AF248767

Gyorgy Kepes

Gyorgy Kepes

Undreaming the Bauhaus

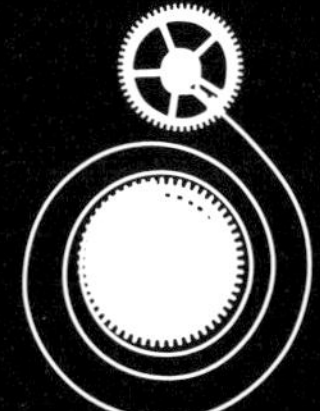

John R. Blakinger

The MIT Press   Cambridge, Massachusetts   London, England

This book was set in Helvetica Neue Haas Grotesk Pro and Dante MT Pro by The MIT Press. Printed and bound in the United States of America.

Library of Congress Cataloging-in-Publication Data is available.

ISBN: 978-0-262-03986-4

10 9 8 7 6 5 4 3 2 1

For my parents

Contents

Acknowledgments

One thing I learned from Gyorgy Kepes is that every intellectual project, even one as individual and sometimes isolating as writing, requires collaboration with a wide community. The support of many made this book possible.

This project began as a dissertation completed at Stanford University in 2016. First and foremost, I thank Pamela M. Lee, whose brilliant teaching, research, and writing provide a model for what art history can be. I have learned greatly from her example. The members of my doctoral committee all supported my work in crucial ways. Alexander Nemerov urged me to think more deeply about my subject. Nancy J. Troy encouraged my very first explorations of the Gyorgy Kepes papers at Stanford. Fred Turner challenged me to see my work from diverse perspectives. I also learned much from other faculty at Stanford, especially Scott Bukatman, Pavle Levi, Bissera Pentcheva, and Bryan J. Wolf. Many colleagues took interest in my work, including David Fresko, Yinshi Lerman-Tan, Gregor Quack, Sydney Skelton Simon, Oliver L. Shultz, Kyle Stephan, Ellen Y. Tani, James Merle Thomas, and Kenneth White. I was lucky to share graduate school with my "academic twin," Kate Cowcher. Even before my arrival at Stanford, it was teachers who inspired; at Wesleyan University, John T. Paoletti's extraordinary teaching pushed me toward the history of art.

A number of institutions provided the funding necessary to complete this study. A Graduate Fellowship at Stanford from 2009 to 2014; a Twenty-Four-Month Chester Dale Fellowship at the Center for Advanced Study in the Visual Arts (CASVA) at the National Gallery of Art in Washington, DC, from 2014 to 2016; and a Provost's Postdoctoral Fellowship in the Society of Fellows in the Humanities at the University of Southern California from 2016 to 2018 allowed me to research, write, and then revise this project. My colleagues at CASVA provided enriching feedback, and I thank Dean Elizabeth Cropper and Associate Deans Peter Lukehart and Therese O'Malley for supporting my work. Special thanks to the many friends and colleagues I made in DC: Joyce Bedi, Monica Bravo, Esther Chadwick, Morten Steen Hansen, Frances Jacobus-Parker, Paul B. Jaskot, Thomas Kren, Marci Kwon, Karen Lang, Lihong Liu, Brendan C. McMahon, Angela Miller, Mary Miller, Barbara E. Mundy, Mauro Mussolin, Lorenzo Pericolo, E. Bruce Robertson, Julia Rosenbaum, Eiren L. Shea, John A. Tyson, Silvia Tita, Iain Boyd Whyte, Mabel O. Wilson, and Kelli Wood. Faculty and fellows at USC helped me reimagine my project. Thank you Rhae Lynn Barnes, Susanna Berger, Andy Campbell, Vittoria Di Palma,

Allan Doyle, Kate Flint, Jennifer A. Greenhill, the late Tom Habinek, Suzanne Hudson, Liying Li, Megan Luke, Jason Nguyen, Amy F. Ogata, Hector Reyes, Monica Steinberg, and Aaron Wile. I must single out Vanessa R. Schwartz for her support; her critical eye pushed me to question assumptions and challenge interpretations. My colleagues at the University of Oxford were enthusiastic about my work during the final stages of this project: thank you Craig Clunas, Anthony Gardner, Dennis Geronimus, Hanneke Grootenboer, Clare Hills-Nova, Ros Holmes, Geraldine A. Johnson, Martin Kemp, Gervase Rosser, and Alastair Wright.

It was my great good fortune that the Kepes papers arrived at Stanford just months before I did, and I am thankful for the help of Glynn Edwards, Tim Noakes, and Bill O'Hanlon in accessing them. Peter P. Blank enthusiastically explored the archives with me and shared his deep knowledge about the collection. Many at MIT also provided crucial assistance: I thank Myles Crowley at Institute Archives and Special Collections, Jeremy Grubman at the Center for Advanced Visual Studies (CAVS) Special Collection, and Daryl McCurdy at the MIT Museum. Caroline A. Jones generously discussed her perspective on Kepes and Leila W. Kinney offered her vision of the arts at MIT.

I am especially grateful to Juliet Kepes Stone, who provided unique insights about her father's work and career, and to the Gyorgy Kepes Estate for supporting this project. I was also lucky to speak with King Collins, Hattula Moholy-Nagy, Kiyoko Lerner, Keiko Prince, and Mya Shone.

An Andrew W. Mellon Curatorial Research Assistantship at Stanford's Cantor Arts Center provided the opportunity to reconstruct Kepes's 1951 *New Landscape* exhibition; Hilarie Faberman, Issa Lampe, Kristen St. John, Wim de Wit, and Connie Wolf helped me realize the show. It was a pleasure working with Isobel Whitelegg on an expanded installation for the Exhibition Research Centre at Liverpool John Moores University in 2015.

Opportunities to share this material have improved it. For invitations to present talks and lectures, I thank Makeda Best, Ben Burbridge, Miguel de Baca, Catherine J. Jolivette, Jennifer Josten, Eik Kahng, Meredith Malone, W. Patrick McCray, Stephanie O'Rourke, Daniel R. Quiles, Sam Rose, Catherine Spencer, Stephanie Straine, Charissa N. Terranova, Amy Tobin, Olga Touloumi, Harry Weeks, and Lily Woodruff. Many in the audience at these presentations shared questions and comments that shaped my scholarship; for especially helpful remarks, thank you Sabine Eckmann, John Harwood, David A. Mindell, Jenevive Nykolak, Alex J. Taylor, and Andrew V. Uroskie. I am grateful to Nikola Jankovic of Éditions B2 for his interest in my work. The participants in a seminar on American art and visual culture at the Newberry Library helped me explore new ideas. I have enjoyed lively exchanges with Márton Orosz about our common interests; Márton generously shared his incredible knowledge and helped with securing images. A workshop organized by Charissa N. Terranova at the University of Texas, Dallas, provided a unique opportunity to discuss Kepes's work. I learned much from Charissa, Márton, Jean-Marie Bolay, Oliver A. I. Botar, Gavin Delahunty, Roger Malina, Twyla Nova, Kate Sloan, and Hadas A. Steiner.

I thank Roger Conover, my editor at the MIT Press, for his commitment to this book. Working with the MIT Press seemed fated to be. During my research, I uncovered a letter from Roger to Gyorgy Kepes expressing an interest in publishing a book "with/by/about" the artist.[1] Such a book arrives a few decades late, but I am happy to supply it—and to have the opportunity to work with MIT instead of only writing about MIT. Others at the Press played key roles in transforming a mere manuscript into a book. Gillian Beaumont greatly improved the text with her meticulous copyediting. Margarita Encomienda brilliantly designed a beautiful object beffiting its subject. Gabriela Bueno Gibbs kept the many moving parts moving properly. Matthew Abbate helped guide the book through final production, and Susan L. Clark assisted with key aspects along the way.

Family and friends are at the center of all academic work. Many have hosted my visits to the archives: The Fogels, Kiplingers, Lashes, LeVees, and Smerlings; John Gatto and the late Art Shirk; and Phoebe Greenwood made my research trips a success. My sister Kate Blakinger shared her writing wisdom. Harvey and Austin Benschoter as well as Linda, Dick, Steven, and Lizzy Schapiro all helped me escape the tedium of writing. Deepest thanks to Martin Schapiro; completing this book would not have been possible without him. I am so lucky that Marty accompanies me so enthusiastically across continents and oceans. My parents, Mary and Chuck Blakinger, have supported my art-historical pursuits over the years, from the homemade readymades and Marcel Duchamp (and Rrose Sélavy) costumes to travels far and wide. This book is for them.

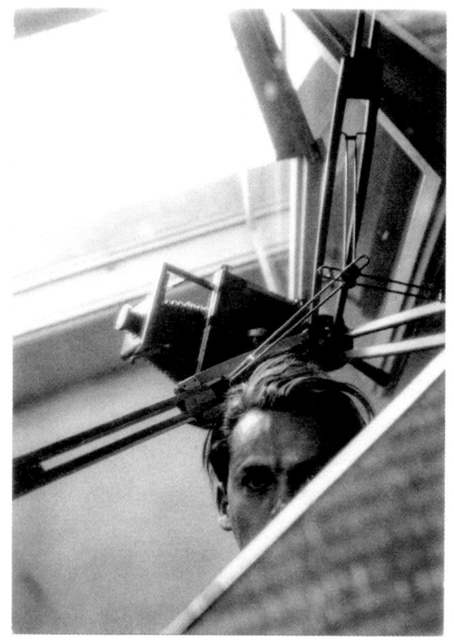

0.1

Gyorgy Kepes, *Self-Portrait*, 1930. Gelatin
silver print, 3 x 2 inches. Courtesy The Kepes
Institute, Eger, Hungary, and Márton Orosz.
© The Estate of Gyorgy Kepes.

Portraits of the Artist

Two photographs depict two versions of Gyorgy Kepes (figs. 0.1–0.2). The first shows a young Kepes, in his mid-twenties, circa 1930, in Berlin. He peers over a mirror placed on the floor below him; a tripod and camera, aimed out an open window, tower above him. His hair falls loosely, his face cropped but still visible and conveying a characteristic intensity of purpose. A beam of light reflected from the mirror's cut edge appears to precisely pierce the pupil of Kepes's left eye, alluding to the eye's reception but also release of light, as put forth in antiquated theories of vision. His right eye is aligned by visual analogy with the camera above it. In this way, the photograph self-reflexively comments on its own formal conditions. It diagrams the science and technology of vision—the optics of light and the human and mechanical apparatuses that transform light into image—and presents the camera itself as a prosthetic extension of human sight. A series of conceptual repetitions within the photograph reiterates this fixation on vision. There are actually two cameras present, the one capturing the picture—unseen, as it is held discreetly in Kepes's hands, and not reflected in the mirror that lies below—and the one captured within it. There are also multiple framing devices: the open window and the mirrored surface, but also the literal limits of the photographic print. The sightlines and viewpoints of camera and eye suggest a series of intersecting and reflecting perspectives. We look up, down, out, and in, all at once. The artist suspends himself within this radically expanded, radically exploded model of vision.[1]

Kepes made this photograph as a new adherent of modernism, and it signifies his belated self-fashioning as a student of the avant-garde—belated because modernism and the avant-garde were not new circa 1930. Kepes is, as he would be throughout his career, late to the scene. Previously, in his native Hungary, he had joined a circle of Marxist intellectuals called *Munka*—the name meant "work" or "labor," indicating the group's social and political ambitions—and it was with these fellow-travelers that Kepes learned of the aesthetic experiments happening abroad. He left Budapest for Berlin in 1930 in order to join them. There, he met Soviet filmmakers Dziga Vertov and Alexander Dovzhenko, and found work with Hungarian compatriot and former Bauhaus master László Moholy-Nagy, known as Moholy—he would become Moholy's most important disciple, though their relationship was often vexed. The art of Aleksandr Rodchenko, Vladimir Tatlin, and El Lissitzky inspired him, and he intentionally imitated it; in its presentation of the visionary seer, for example, Kepes's photograph emulates Lissitzky's 1924 self-portrait as *The Constructor*.[2]

0.2

Portrait photograph of Gyorgy Kepes, 1967. Photo: Ivan Massar. Courtesy Center for Advanced Visual Studies Special Collection, MIT Program in Art, Culture and Technology. © Massachusetts Institute of Technology.

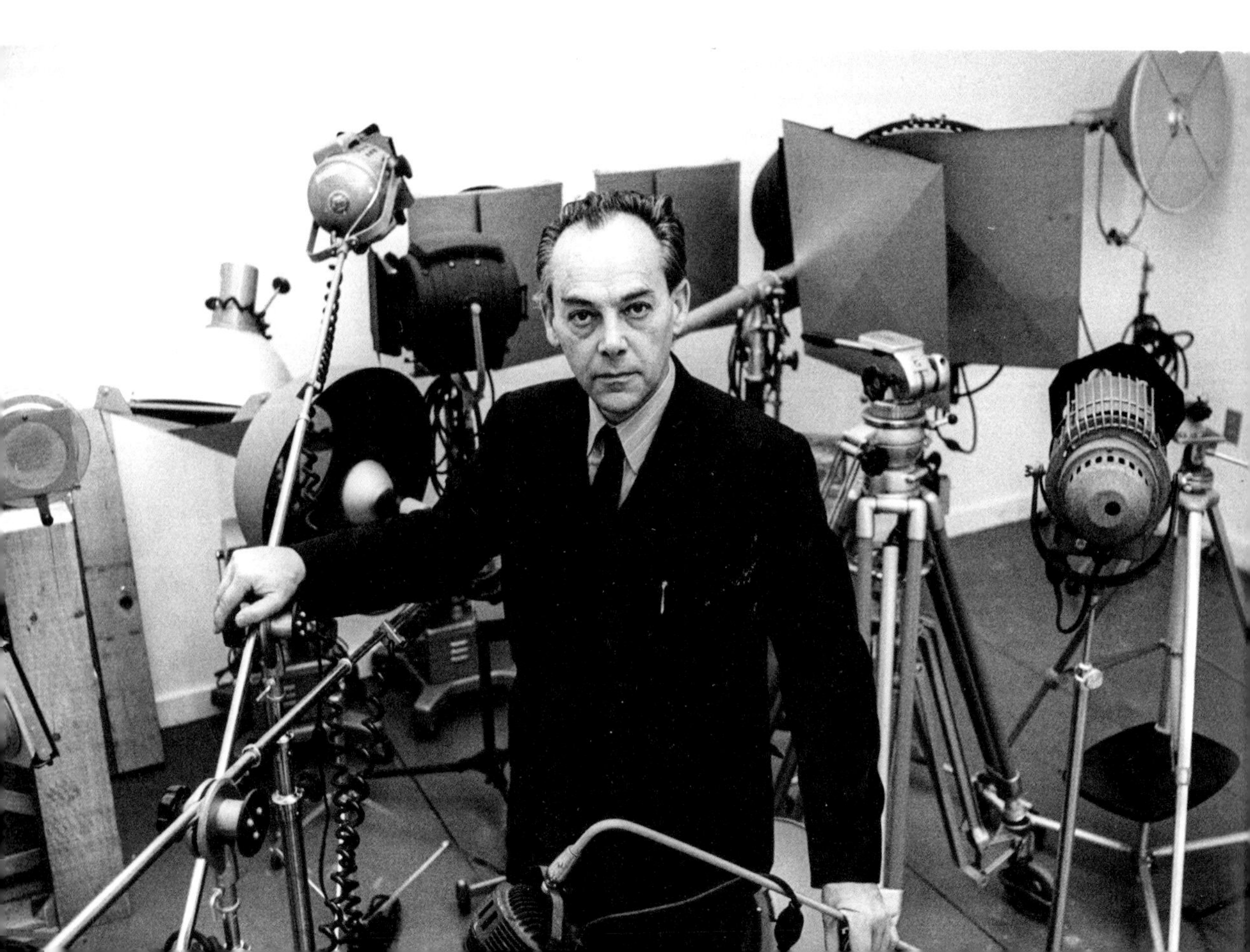

Kepes borrowed from earlier precedents—a strategy that he would frequently employ. In styling himself as an artist revolutionary, Kepes even renounced painting, the subject of his formal training at the Academy of Fine Arts in Budapest, as a reactionary medium. In its place he embraced the "new vision" made possible by the camera, and hoped to unite aesthetic experiments with politics, especially through socially oriented film.

Such aspirations would not come to pass. The second photograph (fig. 0.2) shows a very different Kepes. The photojournalist Ivan Massar of the agency Black Star, on assignment at the Massachusetts Institute of Technology (MIT), took the picture in 1967—as campus conflict over the Vietnam War was intensifying—at Kepes's Center for Advanced Visual Studies (CAVS), the research institute Kepes founded just years prior. MIT's administration commissioned such photographs for use in reports, newsletters, and promotional publications, and Massar made many similar images of MIT's faculty in their laboratories. Like Kepes's earlier self-portrait, this image also depicts the artist with the tools of the trade; Kepes's hands grasp a set of lamps, and he is posed with lighting equipment and assembled tripods, all of which signal the expanded model of human and machine vision that Kepes theorized and practiced—and the association with science and technology that defined his career at MIT. In another photograph from Massar's shoot, Kepes holds a standing softbox light with his right hand—grasped with intention, it becomes an icon of Kepes's obsession with illumination—while a labyrinth of cables and cords surrounds him.

But despite these superficial similarities, Massar's image is very different from Kepes's earlier self-portrait. Nearly four decades have elapsed. We see a changed Kepes. His face conveys that familiar intensity of purpose, but, in a suit and tie, he now looks decidedly more serious. His conservative sartorial statement is matched by a stern facial expression. His posture is stiff. The slightly elevated perspective is imposing. The result is cold, purged of affect. The photograph reads as a generic statement of institutional power; Kepes now appears as the embodiment of the establishment, authority personified. He is not the revolutionary of long ago.

Perhaps the comparison between these two photographs is not fair. Not only are the images separated by time, they are also separated by genre; the first, as a self-portrait, bears the weighty gravitas of artistic convention, while the second is a documentary image, a record prepared for any number of banal and bureaucratic uses. The first is self-conscious, subjective, and meticulously composed; the second is intended to be objective and unambiguous, obvious in its factual presentation. But the comparison is still suggestive, for it indicates a transformation that occurred at midcentury, as circumstance compelled artists like Kepes to carry the familiar project of modernism to new shores and adapt its traditional mission to new ends. It indicates not only two versions of Kepes, the youthful and the worldly-wise, but also two versions of the artist as a creative persona.

This book explores that transformation. It is concerned not with the type of artist in the first photograph, the familiar figure of the avant-garde modernist—a character already well known in the history of art—but with the type of artist he became, this new technocratic figure. With Kepes as protagonist, this book demonstrates how the Bauhaus idea served new purposes in a new world, in contexts that have largely remained unexamined as artistic sites, like the scientific laboratory. In these settings, the artist functioned not as a creator of conventional works of art but as systems thinker, as an aesthetic researcher devoted to the study of vision and the design of the eye. This book examines the forces of displacement and dislocation that created this historical change. It considers how Kepes, in attempting to synthesize art with science and technology, became entangled in the compromising forces of military power during the Cold War.

From the Ivory Tower to the Control Tower

introduction

No More Revolution?

Among the papers Gyorgy Kepes (1906–2001) left behind, the fading manuscripts and boxes of old letters—a life condensed into a vast archive—there is a single white page, blank except for a faint inscription in Kepes's handwriting. In blue ink, the notation states: "No More Revolution?"[1] Posed as both question and answer, the undated and unexplained remark—a stray comment that is without context, unqualified in any way—may be meaningless. But it may also mean a great deal, for it seems to mysteriously distill the central concern of Kepes's career. It hints at self-doubt, maybe possible resignation. If the common conception of the avant-garde as a radical formation with revolutionary potential appeared eclipsed, even exhausted, in the aftermath of the Second World War, what new model became operative in its place? If the role and purpose of the vanguard artist was suddenly irrelevant, what new character emerged from the wreckage? What happened to the art of revolution when there was "no more revolution"?[2]

This study takes seriously Kepes's provocation. It explores the altered fate of the avant-garde and the changed role of the artist in the mid-twentieth century through Kepes's art, design, and visual theory—and especially through the interdisciplinary projects Kepes pioneered. Perhaps more than other artists of the era, Kepes uniquely embodies the dramatically transformed circumstances following the Second World War. Spanning worlds old and new, and paradigms past and future, he personifies the turmoil of this period.

Even a cursory overview of Kepes's youthful ambitions reveals an aesthetic and political orientation significantly at odds with that displayed in his later, and better-known, work. Allow me to elaborate these early positions through Kepes's early biography.[3] Born in the village of Selyp, Hungary, Kepes grew up in the bucolic countryside before moving to Budapest in 1914, where he attended the elite Fasori Evangélikus Gimnázium (Fasori Lutheran Secondary School), and then studied painting with the Impressionist István Csók at the Academy of Fine Arts from 1924 to 1928. The ferment of the era was acute; revolution and counterrevolution left Hungary under the rule of a reactionary military regime. These events prompted Kepes to join a circle of radicals led by the socialist poet and painter Lajos Kassák and named *Munka*, meaning "work" or "labor" in Magyar. The group's name makes clear its Marxist ambitions. Kassák had fled the country after the short-lived Hungarian Soviet

Republic fell in 1920, but he returned in the mid-1920s, after it had stabilized, to promote the aesthetic and political experiments happening abroad. Kepes painted for the *Munka* circle and joined its revolutionary education group, with which he visited factories to recite radical poetry to workers as a means of fomenting insurrection.

It was through *Munka* and Kassák that Kepes learned of Cubism, Suprematism, and Constructivism. Kepes was nearly expelled from the Academy of Fine Arts, a bastion of conservatism, for emulating these dangerous styles; he finished his degree before the charges against him were formally settled. He then renounced painting, the subject of his studies, as impotent ("too static and too limited in terms of social message communication") and made plans instead for a film about the nineteenth-century Hungarian folk revolutionary Rózsa Sándor.[4] Around this time, Kepes began a correspondence with Hungarian compatriot and former Bauhaus master László Moholy-Nagy, with whom he hoped to realize these projects (Moholy had also begun his artistic career in Kassák's circle).[5] At Moholy's invitation, Kepes traveled to Berlin to work for him in 1930.[6] There, he became fast friends with other Hungarian émigrés and aspiring radicals, including the photographers Robert Capa and Eva Besnyö.[7]

As the chaos in Europe accelerated, Kepes was violently uprooted: he returned to Hungary to convalesce from a nearly fatal heart infection in 1932, returned to Berlin in 1934, and then finally escaped to London at Moholy's behest in 1936. His changing name reflects his dislocation: he was, following Magyar word order, "Kepes György" in Hungary, then "Georg Kepes" in Germany, and finally "György Kepes" in the United Kingdom. Once in London, he continued to work in Moholy's design studio and made new contacts with a group of exiles gathered around the anarchist art critic Herbert Read.

Kepes was very much committed to the aesthetic and political program of the avant-garde. His goal was to join the new aesthetic possibilities offered by film and photography with politics—to unite theory and praxis, all in the service of a utopian agenda. He explains: "In this process of examining the analytic self-explorations of the avant-garde, we moved, on the one hand, toward a complete commitment to a kind of formal, structural image-making, but on the other hand we could not close our eyes to the problems of the social structure we encountered."[8] But the opportunities for realizing this agenda were limited. Kepes's work in Moholy's Berlin studio was largely incidental; he prepared the title sequence for *Ein Lichtspiel Schwarz-Weiss-Grau*, the 1932 film of Moholy's Light-Space-Modulator in operation, and designed covers for a German lifestyle magazine called *Die Neue Linie*. Kepes's work in Moholy's London studio was, by necessity, primarily commercial; Kepes designed route maps for Imperial Airways, station posters for London Transport, window displays for the Simpsons of Piccadilly department store, and covers for Shelf Appeal, a British advertising publication. Kepes's plans for a radical film would have to wait.

But Kepes nonetheless took on, whenever possible, selected design commissions that make his commitments clear. He designed covers for *Das Neue Russland*, a Berlin-based journal led by a mix of center-left, and some far-left, intellectuals that promoted the Soviet experiment in Germany (fig. 0.3). Kepes assumed the job from John Heartfield and, before him, Käthe Kollwitz. Though Kepes's designs were blunt and clumsy—minimally collaged images of workers with bright red text—and lacked the incisive critical valence of Heartfield's photomontage, the existence of these projects alone indexes Kepes's priorities. He sought a position designing propaganda because, he explains, "politically I was on the same wavelength. I was interested in doing something where I felt it may help a little to change the mood of history."[9]

He designed covers for a Berlin-based radical press called the MOPR Verlag (a name derived from the Russian acronym for the International Red Aid, a social service organization founded by the Comintern that assisted Communist political prisoners around the world).[10] His commissions included a book cover for the second volume of Soviet revolutionary Alexander Sidorowitsch Schapowalow's memoir, titled *Illegal: Erinnerungen eines Arbeiterrevolutionärs* ("Illegal: Recollections of a Revolutionary Worker") and published in 1932. Kepes also designed covers for a series of booklets called *Rote Reihe: Ausschnitte aus dem Leben der Arbeitsklasse aller Länder* ("Red Series: Excerpts from the Lives of the Working Class of All Countries"). Remembering his experience in Berlin decades later, Kepes minimizes the significance of these projects and their patron: "I didn't have any political background. And I didn't participate in any political party."[11] He would disavow such work throughout his career.[12]

These commitments are most compellingly visualized in an ink drawing by Kepes titled *Against War* and created in response to the Spanish Civil War (fig. 0.4).[13] The image portrays a barren battlefield as a wash of dark ink. A piece of half-eaten bread, a symbol of sustenance, is caught in a tangle of barbed wire from a fence supported not by a wooden post but by the remains of a disembodied limb. A bony hand, stripped of flesh, desperately grasps for the bread. Life is denied. Recalling the now lost drawing decades later, Kepes suggests that it was a generic statement against war, his own personal *Guernica*. It "had no clear political tendency, but it was a protest."[14] But again, Kepes is not honest; he is selectively retelling his past, for he actually submitted the drawing to *The New Masses*, the American Marxist magazine closely aligned

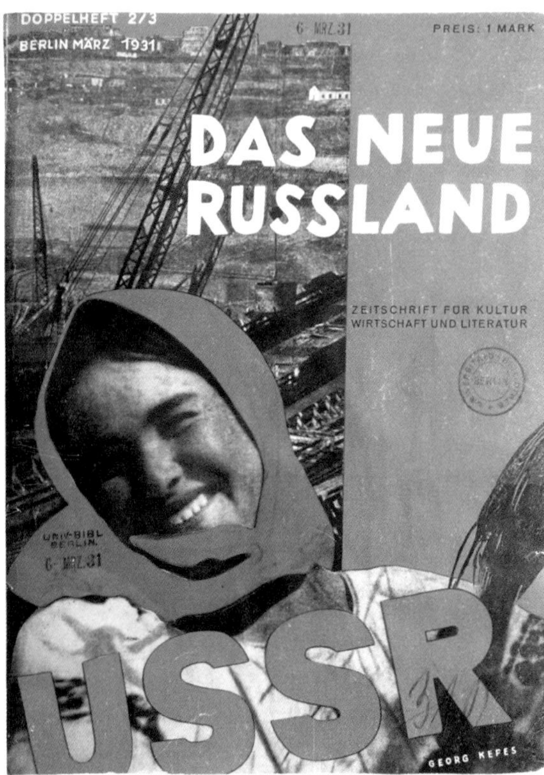

0.3

Das Neue Russland
(March 1931). Cover
designed by Gyorgy Kepes
("Georg Kepes").
Courtesy Márton Orosz.

0.4

Gyorgy Kepes, *Against War*,
1936 (*Bread*, 1938).
Ink drawing (lost). Note
attribution "Ke" in lower
right corner. From
The New Masses, 19 July
1938. © The Estate of
Gyorgy Kepes.

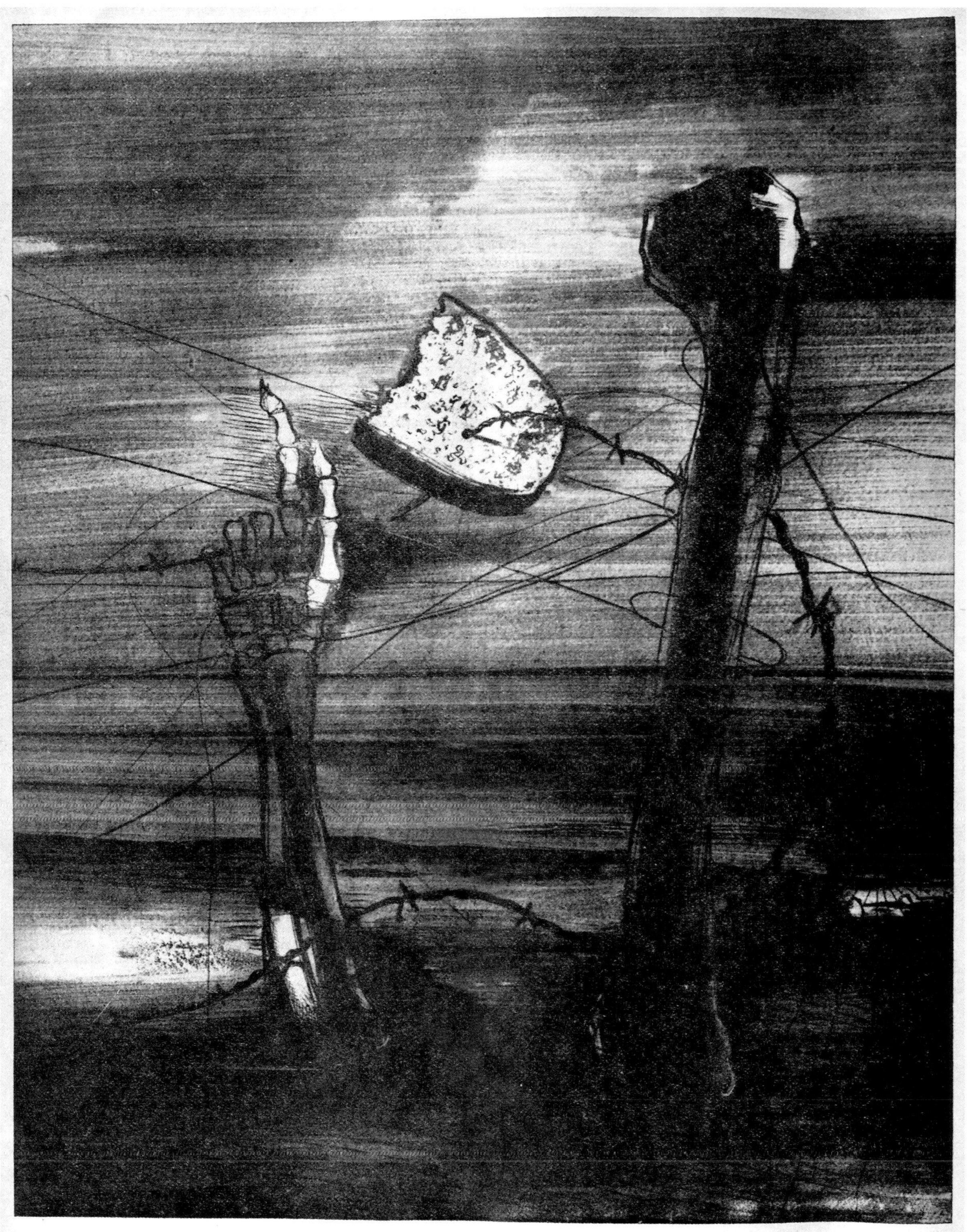

BREAD, 1938

with the Communist Party USA, in 1938. The drawing was printed with pride of place on a single page within the magazine, above an abbreviated attribution pseudonymously listing its creator as "Ke."[15] Alongside the magazine's denouncements of Francisco Franco and Adolf Hitler, not to mention Wall Street, the significance of the drawing's "political tendency" becomes clear. The statement was a final example of Kepes's radical past before his politics were suppressed under a new climate of Cold War conformity.

These commitments were increasingly under pressure in succeeding years. War decimated Kepes's family. His sister was murdered at Auschwitz by the Nazis—although assimilated, Kepes's family was Jewish in origin, and the Nazis began systematic deportations from Hungary after occupying the country in 1944.[16] After the war, his remaining family was further rent apart, physically and ideologically torn between East and West as the Iron Curtain closed. Kepes's brother pledged allegiance to Communism. He fought for the Republicans against the Nationalists in the Spanish Civil War, and later became a functionary and midlevel bureaucrat in Hungary's Soviet-backed postwar regime. He stayed in the East.

Kepes furiously debated whether he, too, should join the fight in Spain with the Republicans; after receiving a telegram from Moholy in 1937 asking him to head the Light and Color Workshop at the New Bauhaus in Chicago, the school Moholy founded that year, Kepes contemplated options. "I wanted to go to Spain, to the International Brigade because that was where my heart was."[17] He hoped to gain entry via France, through the Pyrenees, armed with a press pass provided by Herbert Read for the *Burlington Magazine*, the art history journal Read then edited. But his heart condition made the journey precarious. After much hesitation and delay, he abandoned the cause—though not without a lifetime of doubts. Kepes comments nearly three decades later:

> But with the coming of World War II, my old dream – that there was one group, or the possibility of creating such a group, which could fight for a better life – was virtually destroyed.... [M]y original belief that I or any person could marshal his inner strength to take constructive action began to seem almost hopeless.[18]

Kepes moved to the West. Accompanied by his wife, Juliet Appleby, whom he had met in London—she would study at the New Bauhaus and become an illustrator in her own right—he set out on the *Queen Mary* for America. He took only what he could carry. He left the rest behind.

The Artist as Technocrat and the Cold War Avant-Garde

Perhaps the story of the artist in exile that I tell above is a familiar one, but the story that follows is not. The purpose of narrating Kepes's early biography is to demonstrate how the force of circumstance would, in the years to come, dramatically change his ambitions and the means available to realize them. This book explores the new technocratic order Kepes discovered on the other side of the Atlantic, in the United States, and his struggle to make a place within it.

In the shadow of the atomic bomb, all artists—and perhaps Kepes foremost among them—faced an urgent question: What is the purpose of art in a brave new world dominated by science and technology? Scientific and technological advances wrought destruction not only because of their immense power, but also because of the ideology of enlightenment they supported. Hiroshima and the Holocaust might be understood as direct consequences of such an ideology. Could the power of the arts, as a transformative force, change science and technology for the better? Could the "two cultures"—the phrase British intellectual C. P. Snow proposed as a shorthand for the arts and sciences—collaborate for a common purpose, and a more humane one?[19] Or are they fundamentally incompatible, forever condemned to skepticism of their respective motivations and methodologies?

Confronted with this profound crisis of confidence in the contemporary relevance of the arts, Kepes cultivated collaborations with the sciences at Chicago's New Bauhaus from 1937 to 1943, at North Texas State Teachers College in 1944, at Brooklyn College in 1945, and especially at MIT, where he taught from 1946 until his retirement in 1974. Through his books, exhibitions, courses, and the art-and-science think tank he founded at MIT—the Center for Advanced Visual Studies (CAVS)—Kepes pioneered aesthetic platforms for, as he called it, "confronting, combining, and comparing knowledge." He aimed to encourage "the circulation of ideas, to find channels of communication that interconnect various disciplines."[20] He evocatively termed this mission "interthinking" and "interseeing."[21] This study investigates the impact of Kepes's unusual ideas on the interdisciplinary methods we employ in the art world and academy today. It uses Kepes to establish an unexpected genealogy for contemporary phenomena like what we now call "visual culture" and "new media" (Kepes used both terms before they became common parlance), with the ultimate aim of recovering the continued significance of the Cold War's scientific and technological ethos for the visual arts.[22]

In the first self-portrait he created after arriving in the United States, dated 1940 and published with a small portfolio of commercial design in a trade publication for advertising professionals called *PM*, Kepes clarifies his new focus. He christens himself with science, surrounding his face with a halo-like form created from the lines of a "Lissajous figure"—the mathematically perfect

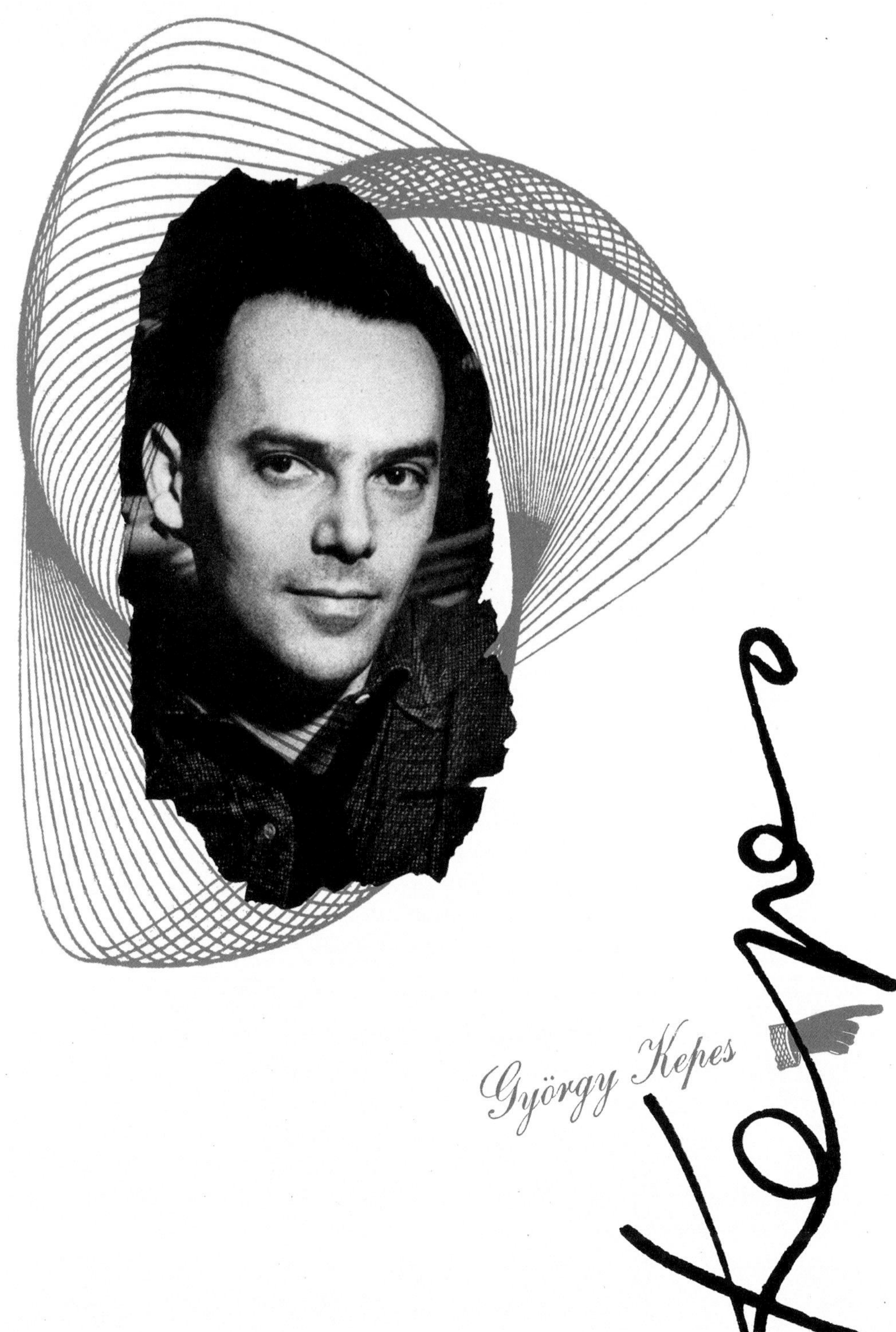

György Kepes

0.5

Gyorgy Kepes, page from *PM: An Intimate Journal for Art Directors, Production Managers, and their Associates* 6, no. 3 (February–March 1940). © The Estate of Gyorgy Kepes.

pattern created by a swinging pendulum's harmonic motion, as recorded in pen on a drawing surface (fig. 0.5).[23] His name is still listed with diacritics—"György Kepes"—but it was only a matter of time before he would remake himself simply Gyorgy Kepes.[24] He purged the past, he grasped at the future. A typographic arrow in the image points to the next page in *PM* but also, we might say, toward this future. Kepes would pursue science and technology with a near-religious intensity in the coming decades.

Indeed, many critics in this period identified the explosive advances in science and technology that began during the Second World War and accelerated in the early years of the Cold War—and the social, political, and cultural changes these advances created—as the beginning of an entirely new society, a technocratic society. They authored major books proposing various models for understanding it. Jacques Ellul called it the "technological society."[25] Alain Touraine called it the "programmed society," a reference to both the rise of the computer and the control it exerted.[26] Daniel Bell described it as "postindustrial," emphasizing the emergence of a new social model based on the "codification of theoretical knowledge."[27] For Bell, a key characteristic of the postindustrial society was the accelerated creation of information through scientific and technological innovation and the rise of a power elite armed with the specialized expertise required to manage it. The postindustrial society was also intrinsically tied to warfare; the "codification of theoretical knowledge" that Bell identified specifically involved the research and development of weapons and the strategies and tactics for using them. All three variations posited by these critics anticipate the computerization of culture so ubiquitous today, the constant blur of ones and zeros that constitutes what we call the "network society."[28]

This new technocratic society also required a new type of artist. Kepes's relationship to technocracy can be understood through the way he captured its peculiar aesthetic in his design projects. Kepes pictures the visual culture of the Cold War, for example, on the book jacket for his 1956 volume *The New Landscape in Art and Science*, for which he illustrated the two cultures through two overlapping images. A radiograph of a rose—an image of beauty representing "art"—is covered with dots, representing "science," from a reel of perforated computer punch tape, the type used for inputting data on early mainframe computers.[29] Editions of Touraine's *The Post-Industrial Society* and Bell's *The Coming of Post-Industrial Society* also employed such an iconography on their own book jackets: Touraine's volume depicts the city of the future as a digital landscape, with towering skyscrapers imagined as giant computer punch cards, while Bell's volume depicts the city of the past—industrial factories with spewing smokestacks—similarly overlaid with a stylized computer punch card.

But Kepes's relationship to this new social organization was not just a matter of illustrating its visual culture, of using its images in his books and exhibitions; he was entangled in technocracy in far more essential ways. MIT, after all, was the center of this new society. It was the archetypal "Cold War university"; its hallowed halls were actually akin to battle-ready war rooms, inextricably enmeshed in the demands and dollars of the US military's research agenda.[30] Here, making weapons and designing ways to use them—the "codification of theoretical knowledge," as Bell puts it—took place in the research laboratories that lined the Institute's famed infinite corridor, the main east-west axis through campus. What happens when the artist enters this corridor and must adapt to this setting? As Heinrich Wölfflin reminds us (and as Wölfflin reminds Kepes, who was fond of the quotation and saved a copy of it in his papers): "Every artist finds certain visual possibilities before him, to which he is bound. Not everything is possible at all times. Vision itself has its history, and the revelation of these visual strata must be regarded as the primary task of art history."[31] Technocracy foreclosed many of the artist's previous possibilities, to be sure, but what new opportunities did it open up?

In exploring Kepes's role in this new social organization, I put forth two major ideas. First, I argue that Kepes developed a new and hitherto unrecognized paradigm for aesthetic practice: the artist as technocrat. This figure operated within rather than against a scientific establishment; seeking refuge, he retreated from the art studio to the relative safety of the research laboratory. As Marshall McLuhan explains, likely with Kepes in mind, "the artist tends now to move from the ivory tower to the control tower of society." In this unusual context, the artist takes on a new role: preparing us all for the increasingly deleterious effects of scientific and technological change. The artist in the control tower, explains McLuhan, observes "technological challenge decades before its transforming impact occurs," and, in response, constructs "models or Noah's arks for facing the change that is at hand."[32] By preserving values from the past, the artist defends society against the effects of the future. This function required oddly antiquated roles. In his personal notebook, Kepes records the Latinate word for prophet: "Latin <u>vates</u> 'soothsayer' foresee + thus promote future."[33] Despite his association with advanced science and technology, Kepes was often described as a mystic or sage—*Time* magazine called him an "oracle."[34]

One might imagine this control tower as a site of limitless panoptic power, but I argue that this technocratic figure was not entirely compromised by or fully complicit with the militaristic logic that governs this setting.[35] He was not just the "organization man," the "man in the gray flannel suit," so characteristic of the era. I explore how Kepes instead navigated the Cold War university through an artful infiltration of its protocols, appropriation of its tools, and transformation of its discourses. Rather than complacent or co-opted, I see him as a subtle operator of this system. Kepes uniquely embodies the complicated negotiations required for agency in a highly pressured, highly charged context. I demonstrate how an individual can work against a dominant ideology even while inhabiting one of the most powerful institutions upholding that very same ideology. Kepes was, to draw from McLuhan, Noah before the Deluge, a prophet for the Cold War and an avatar of its anxieties. (Before the Cold War even began, Kepes would give voice to unease; his 1944 *Language of Vision* opens with the declaration: "Today we experience chaos."[36] Kepes included such language in nearly all of his subsequent writings from the 1950s, 1960s, and 1970s.)

I also put forth a second major idea. I situate Kepes as the foremost artistic actor—he was the first artist to join MIT's tenured faculty, and the only artist ever to attain the prestigious rank of Institute Professor—within a startling network of scientific experts and elites, many aligned with sophisticated war and weapons research. The physicists who designed the first atomic bombs, the engineers who created the earliest digital computers, and the mathematicians who invented novel ways to simulate thermonuclear war all comprise an important but not yet fully explored constellation: the "Cold War avant-garde." The historian of science Sharon Ghamari-Tabrizi uses this term in reference to the creative culture of the think tank, one of the Cold War's most emblematic institutions. The historian of art Pamela M. Lee considers the implications of such a formation for art history.[37] In evoking the phrase, I follow the lead of both scholars. This milieu was on the forefront of advanced science, but also, in a way, advanced art. Innovative research in all fields demands creativity; an intuitive sensibility compels artistic but also scientific insights and imaginings. I use the term "Cold War avant-garde" not to signify a specific roster of individuals but to seize upon the aesthetic dimension inspiring a new scientific vanguard that became socially, culturally, and politically dominant at midcentury. I reveal how the world of science, rather than the art world, became a viable setting for the artist. To understand Kepes, we must shift art-historical attention away from conventional terrain like the gallery and museum and toward new sites like the research laboratory.

Psychology, anthropology, chemistry, biology, crystallography, metallurgy, physics, engineering, linguistics, genetics, mathematics, and more: for Kepes, a vast spectrum of fields became unexpected proving grounds for aesthetics, often through interdisciplinary languages that seamlessly crossed intellectual boundaries, like the cybernetics of Norbert Wiener, the systems theory of Ludwig von Bertalanffy, the information theory of Claude Shannon, and the common study of patterns, models, and symbols. In his collaborative books and exhibitions, Kepes attempted to relate seemingly incommensurate disciplines in the natural and social sciences—despite his lack of scientific training—by finding textual analogies and matching visual metaphors between them, thus translating and fusing disparate knowledge into a holistic "Unity of Science" (such was the name of the philosophical movement that inspired him in the 1940s).[38] In a sense, the academy and its fraught disciplinary map became a microcosm of Cold War conflict, a model in miniature of the "communication crisis" Kepes identified as the cause of countless contemporary problems.[39] Solving this discord in facsimile—by brokering a peace between the warring "two cultures" of art and science—might stimulate, or at least simulate, larger solutions to more global crises. This study takes a sympathetic view of these noble ambitions.

An array of lab-coat specialists thus became very real, if entirely improbable, partners in Kepes's projects; his engagement with science was not a mere trope or theme.[40] Kepes worked with more than a hundred colleagues on his books alone, and they made for surprising company. Many were personally involved in the most catastrophic episodes of the twentieth century. Kepes worked with metallurgist Cyril Stanley Smith, who prepared the fissionable plutonium for the first nuclear weapons test; philosopher of science Jacob Bronowski, who analyzed bombing patterns and officially surveyed the destruction of Hiroshima and Nagasaki on behalf of the Royal Air Force; engineer Jay W. Forrester, who designed the Whirlwind digital computer that powered the Semi-Automatic Ground Environment or SAGE, a vast command-and-control system monitoring the continental US for surprise Soviet aerial attack; and physicist Bruno Rossi, an expert in cosmic rays and interplanetary plasma who helped initiate the space race as an advisor on the Space Science Board, a government committee formed after the launch of Sputnik inspired fear of Soviet military supremacy. These researchers all had genuine interests in aesthetics and longstanding, and often reciprocal, relationships with Kepes.[41] Rossi, for example, participated in Kepes's *Vision + Value* series; Kepes advised Rossi on the preparation of images for the physicist's 1952 textbook *High-Energy Particles*.[42]

Kepes's association with such figures predates his arrival at the Institute. He knew Rossi, for example, from London; like Kepes, Rossi fled his homeland as Europe descended into chaos (Rossi's trajectory took him from Italy to Denmark, the United Kingdom, and finally the United States). Rossi also landed in Chicago before moving to Cambridge, Massachusetts, to a faculty position at MIT. But, remarkably, Kepes's associations with such figures go back still farther. His classmates at the Fasori Evangélikus Gimnázium in Budapest included a generation of brilliant Hungarians who, like Kepes, immigrated to the United States before the Second World War. He studied alongside the polymathic genius John von Neumann (Neumann János Lajos) and the Nobel Prize-winning physicist Eugene Paul Wigner (Wigner Jenő Pál); Edward Teller (Teller Ede), the infamous co-creator of the hydrogen bomb, studied nearby at the Minta Gimnázium.[43]

While many of Kepes's colleagues at MIT, and before MIT, had direct roles in key military endeavors, still others became vocal opponents of nuclear proliferation, defense contracts, and the war in Vietnam—but only because of their own prior involvements. Engineer Jerome Wiesner, who helped develop radar at MIT's famed Radiation Laboratory or "Rad Lab" during the Second World War and eventually ascended to President of the Institute, lobbied for nuclear arms control as science advisor to US Presidents Eisenhower, Johnson, and Kennedy. Philip Morrison, another Manhattan Project scientist, similarly championed disarmament through the Federation of American Scientists and the Union of Concerned Scientists.[44] But the political leanings of this group were hardly cohesive, or even coherent; for just as many became antiwar activists in the 1960s, many others continued to support, and to participate in, military projects. Kepes thus became entwined in the contradictions, and darker commitments, of his many interlocutors and collaborators. In appealing to this vanguard's interest in art and aesthetics, he became mired in the group's competing institutional associations and allegiances. Kepes remained very much "against war," to invoke the title of his 1930s drawing, but he was now forced to appeal to both sides of the debate over science and technology and their relationship to warfare that divided the era, like the arguments at MIT over research with applications in Vietnam.[45] The moral ambiguity of this group, and the ethical questions their activities raise for Kepes, is a central theme in my study.[46]

But I refute the assumption that this technocratic figure necessarily abandoned politics in order to join this milieu. Kepes hoped to forge change not through revolution, but through new models of art and aesthetics as dialogue and discussion, or what he called "interthinking" and "interseeing." His method relied on cooperation between opposite and even actively opposed groups. He was motivated by a utopianism based on the promise of communication.[47] I reveal the possibilities and potential of this approach, but also its pitfalls. I focus on the subtle and subversive ways in which Kepes asserted the value of the arts, but I do not ignore the "reactionary modernism" inherent in any attempt to unite them with the sciences.[48] The larger methodological ambition of this study is to pivot from a completely "paranoid reading" to one that acknowledges nuanced acts of resistance.[49]

Above all, this book aims to use Kepes as a template for understanding the aesthetics of the Cold War—and I use "aesthetics" to indicate not the look of a particular period style but a broad worldview crucial to the era. We often remember the early Cold War in black and white, as the drab Eisenhower Fifties, and we may think of the Cold War's science as governed by a bleak technocratic vision of enlightenment rationality. The historian of science Paul N. Edwards called it a "closed world," a culture of containment and control.[50] But scholars like Fred Turner and Jamie Cohen-Cole have more recently emphasized the period's openness, not its closure.[51] Indeed, the period witnessed explosive creativity. Science captured the imagination. New technological frontiers fascinated experts and elites, but also the public—this was, after all, the era when the Machine Age gave way, in rapid succession, to the Atomic Age, the Space Age, the Computer Age, and the Jet Age. The assumption that science and technology are entirely rational and logical, and thus dull and dry, betrays their intrigue. High-minded methods and methodologies like game theory, information theory, systems theory, and cybernetics had a special allure, an ambience, as new ways of thinking at the cusp of cutting-edge research and advanced forms of knowledge.

Descriptions of technocracy's aesthetics actually abound in the historical record. In his study of protest against this establishment, critic Theodore Roszak refers—not without irony—to the "arts of technocratic domination."[52] Daniel Bell, more sympathetic to these changes, similarly describes technocracy as encompassing a special aura. "The technocratic mind-view is not just a doctrine, one might say, but a temperament."[53] For Bell, advances in science and technology had "all the excitement of an avant-garde movement"; the era's tense geopolitical arrangements, frightening though they were, also sparked "extraordinary revolutions in the art of war."[54] Descriptions of the "avant-garde" and the "art of war" are not just rhetorical flourishes. The confusion between art, science, and warfare is an appropriate evocation of the period's creative sensibility. This confusion is clearly indicated by a cover of the *Bulletin of the Atomic Scientists*, a public policy journal founded by disillusioned Manhattan Project scientists who vigorously supported disarmament and opposed the development of nuclear weapons, especially the hydrogen bomb (fig. 0.6). The cover depicts the journal's signature Doomsday Clock, which recorded the risk of nuclear war as a countdown to midnight. In February of 1959, at the height of Cold War tensions, nuclear midnight was mere minutes away. The specific theme of the issue, surprisingly, was "Science and Art." Kepes is cited within its pages.[55]

We could even describe the Cold War itself as a total work of art. Let me unpack this allusion through an example from Kepes's projects. In an image comparison from *The Nature and Art of Motion*, a 1965 installment of Kepes's *Vision + Value* series, Kepes records the dramatic tensions of the Cold War as a haunting visual comparison (fig. 0.7). He uses a two-page spread to juxtapose images in the classic format of the art-historical slide comparison. On the left, Kepes presents a reproduction of Jackson Pollock's *Number Fourteen*, a web of dripped and poured paint. On the right, he supplies a similar abstract pattern, an electronic pulse of optical signals. A caption describes this second image dryly: "Iterations of non-linear transformations performed by high-speed Los Alamos computers, displayed on the face of an oscilloscope."[56] Produced by Stanislaw Ulam—a mathematician who invented methods for modeling the possible casualties of nuclear war and created, along with Teller, the structural design of the hydrogen bomb—the pattern plots a string of complex numerical equations on a computer screen, encoding digital calculations as a ripple of data.[57] In effect, the abstract expressionist "weapon of the Cold War" now mimics the menacing signs of a literal weapon, the visual traces from a hulking mainframe at the Los Alamos Scientific Laboratory—ground zero for nuclear armament.[58] The "arena in which to act" now mirrors the geopolitical forum of atomic war, the global web of mutually assured destruction— the doctrine known as "MAD."[59] The computer that displays the image was even dubbed "MANIAC" (Mathematical Analyzer, Numerical Integrator, and Computer). The "apocalyptic wallpaper" imagined in paint and canvas is given form on the flickering screen.[60] Kepes sets a quintessential example of modernism's medium-specific materiality against the informational pulse of a doomsday device. His comparison recalls a memorable line by Clement Greenberg: "Modernist art belongs to the same historical and cultural tendency as modern science."[61]

The unlikely visual correspondence demonstrates one of Kepes's essential strategies, one used in all of his interdisciplinary projects: the matching of seemingly unrelated images and ideas through a common visual or intellectual relationship. Kepes employs a form of analogical and metaphorical linking, a concatenation of visual ideas. It may be tempting to reject this particular comparison as false equivalence—or, worse, an embellishment of militarism. I instead understand it as a means for Kepes to visualize the Cold War and its distortions. In aligning these two images, Kepes renders visible otherwise

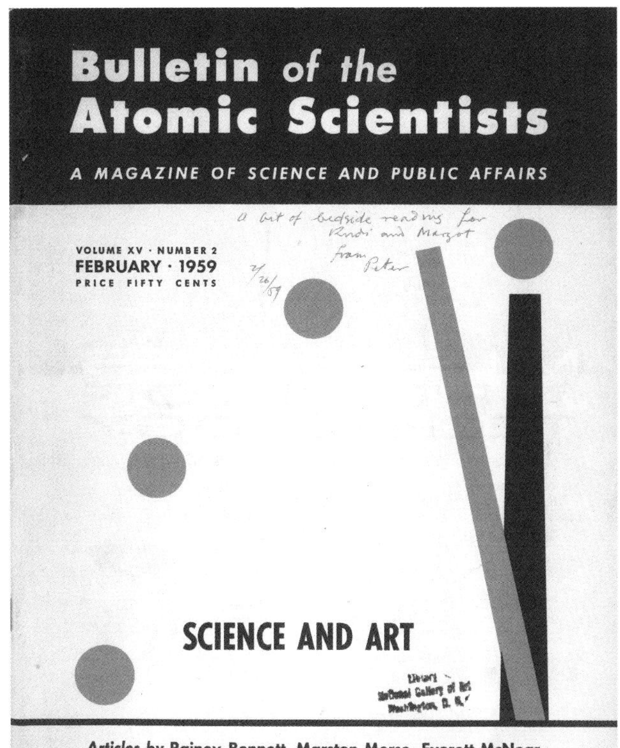

0.6

Cover of the *Bulletin of the Atomic Scientists*
15, no. 2 (February 1959). Copyright
© Bulletin of the Atomic Scientists, reprinted
by permission of Taylor & Francis Ltd,
http://www.tandfonline.com on behalf of
Bulletin of the Atomic Scientists.

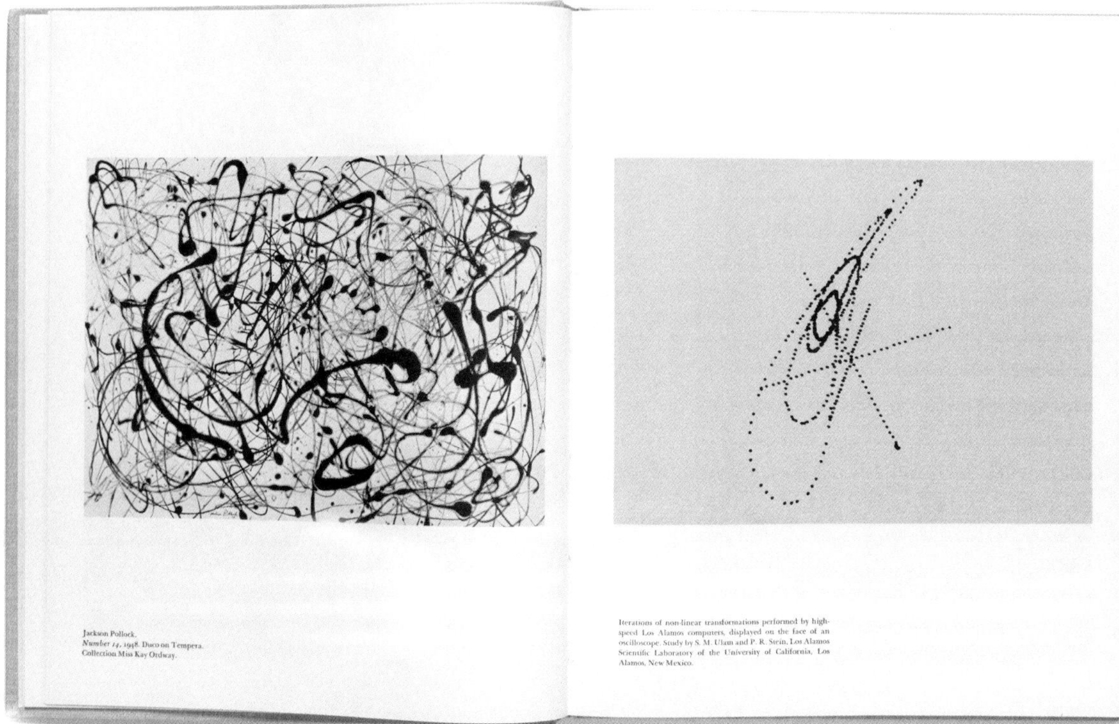

0.7

Two-page spread from *The Nature and Art of Motion*, edited by Gyorgy Kepes. New York: George Braziller, 1965. Courtesy George Braziller.

invisible historical conditions delimiting the visual culture of the period. By suggesting that Ulam's equations are akin to Pollock's splatters, and vice versa, he incisively reappraises both art and science as tied to military power. The infamous "umbilical cord of gold" said to fuel the modernist avant-garde is here envisioned as the mutually parasitic relationships sustaining a new scientific vanguard.[62] For Kepes, to think and see in metaphor—to interthink and intersee—was a means of revealing the deformations caused by the Cold War. This book explores these and many more confrontations between art, science, and militarism.

This book is therefore in dialogue with important biographical and contextual accounts of Kepes's work and career. Anne Collins Goodyear, for example, situates Kepes within the wider context of art, science, and technology partnerships from the 1960s. Goodyear compares Kepes to projects like Experiments in Art and Technology (E.A.T.), the NASA Art Program, the Art and Technology Program at the Los Angeles County Museum of Art (LACMA), and the Pepsi Pavilion at Expo '70 in Osaka, Japan. The first monographic study of Kepes, by Elizabeth Finch, provides a thorough survey of Kepes's artistic output. More recently, Márton Orosz's study deepens the literature with an especially close reading of Kepes's early work and life in Hungary, Germany, and the United Kingdom, before his immigration to the United States.[63]

Kepes has also emerged in complicated debates over aesthetics and politics and the relationship of both to science, technology, and militarism. Architectural historian Reinhold Martin explores how a logic of organizational power consistent with Gilles Deleuze's control society and Theodor W. Adorno and Max Horkheimer's culture industry emerged in Kepes's visual theory, specifically through his use of cybernetics and systems.[64] Martin describes this logic as the "organizational complex," and positions Kepes as its single most significant proponent and perpetrator. He writes with emphasis: "Kepes recognized *and identified himself with* the conditions that underlay the organizational complex."[65] Architectural historian Anna Vallye continues this mode of analysis, understanding Kepes's art-and-science and art-and-technology partnerships as a "version of social engineering," an attempt "to adjust subjectivity to an emergent technoscientific regime."[66] A volume titled *A Second Modernism: MIT, Architecture, and the "Techno-Social" Moment* and edited by architectural historian Arindam Dutta reiterates these arguments.[67] It uses the institutional history of MIT's School of Architecture and Planning, Kepes's home at the Institute, to argue for the rise of a "second modernism" in the postwar period that applied rational, technical, and logical approaches from science and technology to the arts and humanities. It demonstrates how MIT used architecture as forms of control in a wider matrix of power.[68]

These studies are all powerful and persuasive. But they succumb too easily to a narrow repetition of institutional affiliation as ideological orientation. I depart decisively from their version of Kepes. I argue instead that Kepes attempted to change MIT, to transform science and technology by inverting rational, technical, and logical discourse. He offered an alternative model of seeing and thinking, one opposed to MIT's technocratic ethos. He should not be guilty by association; he did not create his context. Rather than frame Kepes as the "agent" or "middleman" of the military-industrial complex (terms used by Martin and Vallye), I consider the nuanced strategies and tactics he employed to change this complex. What lessons can we learn from Kepes's subtle resistance to militarism? In asking this question, this study also moves beyond the entrenched tension in art history between "technophobic" and "technophilic" understandings of art-and-science and art-and-technology projects. I refuse to restage these polarizing perspectives, which have already structured the original reception of these projects. I provide a more contradictory, more conflicted account.

In fact, it may come as a shock, given the common impression of Kepes, to learn that he had a deeply ambivalent relationship to even the most basic scientific and technological devices of daily life. He never operated an automobile, or even a bicycle.[69] According to the curator and critic Katherine Kuh, Kepes "refuses to own or drive a car because he feels he will see less if he moves too fast."[70] He also long refused to purchase a television, even after TV became ubiquitous. Otto Piene, Kepes's successor as Director of the CAVS, writes: "In 1969 I talked the Kepes family into acquiring a TV set. The cat is always there but I have never seen the set turned on."[71] Kepes's home in Cambridge, a traditional New England colonial, was filled with ancient objects and abundant plants, but none of the modern art that one might imagine on its walls. The point is that Kepes was more naturalist than technologist. He had an interest in atavistic traditions, in the mystical and mythological. He was a seer who looked backward as much as he looked forward. This description suggests the fundamental equivocation in Kepes's attitude toward science and technology—one missing from existing interpretations, but one that is the starting point for my analysis. Kepes seemed to be every inch the technocrat, but he also took on oddly antiquated artistic roles; he served MIT's power elite as a medieval mystic and wizardly sage.

Structure

This study does not provide a comprehensive account of Kepes's entire artistic output. I dispense with the conventions of the survey monograph because Kepes is not a conventional artist. He often disavowed traditional labels.

Invited to lecture on photography, he would begin his talks with a caveat: I "do not consider myself a photographer."[72] He cared less about art and more about systems—aesthetic systems, from the systems of visual elements in a single image to the systems of relationships between individuals and institutions in wide-ranging, far-reaching environments. This study focuses on the role and purpose of the artist as a designer of such systems. It reconstructs Kepes's expansive interdisciplinary world through close analysis of archival material: research notes, lecture notes, bibliographies, drawings, sketchbooks, manuscripts, and correspondence. It embraces archive as method. After all, these systems did not take form as proper works of art; they exist only in the archive. In making sense of these materials, I also use Kepes's own approaches—"interseeing" and "interthinking"—as methodologies in their own right, ways to make connections between networks of images and ideas. The artist is but one figure in a wider web; an image is but one object in an endless progression.[73]

I therefore focus on selected chronological episodes from Kepes's career after his arrival in the United States in 1937. Chapter 1, "Camouflage Aesthetics," begins this story. Using previously classified documents, this chapter reconstructs Kepes's collaboration with the US military on camouflage techniques during the Second World War. It presents the first study of the military's wartime camouflage program, held at Fort Belvoir, Virginia, which trained artists, architects, and designers in techniques for disguising ground targets from aerial attack through the application of modernist approaches to the aerial view. Kepes's "language of vision"—the pedagogy he eventually published in a book of the same title in 1944—was uniquely able to address this way of seeing and thinking. This chapter further explores the participation of artists in government bureaucracies and the role of camouflage in boosting "cultural morale." I argue that Kepes's work on camouflage—his first sustained engagement with science and technology—provided a model for self-preservation and self-protection, a way to blur his identity and prior political affiliations.

Chapter 2, "Patterns and Puzzles," investigates the visual design pedagogy Kepes developed at MIT after his arrival in 1946. Using his 1951 exhibition *The New Landscape* and his 1956 book *The New Landscape in Art and Science*, as well as bibliographies, reading notes, and course materials, this chapter explores what Kepes called the "education of vision." Previous scholars have critiqued Kepes's program as a reactionary glorification of the imagery and ideology of science. By contrast, I argue that Kepes used the images he discovered in MIT's labs in subtly subversive ways. He adapted science's images to new ends— ends that were not scientific. He transformed them into tools for cultivating creativity. I claim that Kepes's project therefore had a suppressed—or, using the parlance of the Cold War, "contained"—political agenda.

Chapter 3, "Vision as Value," examines Kepes's set of seven encyclopedic volumes published in the mid-1960s and early 1970s as the *Vision + Value* series. Based on interdisciplinary seminars first organized at MIT, Kepes's books brought together artists and scientists for the study of abstract topics like motion, structure, symmetry, and signs. Decades before communications technology created the social network as we know it, Kepes used the primitive technology of the book to create his own network, one that united more than a hundred colleagues. The series responded to what Kepes identified as a "communication crisis" plaguing the contemporary world; this crisis might be solved through the creation of a holistic visual knowledge.

Chapter 4, "Darkness into Light," explores Kepes's Light Book, his unpublished *magnum opus*. Comprised of hundreds of pages of typescripts, thousands of black-and-white photographs, and countless preparatory notes filled with drawings, the Light Book exists now as a vast archive of textual and visual fragments, not unlike Walter Benjamin's Arcades Project. This chapter reconstructs the Light Book for the first time. It analyzes specific textual passages, like Kepes's meditation on the light released by the nuclear bomb, and speculatively reconstructs the project's visual contents. It situates the Light Book in relation to Kepes's use of light as a creative medium, including his transformation of scientific studies found in treatises on optics into photographic experiments; his creation of "mobile kinetic light murals" using artificial lighting elements; and his interest in the romantic movement and the "painter of light," J. M. W. Turner. For Kepes, light manifests—renders visible—the profound anxieties over science and technology that marked the Cold War.

Chapter 5, "The Military-Industrial-Aesthetic Complex," considers Kepes's Center for Advanced Visual Studies (CAVS), founded at MIT in 1967. The Center was the culmination of Kepes's interdisciplinary ambitions. Its goal was to bring together visiting artists and MIT scientists for collaborative work on civic-scale projects. This chapter considers the first major project fully realized by the Center, a group exhibition titled *Explorations* that Kepes first organized for the tenth São Paulo Biennial in 1969. After participating artists boycotting Brazil's military regime withdrew from the original exhibition, the show was canceled and instead opened at MIT's Hayden Gallery and then at the National Collection of Fine Arts (now the Smithsonian American Art Museum) in 1970. This chapter interprets Kepes's utopian aims, but also the backlash he encountered. Kepes's Center became mired in the politics of Vietnam; while he embraced a philosophy of conversion aimed at changing MIT's military involvements for the better—at cultivating peace and understanding, as he often wrote—he could not escape complicity in such entanglements.

Although the Center's first exhibition was a disaster, the Center persisted in its mission and produced a wide range of imaginative projects in the late 1960s and early 1970s. Chapter 6, "Artificial Natures," considers proposals drafted by Kepes and fellows at the Center for massive environmental installations that could activate the urban landscape, like light towers illuminating Boston Harbor. This chapter uses Kepes's notes and drawings from the period, including his study of phenomena like rainbows and his investigation of concepts from "pollution" to "participation," in order to reconstruct Kepes's and the Center's environmental aesthetics. It presents numerous plans for projects that were never realized, but which nonetheless indicate the utopian dimension of the Center's early activities.

Lastly, an epilogue considers the critical stakes in relating the "two cultures" of art and science, both in Kepes's time and in ours. It contemplates the fallout of Kepes's work in terms of MIT's current commitments to the arts, but also the wider reception of interdisciplinary research and the embattled state of the arts and humanities on university campuses.

This book argues for an interpretation of the artist not as heroic, but as fundamentally flawed, maybe tragic. Kepes is best understood as refugee and exile. "**I am vagabond**," he writes with added emphasis on a page of lecture notes. "Budapest Berlin London U.S.A."[74] In a personal notebook, he describes an evening at home in Cambridge with colleagues—the linguist Roman Jakobson, the poet Octavio Paz, and the architect José Luis Sert, all foreigners in a foreign land: "displaced—uprooted, at home and not at home."[75] But in addition to personal experiences of trauma, Kepes endured professional ones. He was an avant-garde artist who came of age after the avant-garde had disbanded; he was the last of the Bauhaus modernists, resigned to proselytizing the Bauhaus idea after the Bauhaus was no more. He was out of art-historical time, strangely belated. He was also long obscured by the stature and significance of his mentor, László Moholy-Nagy, and constantly threatened by the suspicion that he was nothing more than the failed epigone of the master. Such charges were leveled against him frequently. In a 1970 volume on Moholy, critic Richard Kostelanetz accuses Kepes of copying Moholy's posthumous book *Vision in Motion*, which "has inevitably become a mine for plagiarists and popularizers." He names names: "Gyorgy Kepes, now of M.I.T.'s Center for Advanced Visual Studies, has cribbed freely from it for many of his recent books and anthologies, though he rarely credits Moholy."[76] This charge worried Kepes (he was aware of it; he kept a photocopy of this page from Kostelanetz's book in his papers).[77] I choose to interpret Kepes on his own terms, in his own context, one that is very different than Moholy's. From Kepes's notebooks: "No doubt as the Greek said 'everybody has a father'—but we are more + different than carbon copies of our fathers—important is not only what is similar but what is different."

But there is also truth to the charge. I also choose to take this critique seriously, and to understand why Kepes may have repeated Moholy, and what he hoped to accomplish by doing so. Kepes once referred to his lifelong project as "undreaming" the Bauhaus: as making what were once dreams come true. He attempted to realize the Bauhaus idea in practice and not just in theory. *Gyorgy Kepes: Undreaming the Bauhaus* explores this proposition—both its potential, and the backlash it unleashed.

Camouflage Aesthetics

Protection as petrified terror is a form of camouflage.
These numb human reactions are archaic patterns
of self-preservation: the tribute life pays for its continued
existence is adaptation to death.

Max Horkheimer and Theodor W. Adorno,
Dialectic of Enlightenment, **1947**[1]

Avant-Garde circa 1942

A promotional magazine designed by Gyorgy Kepes opens with a striking diagram (fig. 1.1): pages patterned with the essential formal elements of modern art—points, lines, shapes, and shadows, or what Kepes called "visual fundamentals"—cast rays of light that penetrate a perfectly spherical eyeball. Reflected through the cornea, refracted by the ocular lens, these beams imprint an inverted image on the retinal surface. The illustration outside the eye is mirrored within it.

Kepes's diagram presents an inventory of iconic styles from the historical avant-garde: his design charts depict variations on a Kandinsky, a Mondrian, a Malevich, perhaps a scribble from Klee, even a thumbnail version of Kepes's own commercial advertisement, commissioned by the Container Corporation of America—this last cleverly positioning Kepes as New World successor to a lineage of Bauhaus masters.[2] But in his diagram, titled "Entering the Eye" and dated 1941, Kepes does not focus on the work of art as typically conceived; rather, he dramatizes the act of perception—bluntly emphasized typographically by three pointing manicules—and implies that the principles of modernism are basic to all vision, even its constituent idiom. He suggests a way of seeing in which the formal elements of abstract art are exploded from the scale of the single sheet of paper or solitary square of canvas and expanded to the vast dimensions of our surroundings; amplified to ambient space, dilated beyond the picture plane, these elements turn the environment itself into an immersive optical puzzle. They encode sight as a layered pulse of pictorial signals. They encrypt the visual field as a matrix of sensory data. For Kepes, all the world becomes picture.

The magazine's accompanying text elaborates the astonishing power of human vision, the "almost inconceivable" intricacy of the eye, its muscular systems, its nerve pathways, even the retina's one hundred and thirty-seven million rods and cones. The text reflects Kepes's awe at the eye's technological

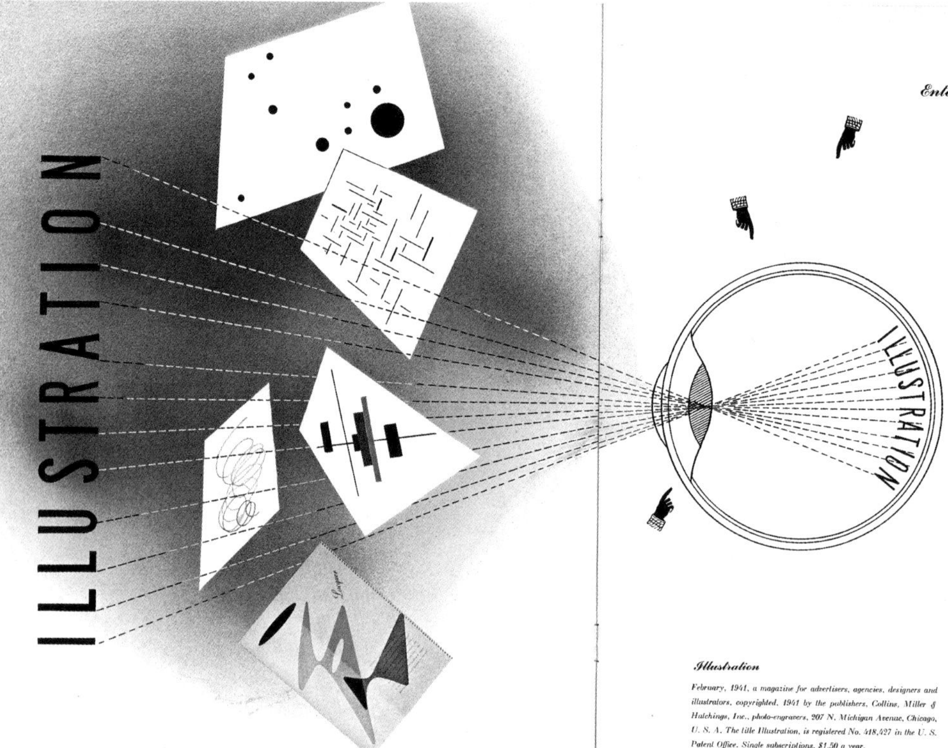

1.1

Pamphlet titled *Illustration* (Chicago: Collins, Miller & Hutchings, Inc., February 1941). Designed by Gyorgy Kepes. IDR_0006_0172_neg27, Institute of Design collection, University of Illinois at Chicago Library. © The Estate of Gyorgy Kepes.

abilities, its exacting capacity to convert photons into electric current, and thus into meaningful images. The human optical organ is "one of the most complex and delicately balanced instruments known to science," it declares. The pamphlet further serves to publicize the largely unknown artist and theorist—Kepes arrived in the United States only in the fall of 1937, and had yet to distinguish himself from László Moholy-Nagy—and it celebrates Kepes's "many experiments" in "art as applied optics" as a teacher at Chicago's School of Design, first founded as the New Bauhaus and later known as the Institute of Design. The magazine concludes with a suggestive comment: Gyorgy Kepes "believes in the coming of a new art that will combine science and aesthetics to the advantage of both."[3]

I open with this diagram because of what it reveals about Kepes's project. Although Kepes designed the pamphlet to advertise the printing services of Collins, Miller & Hutchings, Inc., a photoengraving company, it more clearly serves as a manifesto for his nascent visual theory.[4] It forms an early statement of what Kepes first described in 1938 as the "language of the eye" and would articulate more fully as the "language of vision" in his 1944 book of the same title; indeed, it anticipates a cover design Kepes created for the book (fig. 1.2).[5] He suggests that artistic practice can escape the physical constraints of pigment, paint, and paper, and that light—hypostatized as the very substrate of vision—is a medium in its own right. The grammar and syntax of pictorial organization, developed through Bauhaus and New Bauhaus design pedagogy, would then enable a direct manipulation of optical phenomena. Kepes dreams of total art, of art as absolute as the human eye: a universe of images immaculately distilled in pure light.

But I also open with this diagram because of what it does not reveal about Kepes's project—or, more accurately, because of what it camouflages: his concurrent involvement with the US military. Kepes's radical aspirations for autonomous perception seem separate from the events unfolding circa 1942. They are not. After Japan's surprise strike against the US naval base at Pearl Harbor on 7 December 1941, Kepes's language of vision gained a peculiar expediency. By conceiving of our surroundings as fundamentally visual, Kepes provided a vocabulary that could make sense of a horrific new form of visual experience. Perceiving the world as codes of pictorial signals and arrays of sensory data, seeing in terms of abstract informational patterns that one might quantify and then manipulate, was to perceive like the bombardier engaged in aerial attack—or the *camoufleur* charged with defensive disguise. Kepes's prophecy of a coming alliance between science and aesthetics was therefore surprisingly prescient, but this unlikely arrangement between disciplinary opposites would actually form in response to the terror of air war.

1.2

Gyorgy Kepes, gouache design for the cover
of *Language of Vision*, c. 1944. Gyorgy Kepes
papers (M1796). Dept. of Special Collections
and University Archives, Stanford Libraries,
Stanford, Calif. © The Estate of Gyorgy Kepes.

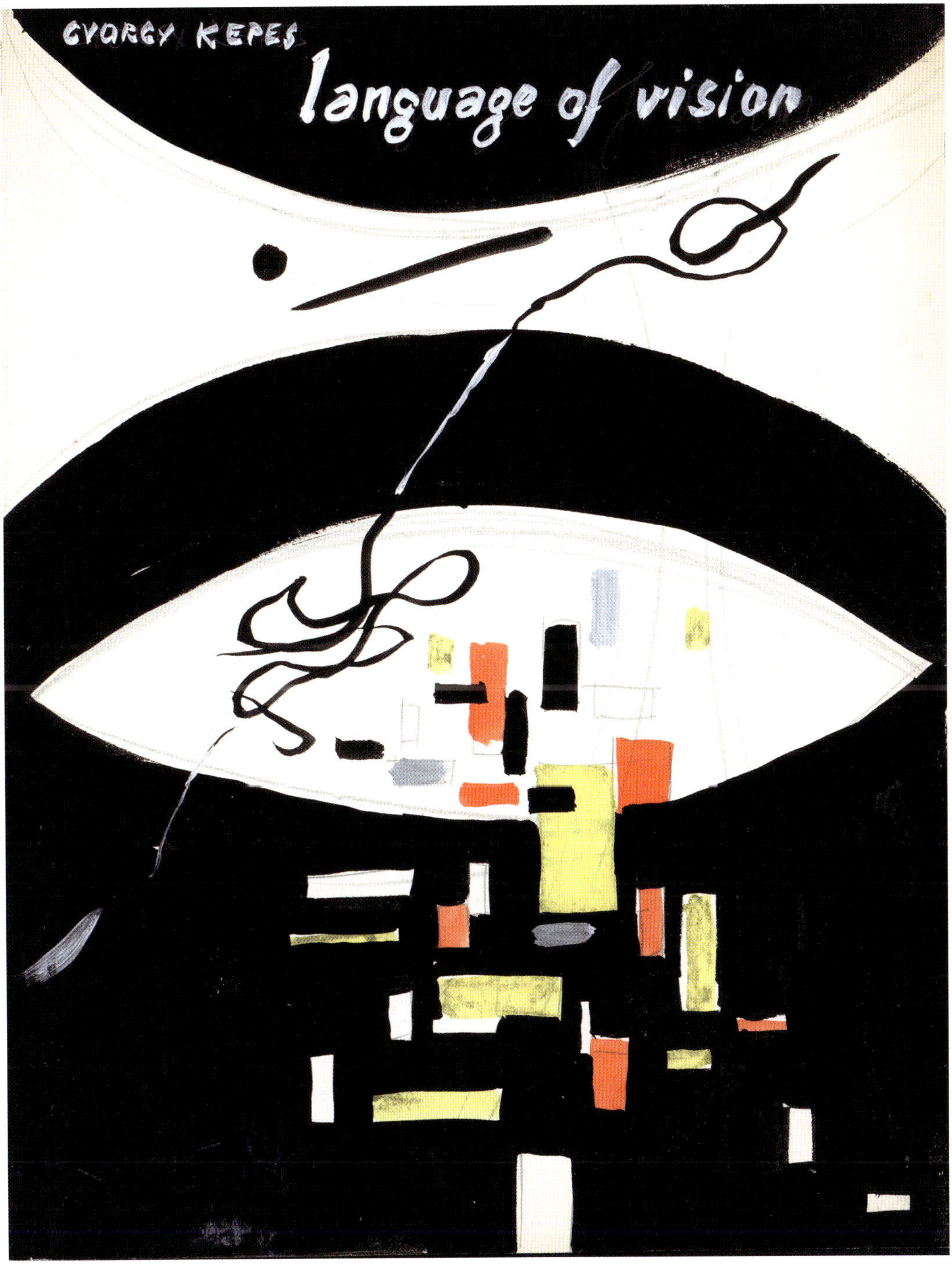

GYORGY KEPES
language of vision
GYORGY KEPES

To explore how Kepes's modernism became entangled with the catastrophic events of midcentury is the point of this chapter. I want to restore a literal militarism to that timeworn but enduring term from the art historical lexicon—"avant-garde"—by returning to the label's etymological origins in warfare. In his genealogy of the term, Mattei Calinescu tracks how it emerged during the early modern period in reference to the advance guard on the battlefield but quickly became rhetorical, migrating to descriptions of artistic production that attacked social, political, and cultural conventions, and for which militarism was an apt metaphor.[6] But suppose we reread this figure of speech letter-for-letter. What would happen if advanced art joined forces with a real military formation? If the avant-garde were not metaphorical but actual? This chapter demonstrates how the strategies and tactics of battle were not only marked semantically, but also literalized; modern art became intertwined with modern war.

In so doing, this chapter recovers institutional adjacencies and interdisciplinary partnerships that have remained concealed, blurred from view. It traces a genealogy for the language of vision back to warfare; it locates the ontology of Kepes's way of seeing in battle. It shows how advanced art moved from an antagonistic position aimed against established culture to a position within that very culture.[7] During the Second World War, the "art world" became continuous with the "real world." Artists could not maintain earlier radical aspirations; they required camouflage.[8] To begin this story, we must look more closely at the system of vision that emerged through the optics of aerial bombardment.

Air War

Imagine the extraordinary view from the Boeing B-17 Flying Fortress or the B-29 Superfortress, the hulking heavy bombardment airplanes that rained terror in campaigns across the European and Pacific Theaters during the Second World War. These enormous metallic leviathans, bristling with machine guns and leaden with bombs, hurtled through the atmosphere at hundreds of miles per hour, thousands of feet up. The extreme velocity and vertiginous altitude created dazzling but also dizzying, probably nauseating, abstractions. The air grew cold and thin and the ground below simply fell away, its recognizable features and familiar landmarks fast dissolving into mosaics of points, lines, shapes, and shadows—captivating pictures not unlike modern art (figs. 1.3–1.4). This was exactly the type of perception Kepes diagrammed in his pamphlet for Collins, Miller & Hutchings, Inc. The height and speed distorted the landscape below, warping the spatial depth of forms and structures. The three-dimensional world flattened and collapsed into an immersive two-dimensional kinetic image. The distance from the ground was thus actual and metaphorical, an erasure of the people and places underneath the aircraft that left only traces of visual information and a cryptic optical blur: patterns in pure light.

Kepes was mesmerized. Though his first flight in an airplane took place over Paris in the mid-1930s, he began considering the implications of aerial experience on the language of vision only after flying over Chicago in 1942.[9] "My imagination was captured," he recalls, "during the Second World War.... It was then I came to appreciate the overwhelming experience of looking down from an airplane on the nightscape of a large city."[10] He would explicitly re-create this view many years later in an electronic mural, or what he called a "mobile kinetic light form," that he designed for the KLM-Royal Dutch Airlines ticket office on Fifth Avenue in Midtown Manhattan (fig. 1.5). The work replicated the cinematic spectacle of the ground seen while moving through the sky in programmed lighting elements that flickered and flashed; points of light suggested street grids while ribbons of neon simulated the fluctuating illumination of the urban environment. In a statement originally drafted for the public announcement of the work's completion in 1959, Kepes describes the exhilarating God's-eye view that the mural intended to capture:

> Flying above a big metropolis in the night offers a visual experience incomparable to anything man experienced before. For no matter how chaotic, blighted, or illegible a large city is, it is transformed when the evening comes and the lights go on. Points, lines, plane figures, and volumes of light, steady and winking, moving and still, white and colored, from windows, signs, headlights, traffic lights, street lights, combined into a fluid luminous wonder – one of the great sights of this or any age. The impressive sight, with its accidental visual wealth, could only be compared with the concentrated, ordered beauty of the great windows of the 13th century cathedrals.[11]

Kepes reproduced similarly resplendent special effects in actual stained-glass windows designed for churches, like Pietro Belluschi and Pier Luigi Nervi's Cathedral of Saint Mary of the Assumption in San Francisco in the 1960s. For Kepes, the luminous perspective from the sky was spectacular, the radiant view from the heavens sublime.

1.3 (following pages)

Aerial view. From Robert P. Breckenridge, *Modern Camouflage: The New Science of Protective Concealment.* New York: Farrar & Rinehart, 1942.

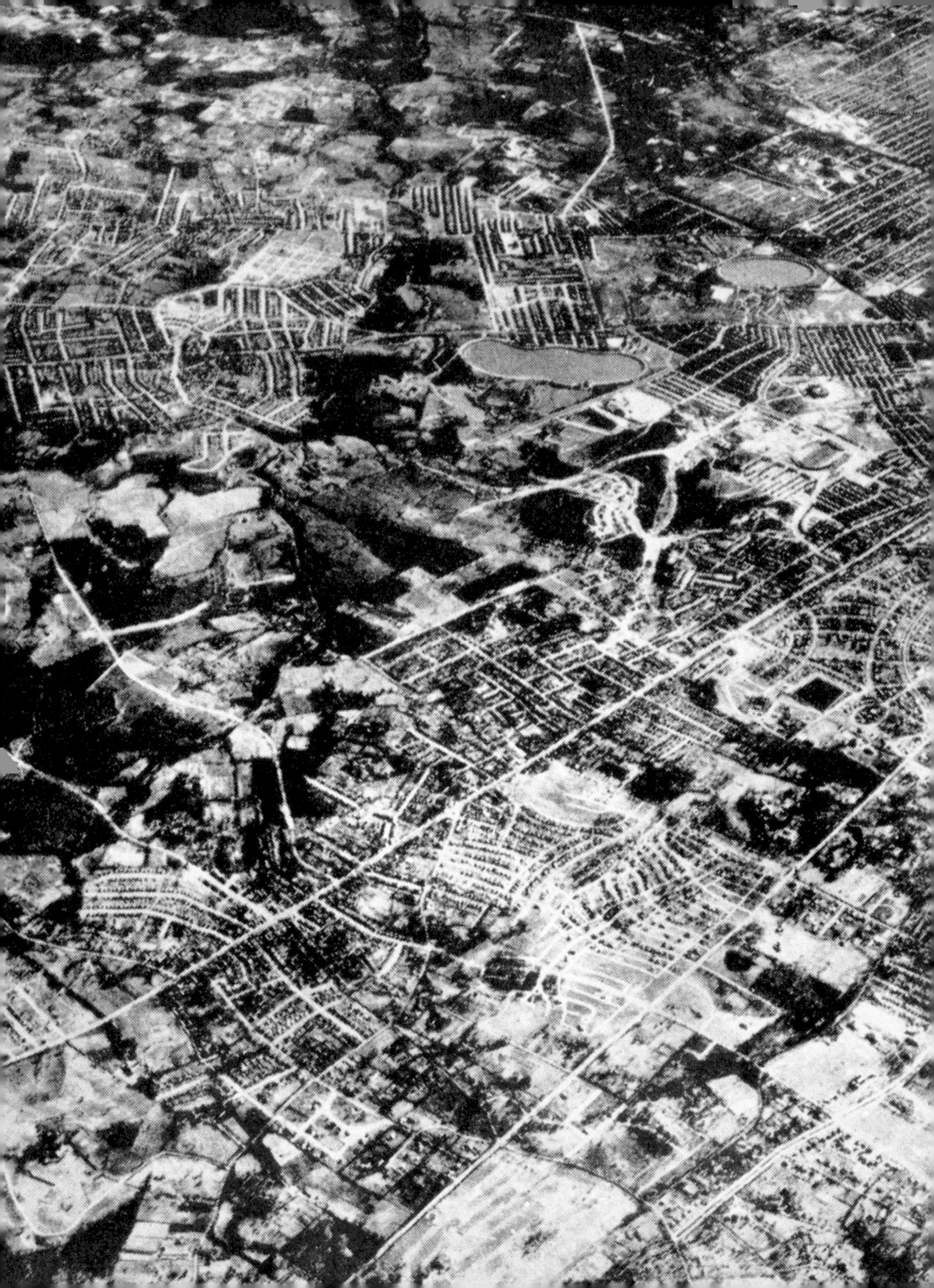

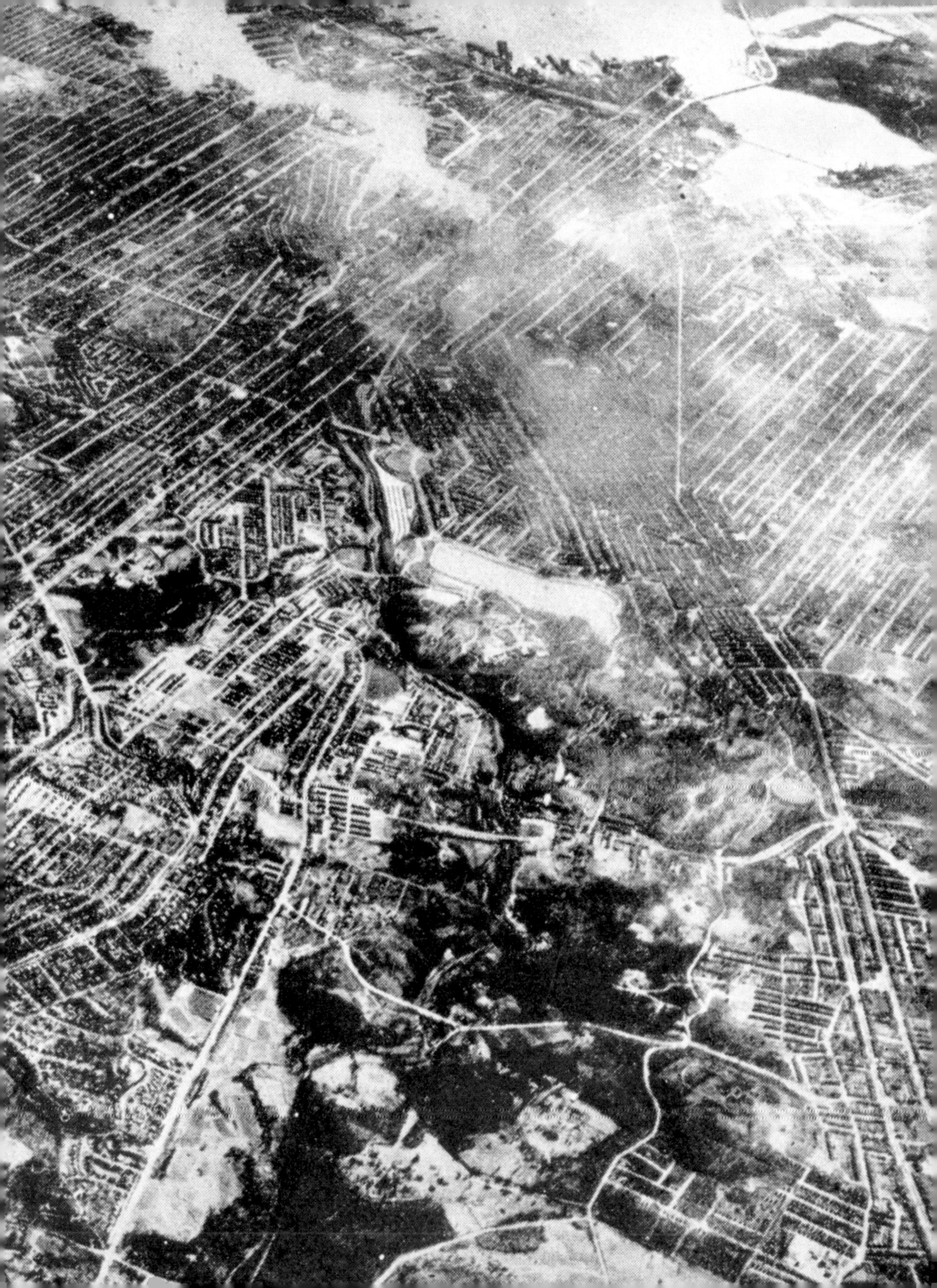

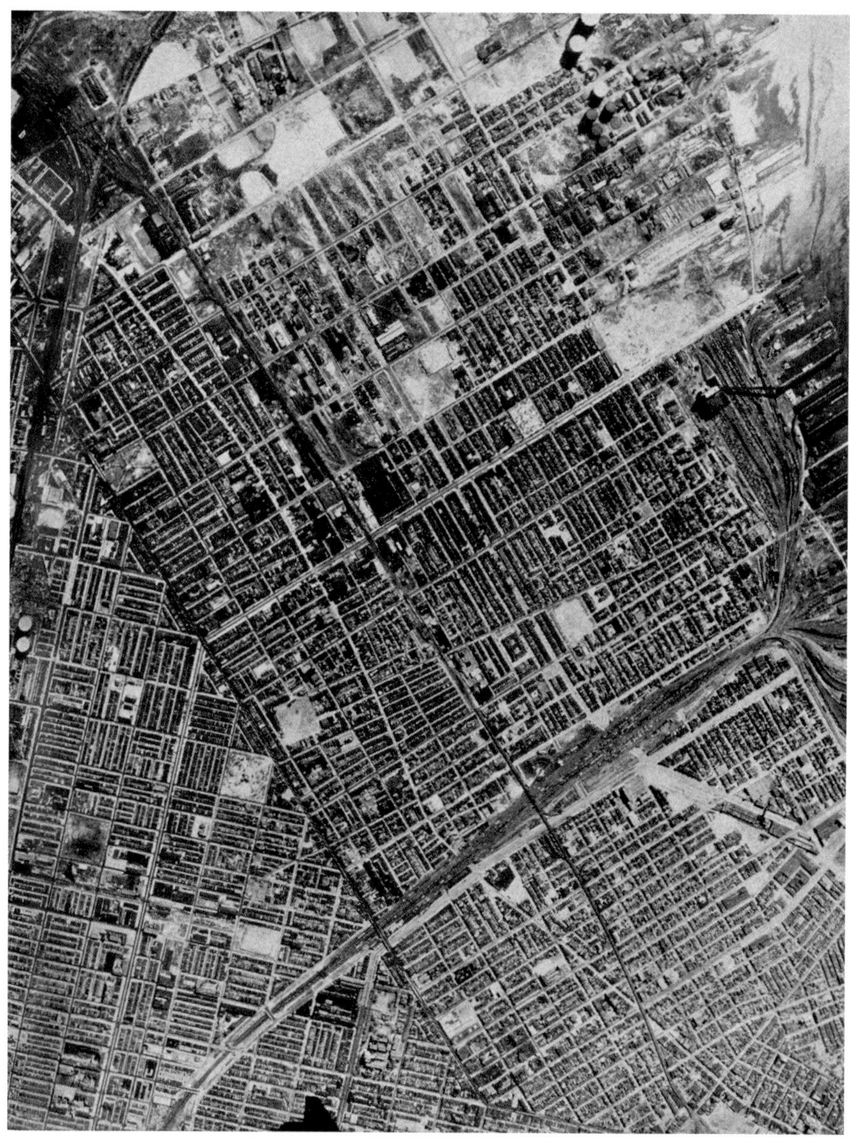

1.4

Aerial view. From Robert P. Breckenridge, *Modern Camouflage: The New Science of Protective Concealment.* New York: Farrar & Rinehart, 1942.

1.5

Gyorgy Kepes, *Programmed Light Mural*, KLM ticket office, New York, 1960. Photo: Nishan Bichajian. © The Estate of Gyorgy Kepes.

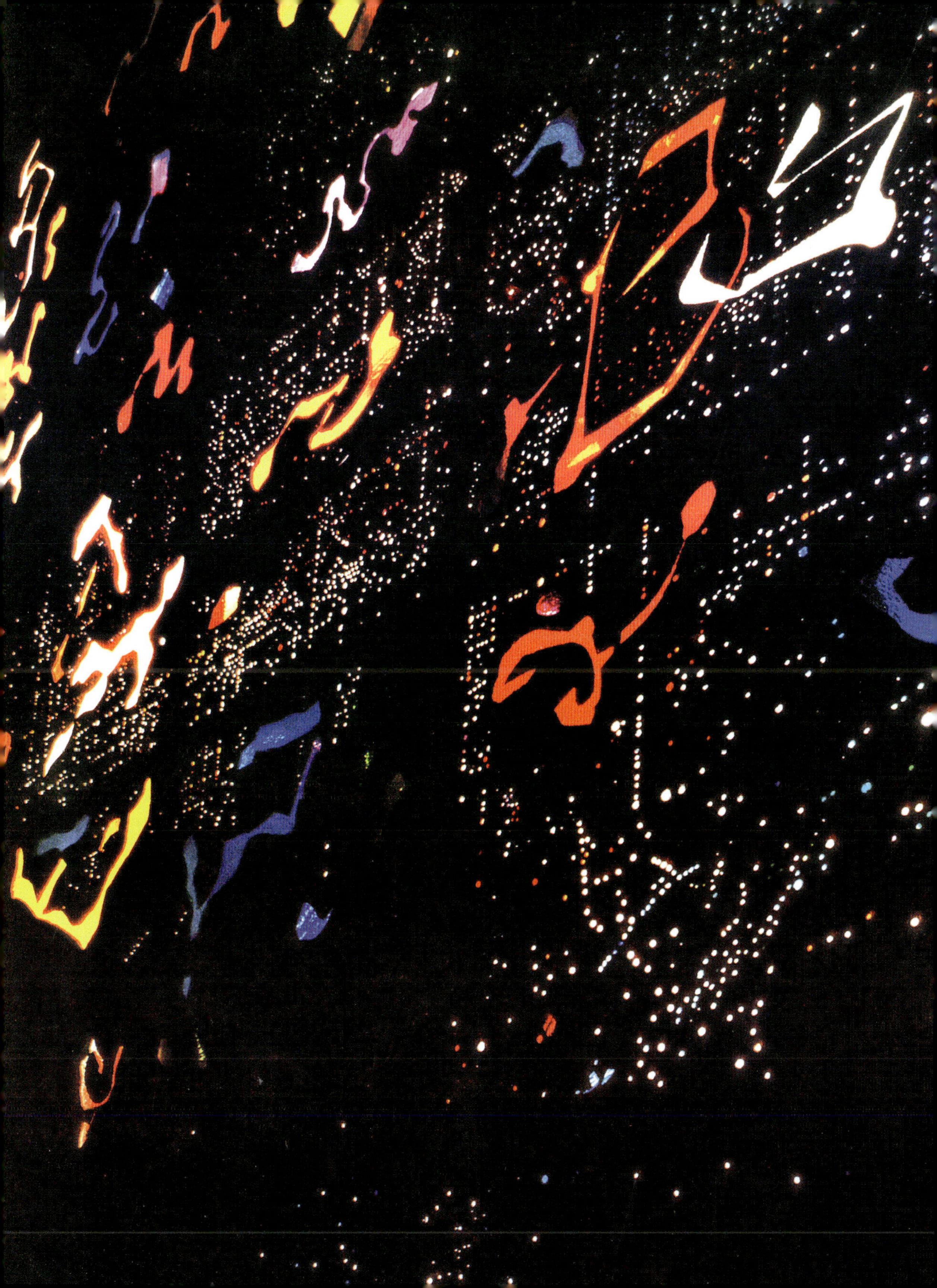

And yet, during war it gained deadly significance. From the B-17 or the B-29, such perception was always-already militarized, the incredible optics transmitted through the most advanced military technics. A series of technologies filtered sight in unusual ways, creating what Moholy had called the "new vision."[12] The flying mechanical fortresses allowed human sight to take to the skies in the first place, and their Plexiglas windows—made of a thermoplastic acrylic—created vistas of the world. But it was a more elaborate apparatus that militarized vision: the Norden bombsight.

The M-series bombsight, invented by the engineer Carl L. Norden and classified top secret by the US military, was a tachometric or gyroscope-stabilized targeting device that employed a series of manually operated telescopes and an analog computer to calculate a bombshell's ideal parabolic trajectory (the device's spinning gyro mechanisms kept its optics aligned to the earth during turbulent bombing runs). The instrument was one of the central components in the US Army Air Force's aerial bombardment strategy. The goal was to use the bombsight's pinpoint accuracy to destroy the nodal points within the enemy's war machine. Identifying these high-value assets also required a new set of interdisciplinary methodologies like systems analysis and operations research, hybrid approaches that allowed the Allies to map the linkages sustaining the enemy's military-industrial complex.[13]

For our purposes, the bombsight is most significant for its dramatically derealizing visual effects, or what Paul Virilio has termed the "logistics of perception." As Virilio explains, "the function of the weapon is the function of the eye."[14] Through the bombsight, the bombardier's sight was mechanically deformed and disturbingly mediated, even virtualized (the device's name etymologically registers this function: bombsight created "bomb sight"). The vertical or oblique perspective, refracted through the instrument's radial viewfinder, was like tunnel vision, reducing the image rushing by to an abstraction, an assemblage of targets hidden within a screen of meaningless visual noise. After the bombardier successfully located a chosen objective, he simply locked it in the sight's crosshairs. The device engaged automatically, even taking over the airplane's flight in order to initiate attack.[15] Bomb bays opened, racks released, and a payload of deadly projectiles fell (fig. 1.6). Allied Forces dropped some 2,700,000 tons of bombs on Europe and some 656,400 tons over the Pacific.[16]

Air war was total war: cataclysmic and apocalyptic. "The war in the air was war pure and undisguised," explains W. G. Sebald.[17] To properly read the aerial image was to know—and ultimately destroy—the enemy; on the other hand, to persuasively deceive through camouflage was to save oneself. Mastering or

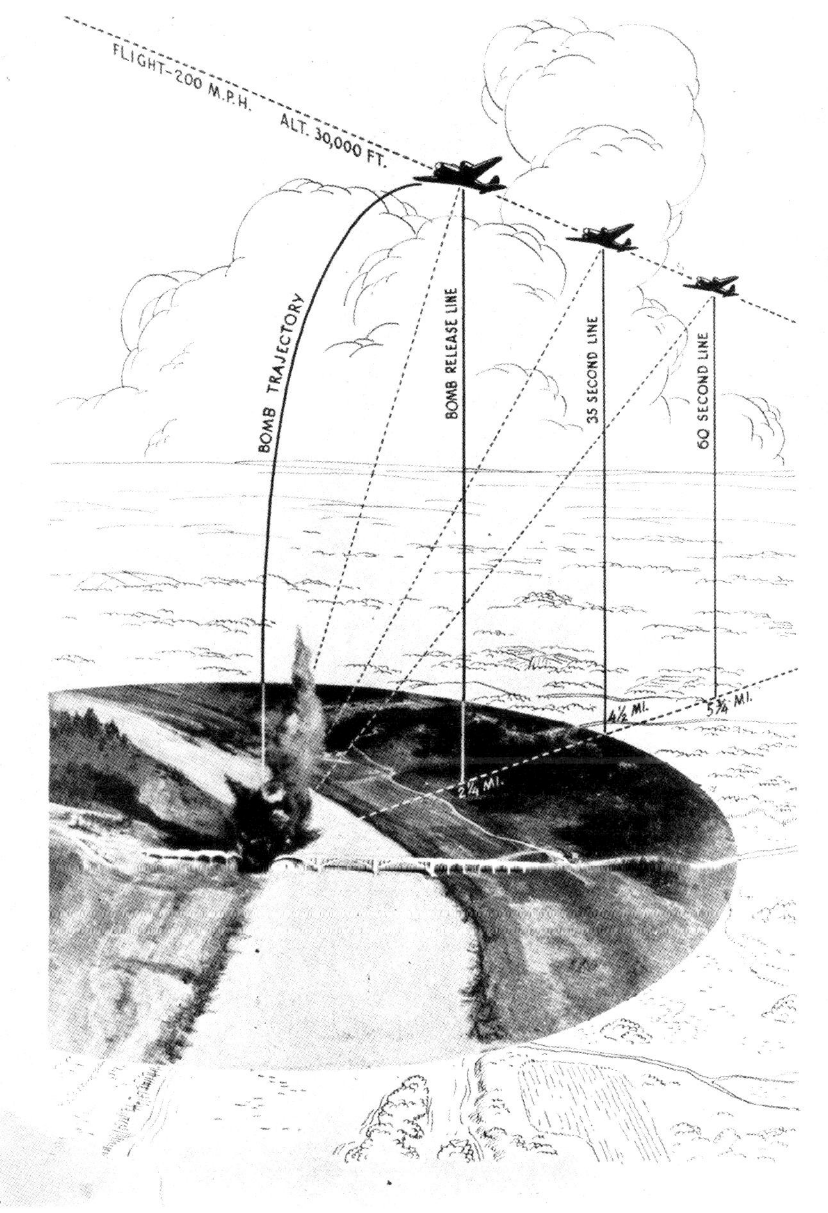

1.6

Bombing diagram. From Robert P. Breckenridge,
*Modern Camouflage: The New Science of
Protective Concealment.* New York: Farrar
& Rinehart, 1942.

misleading a newly militarized visual experience became an urgent matter of life and death. The outcome of war depended on the proper coding or proper decoding of the image of the earth as seen from the sky.

For American civilians, air war was largely understood at a remove, but they became increasingly aware of their visibility and vulnerability after Pearl Harbor. As Peter Galison explains, they began to see themselves reflected in the bombardier's eyes and mirrored in the bombsight's crosshairs.[18] This epistemological shift displaced the frames of self and other, inverted enemy and friend, confused Axis and Allied, and suggested that bombs that could be delivered might also be received.

The frightening imperatives of air war therefore demanded expertise in visual culture, an approach to the image that could make sense of a strange new optical battlefield. The US military required a modern—or, more aptly put, modernist—visual thinker in order to conceptualize techniques of deception. They found this expertise with Gyorgy Kepes, whose rhetoric of the image clarified the exceedingly complex mechanical and human logistics of aerial bombardment, the system of vision built into the bombsight, as an essential question of how to see and how not to be seen. Alongside a surprising network of artists, architects, and designers that spanned the United States—as well as leading scientists, military bureaucrats, government functionaries, corporate executives, and university administrators—Kepes became a key figure in the US camouflage effort. Hannah Rose Shell has argued that camouflage is a unique media phenomenon, and both Jean-Louis Cohen and Jason Weems have revealed the extent of these projects.[19] This chapter focuses particularly on Kepes's unique role in this history.

The Principles of Camouflage

On 13 January 1942, a tabloid newspaper named the *Chicago Herald American* ran an alarming front-page banner (fig. 1.7). The headline blared: "How Chicago May Hide from Bombers!" The image below depicts an ominous formation of what a caption crudely identifies as "German-built Jap bombers" on the approach. But the attack would be thwarted, the city saved: "For if camouflaging experts now feverishly at work have their way, no invading air armada will be able to detect our loop's great skyscrapers." The article elaborates:

> Learning how to conceal Chicago's roofs from possible attack by enemy bombers, a hundred art students are perfecting the art of camouflage by experimenting daily with light, shadows and color mixing. Their training is conducted at the School of Design in Chicago, 247 E. Ontario St., by George Kepes.

A photograph at left shows Kepes, described as a "specialist in camouflage" and a "great camouflage artist," hovering over model airplanes with his student Nathan Lerner. Kepes tells a reporter about his latest research into the city's "light framework," describing how he could distort the urban pattern by covering street lamps with specialized materials, thus breaking "the visual symmetry" of the street grid. Kepes even claims that "it will be possible to change the air view of Chicago's State St. into a lagoon, or the appearance of the Merchandise Mart"—with some four million total square feet, the world's largest building by floor space until the Pentagon's completion one year later—"into a prairie." The newspaper illustrates this latter proposal through a series of absurd diagrams comparing the structure's colossal size against a "Bomber's View From 20,000 Feet!"—what appears to be an aerial view of pine trees—and an "On-the-Roof Closeup!" that merely shows one of Kepes's design charts smudged with pigment. A special coat of "Forest Green" will do the trick. "Just a few daubs of paint—that's all it takes to create this phenomenal optical illusion."[20]

The *Chicago Herald American*'s sensationalism suggests the degree to which world war became media spectacle after Pearl Harbor. The article appealed to the manic urgency that surrounded US entry into the war, but it also registered the public's fascination with the novelty and newness, the uniquely modernist sensibility, of these curious camouflage projects. In framing Chicago as target, the newspaper also presented a problem that most readers had never contemplated. Can the visible world be made invisible? Even the city's streets, its highways and railroads—even its downtown skyscrapers? Not the camouflage of popular conception—little concerned with specific decorative motifs or particular patterns—these proposals instead revolved around innovative forms of visual thinking.

This type of research quickly became an established field of wartime study called "industrial camouflage." Unlike military camouflage, industrial camouflage occurred on the home front, far from the line of battle; it was practiced by Allied and Axis powers alike. The goal was disguise of the total environment through its comprehensive transformation into a cryptographic aerial image—and this meant elaborately concealing airfields, factories, fuel tanks, bridges, canals, and even topographical features that might serve as directional indicators pointing toward high-value assets. Industrial camouflage was highly abstract, requiring the conceptualization of these features at an exploded perceptual scale. The camouflage specialist had to anticipate the bombardier's aerial view, mentally adjusting the standard image of the environment to one miles above through an imaginative envisioning of the world

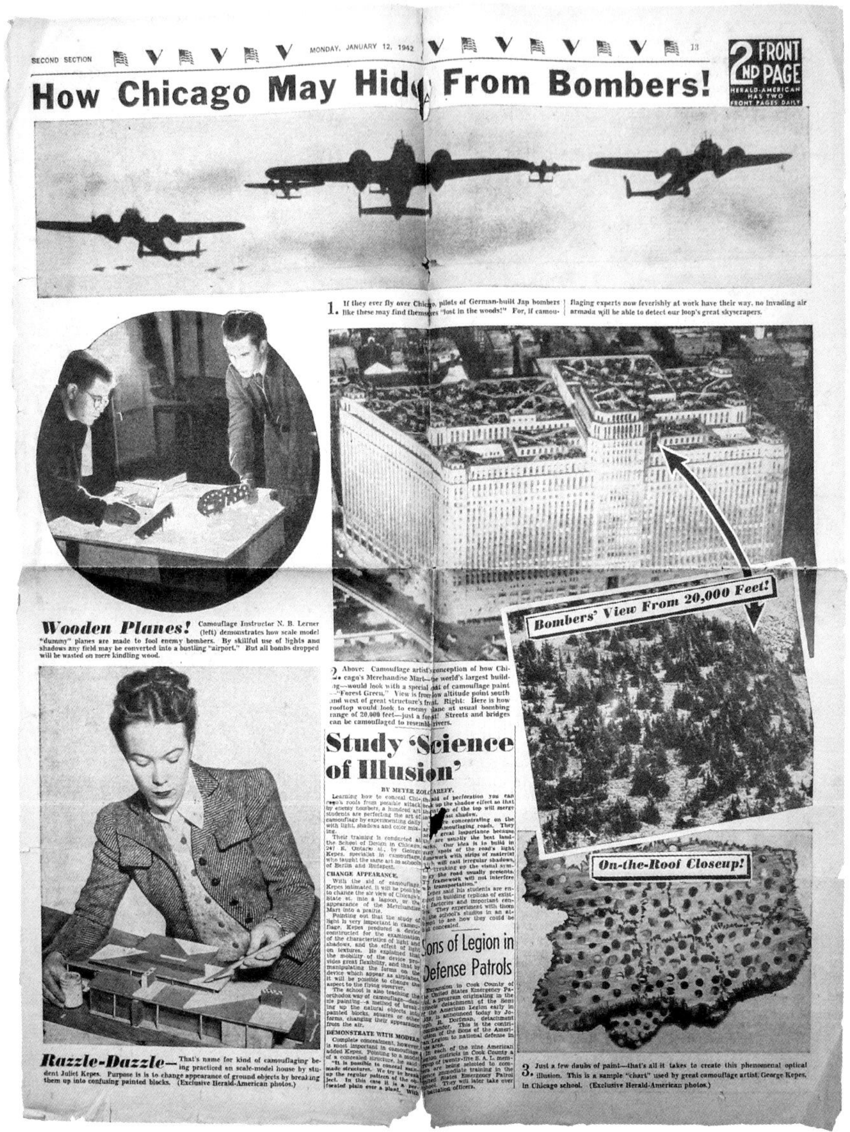

1.7

"How Chicago May Hide From Bombers!"
Chicago Herald American, 13 January 1942.
Courtesy Márton Orosz.

as pictorial. The specialist then subjected this abstract image to a visual game of formal relations: the goal was to collapse a target into the picture plane of the aerial view—to merge figure into ground, to hide signal in noise—by employing effects like pattern interference and gradient distortion.

A series of "camouflage manuals" suggests the wide scope of these projects. The US War Department mass-produced profusely illustrated field books on camouflage for training purposes. But so, too, did the Art School at Pratt Institute; their "Industrial Camouflage Manual" became the definitive guide on the subject. From *The Art of Camouflage* to *Modern Camouflage: The New Science of Protective Concealment*, these volumes suggest how industrial camouflage forged an interdisciplinary ethos between art and science.[21]

In practice, industrial camouflage took the form of all-encompassing engineering projects requiring elaborate constructions—textures plastered over tarmacs, patterns splashed across buildings, and architectural alterations that would disrupt outlines, destroy forms, and blend backgrounds. These constructions often looked ridiculous on the ground, but still deceived the viewer on high. The most bizarre example might be the Boeing Plant II in Seattle (figs. 1.8–1.9).[22] Boeing's massive factories were paradoxically rendered vulnerable to the new vision that its B-17s and B-29s facilitated. In response, Boeing technicians superimposed a deceptively banal simulacrum of suburban sprawl on the factory's roof: tidy yards, parked cars, even ranch houses, all furnished in cheap materials as if the roof was a studio lot or stage set. "Atop the Boeing plant went a 26-acre village made of chicken wire, canvas, lumber, [and] painted chicken feathers," reported *Time* magazine. "The town had 53 houses, stores, a gas station." Seen from the sky, the factory appears to collapse into the city grid, fused into the visual static of the urban array like any other neighborhood. These were hardly isolated projects. The price tag for transforming just 37 US factories was $22 million—more than $300 million when adjusted for inflation.[23]

Industrial camouflage involved constructions on the ground but also required more abstract articulations of visual fundamentals, formalist depictions of the world as picture. In "The Age of the World Picture," a lecture delivered in 1938, Martin Heidegger diagnoses the effects of modernity on phenomenological existence, arguing that science and technology transform our relationship to reality; the world is "conceived and grasped as picture" for the first time. Experience is not immediate but mediated. The world is "set out before oneself" or "set forth in relation to oneself" as an external representation. Heidegger understands this transformation as violent, and declares: "The fundamental event of the modern age is the conquest of the world as

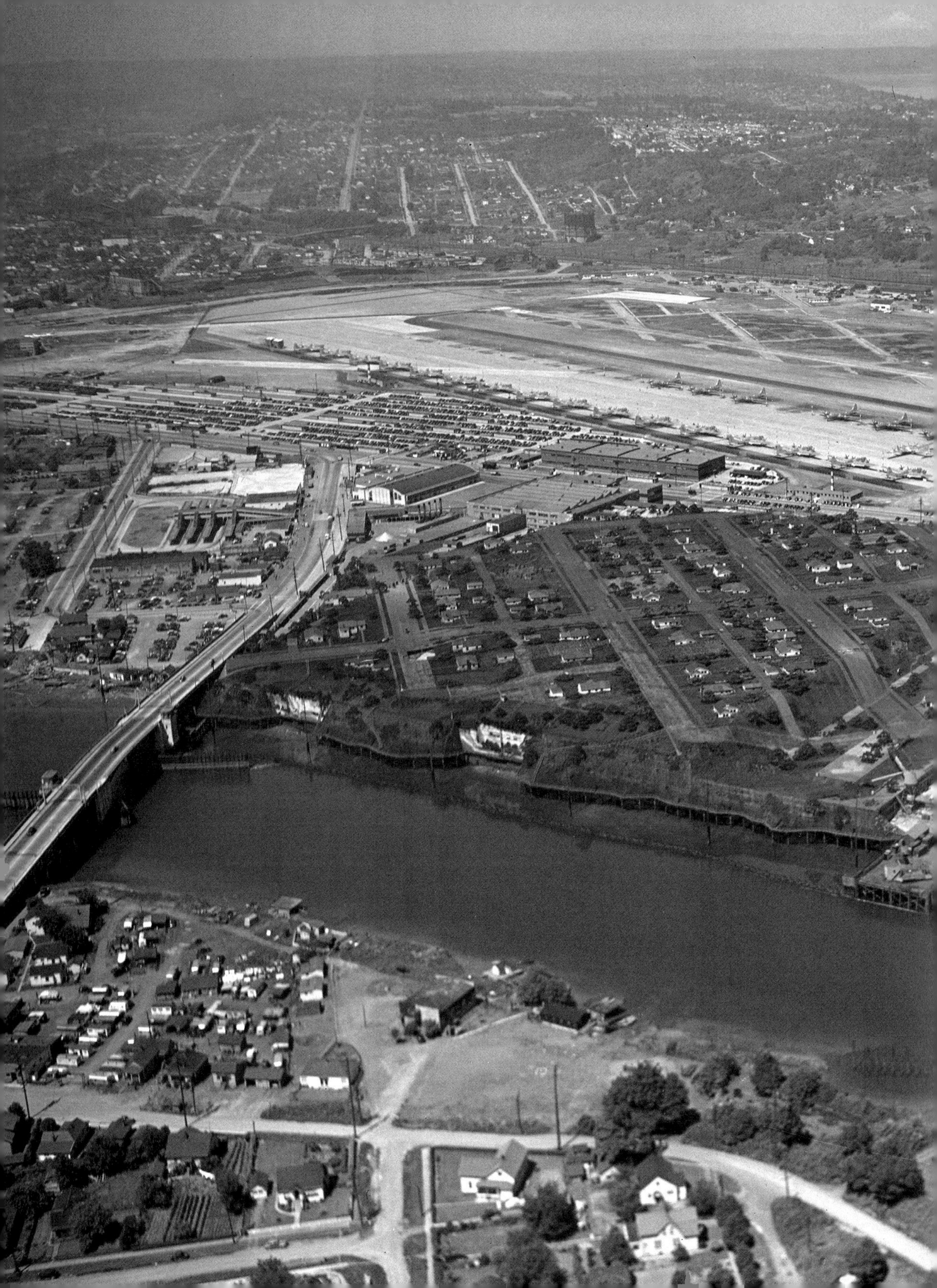

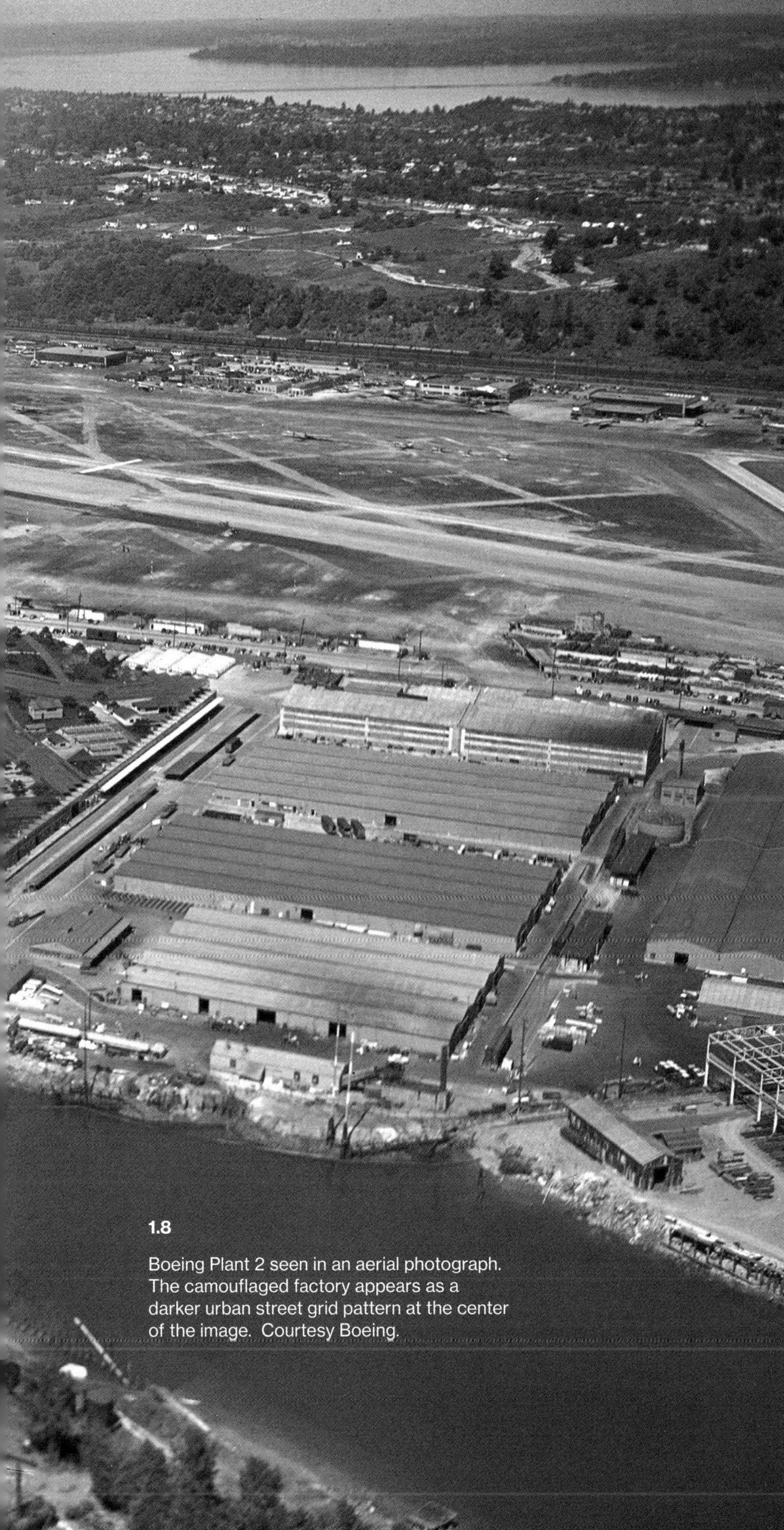

1.8

Boeing Plant 2 seen in an aerial photograph.
The camouflaged factory appears as a
darker urban street grid pattern at the center
of the image. Courtesy Boeing.

1.9

Boeing Plant 2 seen in a vertical photograph.
The camouflaged factory appears as a
darker urban street grid pattern at the lower
left of the image.

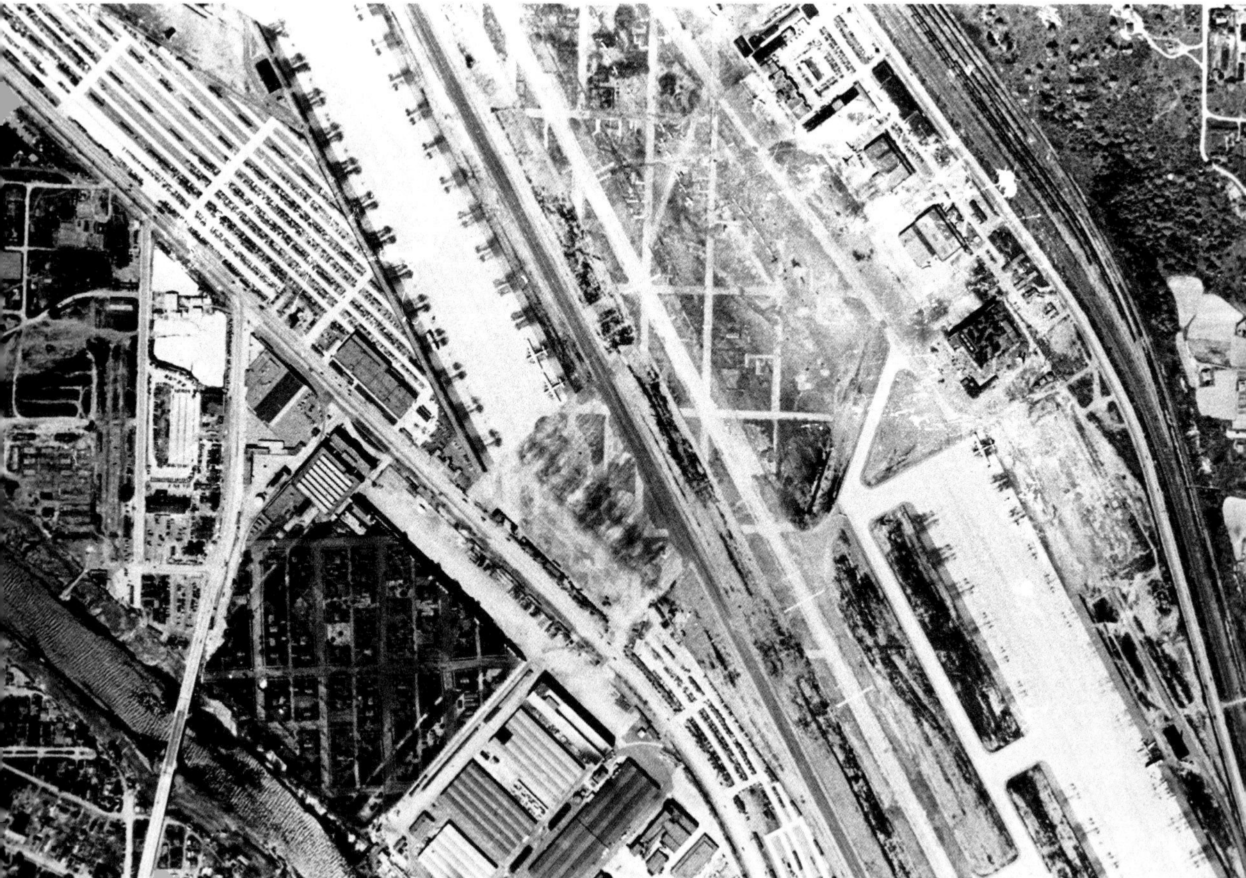

picture."[24] The perception of this picture created a confusion between the work of art and the world itself, a misrecognition decisively evidenced in art-critical commentary. The *Architectural Review* noted that camouflaged installations, when seen from the sky, had "a curious similarity to the conscious creations of modern art." In reference to an aerial view of a German airfield captured by Allied Forces in the North African desert, the magazine clarified: "This is not a Paul Klee picture."[25] Kepes was soon entangled in this confusion.

Camouflage Pedagogy

After Pearl Harbor, the School of Design immediately offered a full program of "National Defense Courses," including "Visual Propaganda in Wartime" and a survey on the "Social Usefulness of Twentieth Century Art and its Relation to a Nation at War."[26] But the "Principles of Camouflage," a course organized and taught by Kepes, would become its most significant wartime offering.[27] A detailed outline for the course reveals how Kepes wielded Bauhaus pedagogy to these unusual ends. All Bauhaus teaching reduces the complexities of design to a set of formal investigations that can be demonstrated through creative experimentation. By applying this approach to the horrifying logistics of air war, Kepes similarly reduced aerial bombardment to an essential problem: one of seeing. "The goal of the aerial observer is to find and fix a target," he explains, and doing so required perceiving the target in a Gestalt relationship, as a figure against a ground. "To grasp an object one needs first to isolate it from its surroundings," to see it articulated through "the basic elements of a visual language."[28]

This language was radically different in the sky. Extended technologically by aircraft, mediated by bombsight, the eye was now emancipated from the earth and liberated from the body, suspended in and propelled through a rapidly shifting matrix. Vision in the air was like that of the transparent eyeball, free-floating and sublime. (Ralph Waldo Emerson: "I become a transparent eye-ball; I am nothing; I see all; the currents of the Universal Being circulate through me; I am part or particle of God.")[29] Kepes elaborates:

> **The aerial observer for whom camouflage has to be largely considered today, is a mobile observer. Every factor involved in his vision is in continuous movement. His eye is moving, the light conditions are changing and the elements of the landscape are moving.**[30]

This experience therefore required a new conception of vision, and how it might be quantified and then manipulated. While standard Bauhaus approaches concerned relatively stable organizations of pictorial elements bounded by a two-dimensional picture plane or three-dimensional space, camouflage demanded

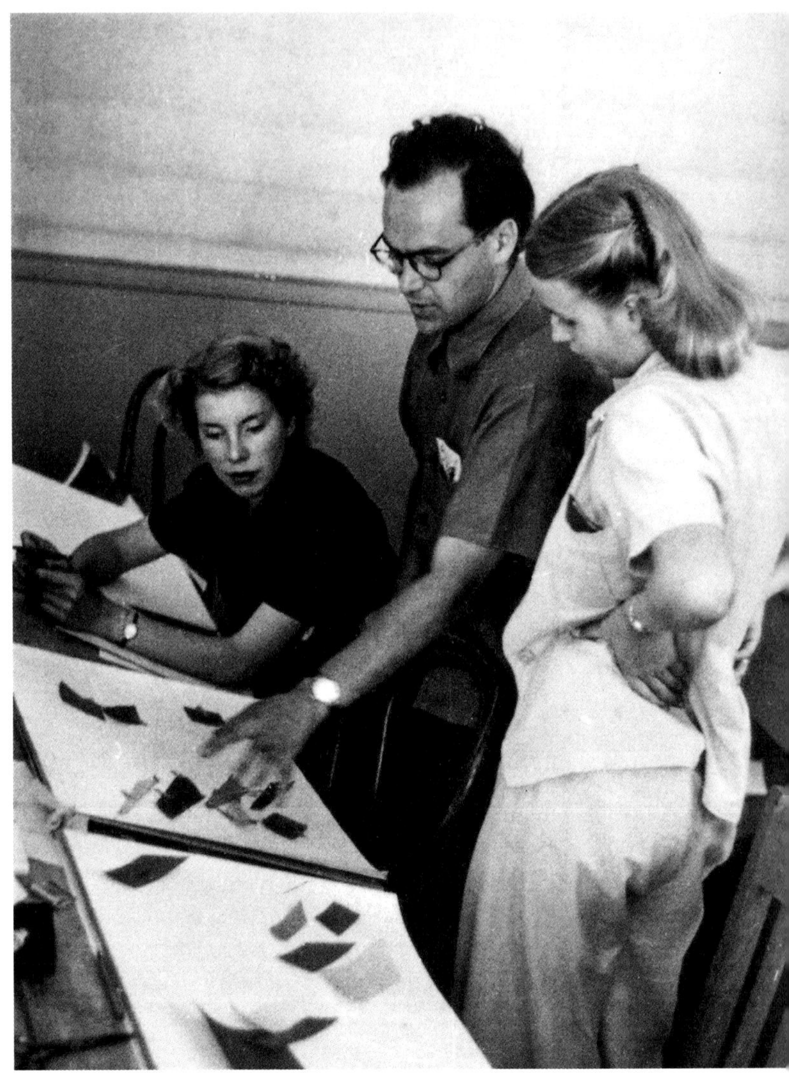

1.10

Gyorgy Kepes with students, c. 1940. Institute of Design Collection, Ryerson & Burnham Archives, The Art Institute of Chicago. Digital File #200702_180720-001.

a conception of vision as a dynamic experience of constantly shifting variables. "This is the reason why at the beginning of camouflage teaching there is a basic investigation of point, line, tone, color, shape etc., put in all their possible interrelationships." Kepes elaborates these interrelationships:

> To mention the most important ones: psychological and physiological phenomena such as figure ground relationship, effects of similarity, closure, inclusiveness, submergence, optical mixture of tone and color values; border, successive and simultaneous contrast, various forms of visual illusion; perspective, size, motion and so on.

In addition, students would "work with light and shadow," investigating "the control of diffusion, reflection, refraction and all other aspects of illumination."[31]

Kepes's existing courses were easily adapted for these purposes (fig. 1.10). Since the fall of 1937, he had conducted lessons in Visual Fundamentals and Light and Color, often through exercises using cameraless photographic techniques like photograms and photomontage and an array of simple devices: mirrors, screens, lenses, and the so-called light modulator. At its most basic, this light modulator was nothing more than a sheet of paper, cut with scissors, folded and rolled, that reflected and absorbed light from its concave and convex surfaces. These tools and the exploratory approaches they facilitated have been extensively studied as formalist artistic investigations, but they acquired a new valence when submitted to the demands of camouflage. Moholy argued that one could use the light modulator "to fit any purpose"—and evidently that included camouflage.[32] In fact, anything could become a light modulator. "In a sense, the world can be conceived as being composed of innumerable light modulating surfaces or substances," Kepes writes.[33] Roads, railways, skyscrapers, even the Merchandise Mart, could be visually shaped and molded by light.

Kepes's student Nathan Lerner constructed another apparatus that enabled the study of essential optical phenomena: the light box. The device, conceived by Lerner just three weeks into Kepes's first course at the New Bauhaus, functioned as a controlled environment that filtered light from outside interference. "I made a box," he explains,

> which was open on one side and with many holes cut into all sides. These holes were used for suspending objects and also served as opening for light to enter. Over these openings, objects (wire screen, etc.) could be placed and projected on the materials inside the box.... This would enable one to study and photograph light in a purer form, as a beam.[34]

Lerner's light box isolated light by removing visual noise. Pages from a series of articles on the School of Design's wartime activities that ran in the magazine *Civilian Defense* show how it could also be used in camouflage research (fig.

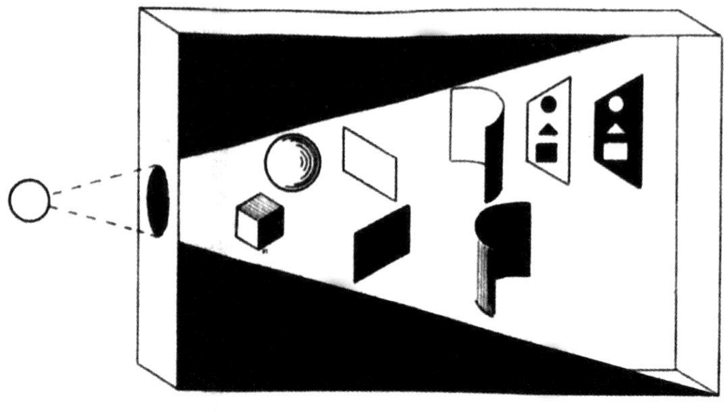

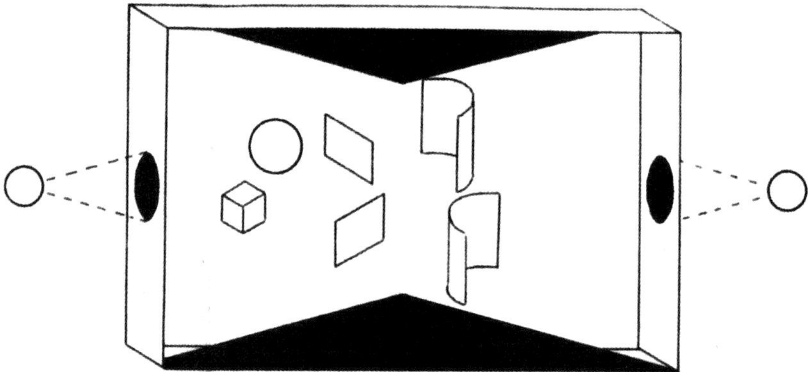

FIG. 25. The diagram at the top illustrates some of the characteristic shadow formations created by a single strong source of light. In the diagram below, another opposing source of light completely counteracts the first, making the objects invisible by blending them into their immediate surroundings

1.11

Diagram from John L. Scott, in collaboration with L. Moholy-Nagy and Gyorgy Kepes, "Materials for the Camoufleur," *Civilian Defense* (September 1942).

1.11). One diagram depicts a cube, a sphere, a surface, and a cutout piece of paper incised with the essential Bauhaus shapes—circle, triangle, and square—hanging in a light box. Illuminated from a single source, distinct countershading reveals these suspended forms in relief. But a double light source renders the forms invisible, camouflaged from view.

Typed and profusely illustrated notebooks compiled by Kepes's students in the fall of 1942 offer more clues about Kepes's teaching. Presumably created in anticipation of a manual to be released by the School of Design ("The findings of the class are being accumulated in a book," one newspaper reported), the unpublished volumes reveal a fantastic imagining of visual experience in the air—what we might call a "bomb imaginary" or "strategic sublime."[35] The opening image in M. Seklemian's set of notes—Seklemian became a prominent graphic designer in the 1950s—registers this fascination as a single, free-floating transparent eyeball (fig. 1.12).[36] Other diagrams from Seklemian's notebook demonstrate how Kepes repurposed standard Bauhaus lessons. One shows how the aerial observer, hovering at the oblique angle of attack, experiences the laws of perspective; others demonstrate how cast shadows reveal a building's function or how common optical illusions might be applied to camouflage (figs. 1.13–1.15).

Other pedagogical exercises from Kepes's course departed from typical Bauhaus teaching; Kepes offered lessons not only on visual fundamentals, but also on the "fundamentals of aerial bombardment," including "typical bombing problems: the relationship to altitude, speed, distance, angle etc., time of bomb release, trajectory, visibility of target; daylight and night bombing; precision bombing, dive bombing; area bombing; demolition bombing."[37] Lectures on the "methods and techniques of aerial attack" and "the equipments [*sic*] and techniques of warfare related to the camoufleur's task," as Kepes puts it in his introductory lecture for the course, are surprising: arts education is presented as military training.[38]

War is Interdisciplinary

At issue in Kepes's camouflage pedagogy is therefore the problem of transvaluation—how the methods and methodologies for making art became commensurate with the strategies and tactics for making war. "Visual Fundamentals" and "Light and Color," courses that sound ideologically blank—focused on the purely formalist investigation of universal design principles—now accommodated very precise military applications.

The question of transvaluation concerns not only the applications of previously neutral classroom exercises, the uses to which Kepes now put his light

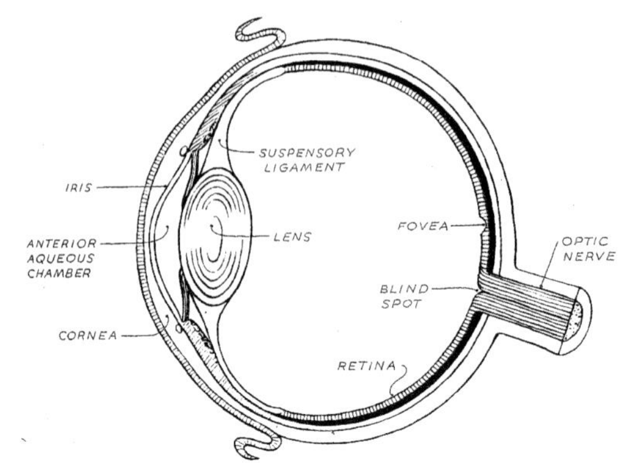

THE HUMAN EYE

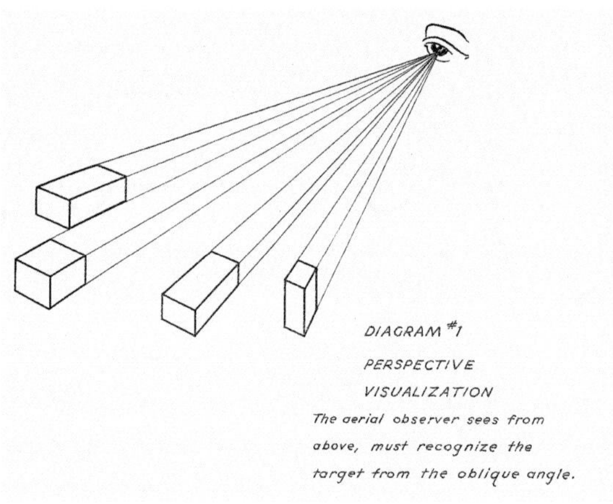

The aerial observer sees from above, must recognize the target from the oblique angle.

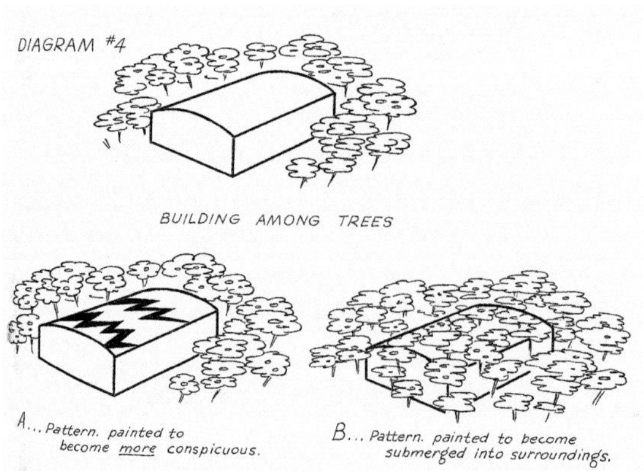

A... Pattern. painted to become _more_ conspicuous.

B... Pattern. painted to become submerged into surroundings.

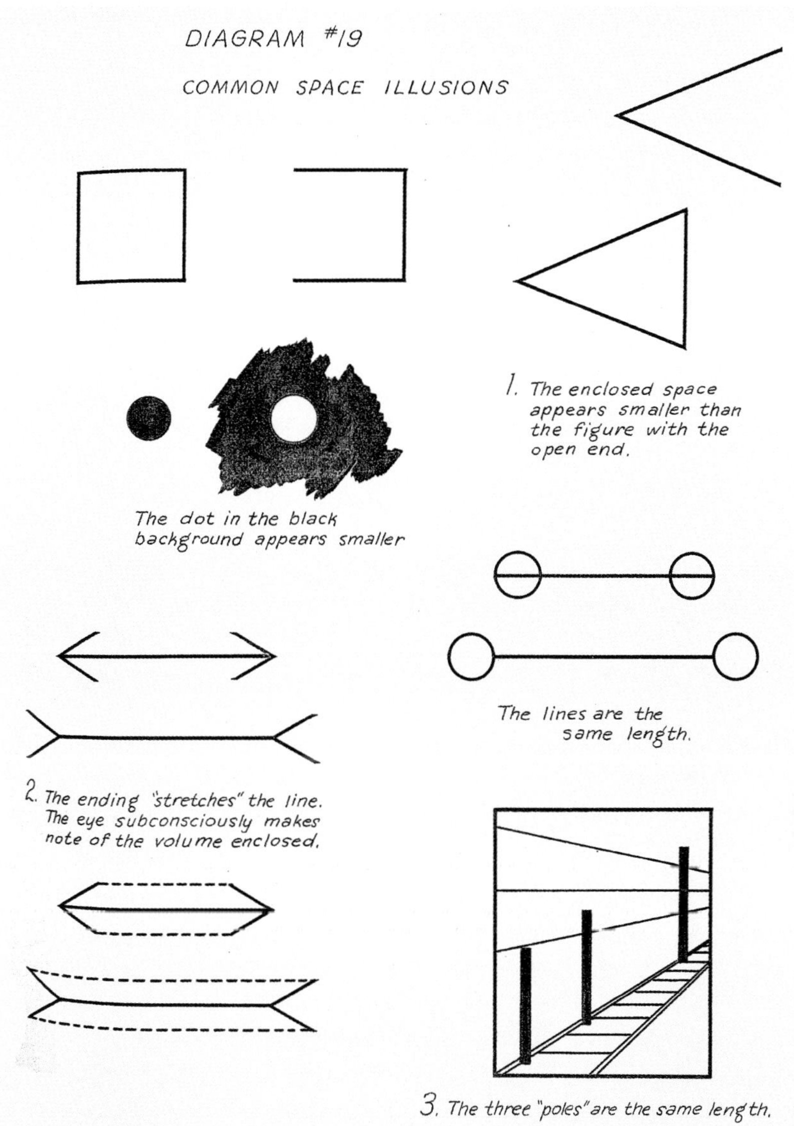

1.12–1.15

Diagrams from M. Seklemian, "A Study of the Principles of Camouflage," set of typed course notes dated September 1942–January 1943. Courtesy Illinois Institute of Technology.

modulators and light boxes, but also the way he taught the course in the first place. Kepes's pedagogy, as war pedagogy, required interdisciplinary exchange; it demanded innovative approaches that could merge knowledge from disparate disciplines in order to solve problems of disguise that defied single fields of inquiry. This dimension of camouflage work fascinated Kepes; his course outline explicitly rejects the "inertia of professional isolation" characteristic of field-specific forms of research and instead embraces procedures that would break boundaries, blur distinctions of discipline, and integrate knowledge, all in order to fight the enemy. Kepes explains this goal in reference to a previous generation of camouflage research during the First World War:

> In [previous] camouflage the painter saw only painting problems, the architect only architecture, the engineer only engineering. They were unable to coordinate their expert knowledge with the necessary flexibility which emerges from a mutual grasp of each others [*sic*] problems.

Kepes's course would, by contrast, embrace such flexibility by creating a "research laboratory" where "mutual cooperation" would unite specialists. This approach would not only offer "better solutions" to specific problems but would also "develop a new creative type of research worker." Innovating through collaboration would produce a visual thinker whose knowledge spanned the expertise of art and science. Bauhaus design pedagogy was ideally suited to this mission. "The School of Design in Chicago[,] because of its past educational policy[,] readily adapted its program to these requirements."[39] Kepes even opened the introductory lecture of his camouflage course with a declaration on the power of interdisciplinary exchange:

> Camouflage requires the combined knowledge of people with a great variety of training – architects, engineers, painters, sculptors, graphic artists. They are finding a synchronization of their divergent knowledge in the fulfillment of this urgent task. This synchronization may be achieved only through a methodic investigation into the purpose and scope of camouflage, and a mutual exchange of knowledge in each particular field.[40]

Kepes also borrowed these ideas from Charles W. Morris, the philosopher and semiotician (and a later collaborator) who taught "Intellectual Integration"—a course organized around the same principle of interdisciplinarity—at the first iteration of the New Bauhaus.[41]

Kepes's rhetoric was sincere; he enlisted an impressive panel of military and civilian experts from a range of institutions to give guest lectures on subjects spanning mathematics, biology, zoology, and medicine, in addition to conventional artistic topics. The original course prospectus proposed a long list of guest lecturers, among them Jack C. Copeland of Riggs Optical Company, who spoke on "Physiological Optics"; ornithologist Rudyard Boulton, who discussed "Camouflage in the Animal World"; and chemist Theodore Ashford, who lectured on the "Chemistry of Artificial Fog."[42] Kepes's approach—what he provocatively terms "flexibility" and "synchronization"—enabled communication between specialists who were unlikely or even unable to communicate with one another under normal circumstances. War demanded pliable methodological tools, and prompted unusual intellectual arrangements. Such methods also had lasting ramifications, as these approaches became standard operating procedure through discourses like cybernetics and information theory. Camouflage offered the aesthetic equivalent to such methods.[43]

This model of interdisciplinarity through camouflage was also put into practice outside the classroom. The city of Chicago took Kepes's camouflage plans, as detailed in "How Chicago May Hide From Bombers!," very seriously. The Chicago Metropolitan Area's Department of Techniques, part of the city's civilian defense initiatives, invited Kepes and Moholy to participate, alongside some 250 experts, in the design of "plans for passive resistance in the Chicago Metropolitan area." Kepes and Moholy represented the School of Design on a special division of "Consultants and Experts" that included figures from "Engineering, Science, Water Navigation, Electrical, Chemical and Laboratory Tests, Construction, Land Transportation, etc." These experts also had a range of professional affiliations: "Steel Industry, Apartment Buildings, Department Stores, Educational Institutions, Electrical Industry, Hotels and Apartment Hotels, Office Buildings, Hospitals, Meat Packing, Railroads, Theaters, etc." Together, the division produced some "2,000 maps, charts, tables, etc." and conducted various "Camouflage Lighting in Motion" tests that involved observation of the city from aircraft.[44]

It was in this context, through these exchanges, that Kepes developed his most imaginative camouflage proposal: an optical simulation of the city itself. "I was on the mayor's camouflage committee in Chicago," Kepes remembers, recalling his work with the Department of Techniques.

> **We tried to re-pattern the light structure of Chicago for camouflage purposes. Areas of dense light were obliterated and lightless areas like parks and the shore of Lake Michigan were brightly illuminated. In a sense, the manipulation of artificial light is a new aspect of nature that must find its own artistic expression.[45]**

A series of powerful lights, strung on cables and floated in Lake Michigan, would illuminate a network mirroring Chicago's gridiron streets. While the city was blacked out, enemy bombers, mistaking these patterns of light for the city grid, would drop their bombs directly into the water.[46] Kepes's student Robert Preusser, who helped him with these "analytical studies" and whom Kepes later helped hire at MIT, recalls this plan for the "camouflaging of Chicago itself": the "enemy would think that the lake was the city."[47]

Disguise by Bureaucracy

Camouflage brought together painters, sculptors, architects, and designers, but also scientists, corporate executives, government officials, military bureaucrats, and university administrators. In the pursuit of concealment, very opposite, even opposing, institutions and individuals became strangely entangled. All succumbed to a generalizing pattern. The artist dissimulated as military scientist, the military scientist dissimulated as artist; both were one and the same, blurred by camouflage. This doubling and confusion occurred through the technocratic protocols of the wartime research laboratory. Fort Belvoir, Virginia, a military installation where the Army Corps of Engineers operated the Army's Engineer School, was ground zero for this study (fig. 1.16).[48] It was here that Kepes would learn the theory and practice of camouflage as officially sanctioned by the US government.

This relationship was set in motion just weeks after Pearl Harbor when Moholy visited Washington, DC, in order to appeal for government funding—the School of Design was close to insolvency after the war began. He approached the US Office of Education, which had established a special commission to administer a program called Engineering, Science, and Management War Training (ESMWT) that subsidized college courses for civilians; the program served as a stopgap for a shortage of trained experts, as its students would then take defense jobs. After his initial meeting failed to generate interest in the school, Moholy wrote to the director of the Office of Education:

> In the middle of January you kindly informed me ... that your Commission is working on war-time speed up plans for higher education, also on the replanning of college curricula in order to adapt these to the requirements of national defense. Since this time I have been working on plans as to how we could offer our services to your Commission.

Moholy argued that the "educational work of the Bauhaus" was an ideal subject for ESMWT courses; "the integration of art, science and technology" could teach all students, even those who were not artists, to be "resourceful and inventive, quick in decisions, [and] courageous in approaching civilian and military tasks."[49] He proposed ESMWT-sponsored Bauhaus courses across the country; his hope was to secure a role for the School of Design in training new instructors to run such courses, thereby giving the school a secure role in the government program (this was the model adopted for ESMWT courses in technical fields like radar, for which elite institutions trained instructors who then taught trainees elsewhere).

The Office of Education declined the suggestion—its sponsorship was strictly for engineering, not art—and routed Moholy's request instead to the Office of Civilian Defense or OCD. Headquartered in Washington, DC, the

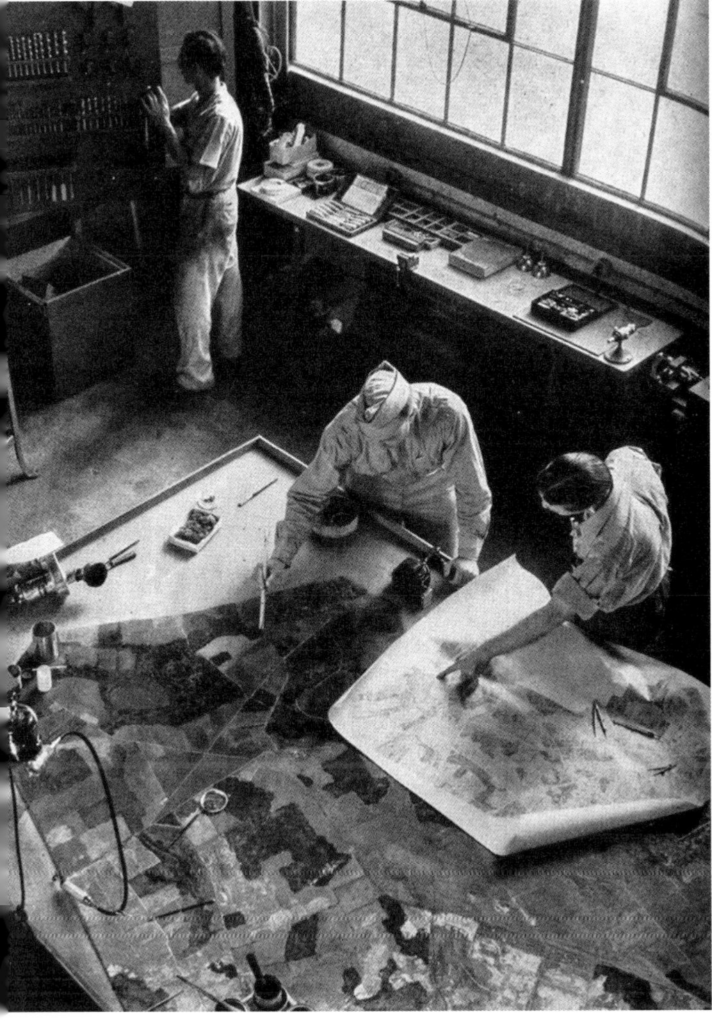

1.16

Camouflage laboratory at Fort Belvoir,
Virginia. From Robert P. Breckenridge,
*Modern Camouflage: The New Science of
Protective Concealment*. New York:
Farrar & Rinehart, 1942.

OCD was a sprawling bureaucratic apparatus with nine designated regions across the country that administrated the mundane procedures of emergency management: blackouts, air-raid shelters, warning sirens, fire protection—and camouflage.

Moholy wrote to Colonel Milton C. Mapes, Chief of the OCD's Engineer Section, explaining the School of Design's financial predicament and asking for "financial assistance."[50] In an attempt to impress the seriousness of the School of Design's wartime commitments on OCD officials, Moholy also mailed the office a copy of Kepes's "Principles of Camouflage" syllabus and a typed sheet of proposals conceived by Kepes and his students that year.

Titled "Research Problems in Camouflage at the School of Design in Chicago," these suggestions reflect the highly experimental dimension of Kepes's camouflage research. Many plans involved the use of huge pneumatic tubes made of rubberized cloth that could be molded into balloons or pillows. Suspended between skyscrapers, these forms would alter the outline of the built environment. Floated on lakes, they would create false shore lines. Oil tanks might be concealed under giant cushions; factory smoke stacks might be hidden among puffed-up columns. Other plans involved "light constructions" mounted on trucks that might simulate illuminated "town units" that could "appear and disappear at different places within brief intervals." False buildings and trees on wheels could be carted onto airfields when not in use. Camouflage specialists could create deceptive railroad tracks by painting fluorescent lines on open ground, or erase rooftops by covering them with reflective pools of water. The most unusual idea involved floating reflective balloons above the city so they could "catch the light from the searchlights creating confusion."[51] This last proposal conjures a vision of pyrotechnics, disguise by light show. These were highly imaginative if frankly bizarre and impractical ideas—all smoke and mirrors.

The OCD's response to Moholy's overtures was coolly bureaucratic. "The contents of your program and letter have been carefully noted with interest," L. D. Gasser, Major General in the US Army and Chief of the Protection branch of the OCD, responded generically. "The camouflage ideas you have submitted will be studied for whatever merit may be seen in them."[52] The OCD was inundated with similar proposals—though the School of Design's were uniquely strange—from hundreds of would-be *camoufleurs* (it received some five hundred letters per month).[53] To all of them, Major General Gasser responded in the bland tone of the government form letter: "You may be interested in obtaining the recently issued booklet 'Civilian Defense—Protective Concealment' from your local defense council."[54]

Around the same time, Greville Rickard—an architect who chaired the Civilian Camouflage Council of New York, a group that lobbied for various camouflage initiatives—also contacted Major General Gasser and Colonel Mapes of the OCD about "the serious lack of enough trained advisors in camouflage in the event that bombing raids might require such service." Mapes found Rickard persuasive—he was, after all, a self-styled camouflage expert—and agreed to join him in meeting with the Office of Education to argue for government-funded camouflage instruction. He even had Rickard appointed to lead the OCD Camouflage Unit. But the Office of Education was still unconvinced; they "could not sponsor, nor pay for, a course that was not given in an Engineering school, for engineers, on [an] engineering subject"—which thus excluded camouflage, deemed an art, from funding. With no other options, Mapes decided that the OCD would have to "commence an educational program in whatever way might present itself."[55]

Such an alternative soon emerged. The War Department rather than the Department of Education agreed to directly sponsor camouflage training for civilians through the Engineer School at Fort Belvoir; after all, the War Department was already offering such courses for army engineers. Rickard collected the names of possible candidates for such instruction by consulting trade publications, and sent form letters to university presidents and art school directors:

> **The Office of Civilian Defense has given careful consideration to the formulation of a national policy with respect to camouflage.... Many institutions throughout the country have urged the inauguration of training courses. In view of these requests, and in order that camouflage instructions may embody approved methods, an educational program has been adopted. An intensive two weeks' course will be conducted by the Engineer Board at Fort Belvoir, Virginia.**

Each school was invited to submit the names of "responsible representatives."[56] The OCD's focus on "approved methods" indicate a fear of the excessively imaginative proposals from artists—precisely the proposals that Kepes had drafted.

The ultimate purpose of training civilians was to establish a network of "Regional Camouflage Schools" across the country; each civilian instructor graduated from Belvoir would return to his home institution to establish government-approved courses under OCD guidance. The major stipulation enforced by the OCD was that all students trained at home institutions must be professionals—"engineers, architects, landscape architects, and people of similar training and qualifications," as Rickard explained in a special memorandum sent to participating institutions. Artists, however, were also qualified. "An artist, for example, is often not practical, yet an artist <u>can</u> be practical as has

1.17

Map depicting regional civilian camouflage
schools. Office of Civilian Defense Records,
National Archives and Records Administration,
Washington, DC.

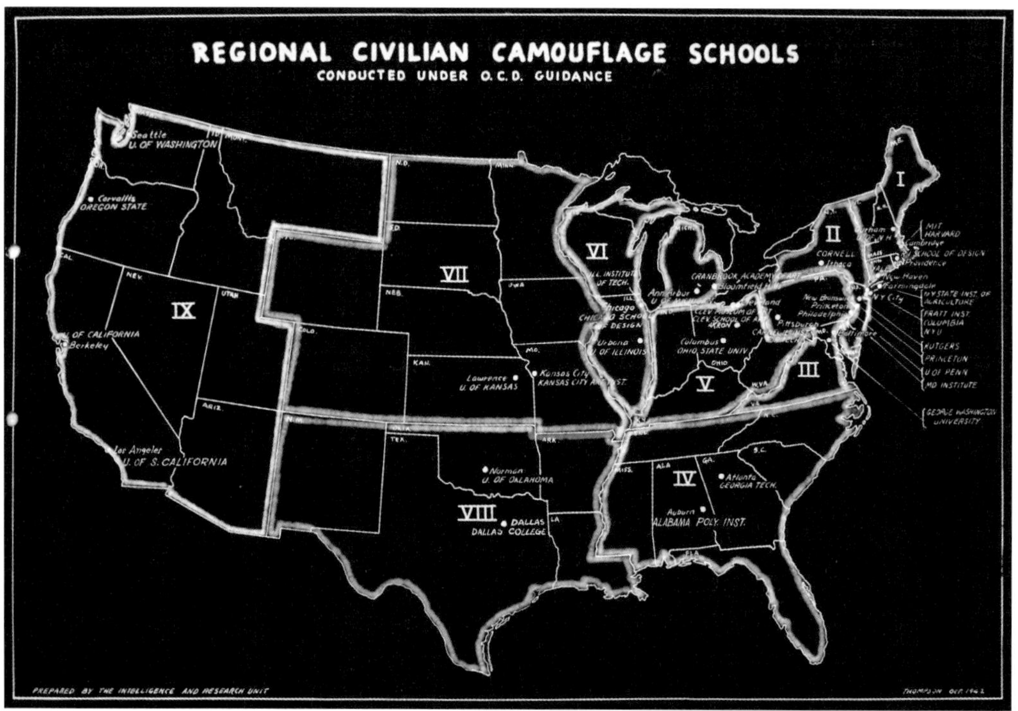

been proved in certain instances, and should not be kept out merely because he is an artist."[57] Students who subsequently completed these courses would receive special OCD certificates, and were then likely to be drafted into military service.[58] They would also be available for domestic camouflage work should the US come under aerial attack—or, as the letters sent to universities and schools euphemistically explained, "should a change in the military situation require an extensive camouflage program."[59] From its graduates, the OCD would also select "Camouflage Specialists" who would administer these officially designated OCD-sponsored courses in each of nine OCD regions across the country. A map from the OCD's official in-house report on the program reveals its comprehensive geographic reach (fig. 1.17).

After each school sent in a list of names, follow-up correspondence from the OCD required the verification of citizenship and loyalty. "The utmost precaution must be taken to select loyal citizens in view of the possibility of introducing subversive elements into the schools that are being conducted on a military reservation"[60] Such warning registers the incongruity of these courses, the way they mixed the radicalism of the arts with the most conservative elements of the military.

Upon receiving his request from the OCD, Moholy cabled an immediate reply: "THANKS FOR YOUR COMMMUNICATION WE SHALL DELEGATE ONE OF OUR PROFESSORS." In a subsequent letter, he clarified: Kepes, now designated somewhat ludicrously as "Head of the Camouflage Department," would attend the government's training course. Moholy had hoped to secure government funding, but the OCD would offer something better: a valuable imprimatur on the School of Design's activities. An officially sanctioned, government-approved camouflage course could only help the school justify its aesthetic mission during the war.

In total, Belvoir's four civilian courses, running through June and July of 1942, graduated seventy-three camouflage specialists. All of the Belvoir graduates, Kepes included, received special certificates upon completion of their training (fig. 1.18). They then established OCD-sponsored courses in thirty-one leading universities and schools, spanning public and private institutions, from the Ivy League to the Big Ten, and including liberal arts, fine arts, and technical and scientific institutions.[61]

At Belvoir, artists, architects, and designers took up the workaday routines of military life: Rickard describes the courses as "intensive," and notes that participants "lived in barracks, [slept] on cots, arose at 6 a.m. each morning, and ate at the Army mess, or in a cafeteria on the grounds."[62] Participants studied subjects like "Tactics of Aerial Attack," but also aesthetic topics like the "Interpretation of Aerial Photography" and "Disruptive Painting."[63] Major

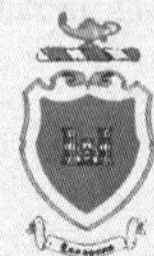

The Engineer School

Fort Belvoir, Virginia

This is to certify that

GEORGE KEPES

SCHOOL OF DESIGN (CHICAGO)

has completed the Camouflage Course conducted for Civilians for the period June 22, 1942 to July 4, 1942. The subjects completed are those outlined on the reverse side of this sheet.

CHAS. J. CRAIG,

Captain, Corps of Engineers

Secretary.

1.18

Certificate for completion of the camouflage course at Fort Belvoir, Virginia, 1942. Note listing of Kepes's name as "George Kepes." Gyorgy Kepes papers, 1909–2003, bulk 1935–1985. Archives of American Art, Smithsonian Institution.

Breckenridge presented an introductory lecture for the course based on his *Modern Camouflage: The New Science of Protective Concealment* (the volume served as the course textbook), and Homer Saint-Gaudens, former Director of the Carnegie Art Museum and a pioneer in camouflage techniques during the First World War, presented a keynote address.[64]

Yet the military experience was by no means mundane. "The interest and 'esprit de corps' was [*sic*] amazing," Rickard recorded in his official report of the program; of the artists and architects, "many had outstanding personalities."[65] In questionnaires submitted to Rickard after completion of the course, participants registered similar excitement. "This is to express my appreciation of the privilege granted me of taking the Camouflage Course at Fort Belvoir," wrote Edward Steese, an architect and poet. "It was a fine and exciting experience, and a privilege. I was especially impressed by the Army's cordiality and cooperation," explained Wallace Mitchell, a painter at the Cranbrook Academy of Art.[66] Participants singled out interdisciplinary exchange—the approaches that Kepes called "flexibility" and "synchronization"—as the course's strength:

> **In addition to classroom lectures and the practical work, the opportunities to discuss informally the problems of camouflage with other students from California to Newfoundland and with members of the staff were invaluable. This would not be possible without a "gathering place" or <u>Camouflage Cultural Center</u>.**[67]

Despite the universal praise, some complained about the lack of actual aerial experience. While the goal of the course was to learn how to create a cryptic aerial image, the participants never took to the sky to see what such an image looked like. "Let the student see the 'Real McCoy' from the air—possibly through at least an obsolete bomb sight—if he is to really effectively fight the bombardier."[68]

Kepes, who attended the second and largest installment of the course (fig. 1.19), was also enamored with its collaborative ethos and aesthetic community. On Fort Belvoir letterhead, he wrote to his student Robert Preusser that he was "quite pleased about the opportunity to take this course"; there were "quite a number of things I can learn here—and nowhere else."[69]

What, aside from interdisciplinary exchange, was unique about those weeks at Fort Belvoir? No doubt access to the military's highly specialized knowledge was thrilling; the chance to consult restricted and confidential training materials and to partake in rare pedagogical experiences conducted on an army base prompted excitement—seduction by military secret. The high value of such knowledge is apparent in an exchange between Kepes and Moholy months later, after Kepes began teaching his OCD-sponsored course.

Moholy requested a report from Kepes about Belvoir; when Moholy did not receive one, he angrily castigated Kepes for "not even showing me the material which you, as stated, brought back."[70] Kepes used this specialized knowledge as professional leverage, a means to distinguish his expertise from Moholy's.

The course also offered a heady mix of artistic avant-garde and military establishment. Kepes had the opportunity to lecture to his colleagues at Belvoir "about visual fundamentals in camouflage." He explained their reactions in a letter to Preusser:

> The general impression was quite good[;] some of the officers/our teachers took notes. But they were still quite cautious with their comments. The civilian group (mainly professors of architecture) were quite interested in what I had to say and they asked for the picture material we have to be made into lecture slides for use in their teaching…. In spite of the fact that the c. course at Belvoir was in the traditional channels I learned quite a lot – because I was able to check up, where our experimental type of work made sense – and where it is only a mental acrobatic for those interested in this type of work.[71]

Kepes records a tension between the artist's approach to the image and the "traditional" approach of the military. Rickard, touring the country to evaluate the teaching methods among Belvoir graduates, jotted down similar notes after visiting Chicago's School of Design: "Saw examples of work done—very thorough course. Gave corrective thoughts on conservative approach in teaching."[72] Rickard's observations about "corrective" measures and the value of "conservative" pedagogy suggest that Kepes's more far-fetched ideas were too imaginative, too fantastical, threatening the OCD's control over the legitimacy of its work.

The School of Design's camouflage course was popular and promotable. The school heavily advertised its OCD affiliation; one course announcement depicts a version of the certificate students would receive upon successful completion of the course (fig. 1.20). The school soon "reached the 200 mark" in enrollment, including "seventy camouflage students," explained Moholy in a letter to Walter Paepcke of the Container Corporation of America, the School of Design's most important benefactor.[73] The course soon became a means to leverage moneys, with Kepes's government training in particular functioning as a selling point in grant applications.[74] Moholy would even consider the year a success—despite the fact that it was the most anxious of the war. "The year 1942 was more positive for the School than we had the right to expect," he explains. "This large increase in the number of students is due of course to the camouflage course which we introduced as an emergency measure."[75]

For all official purposes, the OCD was devoted to its eponymous objective: civilian defense. But was its camouflage program only about protective concealment? Or did camouflage serve other ends—objectives disguised even from those who participated in it?

The biologist Julian Huxley, who served on a committee that advised the New Bauhaus and School of Design, and whom Moholy and Kepes met during their exile in London before their arrival in Chicago, suggests an alternative goal. In a letter to Moholy, Huxley offers ideas about the School of Design's role in the war. He explains that "design clearly must play an important part," and that the school "should be utilized to that fullest extent in the present emergency."[76] To this end, Huxley hoped to help it secure a special relationship with the US government. He wrote a letter to the head of the OCD to offer suggestions on "the problem of employing the arts in general for morale purposes in wartime." He called these purposes "cultural morale" and proposed wielding the arts as a weapon not in combat against an enemy, or even in defense from one, but as a means to boost national confidence.[77] Huxley even suggests that the OCD already served such a purpose:

> There is also what seems to me a complete misapprehension as to the function of the OCD.… Here [in the US], while it is extremely necessary to prepare against air raids, I cannot believe they will ever be serious; and the main job for an organization, such as the OCD, would appear to me to be to keep up morale—both the morale of those concerned with the anti-air-raid organization (for nothing is more demoralizing than waiting about and doing nothing), and still more, that of the civilian population in general.[78]

Huxley argues that the contribution of artists to the war effort may not be their participation in camouflage but rather their cultivation of an approach to the image that, in its very method, would put Americans at ease. As the immediate threat of aerial attack became increasingly remote, camouflage was transformed from a defense activity into a morale activity.

In *The Democratic Surround*, Fred Turner recuperates the history of morale research during the Second World War, revealing how a wide range of figures from the humanities and social sciences became invested in the concept, working to maintain it at home and destroy it abroad.[79] Aerial bombardment was premised on destroying enemy morale as much as destroying enemy targets; similarly, camouflage might be understood as providing defensive disguise not only in a literal sense but also in a figurative one, by maintaining the morale of those on the home front.

Kepes would rethink the purpose of his wartime pedagogy, and its interdisciplinary dimension, under these terms. (Chicago's Mayor Edward J. Kelly

1.19

Photograph of the second camouflage course
for civilians, Fort Belvoir, Virginia, 22 June
1942 to 4 July 1942. Kepes is second from left,
third row. Office of Civilian Defense Records,
National Archives and Records Administration,
Washington, DC.

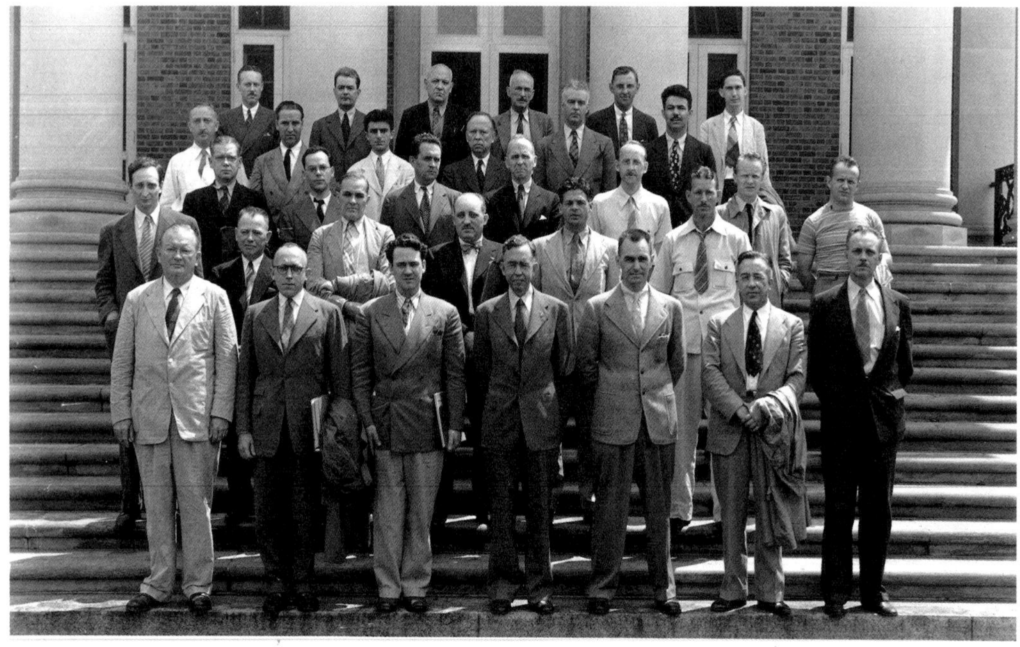

"Principles of Camouflage" course announcement from the School of Design, with official OCD certificate reproduced at left of announcement. IDC_0003_0076_001, Institute of Design collection, University of Illinois at Chicago Library.

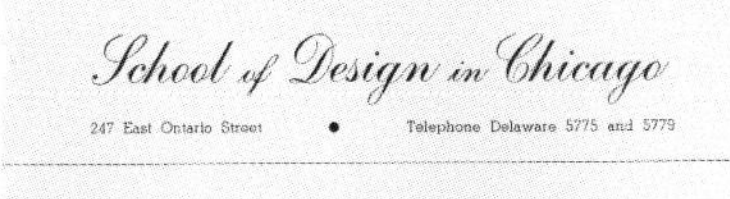

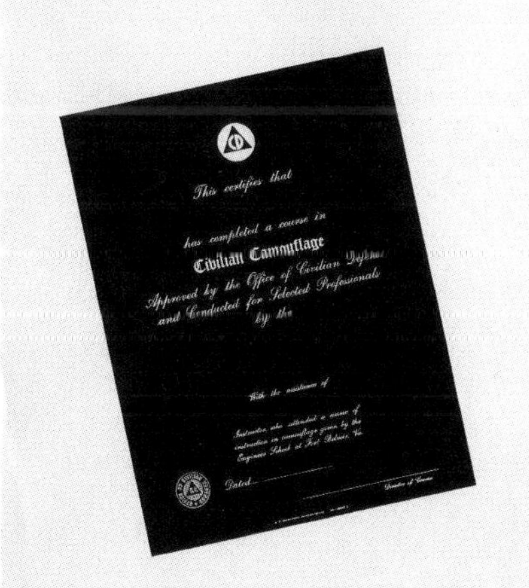

School of Design in Chicago

247 East Ontario Street   •   Telephone Delaware 5775 and 5779

The course in the

Principles of Camouflage

approved by the OCD, Washington, D.C. has opened with 90 students.

Mr. George Kepes, head of the Light and Color Workshop who attended a camouflage instruction course given by the Army Engineer School at Fort Belvoir, Virginia, is in charge of the class.

The Office of Civilian Defense in Washington, D.C., especially encourages "selected professionals"

architects and engineers

to take this course. They are the ones to whom protective concealment problems will best be referred when such decisions are made by the War Production Board. In our Camouflage Course information will be given which will qualify those taking the course to prepare plans, in accordance with principles established by the O.C.D. in Washington, D.C.

We are entitled to give to such participants, after their successful graduation, certificates issued by the O.C.D., Washington, D.C., as shown on opposite page.

Enrolment for the course is still open. New students will receive a transcript of the lecture already given. The second class will meet next Wednesday evening at 8.00 P.M. The panel of lecturers for the course include the following military and civilian experts:

Dr. Theodore Ashford	Instructor in Dept. of Chemistry, University of Chicago
Mr. J. Copeland	Riggs Optical Co.
Prof. Milton Fox	Chairman of Camouflage Committee, Cleveland, Ohio.
Mr. Johns Hopkins	Senior Engineer of Blackout, Camouflage, and Protective Construction, Sixth Regional Office of Civilian Defense.
Mr. Otto Jelinek	Chief of Techniques, Chicago Metropolitan Area
Mr. George Fred Keck	Architect, Head of Architectural Dept, School of Design
Mr. John Moyer	Field Museum, Chicago.
Captain G. G. Preston	Royal Air Force
Mr. R. R. Root	Landscape Architect, author of "Camouflage with Planting"
Dr. Christian Ruckmick	Psychologist
Mr. M. B. Sweet	Northern Pigment Co. Pres. Color Research Assn.
Col. A. D. Tuttle	Medical Director, United Air Lines
An officer from Col. Goddard's Staff	Wright Field, Dayton, Ohio.

The class meets every Wednesday evening from 6:30 to 9:00 P.M. The fee for the Sixteen week course is $30.00.

even asked him to serve on a committee "for developing a program of civilian morale," specifically "among the Americans of Hungarian descent in the Chicago community.")[80] From his notes:

> Not war of techniques but war of human resources. War of morale. The mobilization of morale is the first step in the full mobilization. Morale is a mental state or condition as evidenced in the readiness to endure in spite of the most testing circumstances.
>
> [T]he endurance of [the] individual gets reinforcement from a faith in something outside his personal existence. Morale is a social phenomenon characteristic of social man – [it] is based on recognition that the survival of himself is indivisible from the survival of the group, nation or other forms of social ties. It is the manifestation of belief that his existence is only possible in the components of the others.… Without this identification of the individual with a more embracing dimension … a willingness of supreme sacrifice is unexplainable.[81]

Kepes explicitly argued for the role of visual culture in such a program. "The cultivation and maintenance of morale by propaganda," he notes. "Propaganda uses language—verbal + visual.… Pictures are the most easily understood language."[82] He framed the language of vision as a morale activity; camouflage offered defensive disguise, certainly, but also interdisciplinary experiences that would mobilize Americans.

Research and Development

Camouflage was even an area of research at the highest levels of the military bureaucracy: the National Defense Research Committee, or NDRC, which organized war research that led to developments in radar, submarines, aircraft, digital computers, and the atomic bomb. Just three days after Pearl Harbor, Vannevar Bush and James B. Conant, the NDRC Directors, sent formal letters of appointment for an ad hoc Committee on Camouflage to a hastily assembled group of experts representing the Army, the Navy, the sciences, and also, in a way, the arts; the panel included Major Colonel Mapes, Greville Rickard, and Homer Saint-Gaudens. The purpose of the committee was to investigate the "research and development aspects of camouflage."[83]

After the war, Bush described these programs as "a partnership between groups which one might at first thought consider inherently incompatible," a partnership that required innovative methods—a "pattern of highly effective collaboration"—or what Kepes had already called "flexibility" and "synchronization."[84] A now declassified report elaborates; camouflage demanded the "study of a large gamut of individual problems which are technically dissimilar and have only their major purpose in common."[85]

Most significant is the Committee's explicit consideration of the role of artists in this research. Commenting on one of the earliest camouflage initiatives (known as the "Passive Defense Project") undertaken through the Works Projects Administration, the report states: "This may well have been the first instance in which a Design Department composed of artists, architects, and model makers had the facilities of laboratories of Physics and Chemistry immediately available." Anticipating the tension between art and science that emerged in Fort Belvoir's civilian courses, the Committee on Camouflage ultimately decided against working with artists, architects, and designers:

> Recognizing the present public interest, many art schools and similar organizations have instituted special courses in camouflage…. There will, unquestionably, be opportunities for employment of a few artists and architects in connection with camouflage projects, but the number so employed will inevitably be small. Whether even this small number would profit by a course in camouflage is debatable. For reasons of security, no information of military significance is likely to be learned from these courses; and much of the information may therefore be incorrect. Since the major usefulness of artists and architects on camouflage projects is by application of the skills which they already possess, special instruction does not seem necessary.[86]

The fact that discussion about the arts even appears in a classified document prepared by scientists at the NDRC is remarkable. Not all of the committee members agreed with the report's conclusions, however; Rickard, for one, opposed the report's rejection of artists. In a personal journal, he attacks Bush: "I think he is entirely wrong." Rickard's verdict: "Rubbish."[87]

The NDRC created a complicated bureaucratic web in order to implement these plans; a special section among its various agencies and departments, Division 16.3, handled camouflage. Surprisingly, the head of Division 16 was none other than Karl T. Compton—then President of MIT, a position he held until 1948. The involvement of Kepes's future home institution in camouflage efforts is more than suggestive. The organizational relationships emerging between MIT and the government—the interdisciplinary relationships that would create the military-industrial complex—had already entrapped Kepes, even though he had never even visited MIT. In a retrospective appraisal of the Division's activities, Compton recalls that Division 16 produced "a wealth of knowledge and art"—a telling rhetorical turn of phrase.[88]

The NDRC Camouflage Section contracted research and development across many institutions, from Harvard to the Interchemical Corporation, on studies ranging from the design of special paints to optical devices for camouflage detection.[89] The NDRC also engaged an artists-in-residence program at the Louis Comfort Tiffany Foundation of Oyster Bay, New York. The Foundation offered grants to emerging artists who would then live at Laurelton Hall, Louis Comfort Tiffany's Long Island estate, and were provided with room, board, and studio space, and could peruse Tiffany's art library and paint *en plein air* on the estate's extensive grounds.[90]

But the Foundation assumed a different function circa 1942. "The L. C. Tiffany Foundation is an art school in peacetime, but its facilities were completely engaged during the war years," recalls one researcher who worked in its laboratories.[91] Or, as a confidential NDRC report produced after the war explains: "An art school during the years of peace, the Tiffany Foundation faced the prospect of having its facilities lie unused during the war. It, therefore, offered them first to the Navy and later to NDRC."[92] After meeting with Navy Department officials, the Foundation's Director, Hobart Nichols, put forth a proposal "to place the facilities of the Foundation at the disposal of representatives of the Federal Government for the purpose of carrying on important and secret experiments in the field of art."[93] The Foundation then received government contracts worth some $358,000—more than $5,000,000 when adjusted for inflation.[94] In a 1930 statement, Louis Comfort Tiffany declared that the purpose of his Foundation was to support emerging artists: *"It is my dearest wish to help young artists of our country … in establishing themselves in the art world."*[95] No doubt he never imagined the form this support would take.

The Tiffany Foundation became the government's "central laboratory for camouflage field studies," and was used to test paints and optical devices.[96] It also became the site for elaborate experiments aimed at determining "the perceptual capacity of the human observer."[97] In a purpose-built structure, test subjects observed a white screen illuminated by a test pattern of projected lights; they scanned the projection "at a rate comparable to that employed by lookouts in the military service."[98] More than two million such observations were made, and elaborate statistical calculation determined the visual threshold that rendered a target visible to the human eye.

The goal was to account for vision mathematically, to transform sight into an informational equation. An official summary of the Camouflage Section's activities explains:

> **Unlike most groups in the NDRC organization, the Camouflage Section was not primarily concerned with the development of the instrumentalities of war but rather with techniques for their concealment. By studying the inherent limitations of human vision, and by making proper allowances for the effects of the atmosphere, the concealment aspects of camouflage have been reduced, in most cases, to an engineering procedure.**[99]

These "important and secret experiments" bear a cognate relationship to Kepes's experiments in "art as applied optics," as he called it—to the way of seeing he practiced and theorized, one that turned the world into a layered pulse of pictorial signals and a matrix of sensory data, into patterns of pure light. The NDRC partnership with the Tiffany Foundation also anticipates postwar relationships between the arts and sciences—the very same relationships that Kepes's Center for Advanced Visual Studies later facilitated.

While the clandestine goings-on of NDRC Division 16.3 were strictly confidential, some of the section's general aims, as well as the lessons of Kepes's camouflage pedagogy, were promoted in a series of mass-market volumes. Edwin Boring, a member of the NDRC ad hoc Committee on Camouflage, helped manage the publication of a slim edition titled *Psychology for the Fighting Man* (fig. 1.21). "The human eye is one of the most important military instruments that the armed forces possess," declares one passage in the book, recalling both the NDRC's research at the Tiffany Foundation and Kepes's camouflage pedagogy.[100] The book investigated the topic of camouflage at length, and aligned camera and eyeball in the same way that Kepes aligned eyeball and bombsight: "Your eyes are very fine precision instruments," it explains. "They are in many ways like a camera."[101] Two diagrams from the book illustrate the argument, even depicting a boxlike contraption recalling Kepes's light box.

A Theory of Protective Concealment

By echoing, repeating, imitating his surroundings, by adapting himself to all the powerful groups to which he eventually belongs, by transforming himself from a human being into a member of organizations, by sacrificing his potentialities for the sake of readiness and ability to conform to and gain influence in such organizations, [the individual] manages to survive. It is survival achieved by the oldest biological means of survival, namely, mimicry.

Max Horkheimer, *The Eclipse of Reason*, 1947[102]

Camouflage created a complex network of relationships, almost too hard to pin down, that surrounded and disguised its participants and practitioners. The methods and methodologies for making art provided a set of expedient approaches for making war. Kepes's way of seeing was uniquely suited to this task (he even described the language of vision as "a vital weapon").[103] But my argument also turns on the reverse exchange: the strategies and tactics for making war were also expedient for the artist.

I therefore understand camouflage as a means of self-preservation and self-protection, or what the OCD called "protective concealment." By disguising the environment, artists also gained agency over their own disguise. The War Department's Field Manual 5-20A, distributed to enlisted men—Kepes may have encountered a version of the manual at Fort Belvoir—puts it plainly: "YOU are the target" (fig. 1.22).[104] By blurring identities—and past and present political affiliations—artists found refuge in the relative safety of the military establishment. "Man is a dynamic being struggling individually and socially for survival. To survive he must orient himself in his surroundings," writes Kepes in *Language of Vision*.[105] During war, this orientation took the form of a purposeful disorientation; survival became, paradoxically, an act of destruction. Kepes assimilated into his opposite, ironically destroying himself—or at least the persona he once was—in order to endure.[106]

A tense exchange of letters between Moholy and Kepes discloses these contradictions. The communications reveal the professional disputes that eventually led to Kepes's departure from the School of Design in spring of 1943. The disagreements revolved around an official letter Kepes discovered in the school's administrative files. From Colonel Mapes of the OCD, the letter "expressed dissatisfaction" about the school's camouflage teaching, in particular the school's failure to adhere to guidelines for selecting students (too many admitted students were artists rather than engineers). "Colonel Mapes' dissatisfaction was referred to me specifically," Kepes complained, "the implication being that I had not fulfilled my duties." Kepes worried about his "good or bad standing with the office of Colonel Mapes" and the threat of "being drafted, which could be of major importance for me."[107]

But this initial fear opened up to larger disagreements over the proper relationship between art and war. After Kepes angrily confronted Moholy about the OCD letter, Moholy wrote a lengthy response in which he argued that attracting students—both engineers and artists—was valuable, regardless of the OCD's policies and despite Mapes's criticisms, or the threat they posed to Kepes's chances of being drafted:

> **The consequences of war for a design school such as ours had to be considered.... Also I suggested repeatedly to Bob Wolff [Robert J. Wolff, painting instructor at the School of Design] and you that you should set up a camouflage course. I saw in it a two way service; one towards the country, its future soldiers and for the war industry; secondly, for the school itself. In fact, there are few occasions when altruistic and selfish motives coincide in such a perfect way.[108]**

Moholy suggests the cynical opportunism of camouflage instruction; by instrumentalizing its educational mission, the school made itself relevant, despite the distortion of its primary mission. Moholy was blunt: "Instead of having time to develop quietly the new type of teaching which we visualize," the school must "justify the value of our educational venture [in order] to enlist the interest of prospective students and donors, and the public."[109] Of course, the original Bauhaus had also applied its educational mission in this way, but the applications circa 1942 were more troubling, more militaristic.

It was Kepes's reticence about this cynical opportunism that angered Moholy, who writes:

> I have been, during the past years, sometimes disappointed when I saw that ideas which were obvious in war time were first adopted by Pratt Institute and not by us. I was even more disappointed when I saw that my policy to incorporate the school into the war efforts did not find the approval of the faculty members.... Pratt Institute has no scruples in making a military tactical camouflage course with their normal teaching staff. They do it because such a course is essential, no [*sic*] it would be ideal for the information of people.[110]

In an interview decades later, notably recorded during the Vietnam War, Kepes recalls the School of Design becoming "a vocational school which services the various big industries around Chicago" by "producing people who would be war attendants." Students had entered its programs assuming "that it will be an artistic community [that] will create a revolution of art" when it was actually a training ground for those entering the military-industrial complex.[111]

Kepes's comments are most striking when measured against his earlier, and lesser-known, avant-garde commitments. Consider a cover for *Das Neue Russland*, the German journal from the 1920s and 1930s that promoted the USSR in Germany and was published in Berlin by the anti-Nazi resistance leader, Erich Baron (fig. 1.23). In the early 1930s, Kepes replaced John Heartfield and, before him, Käthe Kollwitz, as designer of the journal's covers; in one, signed "Georg Kepes," the artist depicts a revolutionary soldier armed with a rifle. It makes a striking contrast to the soldier depicted in the War Department's Field Manual 5-20A. Kepes's cover indicates a radicalism that would be highly controversial in the United States a decade later.

And it was. After his falling out with Moholy, Kepes left the School of Design to teach for one semester at Chicago's Abraham Lincoln School, then a target of the House Un-American Activities Committee (HUAC), the infamous government panel that investigated supposed Communists. A 1944 appendix to a HUAC report describes the Abraham Lincoln School as "a streamlined version of the Workers School which was openly run by the Communist Party." The report's language is key: "Although cleverly camouflaged"—that word is crucial—"its staff, sponsors, and curriculum reveal its essential Communist character."[112] Kepes, as well as his wife Juliet, are listed in an accompanying "Exhibit No. 2," which included a transcribed copy of one of the school's course bulletins.[113] That year Kepes taught "Art as a Weapon."

In a letter written many decades later to the art historian Krisztina Passuth, then researching a monograph on Moholy, Kepes was even more explicit in explaining how his falling out with Moholy concerned politics—the politics of teaching camouflage, but also the camouflaging of politics:

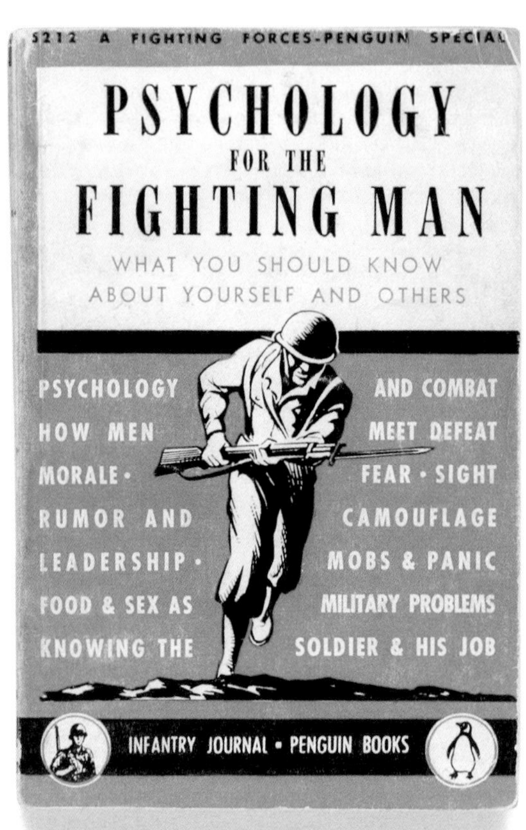

1.21

National Research Council, *Psychology for the Fighting Man: What You Should Know About Yourself and Others.* Washington, DC: Infantry Journal, 1943.

Field Manual 5-20A: Camouflage of Individuals and Infantry Weapons.
Washington, DC: War Department, February 1944.

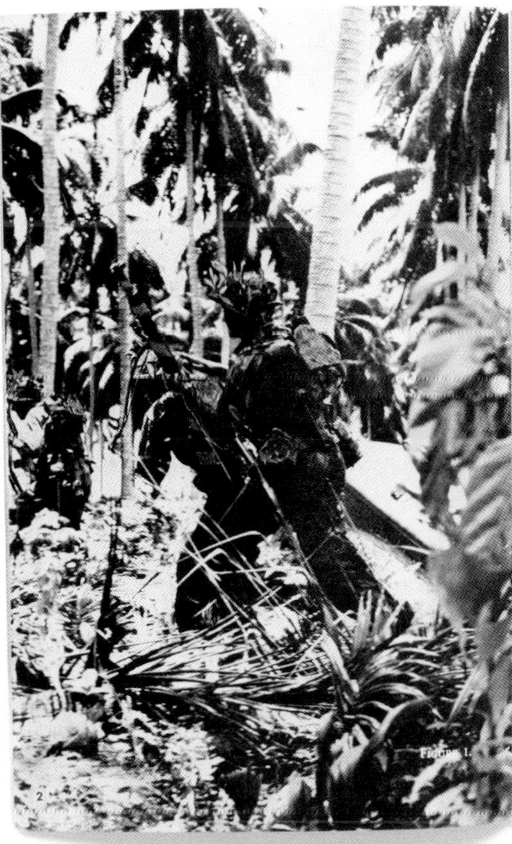

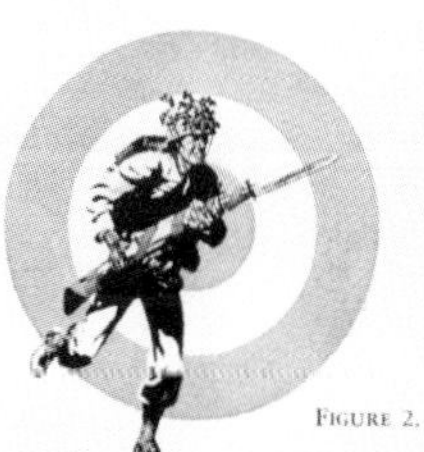

FIGURE 2.

YOU ARE THE TARGET

Individual camouflage is the *concealment* a soldier uses in combat to surprise, deceive, and outwit the enemy.

The ground is the soldier's observation post, jump-off point for attack, route of advance and communication, fortification, protection, and obstacle. He must know how to use the ground for effective concealment. He adapts his dress for best concealment while in the firing positions and for mobility, and carefully selects his routes between positions for such concealment as is possible while he is in motion. Interruptions, crawling (very slow) and running (very fast), aid concealment of motion.

The simple principles in this book have been battle tested. If the soldier learns and practices them continuously in training, he will know what to do about concealment at the right time in battle.

3

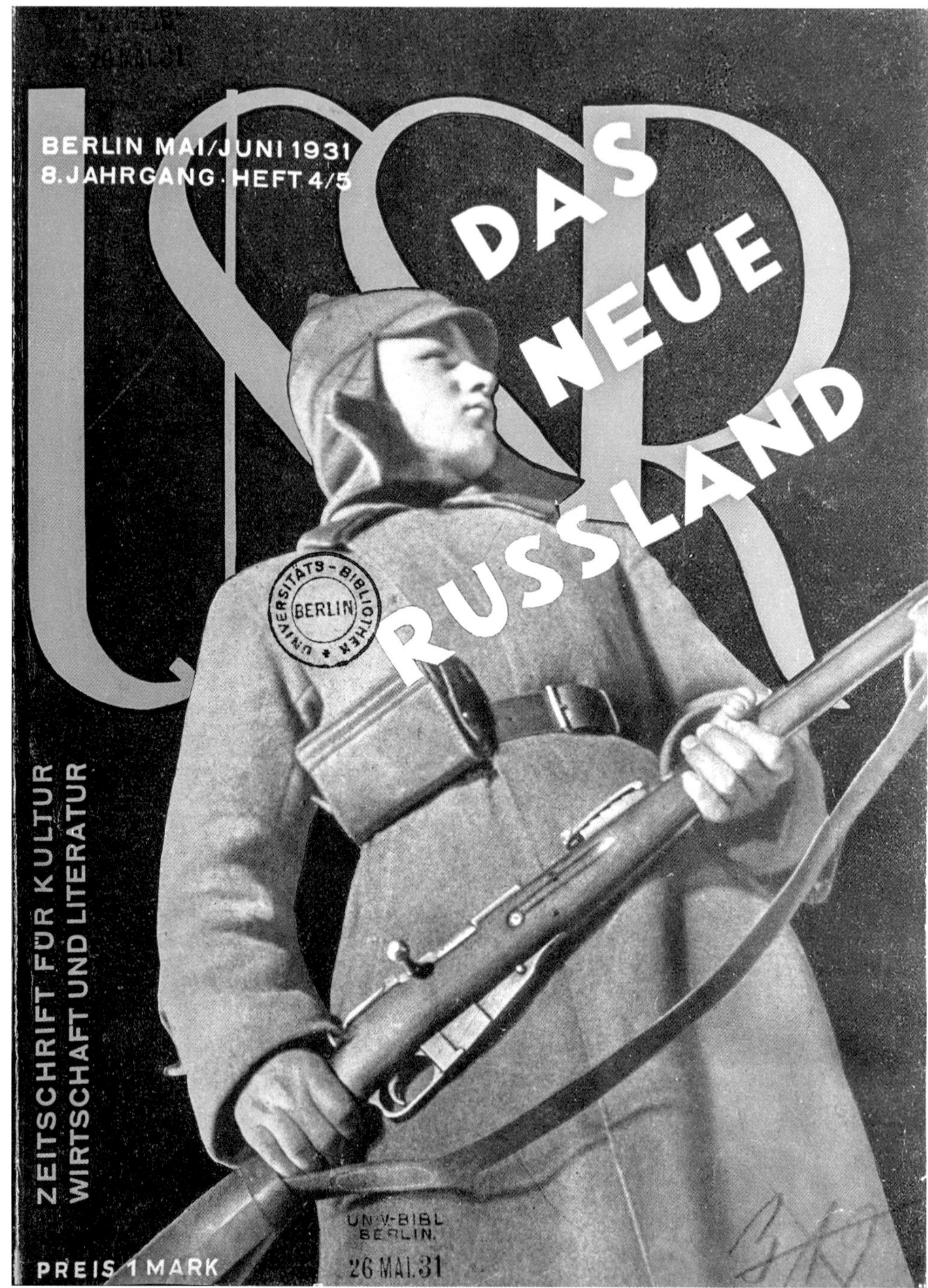
BERLIN MAI/JUNI 1931
8. JAHRGANG · HEFT 4/5
DAS
NEUE
RUSSLAND
ZEITSCHRIFT FÜR KULTUR
WIRTSCHAFT UND LITERATUR
PREIS 1 MARK

> My relationship with Moholy was not a smooth one. We didn't see eye to eye on political/social matters and therefore often had ideological and personal confrontations.... I left the school following a disagreement with Moholy concerning a course organized on civilian camouflage. Though I believed in the importance of using all possible knowledge and strength to combat the Nazi's [*sic*], I thought that to give this particular course had no justification. We were not equipped with the knowledge to teach camouflage nor was it appropriate to have such a course in the context of the declared and believed goals of the school.[114]

Kepes left the School of Design because he did not believe in teaching camouflage—but perhaps also because he did believe in the politics of institutions like the Abraham Lincoln School (such is the implication of his remark that he did not see "eye to eye" with Moholy on political matters).

The necessity of survival put aesthetics into startlingly close proximity with war. These relationships were always equivocal. They enabled artists to endure; Kepes remembers, warily, "it was not easy just to survive."[115] They also permanently mutated the relationship between art and science. The following chapters explore how these interdisciplinary relationships, and the protective concealment they offered, played out during the Cold War that followed.

1.23

Das Neue Russland, May/June 1931. Cover designed by "Georg Kepes." Courtesy Márton Orosz.

Patterns and Puzzles

An individual may be alone in a physical sense for many years
and yet he may be related to ideas, values, or at least
social patterns that give him a feeling of communion and
"belonging." … The kind of relatedness to the world may
be noble or trivial, but even being related to the basest kind
of pattern is immensely preferable to being alone.

Erich Fromm, *Escape from Freedom*, 1941[1]

Revision of Vision

In 1951, Gyorgy Kepes organized an unusual exhibition at the Massachusetts
Institute of Technology (MIT): he suspended scientific imagery—photographs
collected from research laboratories and enlarged onto fiberboard panels—
from a lattice of metal rails in MIT's Hayden Gallery.[2] The pictures appear to
hover and float (fig. 2.1). The immersive quality of Kepes's exhibition drew
from modernist precedents and anticipated defining shows of the period, like
the 1951 *Growth and Form*, which opened at London's Institute of Contemporary Art just months later, and the 1956 *Family of Man* at New York's Museum
of Modern Art.[3] But Kepes wielded these methods of display for peculiar ends.

Titled *The New Landscape*, Kepes's exhibition revealed an entire world
that midcentury viewers had never seen before: his abstract images rendered
visible the inner structure of microscopic minerals, the outer reaches of the
solar system, and everything between the two, from pools of chemical com-
pounds and elastic tissue fibers (seen under 2,000 times magnification) to dense
networks of cells (figs. 2.2–2.5).

The most awesome image was the "Lichtenberg figure," a fractal-like
burst of electric sparks first discovered by and named after the German scientist Georg Christoph Lichtenberg in 1777. By discharging high-voltage electric
shocks onto a resin plate, and dusting the plate's surface with powdered pigment, Lichtenberg exposed, as if by magic, what looked like tiny snapshots
of the universe; "innumerable stars," he notes, even "the Milky Way" itself,
seemed to materialize before the eyes (fig. 2.6).[4] Kepes discovered his Lichtenberg figures in the lab of MIT physicist Arthur R. von Hippel, who created
them by discharging electric shocks directly onto photographic plates (fig.
2.7).[5] Kepes featured these images prominently in his exhibition (one hangs in
the foreground in the show's iconic installation view); he even put one on an
issue of *Arts and Architecture* magazine featuring a report on his show (fig. 2.8).[6]

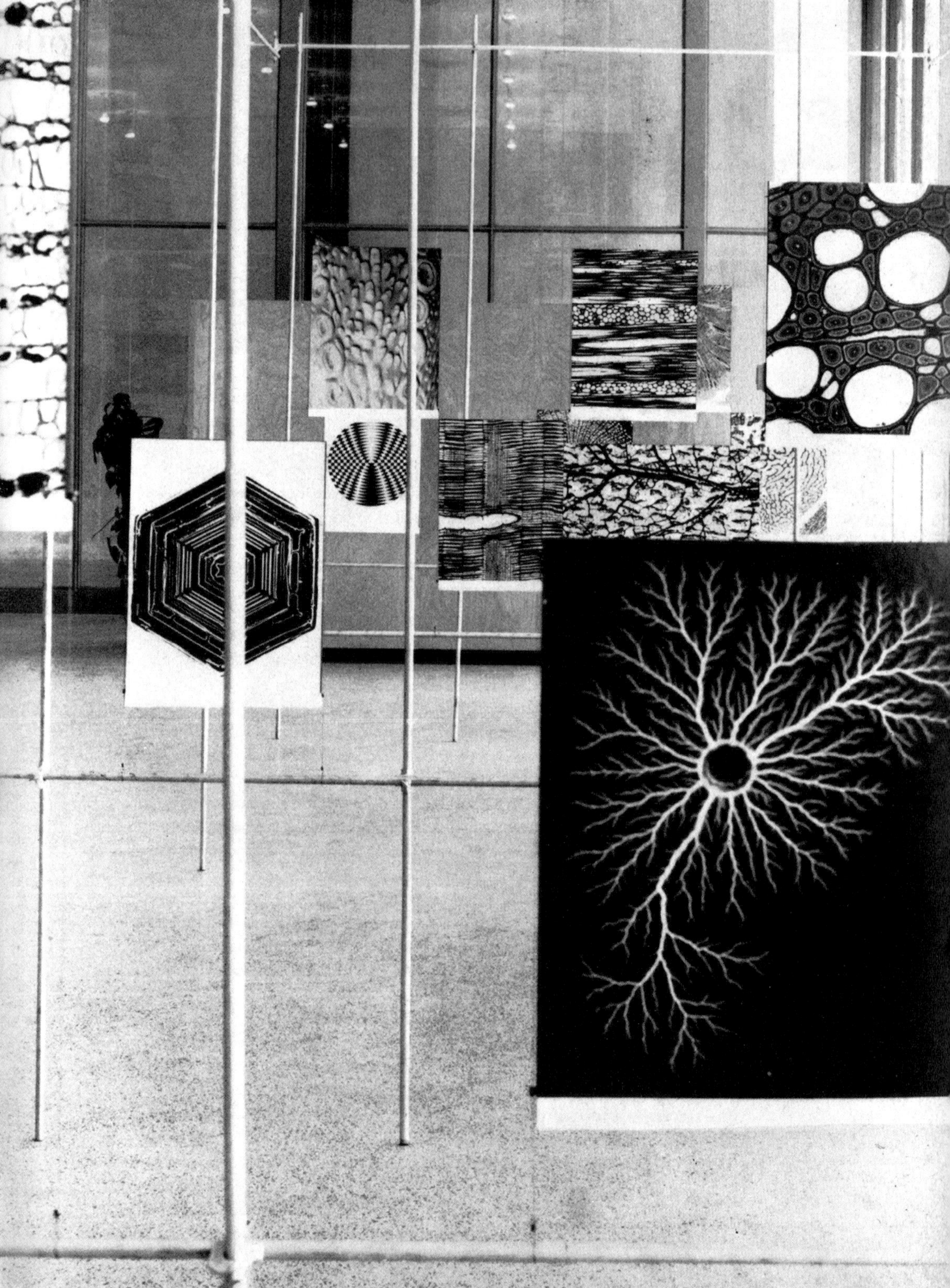

Installation view of *The New Landscape*,
Hayden Gallery, MIT, 1951. From Gyorgy
Kepes, *The New Landscape in Art
and Science*. Chicago: Paul Theobald, 1956.
© The Estate of Gyorgy Kepes.

2.2

Gyorgy Kepes, *Eutectic with island of compound: H. P. Roth*, photographic enlargement on fiberboard panel, 1951. Gyorgy Kepes papers (M1796). Dept. of Special Collections and University Archives, Stanford Libraries, Stanford, Calif. © The Estate of Gyorgy Kepes.

2.3 (following pages)

Gyorgy Kepes, *Elastic tissue fibers: 2000X*, photographic enlargement on fiberboard panel, 1951. Gyorgy Kepes papers (M1796). Dept. of Special Collections and University Archives, Stanford Libraries, Stanford, Calif. © The Estate of Gyorgy Kepes.

2.4 (following pages)

Gyorgy Kepes, *Transverse section of Osmanthus wood: 50X*, photographic enlargement on fiberboard panel, 1951. Gyorgy Kepes papers (M1796). Dept. of Special Collections and University Archives, Stanford Libraries, Stanford, Calif. © The Estate of Gyorgy Kepes.

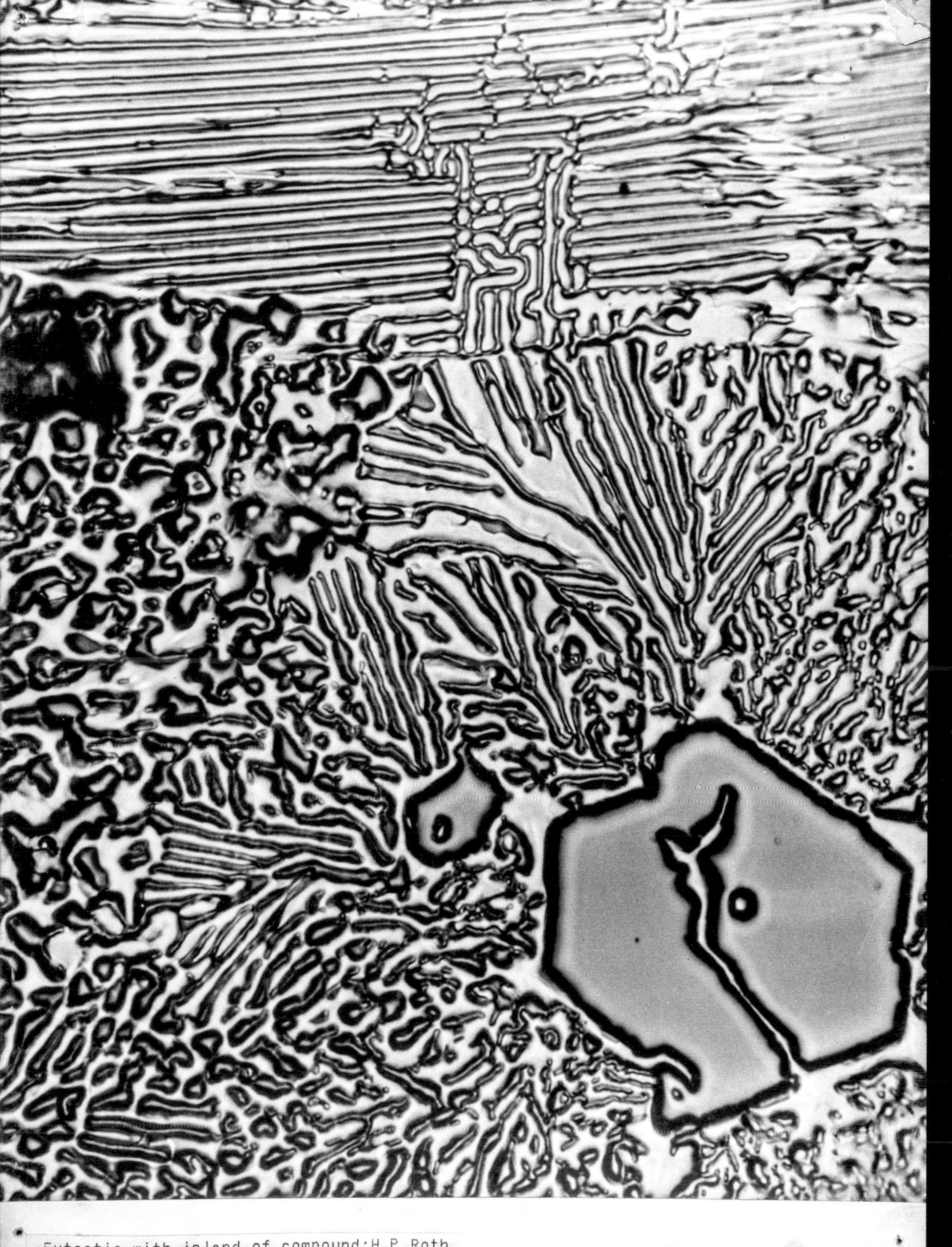

Eutectic with island of compound:H.P.Roth

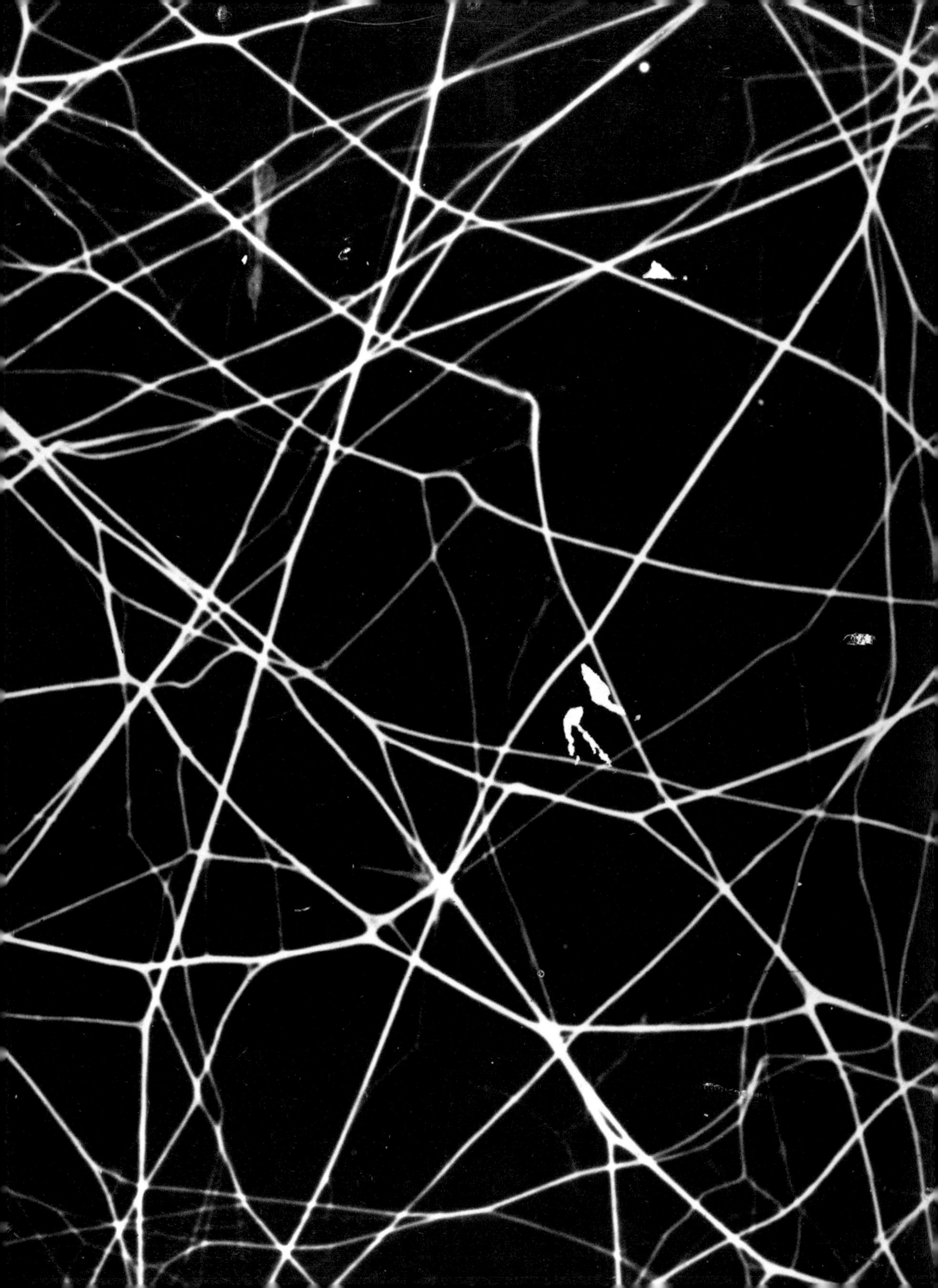

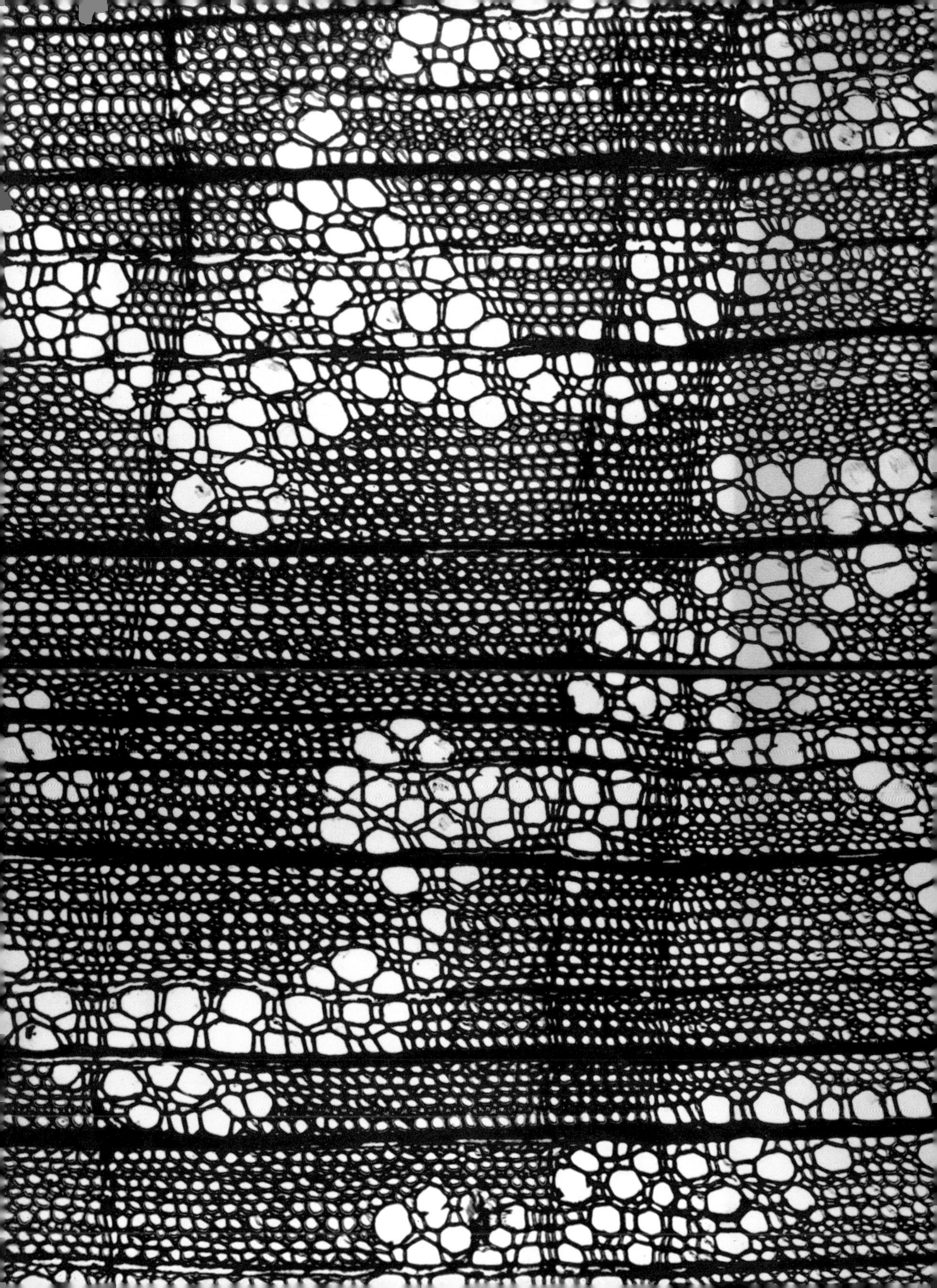

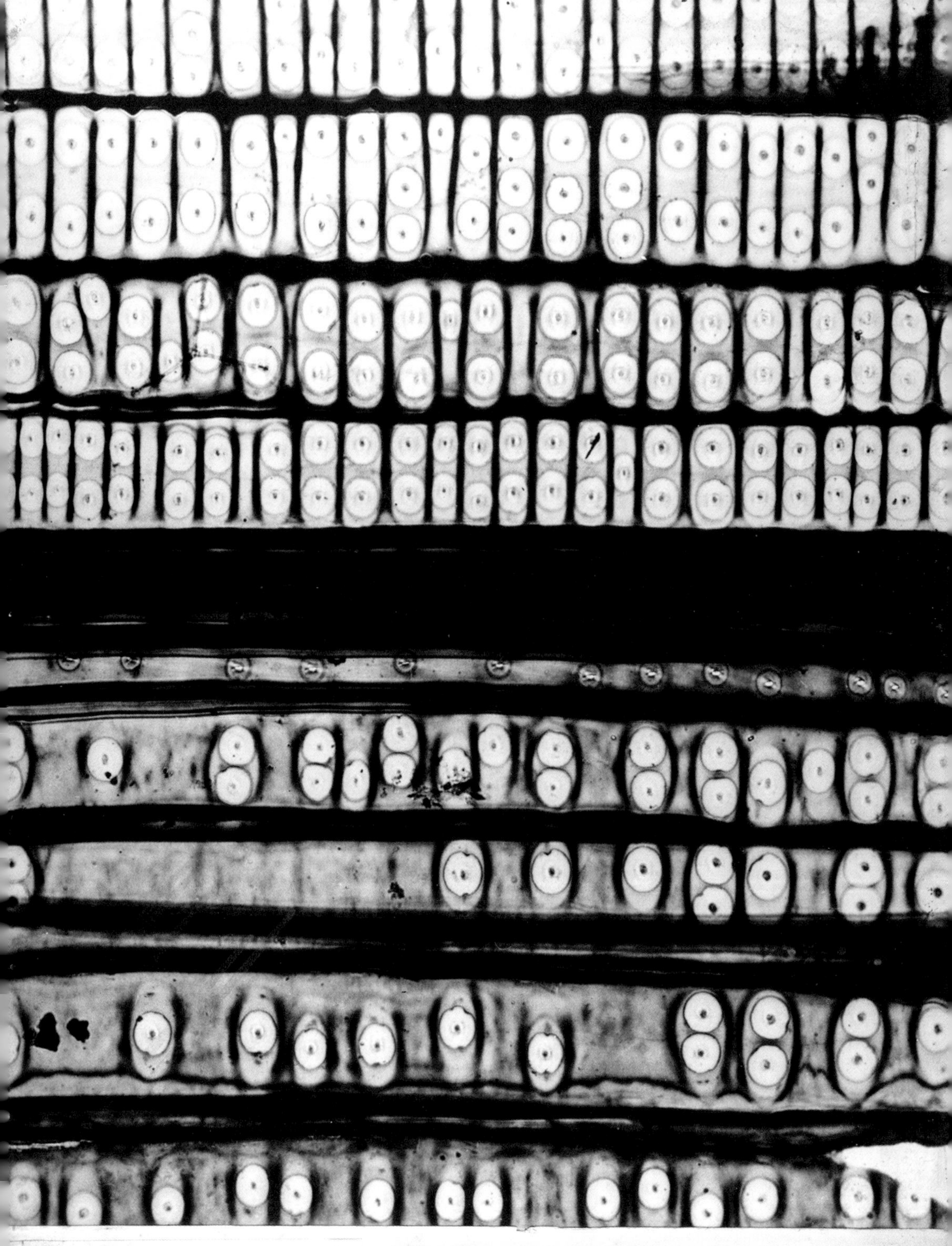

Radial section of redwood: 700X

Gyorgy Kepes, *Radial section of redwood: 700X*, photographic enlargement on fiberboard panel, 1951. Gyorgy Kepes papers (M1796). Dept. of Special Collections and University Archives, Stanford Libraries, Stanford, Calif. © The Estate of Gyorgy Kepes.

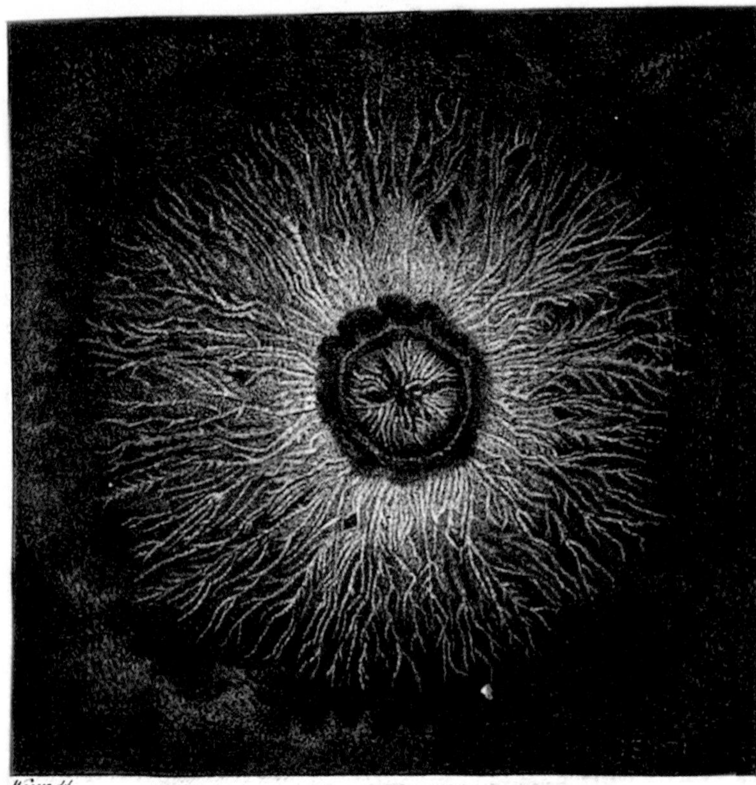

2.6

Georg Christoph Lichtenberg, drawing of electrical figure from "De Nova Methodo Naturam Ac Motum Fluidi Electrici Investigandi" ("Concerning the New Method of Investigating the Nature and Movement of Electric Fluid"), *Novi Commentarii Societatis Regiae Scientiarum Gottingensis*, 1777.

2.7

Gyorgy Kepes, *Lichtenberg figures: A. R. von Hippel*, photographic enlargement on fiberboard panel, 1951. Gyorgy Kepes papers (M1796). Dept. of Special Collections and University Archives, Stanford Libraries, Stanford, Calif. © The Estate of Gyorgy Kepes.

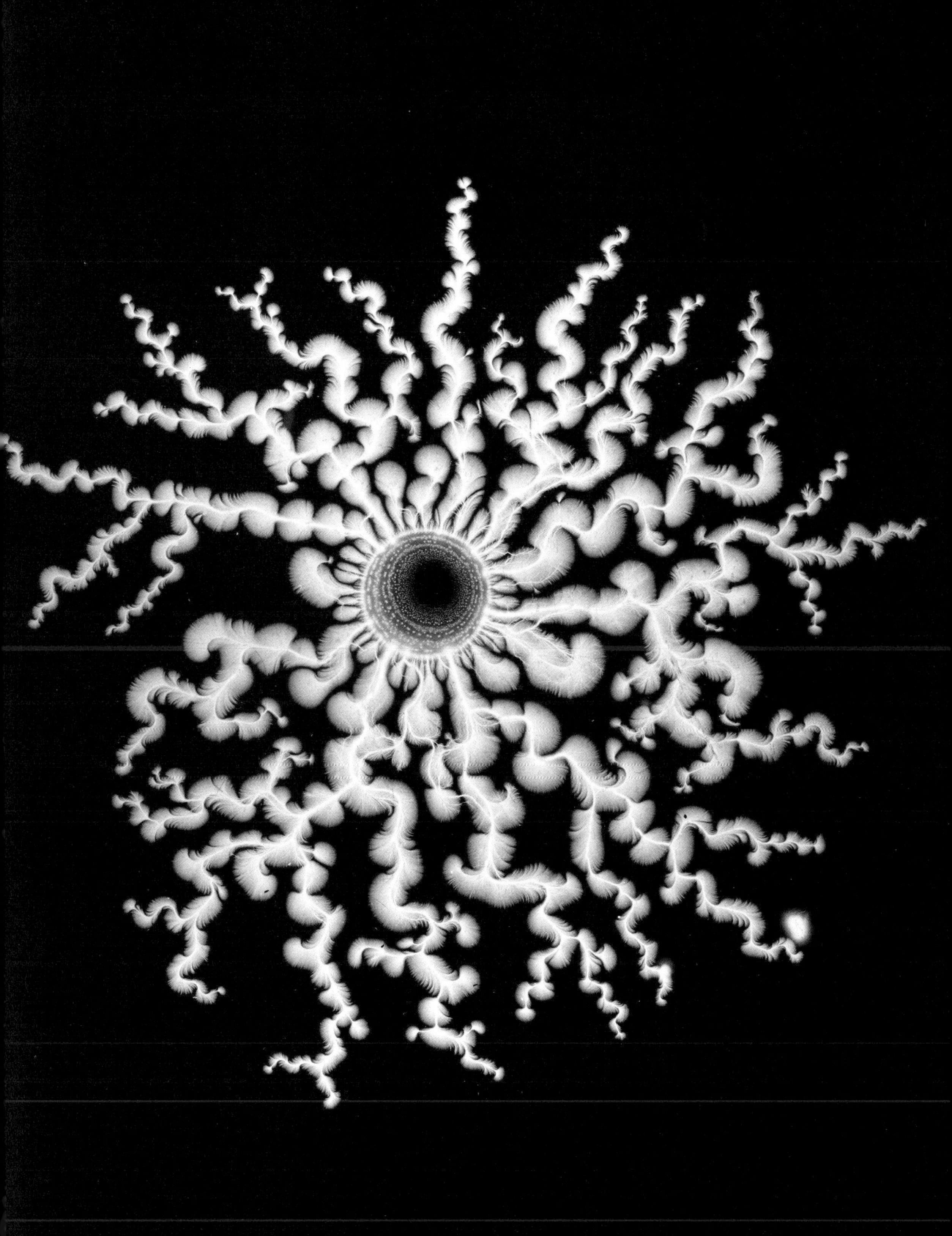

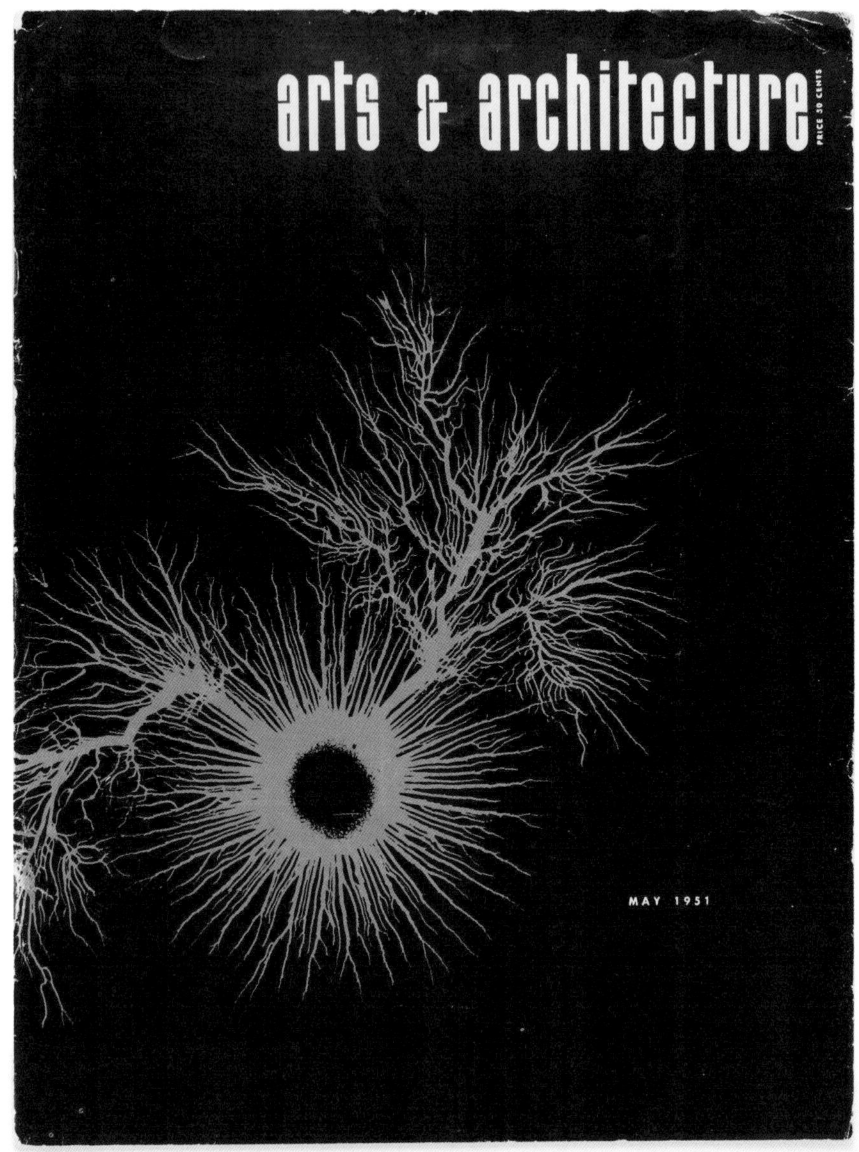

2.8

Arts and Architecture, May 1951. © Travers
Family Trust, used with permission.

Kepes's strange exhibition was preface to an even stranger book. In *The New Landscape in Art and Science*, assembled in 1952 as a catalog of sorts for his exhibition but not released until 1956, Kepes compared and contrasted his scientific imagery to the great masterpieces from art history (fig. 2.9).[7] His book cover elegantly expressed this union of what would become known, following C. P. Snow's formulation, as the "two cultures" of art and science: a radiograph of a rose, symbol of beauty and thus art, is overlaid with dots from a perforated reel of computer punch tape, a very 1950s symbol of science (fig. 2.10). The cloth hardcover below was also embossed with an emblem, originally from a cathode ray screen, that turns two separate strands—one we might read as art, the other as science—into a single spiral, turning toward a center and leading us into the book (fig. 2.11).

Within these covers, Kepes links the disparate visual culture of radically heterogeneous fields through isomorphic similarities, thereby producing bizarre relationships. Sifting through the book's four hundred and fifty-three images, we find affinities through visual likeness that span pages but also places, times, cultures, and even basic object categories and taxonomies: a drifting jellyfish imitates a Gothic rose window, a reticulated dragonfly wing evokes Piet Mondrian's grids, and the ping and ripple of oscilloscope waves from a primitive computer echo the earliest cave paintings. "Every pattern has its own extension and its wider context as well: it contains or is contained by another pattern; it follows or is followed by another pattern."[8]

At their most basic, these arrangements made use of the book's essential format, a format recalling the traditional art-historical slide comparison: two images juxtaposed against one another. They are set on the book's facing pages, with the gutter and spine dividing them (Kepes was fixated on book design; in a 1947 essay, he describes his ambition to "rethink the media" of the book).[9] The black-and-white photographic reproductions make these visual comparisons possible in the first place, erasing distinctions of size and scale, and creating resemblances between otherwise distant phenomena.[10] An exquisite snowflake mirrors, with immaculate perfection, the cross section of a twig—even black and white are carefully inverted (fig. 2.12). One is organic, the other inorganic. Viewing the spread is like seeing a picture card through a stereoscope; two similar but different images must be resolved in the mind's eye—as a unity of opposites. The free-floating panels in Kepes's exhibition functioned similarly, though their associations were more open-ended as one moved around the images; another cross section of a twig, this from the stem of a deciduous tree called the tetracentron, mimics the explosion of another Lichtenberg figure (figs. 2.13–2.14). Organic and inorganic flicker and flash,

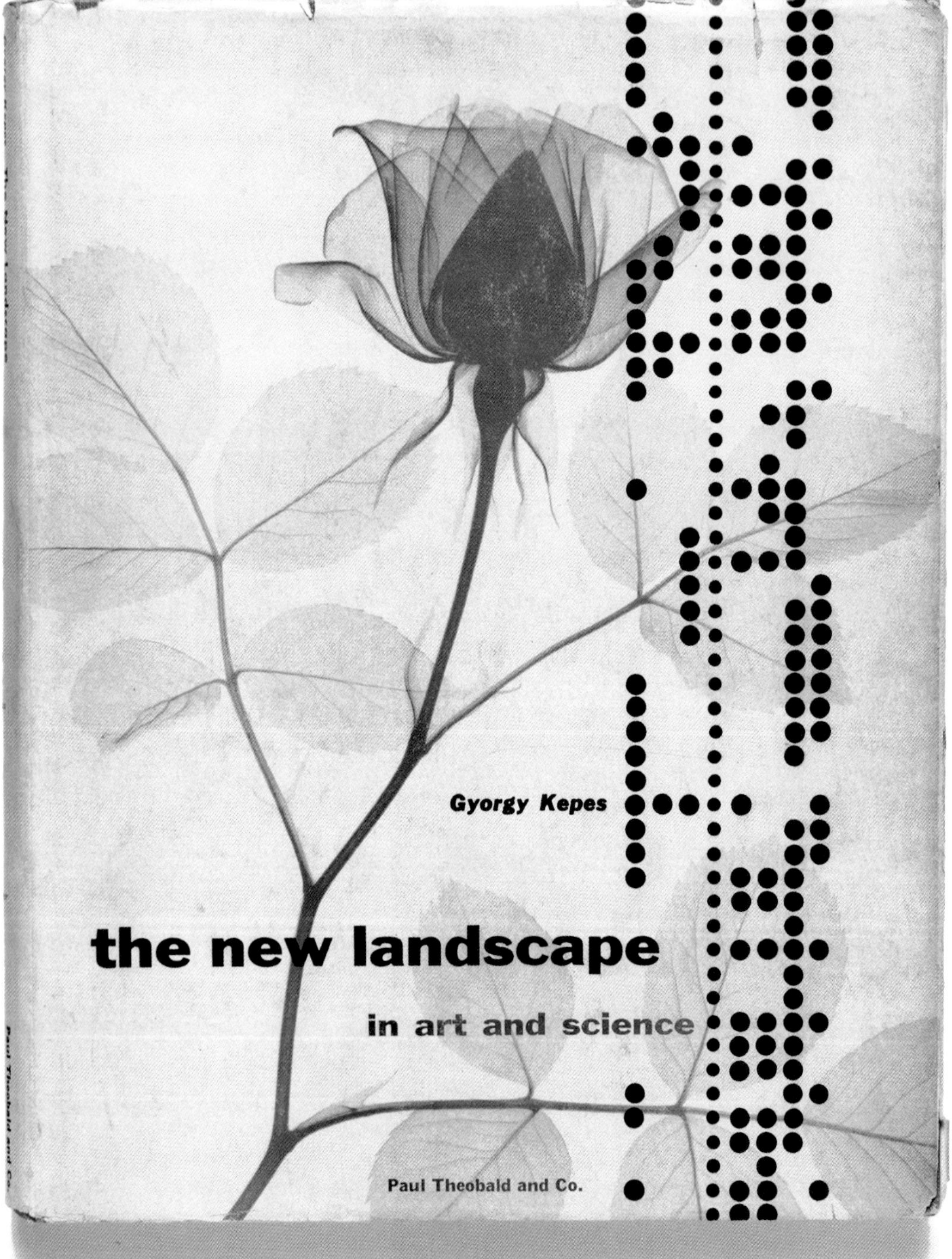
Gyorgy Kepes
the new landscape
in art and science
Paul Theobald and Co.

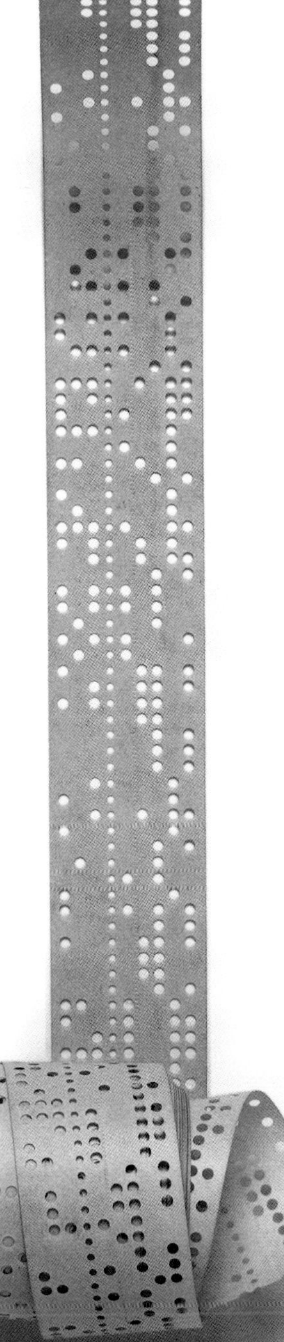

2.9

Cover of Gyorgy Kepes, *The New
Landscape in Art and Science*. Chicago:
Paul Theobald, 1956. © The Estate
of Gyorgy Kepes.

2.10

Computer punch tape circa 1950s.

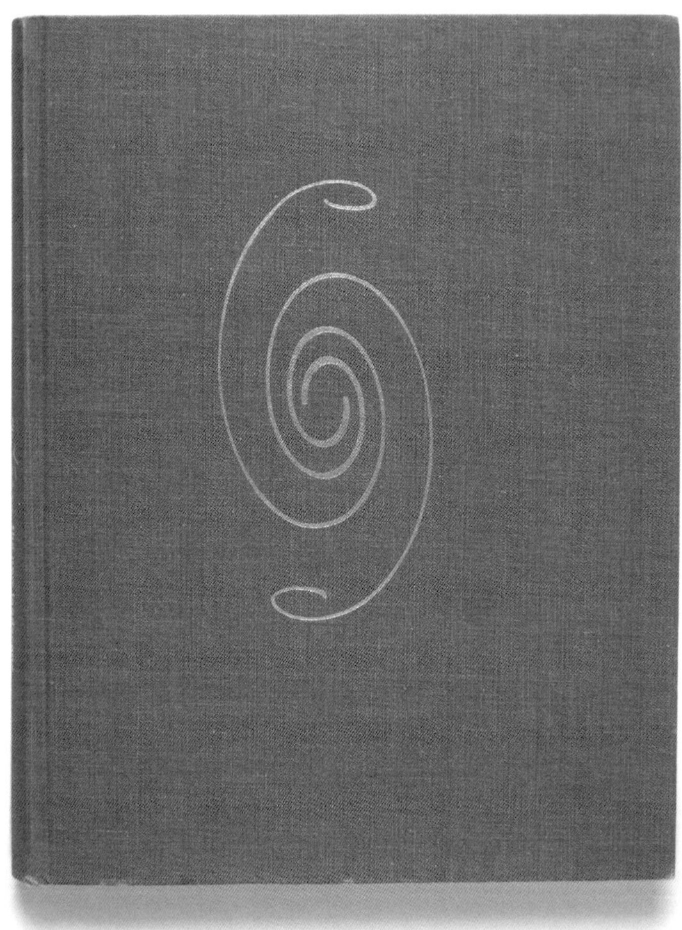

2.11

Cover emblem on Gyorgy Kepes,
*The New Landscape in Art and
Science*. Chicago: Paul Theobald, 1956.
© The Estate of Gyorgy Kepes.

oscillating in an unstable equivalence. The images further suggest the magnified iris of an eyeball, a suitable iconographic reference (it is probably for this reason that Kepes put a tetracentron on the cover of the exhibition's accompanying brochure).

This visual assemblage, from a dizzying list of disciplines, with a staggering array of correspondences and relationships, points to the core question of Kepes's *New Landscape* projects: What is the point of all these patterns? What was Kepes attempting to accomplish through these elaborate puzzles of art and science—and how are we meant to solve his cryptic game?

Before considering these questions, before becoming lost in the labyrinth, allow me to situate the *New Landscape*. These projects are the most significant to emerge from the pedagogical program that Kepes developed at MIT in the 1950s: what he called "visual design"—such was the name of his main course of instruction—or, more generically, the "education of vision," or just "pattern-seeing" and "form-thinking." Visual design was a direct descendent of Bauhaus pedagogy; it built on Kepes's own *Language of Vision* but also adapted these programs to the scientific and technological ethos of MIT, synthesizing fields as disparate as Gestalt psychology, engineering, and biology—any discourse on images and seeing could be absorbed into visual design—and purposefully engaged emerging midcentury fields like Norbert Wiener's cybernetics, Ludwig von Bertalanffy's systems theory, and the information theory of Claude Shannon. Kepes's methodological approach, based on combining and connecting anything and everything, mirrored these omnivorous interdisciplinary languages.[11]

Most critical interpretations of Kepes's *New Landscape* projects are organized exclusively around the theme of instrumentality: the suspicion that Kepes invested aesthetic value into MIT's visual culture in order to give the Institute a humanistic veneer, making it appear intellectually balanced and well-rounded—and to naturalize, while also concealing, the militaristic origins of this visual culture. It has become commonplace to describe Kepes's project as a "reactionary" glorification of the imagery and ideology of science, and to cite his work as representative of "a general faith in technological progress" among the Bauhaus diaspora.[12] Reinhold Martin most forcefully critiques Kepes in this way. According to Martin, Kepes's "'pattern-seeing' was commensurate with the logic of the new machines"—in other words, with military machines then being developed.[13]

By contrast, I argue that Kepes used the images he discovered in MIT's labs in subtly subversive ways. He recalls being met with suspicion while he collected his scientific pictures: "I went like a bulldog who never gives up from

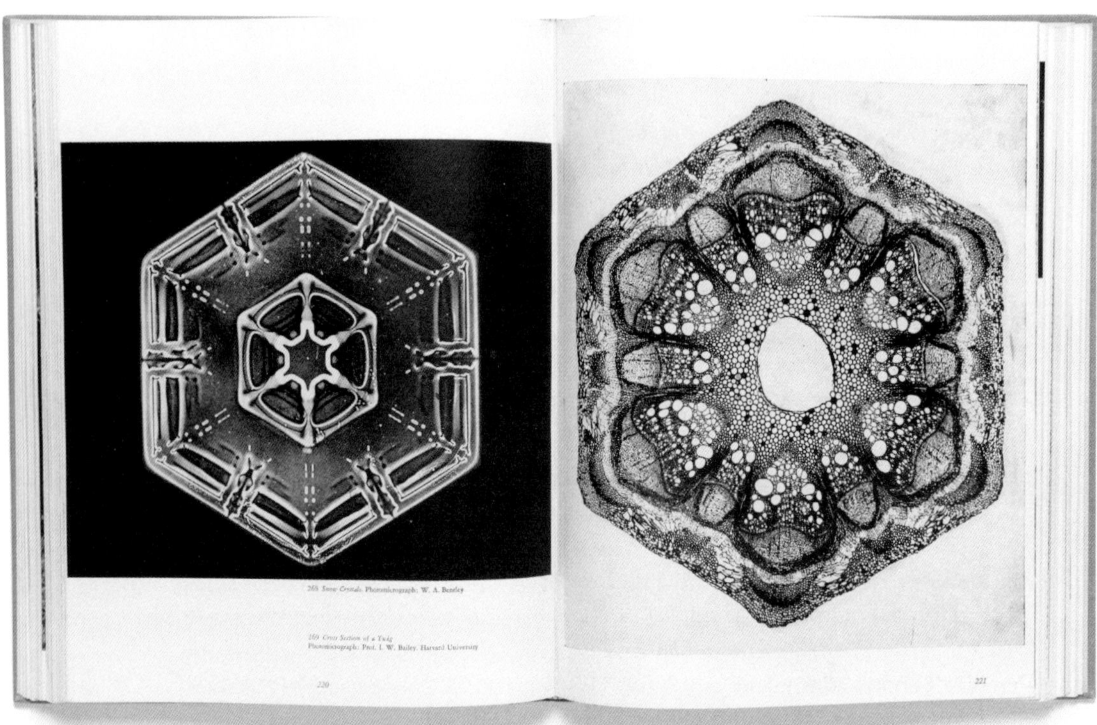

2.12

"Snow Crystal" compared to "Cross Section of
a Twig," two-page spread from Gyorgy Kepes,
The New Landscape in Art and Science.
Chicago: Paul Theobald, 1956. © The Estate
of Gyorgy Kepes.

one laboratory to the other. Sometimes I was almost thrown out by the scientists, because they thought I would distort the meaning of science by trying to use just the optical record."[14] I seize on this description of infiltration, appropriation, and distortion. Kepes adapted science's images to new ends—ends that were clearly not scientific. Through a strange aesthetic alchemy, he transformed them into tools for cultivating creativity: visual patterns, if understood creatively, might teach more creative habits of mind and patterns of being. Sight might provide insight. Kepes's imaginative approach to the image thus opposed rational, technical, and logical discourse, the very discourse that supported research into weapons and warfare. These visual records were created as objective evidence, but in Kepes's hands they became sites for subjective projection. In other words: Kepes provides a counterdiscourse, an alternative model of perception, one structured against the way of seeing and thinking that dominated MIT. A more expansive perception, one freed from the confines of logic and reason, would create a more expansive institution, society, world. To open eyes would open minds. In his notes, Kepes makes this ambition clear: "assumption: a) creativity is value b) creativity can be educated."[15]

In framing Kepes's project as subtly subversive, as potentially progressive and not entirely regressive, I depart not only from existing interpretations, but also from the artist's own stated intentions. Kepes described his project as an expansion of vision through pictures that revealed everything that is "too small or too large, too close or too distant, too dense or too scattered to be seen" by the unaided eye.[16] Such a project reiterates Moholy's "new vision"— indeed, *The New Landscape* is clearly a repetition of *The New Vision*. As Spyros Papapetros states, the new landscape "was already old."[17] The project also has many other precedents. It recalls Aby Warburg's and Fritz Saxl's *Mnemosyne Atlas* and André Malraux's *Musée imaginaire*. Its layouts echo the famous comparison between the Parthenon and the automobile in Le Corbusier's *Vers une Architecture*; it repeats the collections compiled in Siegfried Giedion's *Mechanization Takes Command;* it reiterates the images in Karl Blossfeldt's *Urformen der Kunst.*

And yet, at MIT, such a project—even if it was derivative, even if it was a classic "paradigm repetition"—took on new functions, even functions that Kepes did not openly acknowledge when he described the book.[18] His approach to the image was intended to transform the aesthetic of control that seemed to be inherent in science and technology into a more imaginative form of visual engagement. I examine Kepes's project not only through the lens of the "closed world"—a term used by Paul N. Edwards to signify the militaristic logic of the Cold War, as represented by discourses like cybernetics—but also the lens of the "open mind," an approach that numerous scholars have embraced in creating a more nuanced history of Cold War culture.[19]

2.13

Gyorgy Kepes, *Lichtenberg figures: A. R. von Hippel*, photographic enlargement on fiberboard panel, 1951. Gyorgy Kepes papers (M1796). Dept. of Special Collections and University Archives, Stanford Libraries, Stanford, Calif. © The Estate of Gyorgy Kepes.

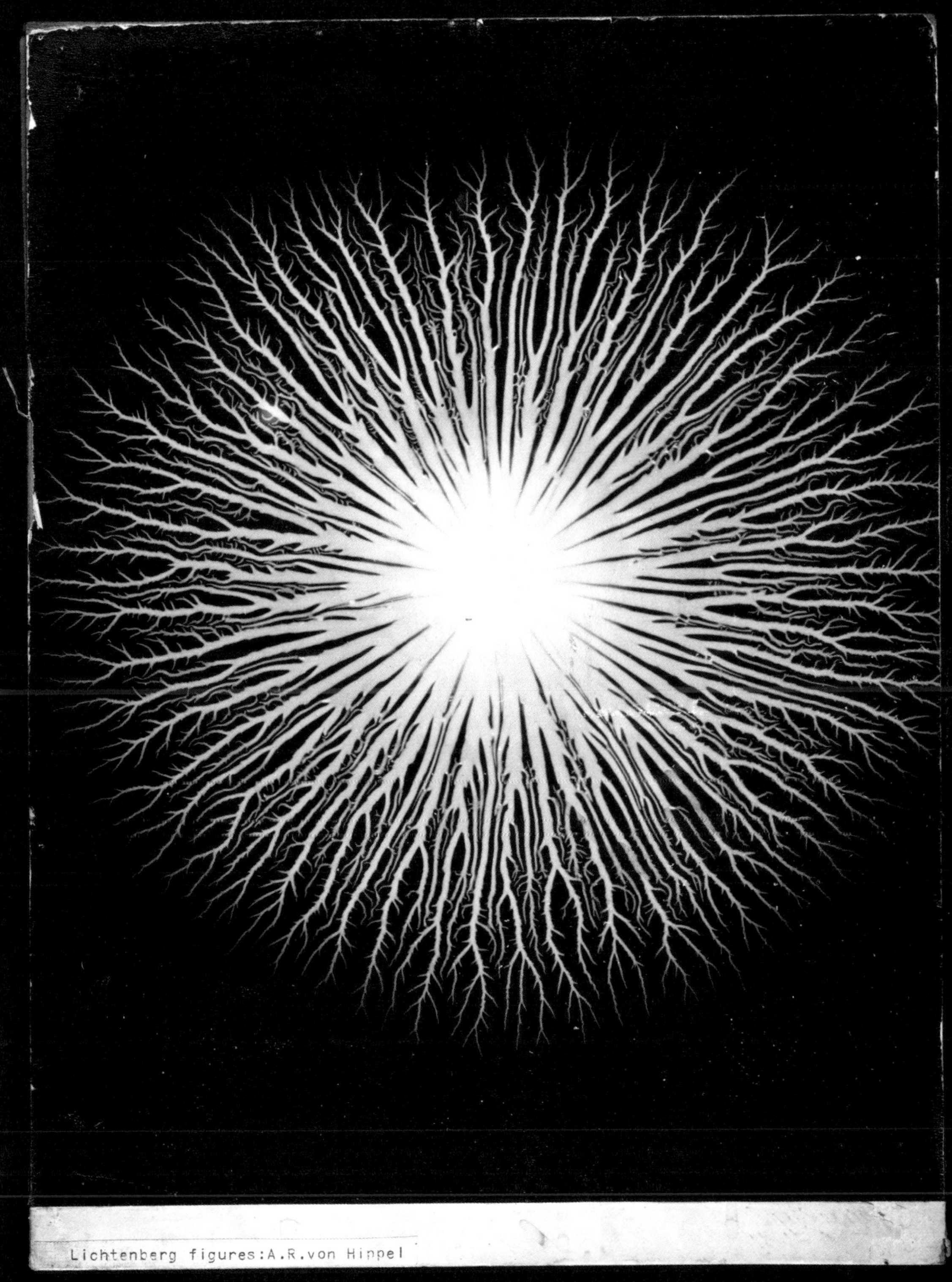

Lichtenberg figures:A.R.von Hippel

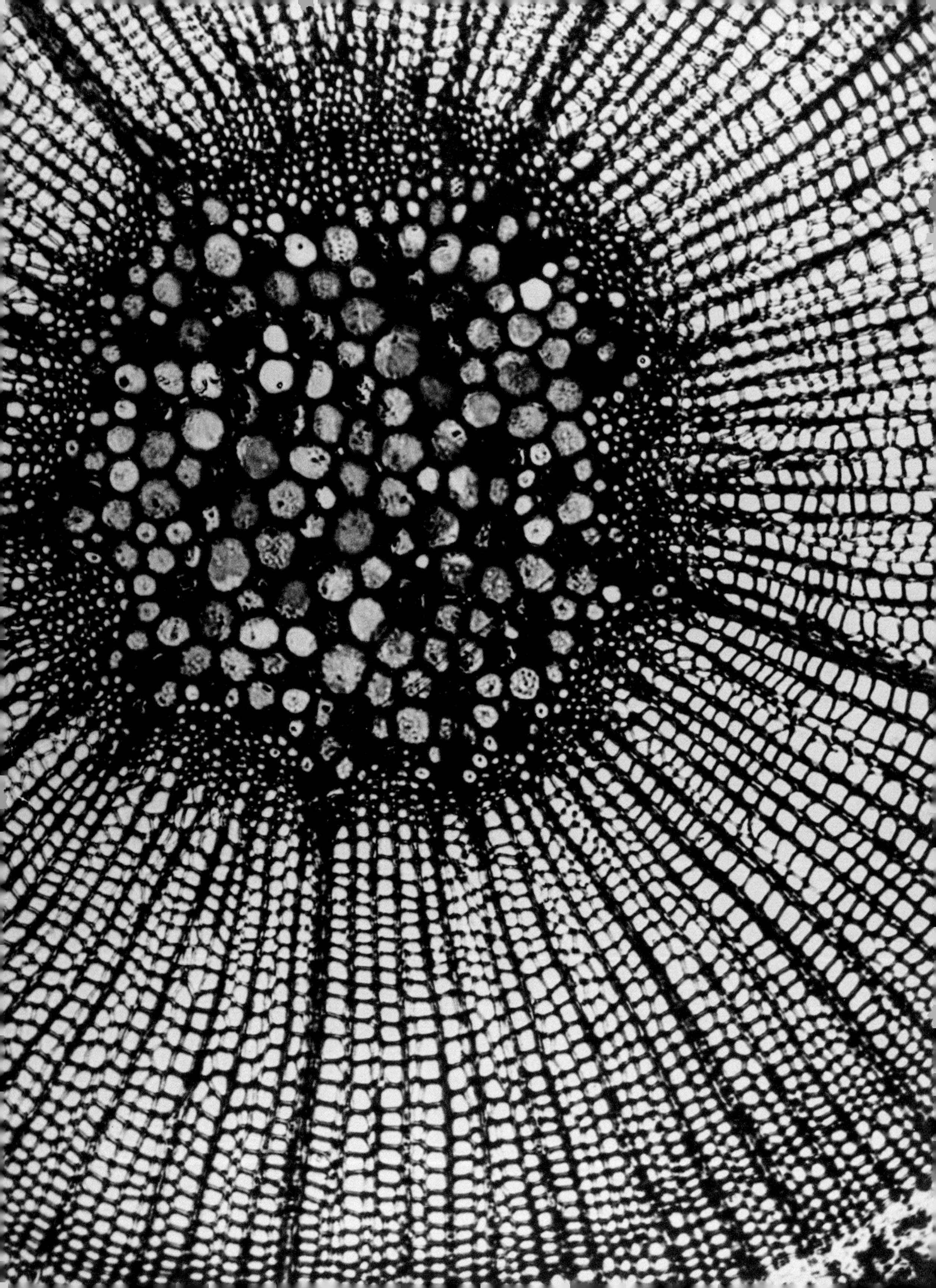

2.14

Gyorgy Kepes, *Cross section of tetracentron: 50X*, photographic enlargement on fiberboard panel, 1951. Gyorgy Kepes papers (M1796). Dept. of Special Collections and University Archives, Stanford Libraries, Stanford, Calif. © The Estate of Gyorgy Kepes.

2.15 (following page)

Gyorgy Kepes, *Camel's tongue: 10X*, photographic enlargement on fiberboard panel, 1951. Gyorgy Kepes papers (M1796). Dept. of Special Collections and University Archives, Stanford Libraries, Stanford, Calif. © The Estate of Gyorgy Kepes.

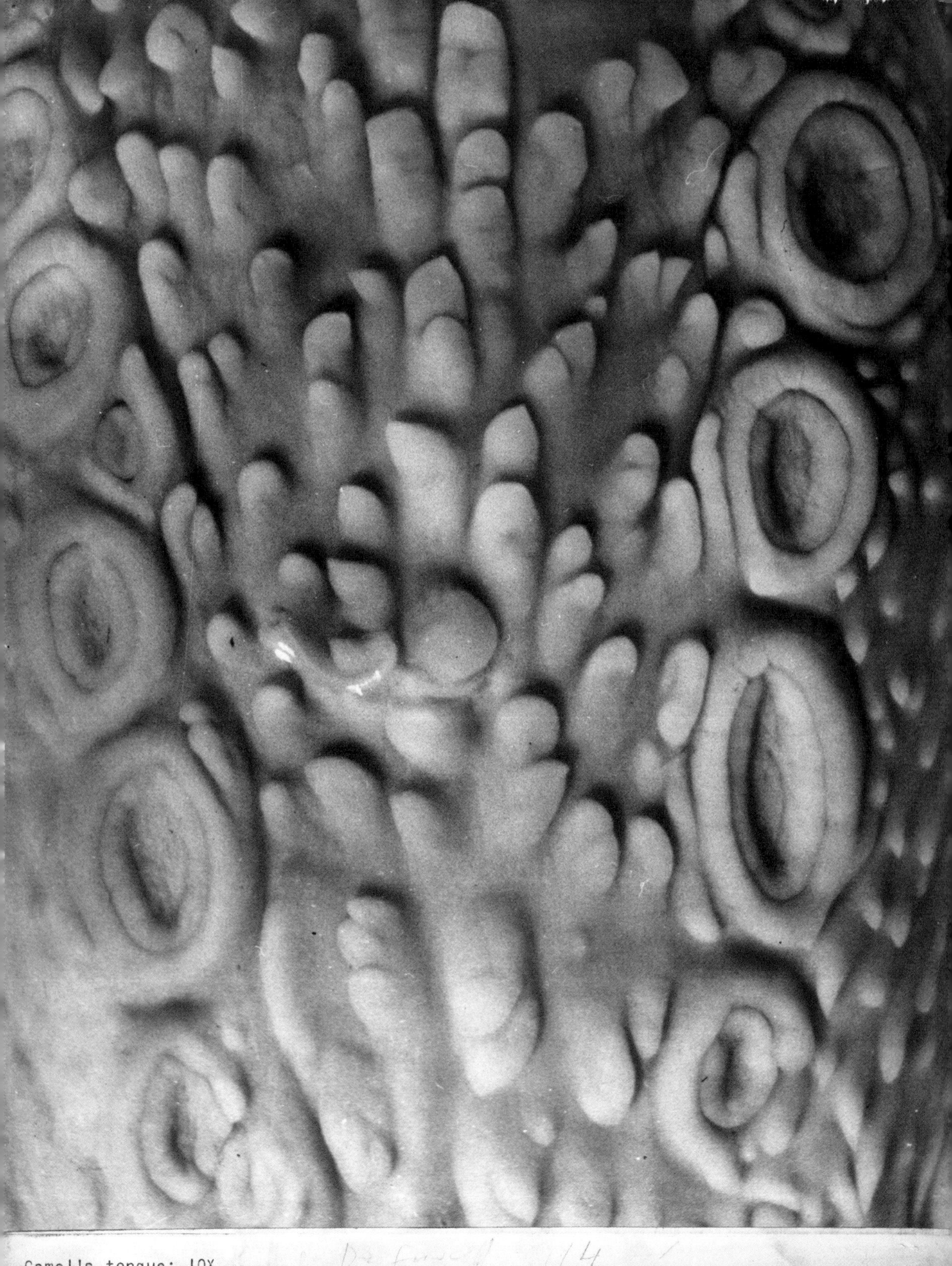

Camel's tongue: 10X

Let us return now to Kepes's image archive. The formalist description with which I opened this chapter was not formalism for the sake of it; my goal was to indicate the fundamentally formalist nature of Kepes's project. While he pulled images widely and wildly—even recklessly—from an encyclopedic spectrum of fields, he selected and arranged them with a formal precision that hints at hidden meaning. I also referred to the relations between images as isomorphic, but they are, more accurately, *pseudomorphic*. Pseudomorphosis is Erwin Panofsky's term for an infamous art-historical fallacy, defined thus: "the emergence of a form A morphologically analogous to, or even identical with, a form B, yet entirely unrelated to it from a genetic point of view."[20] Two things that look alike but are not alike. Yve-Alain Bois has examined the dangerous ahistorical ramifications of such a fallacy.[21] But unlike the art-historical error, the pseudomorphosis is for Kepes a means of generating rich visual experiences. It became his main artistic tool, one replacing the paintbrush, the pencil, the camera.

Pseudomorphic associations occur, for example, within single images. One exhibition panel appears to depict the eerie, crater-strewn surface of a distant moon; according to its textual label, it actually shows a camel's tongue under ten times magnification (fig. 2.15).[22] Fleshy protuberances and taste buds simulate lunar scenery with startling effect. Kepes telescopes viewers between perceptual levels, pulling us from the infinitesimally small (the surface of the tongue) to the infinitely vast (the surface of the moon), and all within a single visual experience. Reinhold Martin argues that it "matters little whether we read the captions provided by Kepes to identify each image."[23] But the interplay of visual and verbal is key. Accurate captions served as proof of Kepes's basic understanding of scientific phenomena—seemingly a prerequisite for his work at MIT. But the text also anchors and releases the meanings of an otherwise meaningless abstraction; it is the textual modifier that causes the image to flicker between associations.

Pseudomorphic associations also occur across multiple images. One book spread compares a leaf sample, prepared with dye so that its venation is visible under a microscope, with a 1938 photograph by Edward Weston of the cracked desert floor (fig. 2.16). Kepes reproduced Weston's photograph exactly as the photographer intended—in authorizing its use, Weston stipulated that "no 'cropping' is done, that my original composition is not changed" in any way.[24] But Kepes manipulated the image of the leaf at will, cropping as necessary so that the two images appear contiguous across the spread, an all-over pattern linking two opposite concepts: the animate on one side, the inanimate on the other.[25]

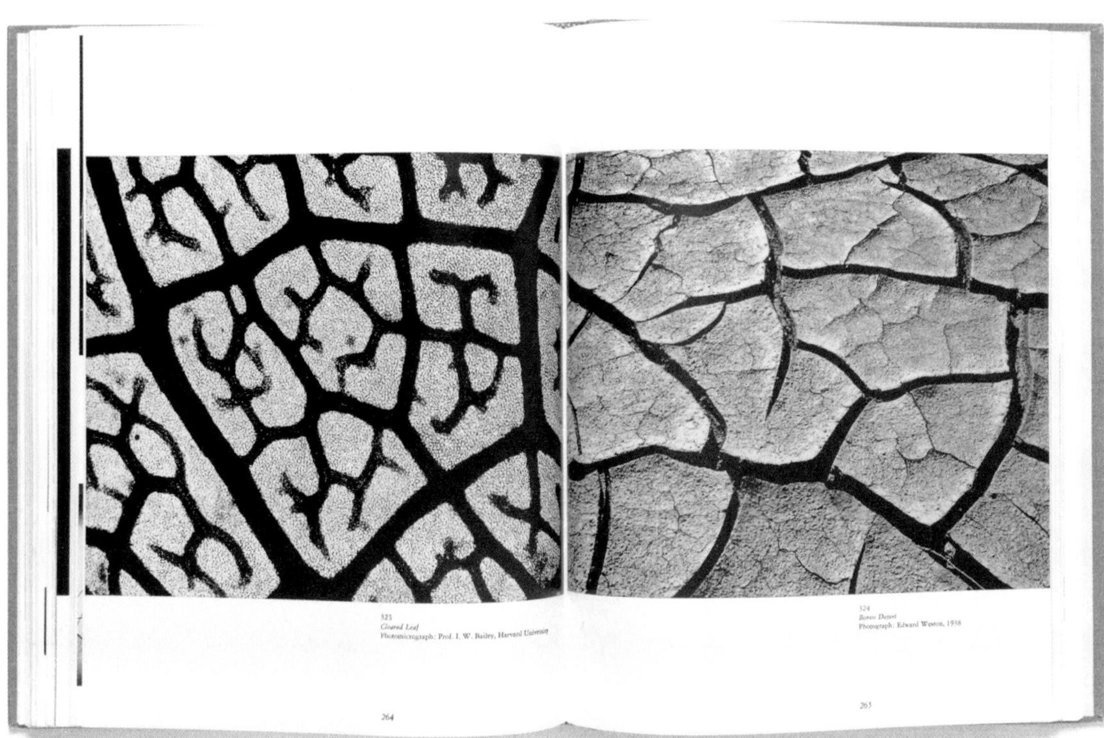

2.16

"Cleared Leaf" compared to Edward Weston,
Borneo Desert, 1938, two-page spread from
Gyorgy Kepes, *The New Landscape in Art
and Science*. Chicago: Paul Theobald, 1956.
© The Estate of Gyorgy Kepes.

Still other pages were more complicated, forming collages with concatenations of images that look as if they might continue well beyond the limits of the book: a piece of Sumerian stone-and-pearl work, a Nasca tapestry from Peru, a Byzantine mosaic, a canvas by Mondrian, et cetera (fig. 2.17). All four images are connected by shared formal traits—the repeating detail of black and white, from interlocking mosaic tesserae to interwoven tapestry fibers—but they also create conceptual inversions. Reading left to right, we move from old world (Sumer) to new world (Peru), from past (Byzantium) to present (Mondrian), et cetera. The staggered arrangement emphasizes these locational and temporal displacements as a chain of dialectical steps, a process of turning visually but also metaphorically. These are patently false comparisons—Byzantium and Mondrian have nothing to do with one another—but they are also purposeful, intended to propel us into the enigmatic realm of imagination. The pseudomorphosis demands that we invent a reason for these visual affinities, making meaningful what is otherwise meaningless.

In these ways, and in many more—there is no limit to the formal relationships one might discover—Kepes encouraged an open-ended interaction with his hanging panels and printed pages that requires viewers to conjure constellations out of the optical profusion. There is no "proper" or "correct" way to connect the dots; these constellations are not fixed in space but chosen at will. This task is not factual, but mystical and mythological, similar to ritual acts of divination like counting stars, deciphering tea leaves, or gazing into the crystal ball (there are starry skies, leaf samples, and all manner of crystals in the book). Kepes recuperated these ancient arts through the photographs from midcentury science. Studying his images is like reading entrails; it is very appropriate that Kepes included a rat's duodenum, the first section of its small intestine, within his image archive (fig. 2.18).

Some of the scientists who contributed to Kepes's project even described visual analysis in similar ways, as mystical and mythological. For them, science required an instinctive interpretation of appearances. Bruno Rossi, an astronomer, wrote in Kepes's book about an "intuitive feeling" that is "as essential to the creative scientists as it is to the creative artist."[26] Von Hippel referred to his Lichtenberg figures as "beautiful spectacles," each one "a riddle akin to the hieroglyphics of the Egyptian script." In interpreting them, the scientist must become a seer or mystic, compelled to "read the signs"—von Hippel's words—of the visual record.[27] It is not surprising that von Hippel approached the image in this way; as a student, he attended Heinrich Wölfflin's lectures in Munich, changing his course of study from art history to physics only after receiving a letter from his father that read: "complete studies in science or starve."[28]

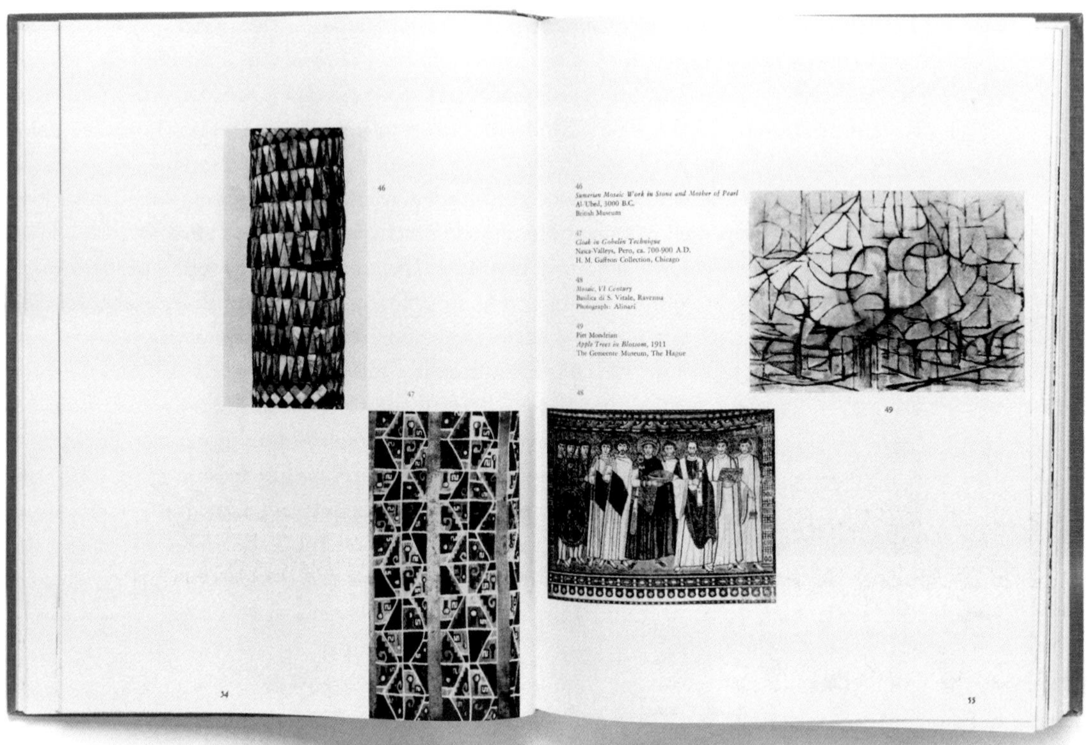

2.17

Two-page spread from Gyorgy Kepes,
*The New Landscape in Art and
Science*. Chicago: Paul Theobald, 1956.
© The Estate of Gyorgy Kepes.

"Duodenum of a Rat," from Gyorgy Kepes,
The New Landscape in Art and Science.
Chicago: Paul Theobald, 1956. © The Estate
of Gyorgy Kepes.

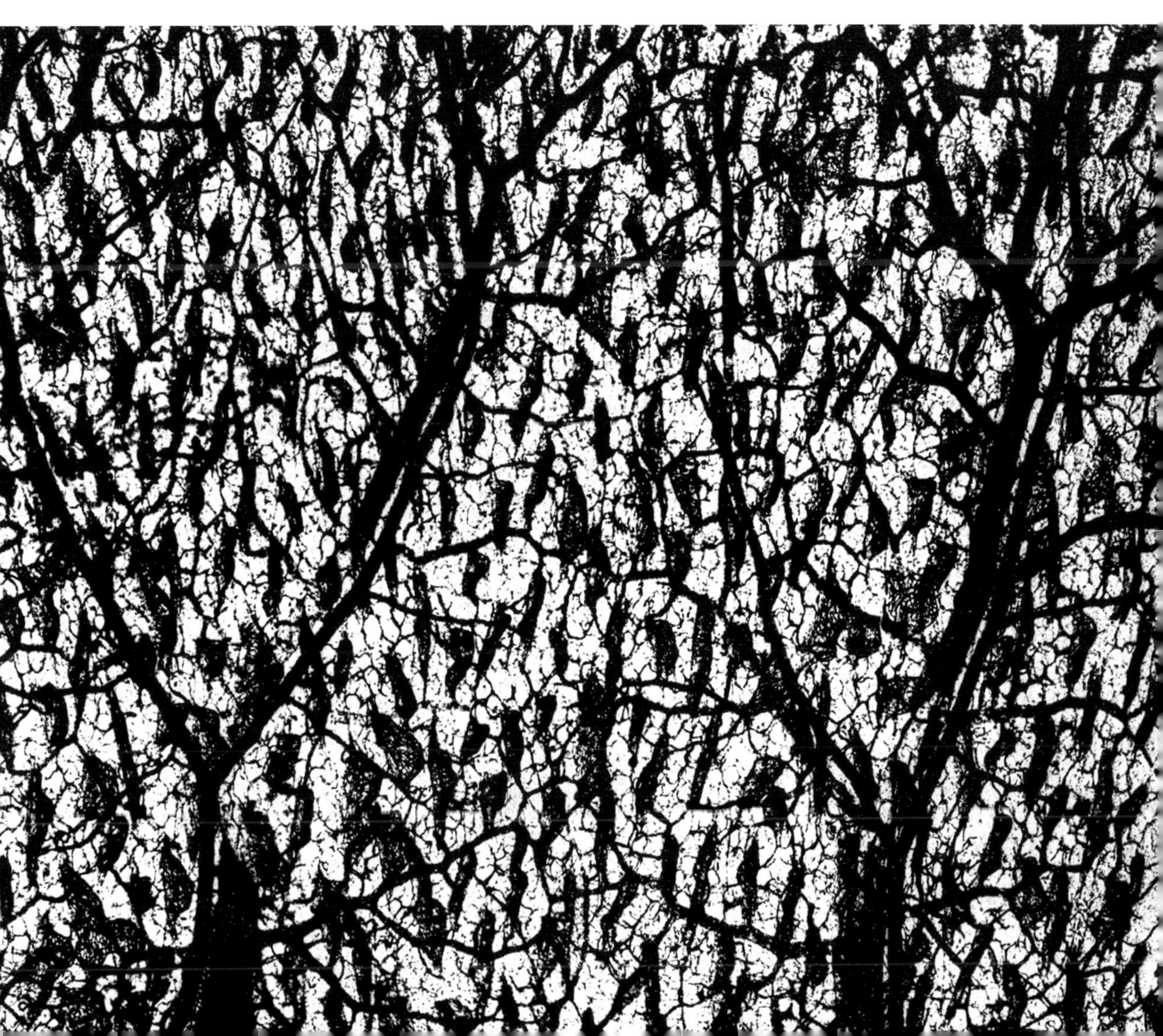

Another page contains another clue: a small but suggestive thumbnail of a Chinese oracle bone from 1700 BCE. The fragment, engraved with a unique script of logographs, was used in pyromancy—divination by fire. Questions were submitted to the gods by carving them directly into the bone; the fire's heat caused the bone to fracture and crack, and a diviner would decipher these patterns as evidence of divine response. This thematic interest in superstition is pervasive in Kepes's notes. Describing pyromancy, he refers to cracks "interpreted by soothsayers" as "magic writing bearing the response of the divinity," a form of "eavesdropping on nature." He similarly references the Taoist tradition of augury: reading omens by studying a bird's path of flight across the canvas of the sky.[29] Kepes aims to recover an ancient enchantment with visual experience, an enchantment eclipsed, negated, by the clinical culture of science and technology.

Art and Science/Art against Science

Actually, Kepes was never so negative. He always framed his ambitions in wholly positive terms, as a peace-making enterprise. In *The New Landscape* he presents the relationship between art and science as disjointed, but hardly antagonistic:

> **Artists and poets on the one hand, scientists and engineers on the other, appear to live in two different worlds. Their common language, their common symbols, do not exist. To develop a vision which brings the inner and outer worlds together, we need common roots once more.... For this we must remake our vision.[30]**

But we need not read Kepes word for word. Indeed, we must not, for Kepes had to frame his ambitions in wholly positive terms—he was employed at MIT, after all—even if the relationship between fields was not so positive. His rhetoric was a delicate balance, intentionally suppressing the hostility between worlds that he actually experienced. Indeed, the volume first took form as an attack against the destruction and devastation wrought by science and technology. An outline for "the projected book, Revision of Vision"—what would become *The New Landscape in Art and Science*—details Kepes's original intentions. The book would have opened with "photographic documents of our time," which Kepes describes as "war photos," "concentration camp photos," "prisoner camp drawings," "children's drawings from the war torn areas," and "disaster images."[31] It would have surveyed the catastrophes of midcentury as a preface to Kepes's scientific image archive, thereby suggesting that these catastrophes were actually continuous with, and maybe even caused by, science and technology.

Kepes abandoned this idea—there are no such disaster images in his book—but he opens the first page of his written introduction to *The New*

Landscape with a similar, if more vague, allusion. In reference to Giorgio de Chirico's 1914 painting *The Mystery and Melancholy of a Street* (fig. 2.19) (a painting evoked but not reproduced in the book), he writes:

> *A child rolls her hoop in a vast, lonely nightmare landscape* ["the new landscape"], *racing toward the shadow where an unknown terror awaits*…. This now familiar image in modern art distills for the spectator the pervasive emotional disaster of the twentieth century – the sense of being lost in an alien, menacing world.[32]

Kepes begins with this painting because it depicts what he referred to as the "blind alley"—a figure of speech indicating his understanding of contemporary crisis as a problem of foreclosed vision, of a blindness perfectly illustrated in de Chirico's skewed perspectival distortion, like being lost in a labyrinth. Kepes's description of the blind alley continues the argument that he would have made with his assorted "disaster images," but does so in a more contained way. As his working title "revision of vision" indicates, this blindness was treatable; Kepes would restore sight to an eye damaged by such violence.

Kepes felt this acute hostility between fields after he arrived at MIT. He offers the most honest assessment of these relationships not in published writings but in less guarded lecture and interview statements. "I must confess I was utterly scared," he states in a lecture delivered at Yale in the 1960s. He recalls discovering himself within de Chirico's painting, experienced as MIT's famed infinite corridor—the blind alley. "If you have ever been to MIT, there's this terrific long corridor and each time you look into any kind of open window"—he refers to the windows of laboratories lining the hallway—"you just see all these kinds of fear instruments." Kepes had doubts about his role in this context. "I felt that science would just etch in into [*sic*] my personality…. I will be corroded…. I just will be deflated within. But I came there, and had to face it."[33] In remarks made years later, he is more candid about the cause of this anxiety: "I was practically shocked." MIT was "the arsenal of the world. It had a tremendous competency in producing intelligent weapons," it was "without heart."[34]

There is even evidence of how tense interdisciplinary relationships actually were directly in *The New Landscape in Art and Science*, through the words of Kepes's many contributors. John E. Burchard, Dean of MIT's School of Humanities and Social Studies, points to the divide in his foreword to Kepes's book: "It has been the nature of our age to place the word above the picture, the prose above the poetry … the search for truth by the methods of science above the search for truth by the intuitive methods of the artist."[35]

Burchard was even more blunt in an initial, unpublished submission for Kepes's book; he first suggested a preface to his foreword that would have

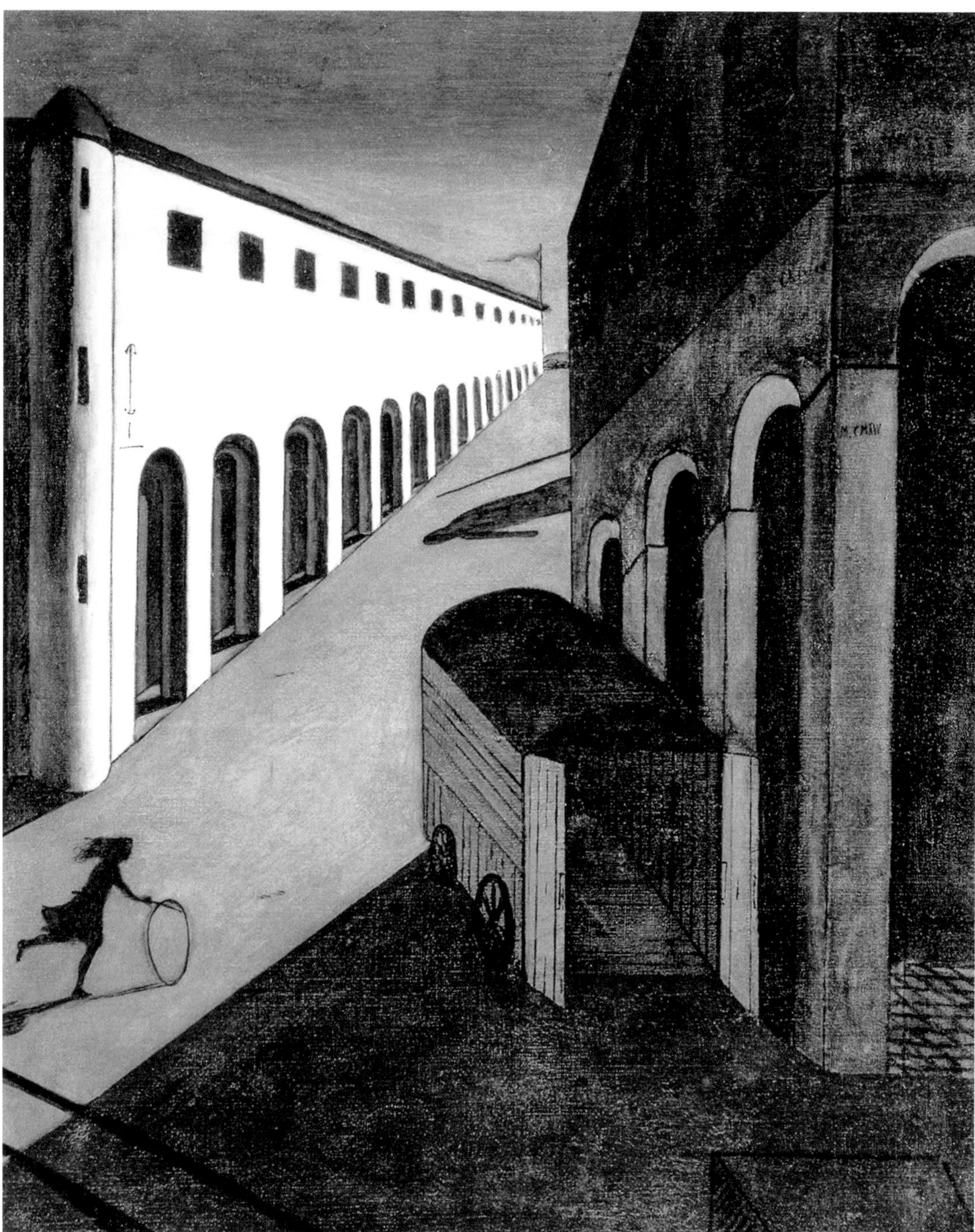

Giorgio de Chirico, *The Mystery and Melancholy of a Street*, 1914.
© 2018 Artists Rights Society (ARS), New York / SIAE, Rome

positioned Kepes's project within public policy debates over the funding of higher education: "I enclose on a separate sheet a few paragraphs which I believe could profitably be put at the head of the foreword if you agree, although I realize they will date it. I leave it to you."[36] These paragraphs—not included in the final book—frame Kepes's volume as a response to contemporaneous debates over the value of specific fields of study. Burchard refers to a report released by the US House of Representatives in September 1954 regarding proposed federal grants for the fine arts. He quotes from the report's majority statement, written by a Republican congressman, which rejects federal sponsorship and reads:

> [W]e cannot conceive it our duty to dip further into his [the American taxpayer's] threadbare pocket to indulge in the luxury of subsidizing an endless variety of programs and projects supposedly related to the so-called finer arts.... We think, instead, that young American artistic talent will overcome the hardships it encounters, just as outstanding engineering, medical and related talent surmounts the obstacle in its way to success.[37]

Burchard lambasts the congressional commentary, arguing that its comparison was false, based on ignorance of the massive investment in scientific and technological fields and correspondingly meager investment in the arts.

> This report was made in the face of testimony in the other direction by most of our national leaders in music, painting, sculpture, theatre, and the other fine arts. It was made in the face of the billions spent by the United States Government on its scientific research and development programs, on the millions available in support of research and fellowships in science provided by the National Science Foundation, in the face of the evidence that there are no more secure and lucrative careers in our country today than those of doctors and engineers.
>
> I do not cite this example to chastise the subcommittee because I believe that they came very near to expressing the common will of the people of our country. We have become a people who care more for the practical arts than the visual arts, if indeed we were ever anything else. This book [Kepes's *New Landscape in Art and Science*] comes to us as a people at a time when our worship of the pragmatic and the verbal and our neglect of the poetic and the visual are at a crisis.[38]

Burchard here presents Kepes's project not just as a veiled critique of science and technology but as a critique of vocational education. By promoting visual thinking as a valid intellectual project, Kepes offered a rebuttal to the narrow focus on the practical and pragmatic, on scientific and technical skills and their instrumental application in specific careers.

While Burchard wrote in sweeping terms about national trends, other contributors to Kepes's book brought this debate directly home to academic life in Cambridge, Massachusetts. Walter Gropius, the Bauhaus founder then teaching at Harvard University's Graduate School of Design, is even more specific. Though his remarks are buried in Kepes's volume—not highlighted in any way, and therefore easy to miss—they still make a provocative claim about the actual experience of interdisciplinarity in the academy:

> In today's universities we keep large departments under the name "Arts and Sciences," but when we scrutinize their activities, we find that science has everything.… But what about the Arts? … This is the century of science; the artist is the forgotten man, almost ridiculed and thought of as a superfluous luxury member of society. Art is considered to be something which was accomplished centuries ago and is now being stored up in our museums from which we may tap as much as is needed. As science is supposed to have all the answers for our predominately materialistic period, art is doomed to languish. Which so-called civilized nation today honestly promotes creative art as a substantial part of life? This disintegrating society needs participation in the arts as an essential counterpart to science and its atomistic effect on us.[39]

Gropius's remarks are striking—he really could be talking about *today's* universities. He could be talking about MIT—or Harvard, or Stanford, or Columbia, or just about any other institution—in the twenty-first century. Even generous investments in the arts and humanities do not allay my suspicion that science still "has everything," that creative pursuits on most campuses are supported for their compensatory role, as decoration, maybe even as meaningless decoration, something intended to make the university appear balanced and well-rounded.

But, going back to the 1950s, Gropius's remarks are also striking because they set the stakes of Kepes's project in high relief, as an antagonistic project, more civil war than peaceful union. At issue for Burchard and Gropius—and thus also for Kepes—is not only the disparity in money, in funds and facilities, but also the difference in prestige and power, the sense that science and technology are legitimately valued while the arts are viewed as irrelevant. In this context, Kepes's project takes on a particular charge. Rather than permitting the sciences to dictate a diminished role for the arts, how could the rapprochement between the "two cultures" play out in terms set by and beneficial for the arts? There is a famous dictum that states: *Ars sine scientia nihil est*, or, art without science is nothing.[40] Kepes set out to prove this claim, but also its inverse: science without art is nothing.

Designing Vision

Let us move further into Kepes's program. He called it "visual design." Although the name may sound generic, encompassing all types of artistic practice, Kepes used it with specificity. He referred not to making art—not to instruction in how to paint, how to draw, or how to sculpt—but rather to the design of vision itself: *how to see*. In visual design, the eye was a medium to be molded and formed, trained to see more freely.

MIT's School of Architecture and Planning recruited Kepes expressly to teach this subject, though they evidently did not understand what it was. A letter of appointment to Kepes from the architect William W. Wurster, then Dean of the School and a key figure in Kepes's hire, indicates the Institute's interest in but also puzzlement over Kepes's pedagogical mission: "We are eager to make 'the Drawing' (for want of a more complete word) a strong and integral part of the school."[41] Wurster similarly announced Kepes's position as "Associate Professor in charge of Freehand Drawing" in an annual statement about new faculty joining the Institute (Kepes would soon adopt "Professor of Visual Design" instead).[42] Despite this confusion over Kepes's program, his hire was a major part of the School of Architecture and Planning's midcentury transformation, which was intended to align MIT more closely with modernism. The School embarked on a period of "recruitment and reorganization" in preparation for a new "postwar orientation."[43] Visual design, centered on interdisciplinary methods devoted to vision, was the core of this new orientation.

But Kepes was little concerned with pencils or other drawing implements—or, indeed, with any conventional artistic tool or medium, even when he utilized them in his teaching. His course offerings reflected this new approach. He immediately replaced the traditional "Freehand Drawing" curriculum, previously taught by sculptor John Selmer-Larsen and primarily based on "drawing direct from the human figure," with a series of courses reflecting Bauhaus methods.[44] These continued the curriculum he had first developed at the School of Design and included "Visual Fundamentals," "Light and Color," and "Graphic Presentation."[45]

Visual design was a hybrid discipline, encyclopedic in its reach, based on combining and connecting a vast array of sources. In his notes, Kepes explains how visual design departed from more familiar models of arts education (he refers to it with a number of interchangeable terms). "Form thinking—structural thinking, configuration thinking, is different from [the] simple making [of a] little painting in an art department of the university. Form thinking must penetrate all disciplines of education. How to do it?" He offers a number of solutions, including teaching methods that emphasize the "active work of forming" and also compare the "structural correspondence between one and other types of forming." But he also suggests "making form thinking an education subject in itself." Such a subject could be taught by faculty within a department "that specializes in this new approach" or faculty who "pool concrete materials from other departments—biology, physiology, psychology, art, painting, architecture, music, etc." Kepes explains the goal of this instruction: "not the medium but the nature of organization, the common laws of configuration should be taught."[46] His ultimate ambition was to make these interdisciplinary methods part of MIT's general education curriculum.

In another note, he summarizes this pedagogical mission with a suggestive proposal for a "university of vision," the purpose of which would be the analysis of visual culture from "history, sociology, political science, geography, economics."[47] Yet another puts forth an early, still inchoate "suggestion" for a "Research lab" dedicated to "various solutions" to the problem of vision, including "training for color engineering" and the production of "pamphlets, exhibitions," and other interdisciplinary platforms addressing images and seeing.[48] These hopes would be fully realized only decades later, when Kepes established his Center for Advanced Visual Studies. But even now, Kepes became widely known for this obsessive focus on the visual; his name was synonymous with the eye. Alexander Calder, for example, addressed a postcard to Kepes with the name "Gyorgy Kepes" reduced to this single pictographic icon (fig. 2.20).

In order to prepare this program, Kepes began an intense period of research, recorded in a series of detailed bibliographies labeled "Education of Vision," another name for his pedagogy, and "Visual Fundamentals," the specific title of one course of instruction. These wide-ranging, all-encompassing reading lists suggest how Kepes developed visual design by working through books and articles from fields far and wide. He also produced an extensive set of research notes—many hundred pages in total—filled with illustrations, quotations, and yet more citations of additional sources; the notes point to an even wider set of references. These documents, taken as a whole, reveal the expansive intellectual world Kepes inhabited in the 1940s and 1950s.[49]

It was, to be sure, hardly cohesive or even coherent; Kepes studied disciplines not typically considered commensurate with one another, and often cited authors from a single field who were explicitly at odds. Many of the texts he studied were also beyond his expertise; he often puzzled over diagrams and then copied them into his notes, changing meaning in the process. In effect, he reinterpreted and reimagined a wide body of scientific and technological research as aesthetic, comparing and contrasting ideas and images from various fields subjectively, even impressionistically.

Some of Kepes's sources were predictable. Kepes closely studied classic works of art history and aesthetics, like Wölfflin's *Renaissance and Baroque* (1888), Gottfried Semper's *Style in the Technical and Tectonic Arts* (1860–1862), Erwin Panofsky's *Perspective as Symbolic Form* (1927), and John Dewey's *Art as Experience* (1934). Some of the art historical titles he consulted were more obscure. Mary Hamilton Swindler's *Ancient Painting* (1929) and Miriam Schild Bunim's *Space in Medieval Painting and the Forerunners of Perspective* (1940) are hardly required reading today.[50] Nineteenth-century painter and printmaker John Burnet's 1837 *Treatise on Painting*, a basic primer on draftsmanship, was

2.20

Postcard from Alexander Calder to
Gyorgy Kepes, 8 April 1950. Gyorgy Kepes
papers, 1909–2003, bulk 1935–1985.
Archives of American Art, Smithsonian
Institution.

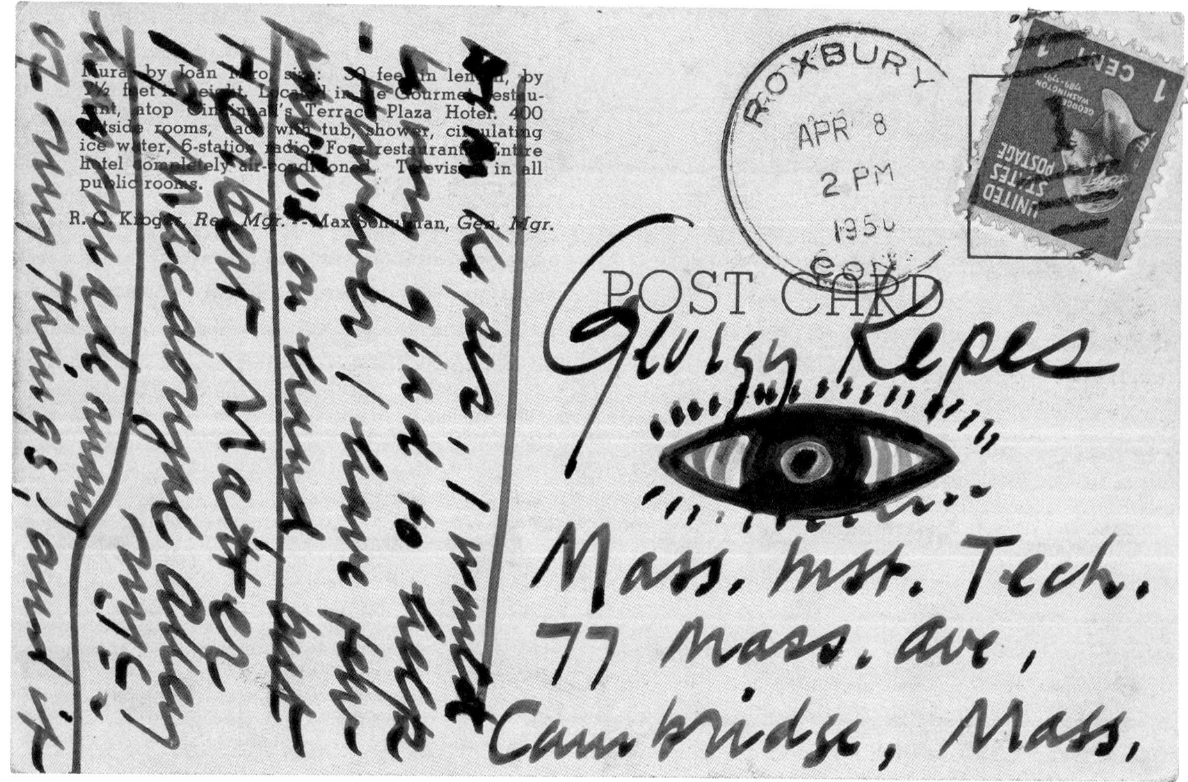

not a significant text for other artists at midcentury—but it evidently served as the source for Kepes's expression "the education of the eye." Such was the title of one of Burnet's essays.[51]

Kepes also turned to the "pioneers" of the "education of vision," among them the art critic Herbert Read; the art educator Viktor Lowenfeld; and Bauhaus figures like Johannes Itten, Josef Albers, and of course Moholy. He turned to innovators like Swiss education reformer Johan Heinrich Pestalozzi and his student, the German pedagogue Friedrich Froebel, who invented kindergarten education.[52] Pestalozzi advocated learning through participatory activities; he also distilled his holistic approach as the education of head, heart, and hand, a maxim that Kepes would reformulate as an integration of "the scientist's brain, the poet's heart, the painter's eye."[53]

As intriguing as these references are—and they are intriguing; it is hard to think of other artists engaged in such an extensive, if eclectic, study of ancient, medieval, Renaissance, and modern art—I find the most obscure sources, those from the intellectual margins, of special interest.[54] Kepes was a keen reader of the biologist Karl von Frisch's 1950 study *Bees: Their Vision, Chemical Senses, and Language*.[55] In his notes, Kepes explains how von Frisch "lucidly" demonstrates that perception "depends upon the way the apparatus such as the eyes modulates the impacts coming from the outside."[56] That is to say: even if seeing is objective, perception—the interpretation of what one sees—is subjective. A diagram from the book, used in an experiment conducted by von Frisch, demonstrates the claim; visual patterns that are distinct to the human eye are perceived as all the same by the eye of the bee (fig. 2.21). Kepes elaborates in his notes: "In shapes only the contrast between solid and broken patterns are perceived but [there are] no differences within these groups."[57] He would have been especially interested in this diagram, as it contains the essential Bauhaus shapes—circle, square, and triangle. The point is that perception is mediated, it occurs in the mind's eye—and that it therefore can be educated.

Kepes was also intrigued by the writings of perceptual psychologist Adelbert Ames, Jr., of the Dartmouth Eye Institute, who created a set of experimental displays that came to be known as the Ames demonstrations. Kepes first viewed them in 1947; they were a widespread fascination in the period (Nelson Rockefeller and Alfred H. Barr, Jr., of the Museum of Modern Art, also viewed them). The most famous demonstration was the distorted room: a structure built with extreme exaggeration that made its occupants appear ridiculously small or large (fig. 2.22). The lesson of the Ames demonstration was, again, that perception is subjective, not objective; that what we see lies in us as observers, not just out there, in the world beyond—and that perception can therefore be educated.[58]

Kepes's approach with these sources—how he borrows ideas from the sciences and changes meaning in the process—is especially clear in the way he used crystallographer Kathleen Lonsdale's 1949 study *Crystals and X-Rays*, an investigation of X-ray diffraction techniques applied to crystallography. Kepes read the book not as a scientific text at all, but as an aesthetic one. In her book, Lonsdale describes how crystals form on imaginary axes extending in all directions; these axes "define a framework or *lattice*" (Lonsdale's italics) which functions like a "scaffolding, as it were, upon which the structure is built." This "steel scaffolding," as she terms it, structures the arrangement of the atoms and molecules within the crystal.[59] Kepes applies this concept to the design of his exhibition: his lattice, made of actual steel pipes affixed with simple vise clamps, replicates a crystalline framework. The individual photographic panels could then hang within the lattice like the atoms or molecules forming the crystal. Kepes emphasized these allusions by putting classic ball-and-stick models of crystalline structures on view with his show.

The specific arrangements of panels in Kepes's exhibition design also derived from Lonsdale's research. She explains how a "repeat unit" or "unit cell" within a molecular lattice is arranged in special "space groupings." She illustrates this principle with a diagram depicting an ampersand, functioning as a generic placeholder, organized in various standard formats; Kepes copied Lonsdale's diagram into his notes, but in doing so changed its meaning (figs. 2.23–2.24). Instead of arranging atoms or molecules, such a framework could organize what Kepes describes as "the same fundamental unit of thought or vision"—in other words, an idea or image—which "can be related into symmetrical structure by simple space grouping." Kepes copies the particular crystallographic relationships Lonsdale explores—"mirror image, reflection, rotation, inversion"—that might connect these fundamental units of ideas and images into a larger structure.[60] The way he hung panels suspended corner to corner, reiterated in his book layouts, derived from the types of relationships between units that Lonsdale articulates. These sources reveal the basis of visual design as an open-ended appropriation of discourses.

Dialectics of Seeing

Kepes's visual design required a dialectical way of seeing, a relational approach that would allow one to see similarity and difference—recall all the comparisons between the animate and inanimate, the organic and inorganic, on the pages of his book and the hanging panels of his exhibition. It required synthesizing opposites, forming constellations between sets of reflected, inverted, or rotated images and concepts.

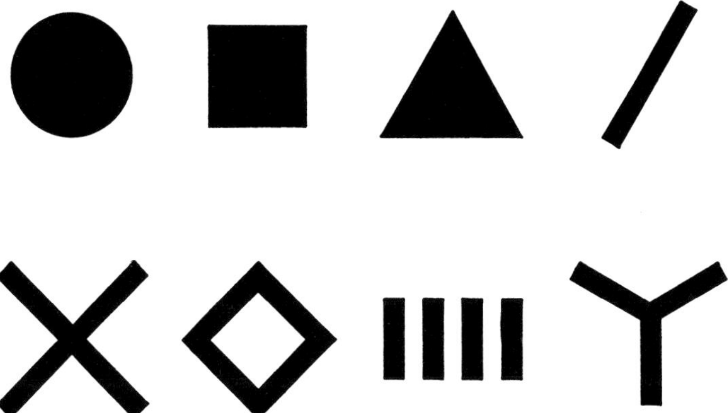

Figure 14. Bees do not learn to distinguish between the shapes in the upper row of the figure or between the shapes in the lower row. But for the eye of a bee each shape in the upper row is distinctly different from every shape in the lower row.

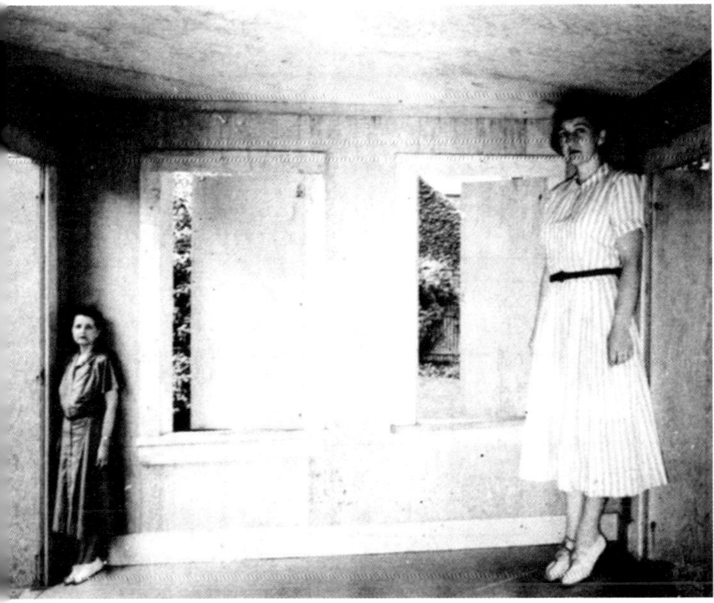

2.21

Reprinted from Karl von Frisch, *Bees: Their Vision, Chemical Senses, & Language*, revised edition. Copyright © 1950, 1971, by Cornell University Press. Used by permission of the publisher, Cornell University Press.

2.22

Ames distorted room demonstration. © The Estate of Adelbert Ames, Jr.

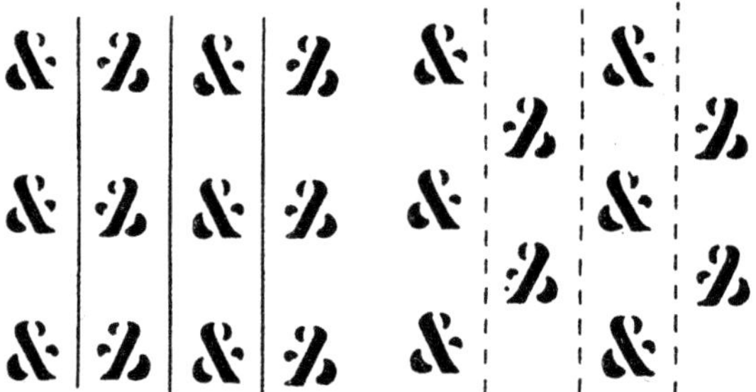

FIG. 47. Mirror plane (010).

FIG. 48. Glide plane (010), with translation $\vec{a}/_2$.

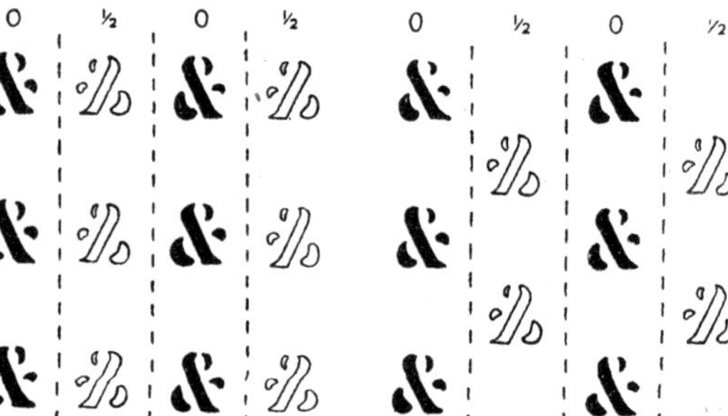

FIG. 49. Glide plane (010), with translation $\vec{c}/_2$.

FIG. 50. Glide plane (010), with translation $(\vec{a} + \vec{c})/_2$.

2.23

Space grouping diagrams.
From Kathleen Lonsdale,
*Crystals and X-Rays.*London:
G. Bell, 1948.

2.24

Gyorgy Kepes, note labeled
"Education of Vision." Gyorgy
Kepes papers, 1909–2003,
bulk 1935–1985. Archives
of American Art, Smithsonian
Institution.

Education of vision

The same fundamental unit of thought or vision
can be related into symmetrical structure by simple
space grouping

mirror image, reflection, rotation, inversions.

Illustration photo, Kaleidoscope, mirror.
related to a central axis

Crystals and x-ray   K. Lonsdale   London. 1948
Also Bernal.

Kepes believed this dialectical method made his program distinct from that of other figures in the Bauhaus diaspora. In a 1948 letter to Moholy's widow Sibyl Moholy-Nagy, who was then writing a history of the New Bauhaus, Kepes argues that what distinguishes the "education of vision" from Moholy's program is the dialectical method. He suggests that Moholy-Nagy refer to his 1944 *Language of Vision*, specifically the last chapter on "Dynamic Iconography," for elaboration. In that text, Kepes explains his understanding of perception as a clash of forces. When a single image contains "statements which seem counter" to one another, the "spectator's attention is forced to seek out the possible relationships" that might reconcile these opposites. This montage-like process (it is no surprise that Kepes was an avid reader of Sergei Eisenstein's montage theory) could continue indefinitely:

> **The association-fields of the representation … can reinforce one another or clash with one another, creating strains, stresses, tensions. Each tension is resolved into a meaningful configuration. These configuration in turn serve as a basis for further tension; consequently for further configurations.**[61]

In *Language of Vision*, Kepes applied this "dynamic" approach to the arrangements of discrete visual elements—points, lines, shapes, and shadows—in single compositions; but visual design could expand this process to broad categories of visual experience.

His notes from the 1940s and 1950s illuminate the program. Kepes created extensive lists of opposite terms—he referred to these lists as "balance sheets" and often labeled the columns on a single page with a plus or minus sign—that recorded positive and negative associations.[62] One such balance sheet reads:

> Conflicts:
> familiar, hackneyed← →unfamiliar, new, fresh
> abstract, schematic, general← →concrete, actual, unique
> organized, structural, formed← →chaotic, ambiguous
> stabilized, freeze← →fluid, changing
> named← →unnamed
> meaningful← →meaningless

The list maps experience as a series of tensions that could be held in suspension, like the montage image. Seeing dialectically was not strictly pictorial; for Kepes, it had a philosophical resonance. The greatest "strains, stresses, tensions" occurred between a pair of terms that he does not list but which he must have had in mind: "art← →science" or "science← →art." The purpose of Kepes's exercise was to train a "correct proportion between the two poles," a creative equilibrium between alternating forces.[63]

The dialectics of seeing has diverse origins. Kepes was reading canonical literature in art history, like Wölfflin's *Principles of Art History* (*Kunstgeschichtliche Grundbegriffe*), which famously establishes a series of tensions that organize visual expression and create a balance between extremes that push and pull the history of art. From Kepes's reading notes:

> **Wolfflin [*sic*] in Grundbegriffe 1915 set up series of opposite concepts felt by hand followed by eye:**
>
> linear – painterly
> plane – depth
> multiplicity – unity
> closed – open form
> clearness – unclearness

Kepes understood these oppositions as existing not only within single images or between multiple ones—between sets of two, as in the traditional art-historical slide comparison—but also "in cultural history," as the motor propelling the past to the future. He then writes: "scientific thinking against mystical," indicating, again, that art and science, science and art, form the greatest tension of all—that rationality might be balanced with irrationality, logic with illogic, order with a measure of disorder.[64]

Kepes was also reading widely in the sciences, and this literature also informed his dialectics of seeing (it is fitting that Kepes consulted two divergent bodies of scholarship). He related his model to Gestalt psychology, specifically to the figure-ground relationship; he understood the visual clash of interlocking relationships between positive and negative fields as a basis for a much larger set of interlocking associations, ones with philosophical resonance:

> **Figure ground relationship … has correspondence to regions of broader human experiences.**
>
> knowledge← →ignorance
> love← →hate
> attention← →neutrality
>
> **On the boundary line of complementary concepts one gets the fluctuation, oscillation, the activities enhancing growth, development, change.[65]**

Theories of Everything

The idea of seeing in opposites could also be expanded indefinitely to encompass vast arrays of similarities and differences, constellations much larger than a single set of two images or pair of two ideas. Kepes obsessively traced these patterns across whole taxonomies and categories of being. His was a theory of everything.

These encyclopedic ambitions become clear as we continue to page through Kepes's sources. Among his bibliographies are citations to the *Book of Signs*, a catalog of symbols used in medieval Europe; maybe a page of astrological symbols caught Kepes's eye, given his propensity for mysticism and magic (fig. 2.25). We find physicist Hermann Weyl's *Symmetry*, which analyzes the concept through such mundane visual culture as wallpaper patterns, and aesthetician Matila Ghyka's *The Geometry of Art and Life*, which examines mathematical proportions in art and architecture.[66] Kepes was especially captivated by morphogenesis, or the development of biological shape across the natural world. He studied biologist D'Arcy Wentworth Thompson's *On Growth and Form*, a well-documented source for many artists and designers; he copied specific diagrams from Thompson's volume, like a chart demonstrating that all fish share a common pattern (Kepes turned this principle into a photographic exercise) (fig. 2.26).[67]

I am drawn to some particularly odd sources from Kepes's research notes, like a long-forgotten 1908 treatise by James Bell Pettigrew bearing an extraordinarily lengthy title, one I will repeat here in full: *Design in Nature, Illustrated by Spiral and other Arrangements in the Inorganic and Organic Kingdoms as exemplified in Matter, Force, Life, Growth, Rhythms, &c., especially in Crystals, Plants, and Animals. With Examples selected from the Reproductive, Alimentary, Respiratory, Circulatory, Nervous, Muscular, Osseous, Locomotory, and other systems of Animals.* Actually, the epic title is appropriate for the project's epic ambitions: the volume contains some two thousand images with which Pettigrew aimed to capture—as if it were really possible in the format of a book—"the cosmos as a whole." His study concerned all fields of knowledge: "Physics, Chemistry, Botany, Zoology, Anatomy, Physiology, Psychology, and Paleontology more or less in detail."[68] It provides a comprehensive compendium of life; as Pettigrew explains, "like produces like in endless sequence."[69]

2.25

"Astrological Signs." From Rudolf Koch, *The Book of Signs, Which Contains All Manner of Symbols Used from the Earliest Times to the Middle Ages by Primitive Peoples and Early Christians*. New York: Dover, 1955. Courtesy Dover.

2.26

Two-page spread from D'Arcy Wentworth Thompson, *On Growth and Form*, 1942, © Cambridge University Press, reproduced with permission.

9. Astrological Signs.

These are used for casting clearly defined horoscopes, in which the planets and the signs of the Zodiac find themselves in direct relation. The casting of the horoscope depends, amongst many other considerations, upon the way in which they stand in relation to each other. The planets stand to one another:

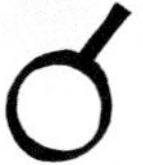

In conjunction, that is at a distance of 0 degrees.

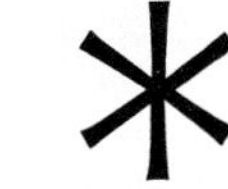

Semi-quadrate, or 45 degrees

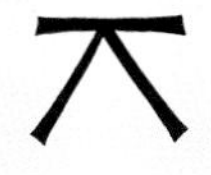

Quadrate, or 90 degrees

Quincunx, or 150 degrees

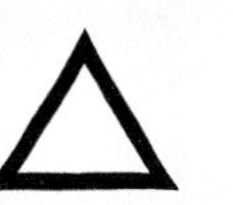

Trigon, or 120 degrees

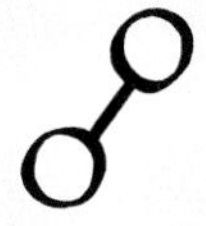

Opposition, or 180 degrees

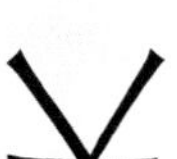

Semi-sextile, or 20 degrees

60

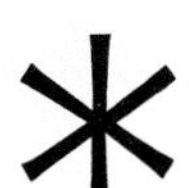

Sextile, or 60 degrees

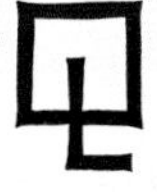

One and a half quadrates, or 135 degrees

Constellation in a state of retro-gradation

61

1062   THE THEORY OF TRANSFORMATIONS   [CH.

which fossils are subject (as we have seen on p. 811) as the result of shearing-stresses in the solid rock.

Fig. 519 is an outline diagram of a typical Scaroid fish. Let us deform its rectilinear coordinates into a system of (approximately) coaxial circles, as in Fig. 520, and then filling into the new system,

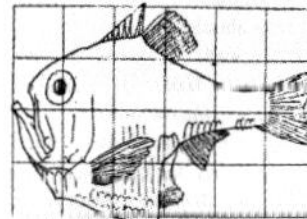

Fig. 517.   *Argyropelecus Olfersi.*

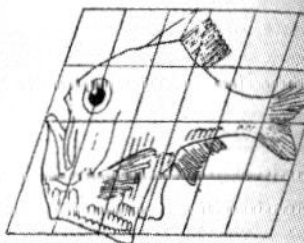

Fig. 518.   *Sternoptyx diaphana.*

space by space and point by point, our former diagram of *Scarus*, we obtain a very good outline of an allied fish, belonging to a neighbouring family, of the genus *Pomacanthus*. This case is all the more interesting, because upon the body of our *Pomacanthus* there are striking colour bands, which correspond in direction very closely

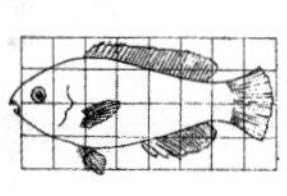

Fig. 519.   *Scarus sp.*

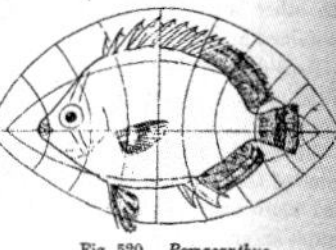

Fig. 520.   *Pomacanthus.*

to the lines of our new curved ordinates. In like manner, the still more bizarre outlines of other fishes of the same family of Chaetodonts will be found to correspond to very slight modifications of similar coordinates; in other words, to small variations in the values of the constants of the coaxial curves.

In Figs. 521–524 I have represented another series of Acanthopterygian fishes, not very distantly related to the foregoing. If we

XVII]   THE COMPARISON OF RELATED FORMS   1063

start this series with the figure of *Polyprion*, in Fig. 521, we see that the outlines of *Pseudopriacanthus* (Fig. 522) and of *Sebastes* or *Scorpaena* (Fig. 523) are easily derived by substituting a system

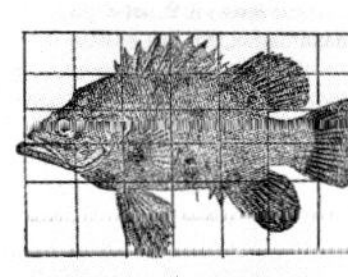

Fig. 521.   *Polyprion.*

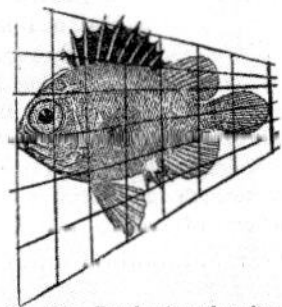

Fig. 522.   *Pseudopriacanthus altus.*

of triangular, or radial, coordinates for the rectangular ones in which we had inscribed *Polyprion*. The very curious fish *Antigonia capros*, an oceanic relative of our own boar-fish, conforms closely to the peculiar deformation represented in Fig. 524.

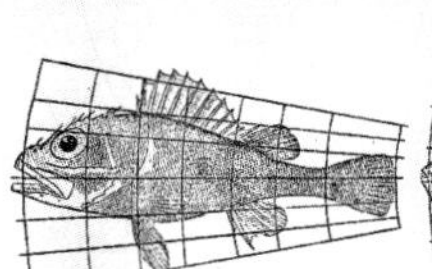

Fig. 523.   *Scorpaena sp.*

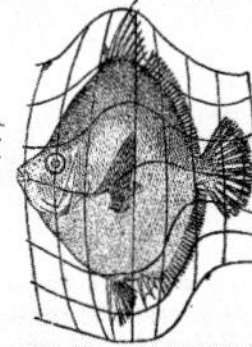

Fig. 524.   *Antigonia capros.*

Fig. 525 is a common, typical *Diodon* or porcupine-fish, and in Fig. 526 I have deformed its vertical coordinates into a system of concentric circles, and its horizontal coordinates into a system of curves which, approximately and provisionally, are made to resemble

Pettigrew claims this wealth of pattern as evidence for intelligent design by a divine creator. While such arguments would have had little appeal to Kepes, the aesthetic dimension of these endless sequences certainly would. *Every pattern contains or is contained by another pattern; it follows or is followed by another pattern.* In his notes, Kepes made a map of this sublime view, of a Great Chain of Being, of a system linked through concentric circles (a notion he is also borrowing from Thompson, as he indicates on the drawing) (fig. 2.27): we move outward from "particles" to "atoms," "molecules," "colloidal aggregates," "cells," "organs," "bodies," "society," and finally "nature cosmos."

I am also drawn to another odd volume: Theodore Andrea Cook's 1914 book *The Curves of Life*, which examines spirals of all and every variety, from "the microscopic foraminifer, and the even smaller bacilli to the enormous nebulae in the firmaments of space."[70] Between these extremes are many spirals: the umbilical cord, the inner ear, the whorls of a fingerprint; horns and tusks; watch springs and the pendulum's swing; sunflowers and twisted tree trunks. Cook collected "a larger number of instances of the spiral in Nature than have ever been similarly collected before."[71] He took special interest in the Fibonacci sequence, the golden section, and the spiral of Phidias, which was named after the Greek sculptor who supposedly employed the golden section in proportioning the sculptural decoration at the Parthenon. The Greek letter Phi represents this divine proportion in his honor.

2.27

Gyorgy Kepes, "The New Landscape,"
undated note. Gyorgy Kepes papers
(M1796). Dept. of Special Collections and
University Archives, Stanford Libraries,
Stanford, Calif. © The Estate of Gyorgy Kepes.

The new landscape.

Each level has its own laws — and
can only be understood by applying the concept
tools appropriate to the level.

But the understanding can only attain full
meaning, if the levels are observed in their interrelationships.

Biological facts as chemical cannot be understood by
physical laws alone — but without physical laws
as f.i.c. psychosomatic relationships & social
great amount of biol. fact would escape
our grasp.

Higher and coarser — can not be understood
in terms of lower minute — or vice versa — but
their interrelationships of levels offers significant key
to interpretation.

one must understand both similarities as well
differences.

organisation levels are envelopes + successions

integrative levels have hierarchy.

see. D'Arcy Thompson levels
Fortune levels.

vision has its own levels of organisation

Cook's project was based on both art and science, and the role that vision plays in relating the two. In the opening pages of the volume, Cook describes how his unusual study emerged from an encounter with a biologist. Presented with an image of a spiral staircase's central column, the biologist immediately recalled the columella or central axis of a seashell (fig. 2.28).[72] Fascinated by this peculiar relationship, a pseudomorphological relationship, Cook began collecting more comparisons between spiral staircases and the interior chambers of seashells. The "Scala del Bovolo" (literally, staircase of the snail) at the Palazzo Contarini in Venice, for example, appeared to exactly match the shell called the *scalaria scalaris* (the staircase shell)—the Scala del Bovolo was even decorated with shells in its interior design. Cook explains that he was "not so much concerned with origins or reason as with relations or resemblances."[73]

The point of my journey into the depths of Kepes's library is to locate the approach to the image that he used at MIT. These sources may all seem to suggest a highly rational, highly technical way of seeing, a visual methodology for which the entire world could be quantified and measured (consider the numerically perfect Phi, which had an approximate value of 1.61803398). But I argue the opposite: that it is the irrational, the surreal quality in these texts and images,that captured Kepes's eye. Skimming though Cook's volume, we find more unexpected affinities: pinecones, violins, horns, and yet more staircases and seashells. Even Cook's obsessive fixation on Phi suggests a secret numerology, as if this magical number might unlock the mysteries of the universe. The scientific analysis of pattern paradoxically provided a very unscientific, even fantastical way to connect and combine micro and macro, distant and near, foreign and familiar. In a note labeled "pattern vision," Kepes describes "creative thinking" as the "introduction of a new pattern."[74] Against the oppressive conformity of facts and figures, Kepes offered such a pattern.

Mind-Pictures

This type of intuitive visual thinking recalls the famous Rorschach inkblot test. And indeed, Kepes's bibliographies include Hermann Rorschach's *Psychodiagnostics: A Diagnostic Test Based on Perception*, the 1942 English translation of Rorschach's 1921 book on the famous projective test bearing his name.[75] Rorschach employed a standard set of ten inkblot cards to evaluate a subject's apperception—what an individual projected onto an ambiguous, accidental visual stimulus. In this way, the inkblot was a window into the mind. It did not matter what one saw in the blots (there was no proper or correct image to decipher) but rather how one saw it; the test revealed "the formal principles (pattern) of the perceptive process."[76] Rorschach's meaningless images (fig. 2.29) could become anything at all: subjects saw bats, birds, beetles, crabs, by the unequal thicknesses observable along the rim. This method was well known to the classical Greek medallists and to the best of the Italians ; and in such modern work as Bertram Mackennal's medals for the Olympic Games of 1908, or in the magnificent head of the French Republic by George Dupuis, you see it still. The exigencies of modern coinage may perhaps make it difficult to use the system when money has to be capable of being stacked in regular piles of similar denominations, but this necessity does not exist in

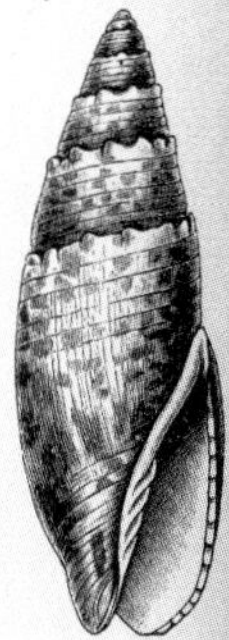

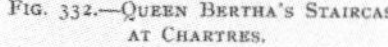

FIG. 332.—QUEEN BERTHA'S STAIRCASE AT CHARTRES.

FIG. 81. MITRA PAPALIS.

modern medals. Yet if we take such a recent example of the official coronation medal of Edward VII., the "floor" of the design seems so flat that the whole thing looks very like a postage stamp stuck on a thin disc, and has as little light and shade. But I am wandering from our immediate subject, and must return to it at once.

An even more delightful example of the close connection between a good architect's plans and the exquisite lines of Nature is to be found in the stairway called " Escalier de la Reine Berthe " at Chartres (Fig. 332). It exhibits the delicate exterior ascending dextral helix, and even the top of *Mitra papalis* (Fig. 81, Chap. III.) with extraordinary faithfulness, and the parallel becomes even more complete when the position of the darkened doorway is compared to that of the shadowy orifice of the shell. The staircase is contained in an external turret, and the internal spiral is expressed externally by curved timbers and the outer side of the string-course. The surface is also divided by numerous vertical beams, most of which rest on the string timbers, but the three main beams go from top to

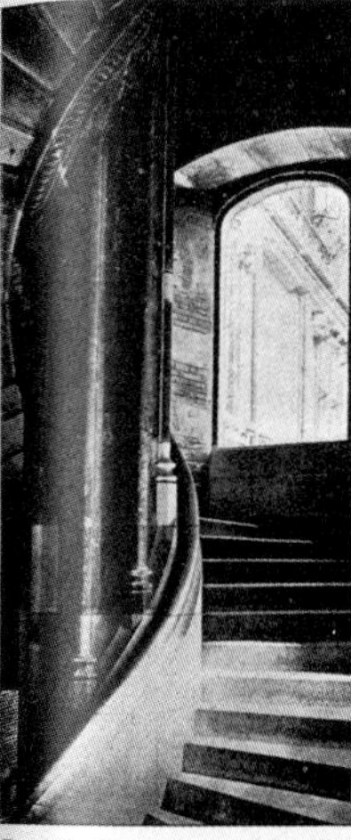

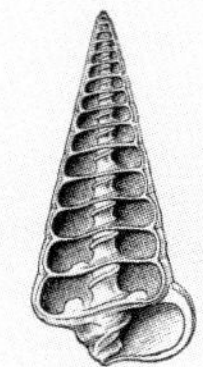

FIG. 333.—STAIRCASE IN THE OLD WING OF THE CASTLE OF BLOIS.

FIG. 73.—TELESCOPIUM TELESCOPIUM.

bottom of the thirty-six steps which swing round the central newel.

In *Mitra papalis* at this orifice you will observe the beginnings of three internal spiral lines, which suggest that the internal arrangements of a shell have as much to teach as its exterior forms ; and a very beautiful spiral may be seen by the aid of the X-rays continuing throughout the whole length of the long

Y 2

2.28

Two-page spread from Theodore Andrea
Cook, *The Curves of Life*. London:
Constable, 1914.

snakes, or monkeys; daily objects like lamp shades or flower vases; and landscape features like mountains or clouds.[77] This inventory is very similar to the collection of images Kepes uncovered in books by James Bell Pettigrew and D'Arcy Wentworth Thompson—or the collection he compiled in his *New Landscape* project.

Those who displayed a high originality of visual responses to Rorschach's inkblots were often artists.[78] In the test samples Rorschach provides in his book, the "above average" responses are from an individual who "has a well rounded education," is "flexible," and is "able to draw and paint."[79] These methods have long intrigued artists. Leonardo da Vinci famously suggested that artists might study "certain walls stained with damp" in order to see "an infinity of things," a process he likened by analogy to hearing distinct words in the indistinct ringing of bells. He described how a sponge full of paint thrown against the wall would leave a blot rendering visible "whatever you desire to seek in it."[80]

Rorschach's inkblots are visually similar to at least some of Kepes's scientific images; one test card looks to me like an image of plant cells under a microscope labeled *Transverse section of a stem of Kadsura: 200X* (fig. 2.30). But such similarity is really beside the point. It is the projective method, not the look of the blot, that interested Kepes. He writes in his notes:

> **Rorschach has shown that the way a person sees is essentially the character of the person.**
>
> **The structure of his vision – the rhythm, form configuration of his perception is the microcosms of the man's attitudes, concepts, conflicts.**
>
> **Not what one sees – how one sees is today significant.**
>
> **The objective reality – the physical stimulus is only the basis (also Ames, gestalt, etc. Leonardo, Freud).[81]**

In other words, the objective record—the Kadsura tree—is only a stimulus for a very subjective process. Kepes understood the images of science as surfaces for imaginative projection, therapeutic tools with which to enrich an individual's perceptual capabilities. He maintained a fascination with their power; he was still contemplating the process in the 1970s, as revealed in a series of notebooks: "art expression not unlike Rorschach test." Both art and inkblots form "chance figures [that] induce imaginative excavation of hidden desires [and] latent wishes."[82] They create "optical puns."[83] The fact that Kepes appropriated images serving this purpose from MIT's laboratories was to the point: these empirical records could teach a very unempirical visual process.

Kepes's method becomes yet more fantastical. He studied not only Rorschach's *Psychodiagnostics* but also psychologist Frederic C. Bartlett's 1932 book *Remembering*. In a section titled "Experiments in Imaging," Bartlett explains a test in which he provided subjects with a series of inkblots that "represent nothing in particular, but might recall almost anything," generating the same wild associations as Rorschach's cards (fig. 2.31). [84] And Kepes consulted Herbert Read's 1942 book *Education through Art*, which describes an exercise in which students closed their eyes and relaxed until seeing subjective pictures, or what are known as "eidetic images": lucid visualizations that exist only in the mind's eye. The students recorded these images in watercolor on paper.

Read called them "mind-pictures" (fig. 2.32). To him, they looked like the visual effects known as entoptic images that result not from vision per se but from physiological simulation of the retina, like the spots seen when applying pressure against the eye—the name "entoptic" indicates that these images originate within the eyeball.[85] He further compared them to an electron density map showing an arrangement of atoms in a molecule of platinum phthalocyanine (fig. 2.33). Read believed that the children's mind-pictures looked like molecular patterns because they actually were images of molecules in the brain, as if the mind's eye were a microscope aimed inward, pointed at brain tissue; "the form assumed by this pattern is determined by the chemical structure of the molecules which are the material basis of the brain," he explains. He also connected these mind-pictures to the mandalas drawn by patients of Carl Gustav Jung, suggesting that they were also sublime spiritual symbols, a macrocosm collapsed into a microcosm (fig. 2.34). I do not think it is a coincidence that the very same type of mandala also appears on one of Kepes's photographic panels, purposefully cropped and framed so it would take on these meanings (fig. 2.35). Kepes also owned a photocopy of the plates from a famous study of entopic images and other subjective visualizations, like phosphenes, dated 1819; these images, too, look like mind pictures and miniature mandalas (fig. 2.36).[86]

This type of visual thinking—shifting from science to spiritual symbol and back again—is actually described in Kepes's *New Landscape in Art and Science*. A text by philosopher and semiotician Charles Morris recalls an encounter with Kepes's exhibition as displayed in MIT's Hayden Gallery. Morris was enthralled with the way images spanned perceptual extremes; one could "imagine oneself as small as an insect" but also "as large as a mountain." Some of the images looked like metals but "turned out to be records from plants," while those that looked like plants "were captioned with the names of metals." By comparing and contrasting, the spectator could "look from below and above," could be "inside and outside simultaneously."[87] Morris explains how this process could ultimately be used for "spiritual exercises," the type of affective experience apparently missing from MIT's blind alleys:

2.29

Plate 4 of Hermann Rorschach's
Psychodiagnostics, 1948.

2.30

Gyorgy Kepes, *Transverse section of a stem
of Kadsura: 200X*, photographic enlargement
on fiberboard panel, 1951. Gyorgy Kepes
papers (M1796). Dept. of Special Collections
and University Archives, Stanford Libraries,
Stanford, Calif. © The Estate of Gyorgy Kepes.

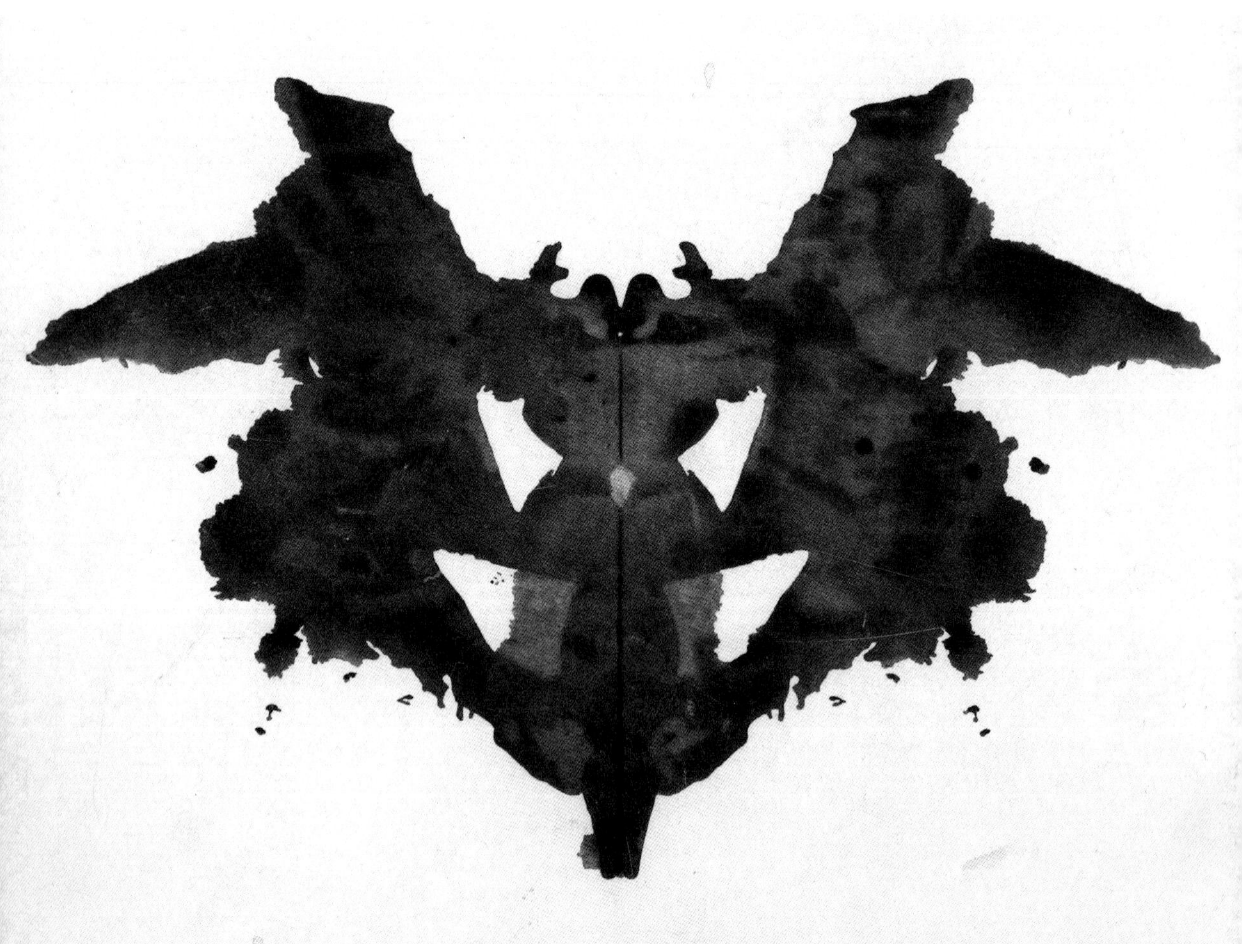

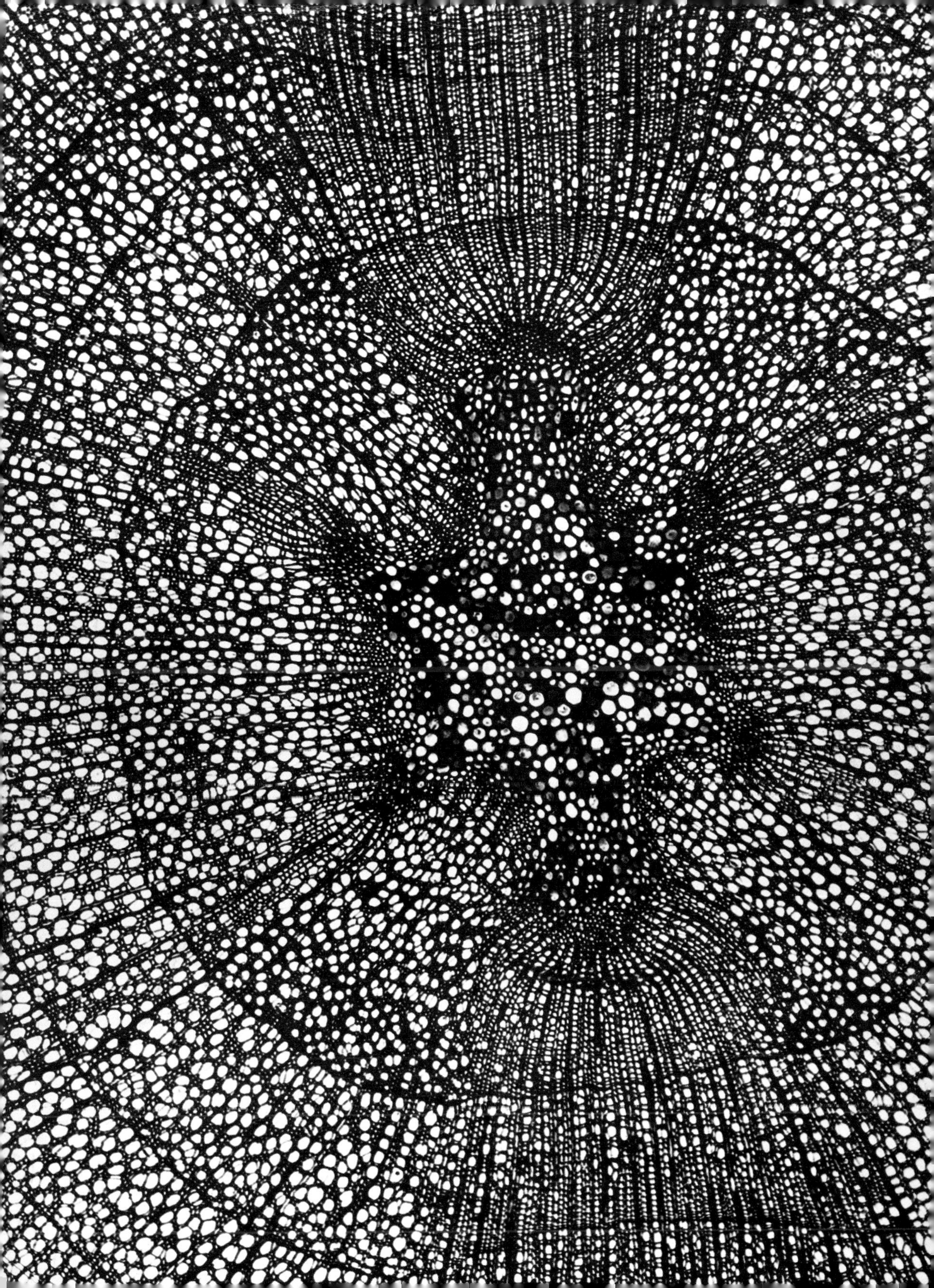

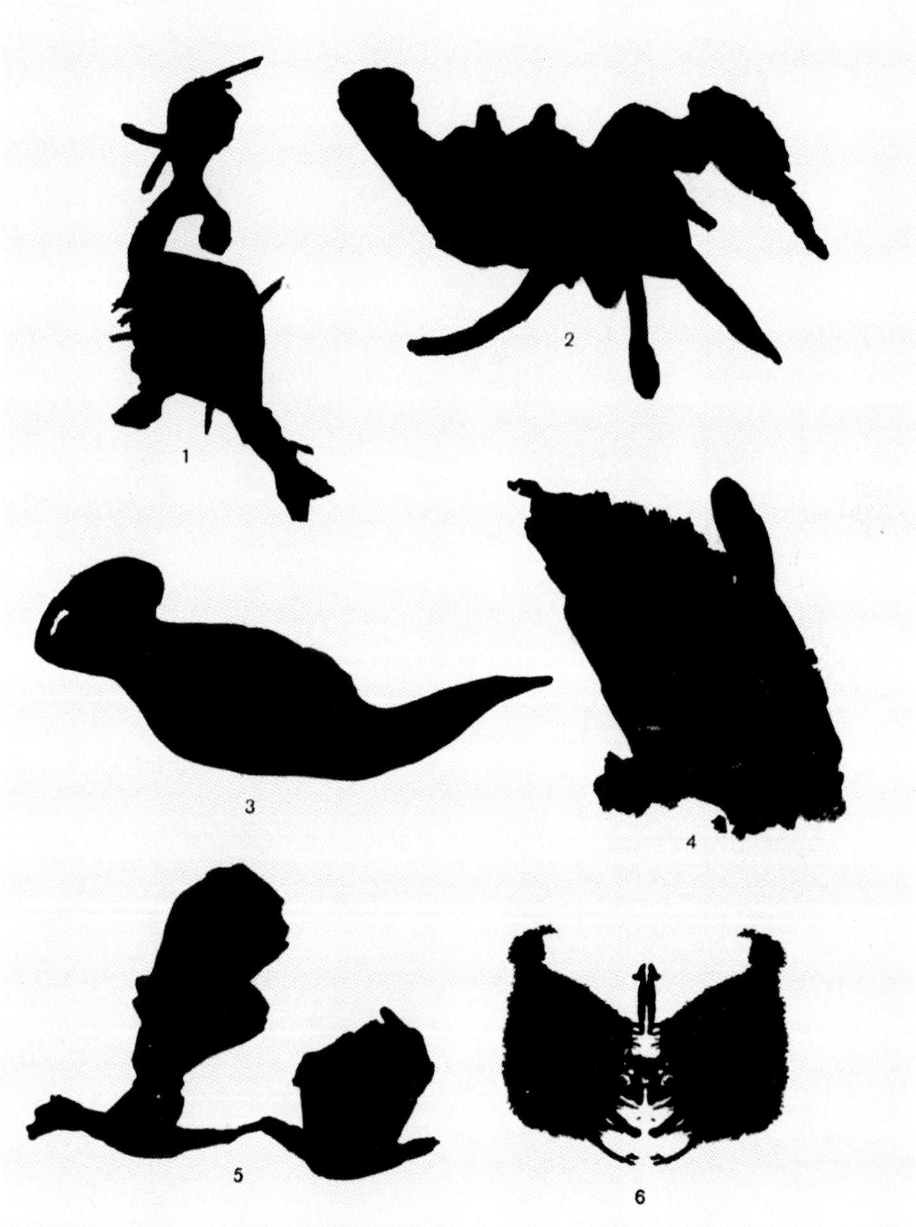
1
2
3
4
5
6

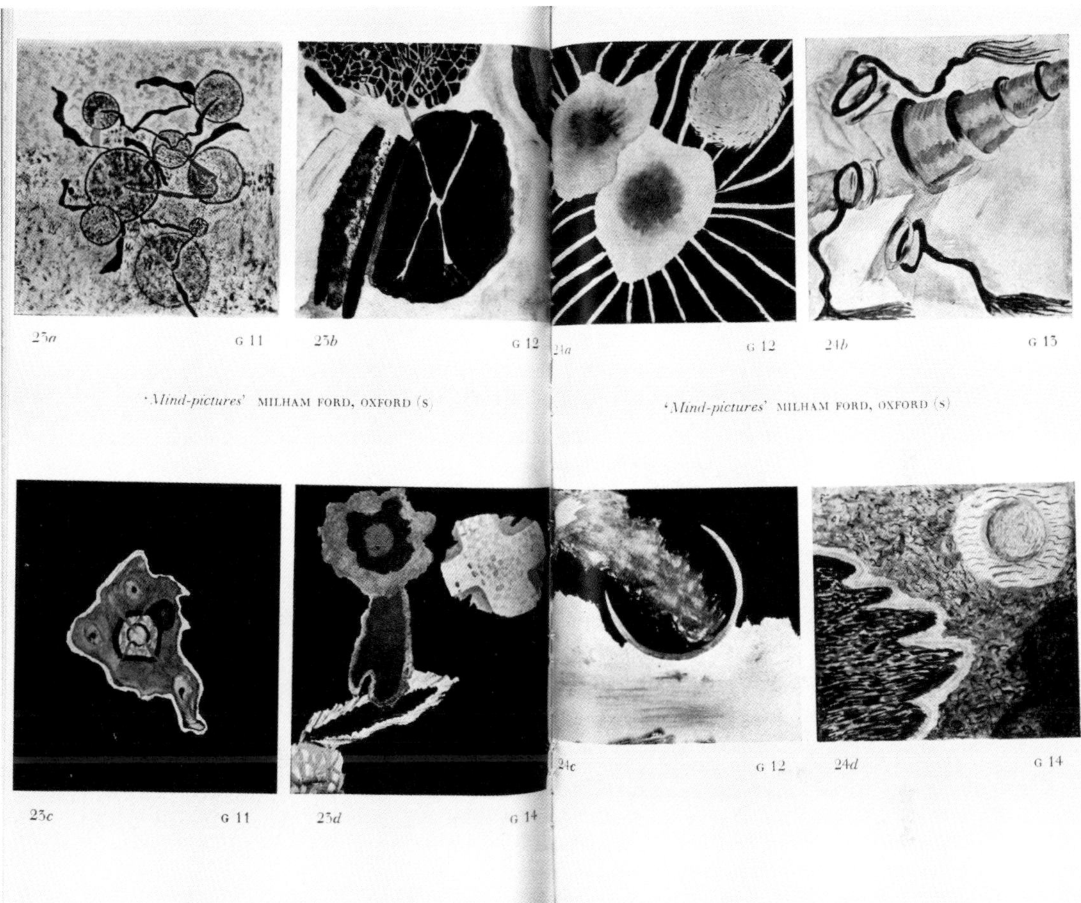

2.31

Ink blots figure from: Frederic C. Bartlett, *Remembering: A Study in Experimental and Social Psychology*. © Cambridge University Press 1932.

2.32

"Mind-Pictures." From Herbert Read, *Education through Art*. London: Faber and Faber, 1943.

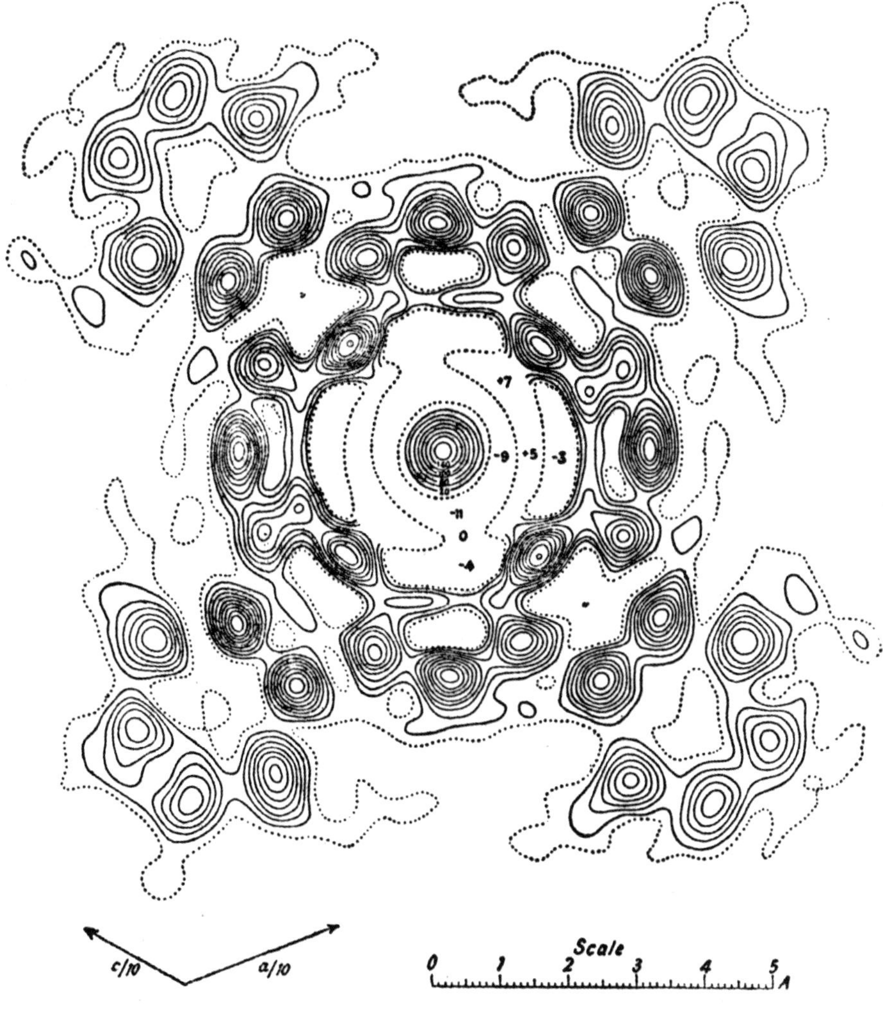

+7
-9 +5 -3
-11
0
-4
c/10
a/10
Scale
0 1 2 3 4 5
A

Fourier projection of a molecule of platinum phthalocyanine. From Herbert Read, *Education through Art*. London: Faber and Faber, 1943. Image originally published in J. Monteath Robertson and Ida Woodward, *J. Chem. Soc.,* 1937, 0, 219–230. Reproduced by permission of The Royal Society of Chemistry.

2.34

C. G. Jung, mandala. From C. G. Jung, *The Red Book*, edited by Sonu Shamdasani, translated by Mark Kyburz, John Peck, and Sonu Shamdasani. Copyright © 2009 by the Foundation of the Works of C. G. Jung. Translation copyright © 2009 by Mark Kyburz, John Peck, and Sonu Shamdasani. Used by permission of W. W. Norton & Company, Inc.

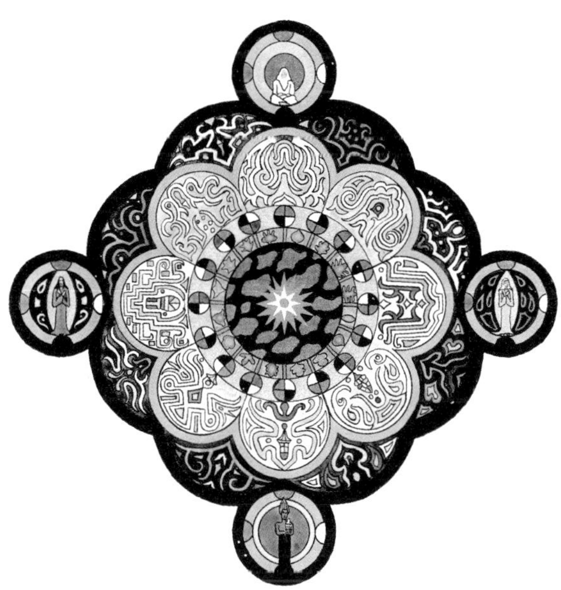

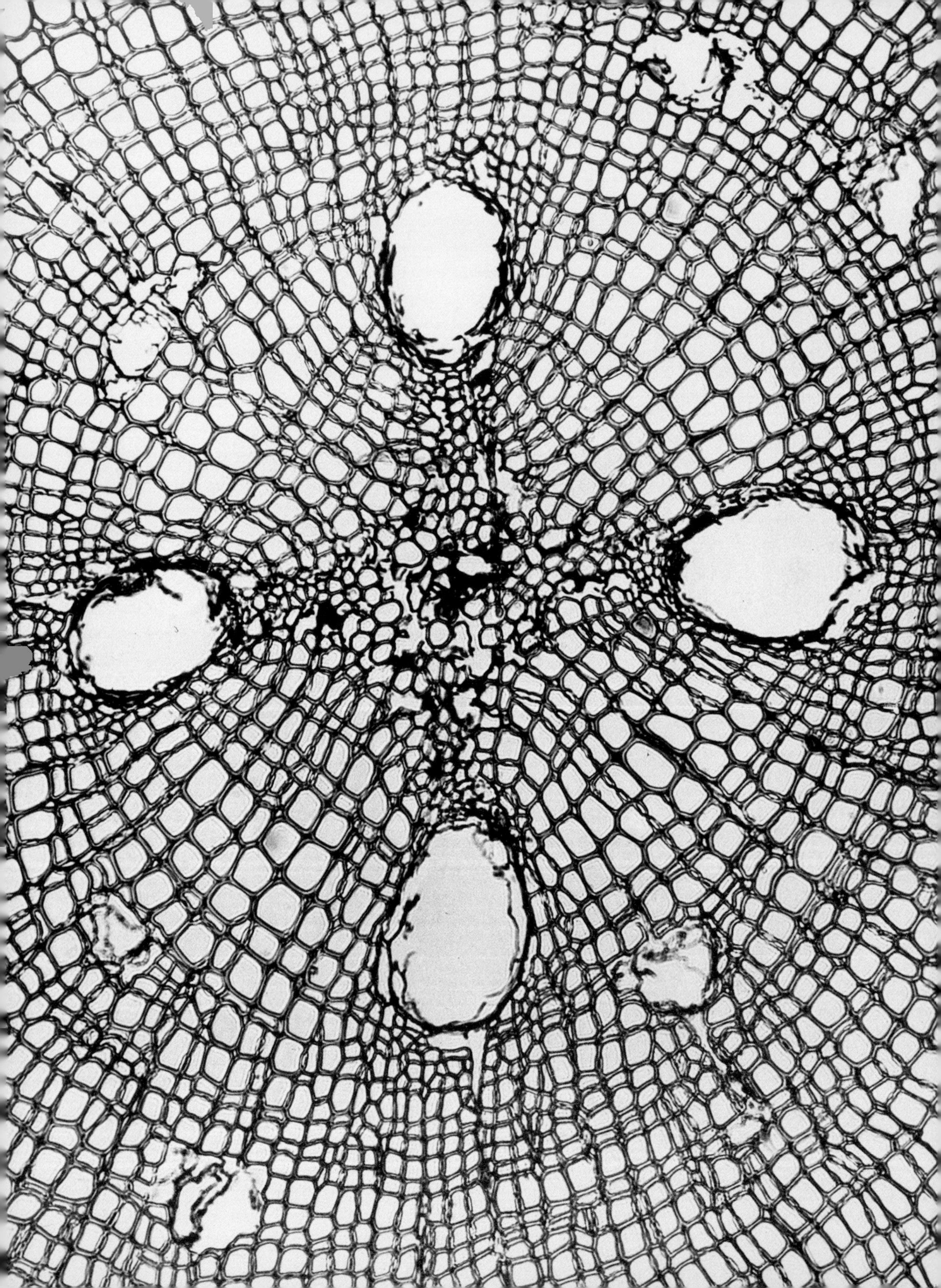

2.35

Gyorgy Kepes, *Transverse section of root of pine: 100X*, photographic enlargement on fiberboard panel, 1951. Gyorgy Kepes papers (M1796). Dept. of Special Collections and University Archives, Stanford Libraries, Stanford, Calif. © The Estate of Gyorgy Kepes.

2.36 (following pages)

Kepes's copy of a plate depicting patterns of "subjective vision," from Jan Evangelista Purkyně, *Beiträge zu Kenntniss des Sehens in Subjectiver Hinsicht* [Contributions to the Understanding of Vision in Its Subjective Aspects], 1819. Gyorgy Kepes papers, 1909–2003, bulk 1935–1985. Archives of American Art, Smithsonian Institution.

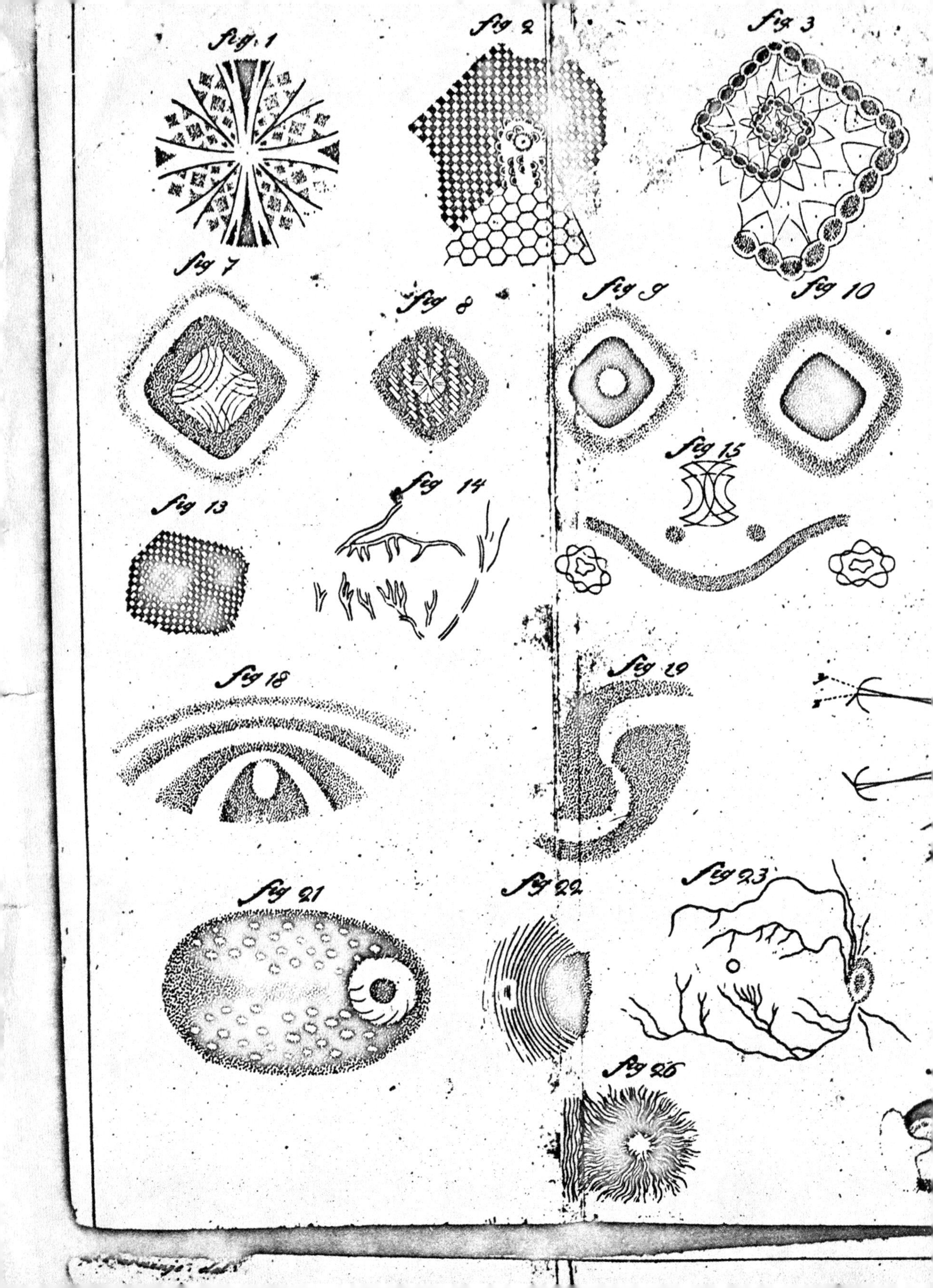

fig. 1
fig. 2
fig. 3
fig. 7
fig. 8
fig. 9
fig. 10
fig. 13
fig. 14
fig. 15
fig. 18
fig. 20
fig. 21
fig. 22
fig. 23
fig. 26

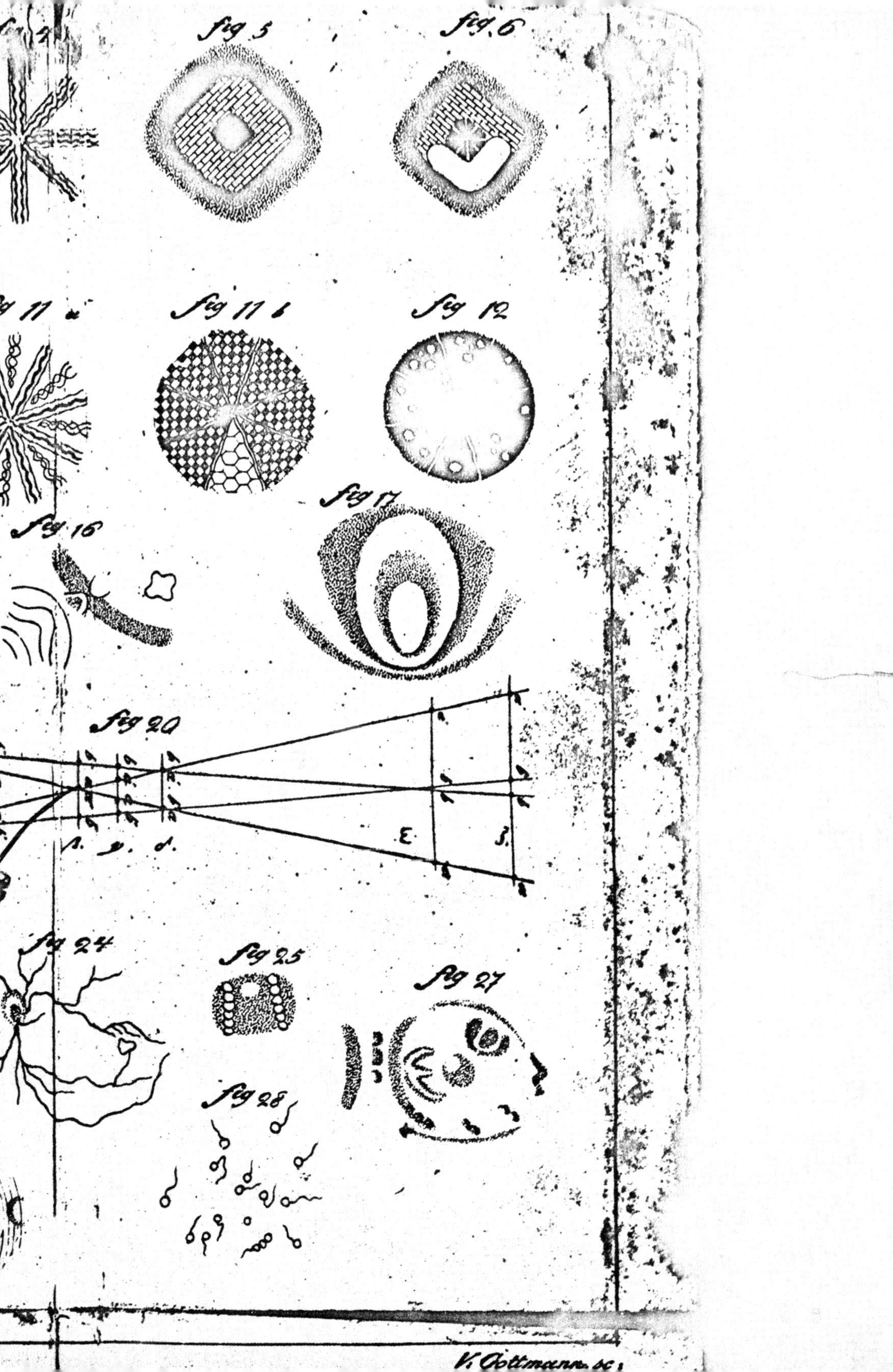

fig 5
fig 6
fig 11 a
fig 11 b
fig 12
fig 16
fig 17
fig 20
E.
3.
fig 24
fig 25
fig 27
fig 28
V. Oottmann sc.

> [T]he photographs can be used in "spiritual exercises" for the unceasing task of man-making, used for jarring the self from an ever-settling rigidity, breaking its egocentric boundaries, renewing its vitality by connecting it with what is vital, offering it something fine enough to lure it into losing itself, and so into finding itself. Contemporary man must be able to move among and between diverse perspectives, cultural perspectives on the earthy, spatial and temporal perspectives in the cosmos. He must become flexible.… He must open himself.… Oddly enough, these photographs from science, though made solely for the purposes of science, can aid man to live in this impersonal-yet-human West-yet-East dimension.[88]

Kepes placed Morris's text immediately preceding the iconic installation view of his original exhibition, suggesting its relevance as a guide for experiencing *The New Landscape.* The terms Morris uses—flexibility and openness, the breaking of mental rigidity—suggest how patterns activate a new mode of perception, one that erupts with hallucinatory vitality.

Let me pause for a moment. Maybe I have wandered too far, becoming lost in Kepes's library, in his labyrinth of patterns and puzzles. Perhaps my interpretation of Kepes's images is also just projection, just fantasy, just my attempt to make meaningful what is meaningless. Maybe I am seeing more than is really there, reading too much into Kepes's bibliographies and stray notations. Maybe these visual riddles cannot be solved: conundrums with no answer. But I do not think so. It is the point of Kepes's project to encourage this creative projection, this imaginative synthesis of images and discourses on images from across all fields of art and science. "Creativity is value," as he notes.

So, let me continue: Hypotonic lines, vertiginous webs, spirals of concentric circles—these pictures seem to come alive—to scintillate and pulse. The psychologist Heinz Werner called this vivid mode of cognition "physiognomic perception," the perception of liveliness, of animation, in abstract formal qualities. In his 1940 *Comparative Psychology of Mental Development,* a study comparing perception among so-called "primitives," so-called "psychotics," and the young through such visual culture as prehistoric cave paintings and children's drawings, Werner reveals the propensity among all three groups for syncretism (seeing unrelated phenomena as related), synesthesia (experiencing visual sensations as auditory or tactile), animism, and magic—all tendencies suppressed in the "normal" adult.[89] Werner explains physiognomic perception with a diagram of lines, each of which conveys an affective charge through form alone (fig. 2.37). One group of test subjects drew lines to evoke particular concepts—"iron," "silver," or "gold"—while the other test subjects matched the lines back to the metal. The responses were consistent, suggesting that abstract qualities innately call to mind particular concepts.

Werner similarly asked children to write various words in such a way that he could "see what they were," with the word's appearance taking on its meaning—synesthesia by visual onomatopoeia. The word *ängstlich* (scared) was produced in "cramped, narrow" forms, while the word *froh* (happy) had "swinging, rounded" forms.[90] The same concept was popularized in an arts education textbook by Viktor Lowenfeld, a pioneering arts educator; it shows how the physiognomy of words, rather than their definitions, expresses meaning (fig. 2.38).[91]

Werner associated this type of perception not only with "primitives," "psychotics," and children, but also with artists ("The self-descriptions of artists often reveal that it is normal for them to perceive things physiognomically"). By way of example, he cites Wassily Kandinsky's autobiographical account of the "physiognomies" of various mundane items, including a set of oil paints in which each color seemed to come to life with a distinct personality (different colors appeared to Kandinsky as uniquely "festive," "thoughtful," "sorrowful," or "stubborn," each one taking on an affective charge).[92]

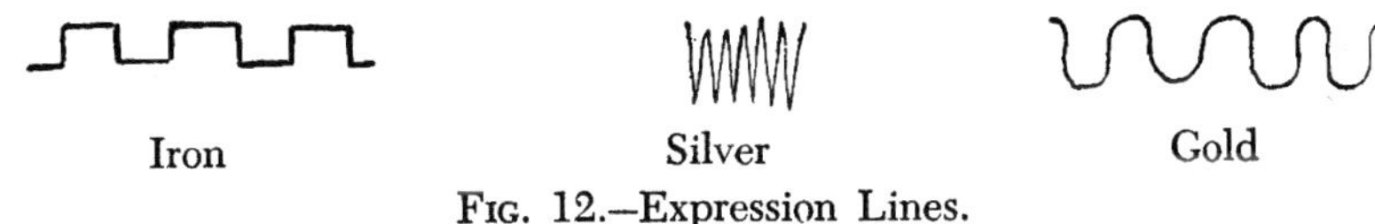

Fig. 12.—Expression Lines.

2.37

"Expression Lines." From Heinz Werner, *Comparative Psychology of Mental Development*. New York: Harper, 1940.

2.38

"Word Illustrations." From Viktor Lowenfeld, *Creative and Mental Growth*, 5th ed., © 1952. Reprinted by permission of Pearson Education, Inc., New York.

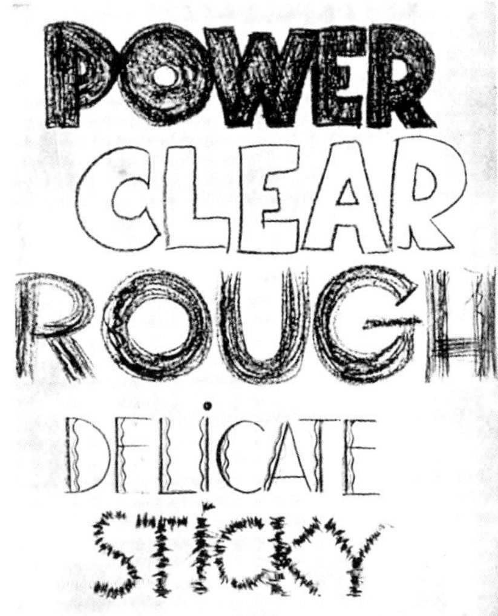

"Physiognomic perception," as an artistic way of seeing, was explicitly motivated against the rational, technical, and logical worldview common at MIT; it was an alternative to the deadening effects of science and technology —a counterdiscourse. In his foreword to Werner's book, psychologist Gordon Allport says as much, explaining Werner's study as a celebration of "the kind of mental life that scientists do not have, and, not having, tend to overlook."[93] Werner explains the oppositional potential of physiognomic perception in a text actually published in Kepes's *New Landscape* volume:

> Though physiognomic experience probably is the primordial manner of perceiving, it grows, if given the chance, by its own rules and laws to a level, not below but on par with that of "geometrical-technical" perception and logical discourse.... [T]he richness of a culture and its impact upon individual experience will forever depend on the growth and coexistence of both modes of cognition.[94]

This mode is precisely what Kepes offered MIT. It strikes me as quite appropriate that the very first images in his *New Landscape* book are not scientific at all, but instead index physiognomic perception: the first spread depicts a prehistoric cave painting on the left and a drawing of a dog by little Imre Kepes, age three and a half, on the right (fig. 2.39). These two images serve as a lens through which we are meant to view the book's remaining contents.

A notation in Kepes's papers on the "physiognomy" of "visual forms" demonstrates how one might understand an image, any image—even a scientific image—in terms of these lifelike qualities:

> Feeling, color, shapes, lines are dancing, exploding, balancing. They are riveted, dovetailed, bond molded, and blended. They have hydraulics – they are pregnant, they attract + repel – budding flowering hovering.... They have weights, density, they sit calm or explode, they are shy or brash, audacious or retreating, beating in or out [with a] heartbeat rhythm.

Kepes describes the complicated personalities he saw in visual forms; some appear "like [a] lion tearing the weaker up" or "like [a] mother caressing their [*sic*] child." Others are "sometimes friends, or sweethearts." Comparing the visual to the verbal, I, too, see blobs and clumps come to life; shapes in a slice of magnified wood are "pregnant," each circular cell birthing a new form (fig. 2.40). The horizontal band in another photographic panel depicts the "hydraulics" of the visual image, the "weights" and "density" that push and pull at the picture plane (fig. 2.41).[95]

We need not focus on any single bibliographic reference to understand Kepes's project. It is too easy to become lost in his library, confused by the patterns and puzzles. In bringing some of his sources together here, my goal is to demonstrate how visual design became a means to encourage ways of seeing and thinking that defied the scientific and technological ethos at MIT.

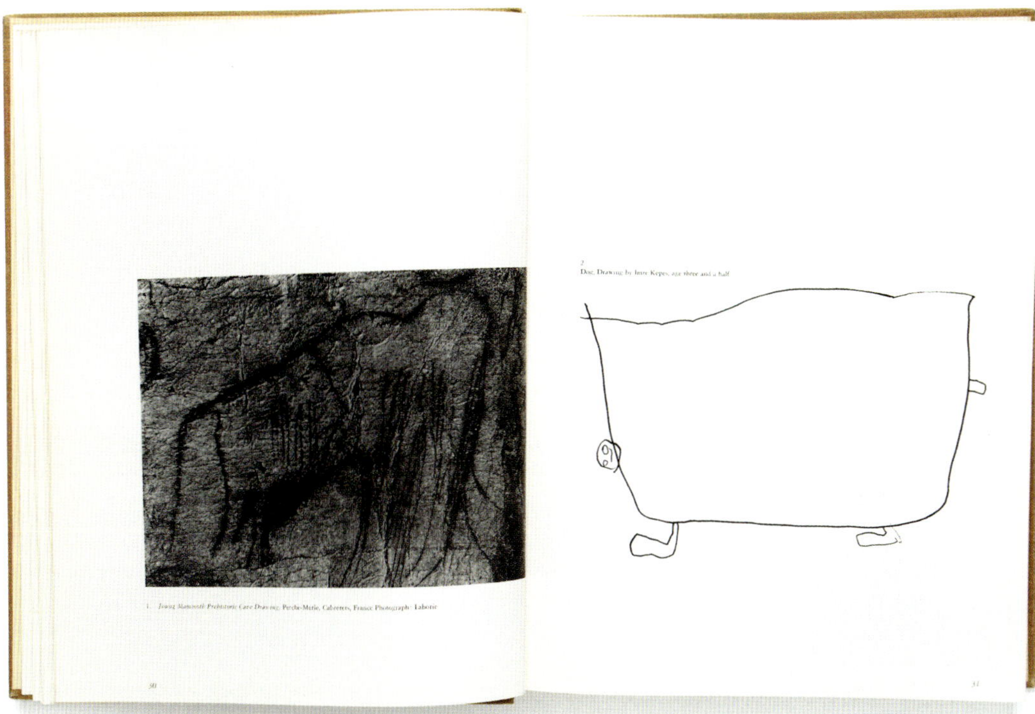

2.39

"Joung Mammoth Prehistoric Cave Drawing"
compared to "Dog," two-page spread
from Gyorgy Kepes, *The New Landscape
in Art and Science*. Chicago: Paul Theobald,
1956. © The Estate of Gyorgy Kepes.

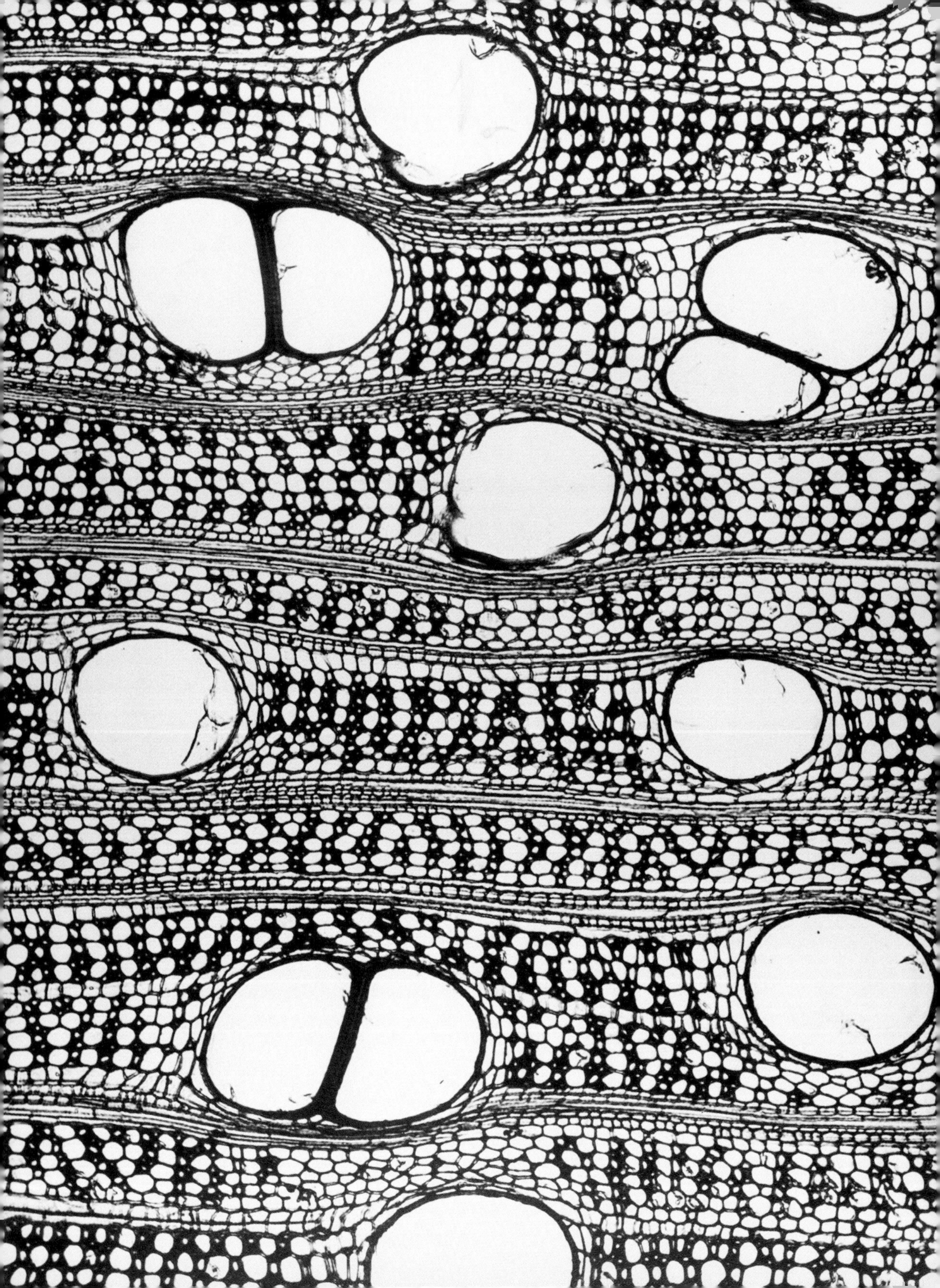

2.40

Gyorgy Kepes, *Transverse section of wood: 250X*, photographic enlargement on fiber-board panel, 1951. Gyorgy Kepes papers (M1796). Dept. of Special Collections and University Archives, Stanford Libraries, Stanford, Calif. © The Estate of Gyorgy Kepes.

2.41 (following page)

Gyorgy Kepes, *Yew tree, junction of normal and compression wood: 50X*, photographic enlargement on fiberboard panel, 1951. Gyorgy Kepes papers (M1796). Dept. of Special Collections and University Archives, Stanford Libraries, Stanford, Calif. © The Estate of Gyorgy Kepes.

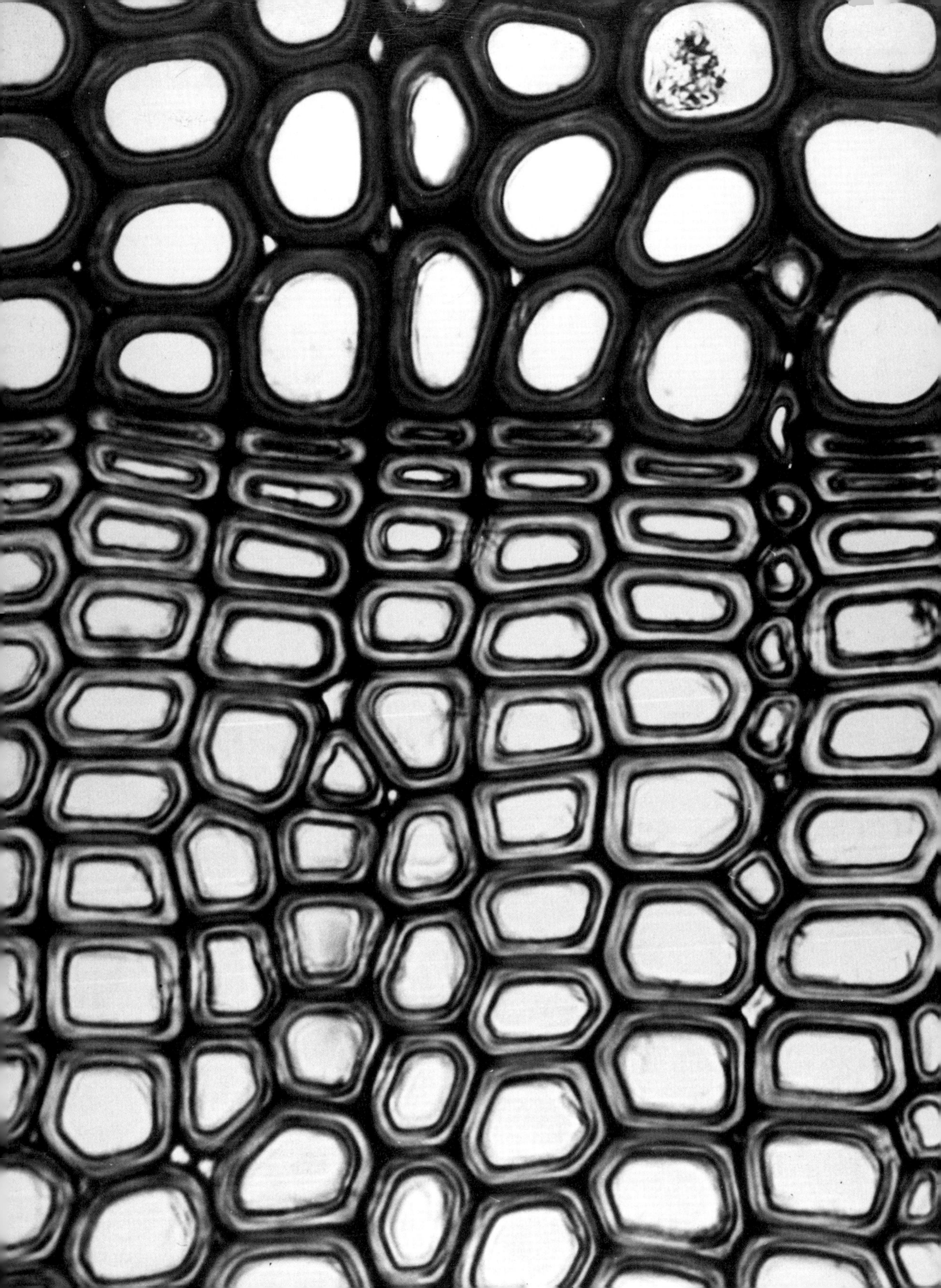

Applied Visual Design

The splinter in your eye is the best magnifying glass.
Theodor W. Adorno, *Minima Moralia*, 1951[96]

What happens when we put all that we have learned about Kepes's visual methodology into practice? How does his approach to the image actually work, for example, with the Lichtenberg figure—especially once we know more about this image's provenance, about the scientist who first created it and the purposes it originally served? The MIT physicist Arthur R. von Hippel produced his patterns for research into dielectrics, the study of electrical insulators, or materials that inhibit the transmission of electricity. In his experiments, von Hippel and his graduate student Fred H. Merrill discharged high-voltage electric shocks directly onto dielectric materials covered with photographic film; each image records the breakdown of the electric charge on a dielectric surface. In *Dielectrics and Waves*, his 1954 textbook on the subject, von Hippel describes how the charge effectively "photographs itself" as light is emitted in this breakdown process (in other words, these images are photograms; this is no doubt part of their appeal to Kepes).[97] Visual analysis of the image's fractal-like burst would then reveal particular attributes about the underlying material's insulating properties under a given set of environmental conditions.

But von Hippel's research was not a neutral endeavor. In *Dielectric Materials and Applications*, a companion volume to his 1954 textbook, von Hippel explains that dielectric materials "are especially important to the Armed Services." Statements in the book from representatives of the Air Force, Army, and Navy elaborate applications of electrical insulators in an aircraft's gunfire control system, a soldier's Geiger counter, and a submarine's sonar equipment.[98] Von Hippel was also a principal developer of radar—a military technology created at MIT under clandestine circumstances during the Second World War—and he later established MIT's Laboratory for Insulation Research, a lab funded by defense contracts through the Office of Naval Research, the Army Signal Corps, and the Air Force.[99] Von Hippel's Lichtenberg figures cannot be separated from military power.

If this lineage, a chain of signification that traces so clearly back to warfare, structures the image, then does Kepes's counterdiscourse allow us to see through or past such meanings? Is eyesight alone enough to change the ideological context the image records—the military funding that enabled the image's creation, the military applications the image then enabled? Of course, the answer is no; simply looking at an image in a new way is not enough to actually purge the associations that have already accrued in that image. But for me, the mysticism and mythology are not only in seeing in a more "mystical" or "mythological" way, more creatively, more physiognomically, more pseudomorphically—not just in seeing more than is really there—but also in believing that seeing is enough to make the aesthetic transformation happen, to symbolically alter the militaristic resonance that the picture originally obtained. The Lichtenberg figure illustrates this perfectly; it records how a deadly bolt of electricity becomes a captivating visual record. Kepes explains its transformative function in his notes: "Lichtenberg found that positive electricity conducted through metal plates covered with a powder could produce figures.... Art as miniature Lichtning Rod [*sic*] avert inner lethal thunderbolts."[100]

Other pictures in Kepes's book are even more directly implicated in warfare; one shows the Earth receding from a camera carried by a V-2 ballistic missile as it hurtles through the diaphanous outer limits of the atmosphere (fig. 2.42). The ethereal perspective is sublime, the view transcendent. But what about its implacably militaristic origins? Such origins are pervasive; in the book's acknowledgements, Kepes recognizes institutions that provided photographs for his project, including military installations like the Ballistic Research Laboratory at Aberdeen Proving Ground; research-and-development centers focused on weapons like the Applied Physics Laboratory at Johns Hopkins; and numerous companies that were or would soon become major defense contractors, from General Electric to General Motors.

Moreover, these images had highly charged public functions at MIT. The Institute's Lincoln Laboratory, a notorious center for weapons research, identified itself with a logo that placed two interlocking letters "L" around the looping pattern of a Lissajous figure. Created two years after Kepes's book was published, the logo may well have used an image from his volume as a design source (figs. 2.43–2.44). The Institute typically ignored the ideological context of its visual culture. The December 1952 issue of the *Technology Review*, MIT's alumni magazine, carried a cheerful "Lichtenberg Christmas Tree" frozen in a block of Lucite on its cover; there is no explanation of the pattern's role in von Hippel's laboratory (fig. 2.45).[101]

2.42

"Aerial Photograph, Made from a V-2 Rocket Fired from White Sands Proving Ground, New Mexico," from Gyorgy Kepes, *The New Landscape in Art and Science*. Chicago: Paul Theobald, 1956. © The Estate of Gyorgy Kepes.

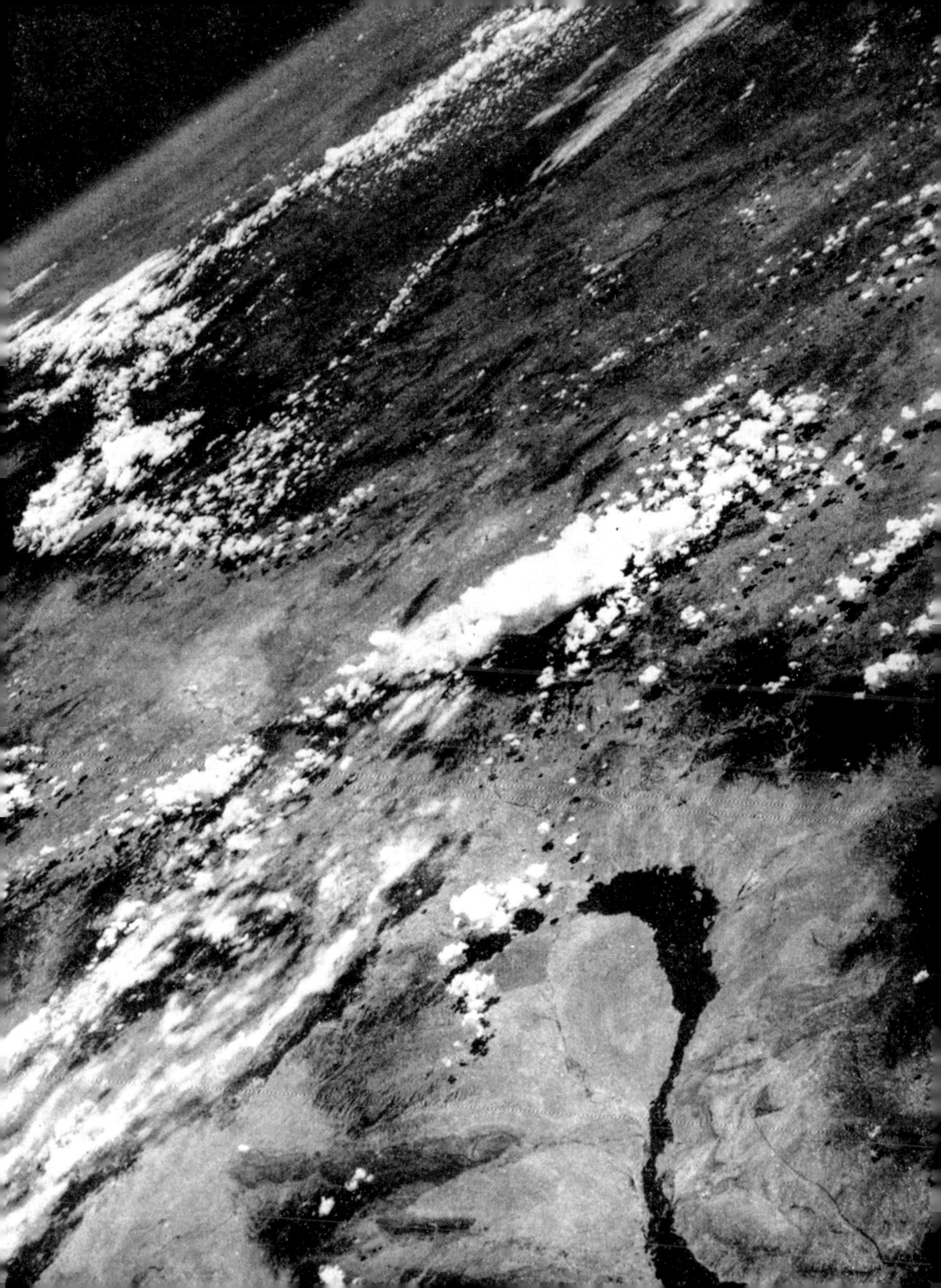

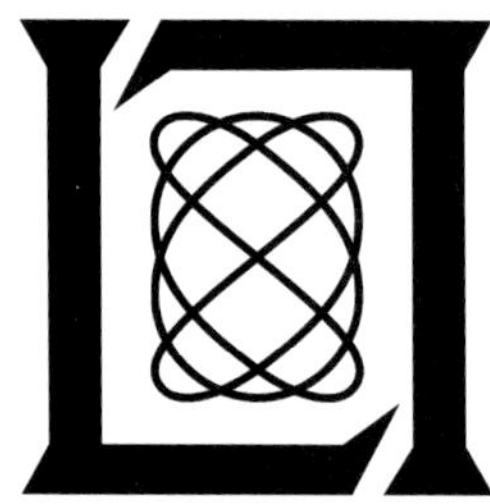

2.43

Lincoln Laboratory logo, designed in 1958.
Courtesy Lincoln Laboratory.

2.44

"Lissajous figure," from Gyorgy Kepes, *The New Landscape in Art and Science.*
Chicago: Paul Theobald, 1956. © The Estate of Gyorgy Kepes.

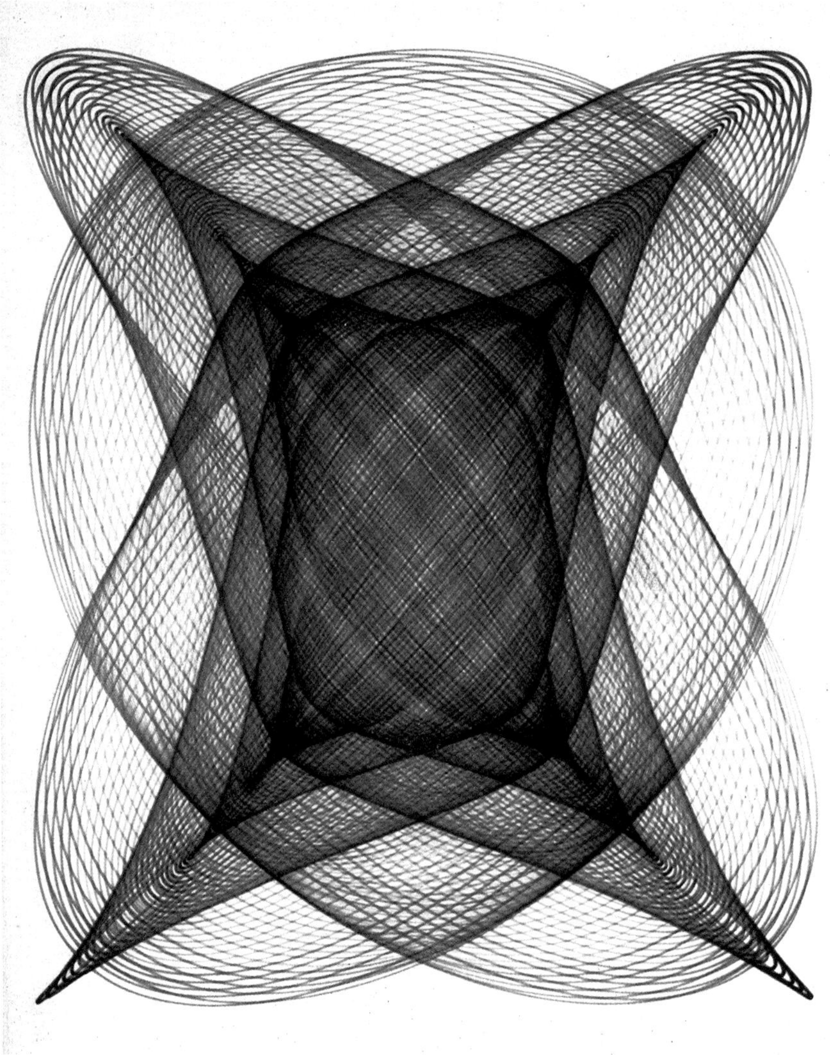

TECHNOLOGY
REVIEW *December* 1952

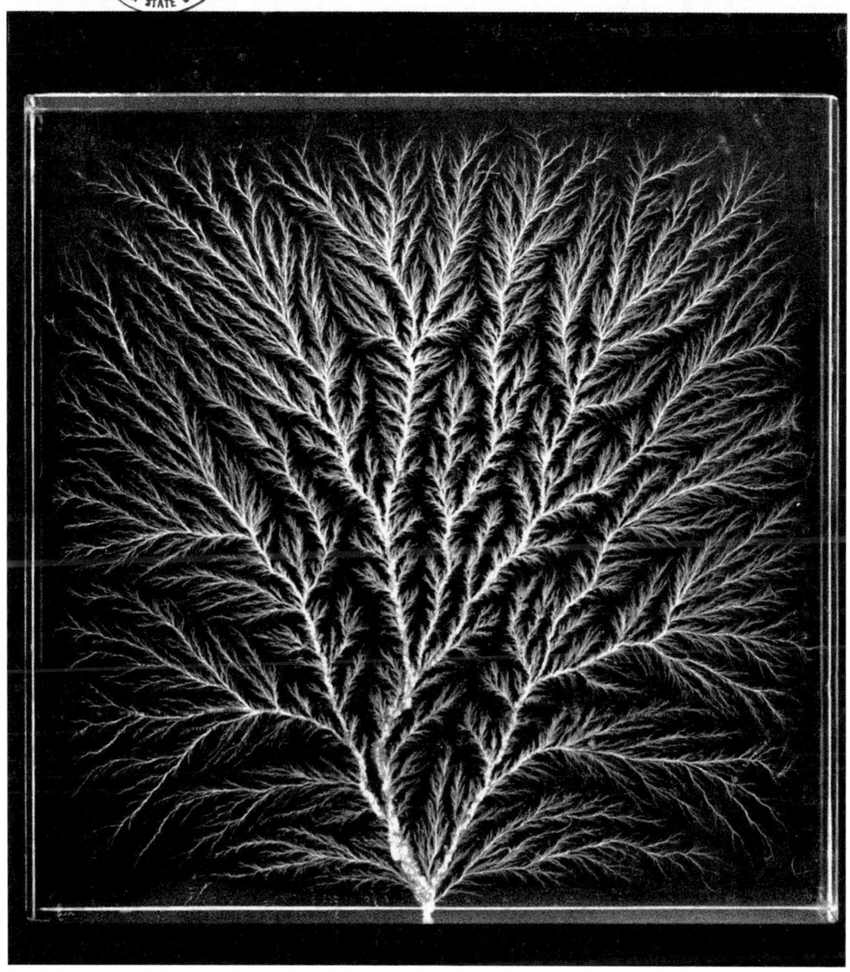

2.45

Technology Review, December 1952.
Cover depicts a "Lichtenberg Christmas
Tree." Copyrighted 1952. *Technology
Review*. 280537:0818SH.

For Kepes, addressing these militaristic origins was the very purpose of his project. Again, it is not possible to purge these menacing associations, but that does not mean Kepes cannot recontextualize these images, looking at them in ways that defy their scientific and technological origins. In his notes, Kepes lists images of death and disaster that might be included within his *New Landscape* collection (these suggest the types of pictures he originally envisioned for the opening of the book): "Hiroshima atom bomb," "cancer cell micro photo," "flak photos (war) light play"—this last a reference to the night sky illuminated by aerial warfare.[102] Kepes was not celebrating such images but attempting to gain power over them through a form of apotropaic magic. These notes offer strategies for such a transformation: integrating the "unity of opposite[s]," comparing "quantitative vs. qualitative," creating the "negation of negation."[103]

There is one spread in Kepes's book that holds special significance, distilling the logic and illogic of his ambitions (fig. 2.46). It is the only spread in color. It falls at the approximate center of the book.[104] On the left, two images of penicillin. On the right, an explosive chemical compound called hexanitrodiphenylamine. During the Second World War, the Axis powers mixed this chemical with TNT for use in high-power incendiary weapons, including torpedo warheads, depth charges, and underwater mines. The chemical is so toxic that it burns and blisters skin on contact. By placing hexanitrodiphenylamine against penicillin, Kepes attempts his strange aesthetic alchemy, a transformation, a transubstantiation, of this elemental form of evil into something good, a miracle drug. In his notes, he lists the Latin phrase *similia similibus curantur:* "curing like with like."[105] The dictum, known as the law of similars, forms the basis of homeopathy, in which toxic substances that cause the symptoms of disease are used as a form of treatment for that very disease. This tautological principle suggests how Kepes might inoculate the eye against these disturbing artifacts of visual culture through exposure to them, by vaccinating vision. He compares and contrasts an agent of death with an agent of life in an attempt to "unify opposites," to "negate the negative," to balance war with an inverse image of peace.

Containment Strategies

Kepes's visual methodology is, to be sure, impossibly complicated and convoluted. On the one hand, I cannot imagine Kepes choosing his images randomly, carelessly, without meditation on their meanings, and how these meanings might be changed or altered. He would not have selected pictures without knowing what they depicted, how they were used, and why they were first

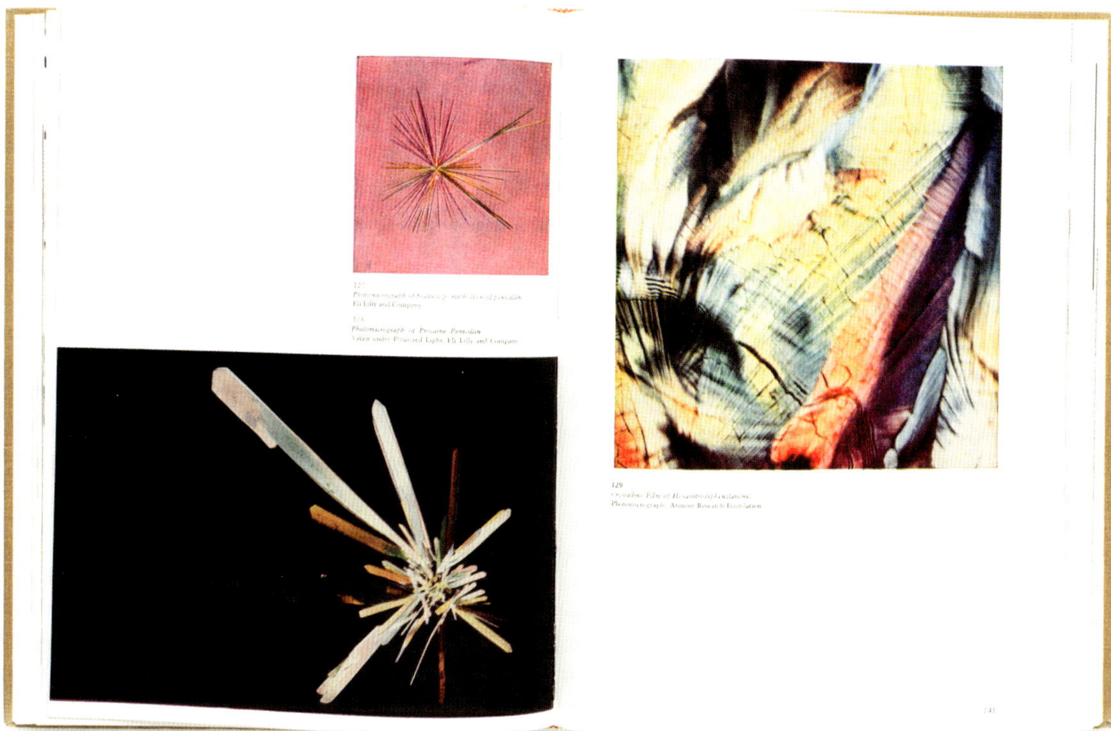

2.46

Two-page spread from Gyorgy Kepes,
The New Landscape in Art and Science.
Chicago: Paul Theobald, 1956. © The Estate
of Gyorgy Kepes.

created. He was a master of such detail. And yet, I also cannot imagine his audiences actually being able to decipher it all, to fully grasp the symbolic transformations. To wit: the uses and abuses of hexanitrodiphenylamine are hardly common knowledge; could anyone recognize the meanings attached to this chemical? Could even an informed reader or viewer grasp Kepes's arcane method?

I understand this complexity as a result of Kepes's attempt to continue a disguised form of political engagement—a politics so hidden that it would not be noticed as such. In developing visual design, Kepes introverted his former political and aesthetic goals, sublimating them into elaborate mental games for the mind's eye. In his notes, he linguistically dismantles the terminology of revolution, offering "re-volution: turning of vision," and "radical: from the root, to understand the very basis of vision."[106] He rephrases social and political action as purely visual concepts, as visual studies. He reimagines change in the world as change solely in the eye, in the way we see the world.

The fact that these agendas are so latent that they are rendered unrecognizable—camouflaged—is completely consistent with the ethos of the era. In a 1958 lecture at the Cooper Union, Kepes reflected on the sharply restricted potential for political art in the 1950s, as historical circumstance seemed to prohibit the type of engagements that had defined the historical avant-garde:

> **What are, then, the basic characteristics of our art situation today? First of all, we could observe a revulsion against ideology. The last generation was the generation of ideologies, manifestos, political, cultural manifestos, clearly defined tasks they [were] characterizing. Today, if one has an ideology it is the absence of the ideology as a creed, an anti-ideological ideology.**[107]

In preparatory notes for his lecture, Kepes explains that this "fear of ideology" came about when an earlier generation aligned with the "Marxist assumption" was supplanted by one that believed instead in "Calvinist determinism," in a vision of the social world as predestined and preordained—in a complacent resignation to the status quo.[108] Kepes's observations are surprising given the reality of explosive revolutions then occurring around the globe, from decolonization movements to the failed Hungarian Revolution of 1956 in his native homeland, to say nothing of Civil Rights and social movements in the United States—but they nonetheless indicate Kepes's understanding that a particular type of Marxist engagement was, at least at MIT, at least for him, now impossible.

Kepes's comments align with what sociologist Daniel Bell diagnosed as the "end of ideology" in his 1960 book of the same title (fig. 2.47).[109] For Bell,

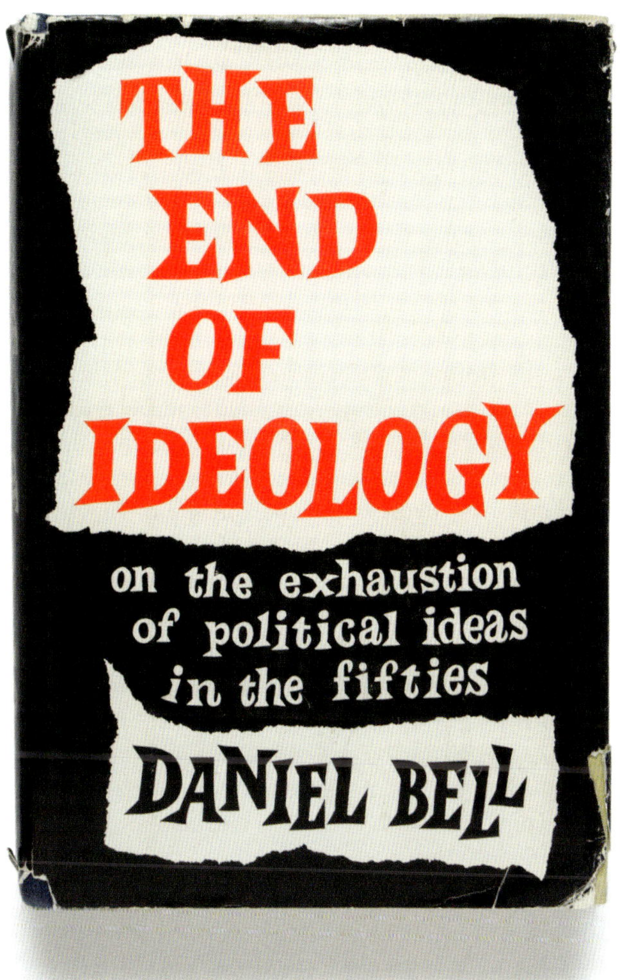

2.47

Book cover from Daniel Bell, *The End of
Ideology* (New York: Simon & Schuster, 1960).
Reprinted with the permission of The Free
Press, a division of Simon & Schuster, Inc.
All rights reserved.

the 1950s witnessed the exhaustion of ideological struggle and the ultimate triumph of a new technocratic order, one free from the clash between left and right. Bell's title might seem to be an appropriate description of Kepes's defection to MIT and abandonment of the historical avant-garde. And yet, Kepes was not praising the end of ideology—that much is clear in his Cooper Union lecture—but observing the limitations that seemed to foreclose previous ideological possibilities. Visual design reflects not the exhaustion of ideological struggle so much as its continuation by another name, in a repressed form. We might term this, to use the parlance of the Cold War, the "containment" of Kepes's prior radicalism.

This containment strategy constricted Kepes's projects in specific ways. His tentative outline for the "Revision of Vision"—recall that this was Kepes's original title for *The New Landscape in Art and Science*, and a title we can now read as a reference to the revolutionary turning of vision, its "re-volution"— lists possible contributors to his book, many of whom had past or present affiliations with radical politics. Kepes includes "Jan Struik," properly known as Dirk Jan Struik, a Dutch mathematician then teaching at MIT who was also a card-carrying Communist. In 1951, the House Un-American Activities Committee (HUAC) subpoenaed Struik; needless to say, Kepes did not include him in his published book. The "Revision of Vision" lists other Communists, too, including J. B. S. Haldane, the British biologist, and Harlow Shapley, the Harvard astronomer. Kepes met Haldane in London but now, in the 1950s, his association with Haldane had become a liability; so had his association with Shapley, who also testified before the HUAC and was a frequent target of suspicion. Neither Haldane nor Shapley appears in Kepes's book.[110]

In purging Communists from his project, Kepes was not opposed to their real or imagined political positions, but fearful that these associations would implicate him (after all, Kepes's name had already appeared in an HUAC report in 1943 due to his position at Chicago's Abraham Lincoln School). A series of letters from 1966 between Kepes and Julius Stratton, then retired as President of the Institute, alludes, in the vaguest of terms, to Kepes's anxieties about being discovered as a former Communist in the age of McCarthyism. The correspondence is suggestive but hardly conclusive, referencing Kepes's worries about his political past and its ramifications only obliquely, as abstract "threats and dangers." Stratton writes to Kepes in 1966:

> I remember so very well the first time you and I met – a long conversation in my office about somber threats and dangers that had nothing to do with art. That was fifteen years ago, and from that first afternoon I have never altered in my respect for the principles which guide your life, or ceased to be grateful for all that you have contributed to the life and quality of MIT. No one could understand better than I your own hopes for the ultimate place of visual art at the Institute.[111]

Stratton recalls that his initial conversation with Kepes was "fifteen years ago"—specifically dating the encounter to 1951, the same year Stratton was appointed Vice President of the Institute. That was also the year that Kepes assembled his *New Landscape* exhibition, and the year that the HUAC subpoenaed MIT's Struik to testify during a series of hearings on the "Exposé of Communist Activities in the State of Massachusetts."[112] At the hearing, Struik refused to answer a single question.[113] In 1953, a second committee devoted to "Communist Methods of Infiltration in Education" continued the witch hunt, searching for potential sleeper cells among the professoriate—a secret "MIT Communist cell," as the House Committee refers to it. The presence of so many MIT faculty brought before the Committee during these hearings prompted the Committee's counsel to ask MIT mathematician William Martin "how you can account for what would seem an abnormally large percentage of Communists at MIT?"[114]

When FBI agents visited the Kepeses in Cambridge to investigate possible connections to this MIT Communist cell, Gyorgy and Juliet hid their books, wrapped in brown paper to disguise them, in the basement of their house.[115] Given their personal history, and their intellectual interests, they were scared. Kepes's daughter recalls a "sense in the home of a kind of paranoia and fear."[116] Kepes's reply to Stratton in 1966 reflects on Stratton's support of his career at MIT, despite whatever earlier anxieties Kepes disclosed during that "long conversation" in Stratton's office in 1951:

> I don't think I have told you how grateful and moved I was by your letter, in particular with your remembrance of our first close contact, which was in my life a very gloomy moment. Your encouragement and understanding meant more than I can ever tell you. And your human and courageous support gave me a new confidence. The fact that you remembered it and mentioned it in your letter added still more to my feeling toward you.[117]

Stratton had suggested that these unnamed events—these "somber threats and dangers"—had "nothing to do with art." But they had everything to do with art: the fear of exposure is what compelled Kepes to suppress what was once an explicitly political project into one that seemed only aesthetic—a politics contained as visual design.

World War III

The introversion of a political project into visual design was, of course, still impossibly complicated and convoluted. Viewers and readers could not grasp Kepes's goals; most had no idea what he was trying to accomplish.

When Kepes first attempted to secure permission for reproducing Wassily Kandinsky's paintings in his book (specifically for Heinz Werner's text on

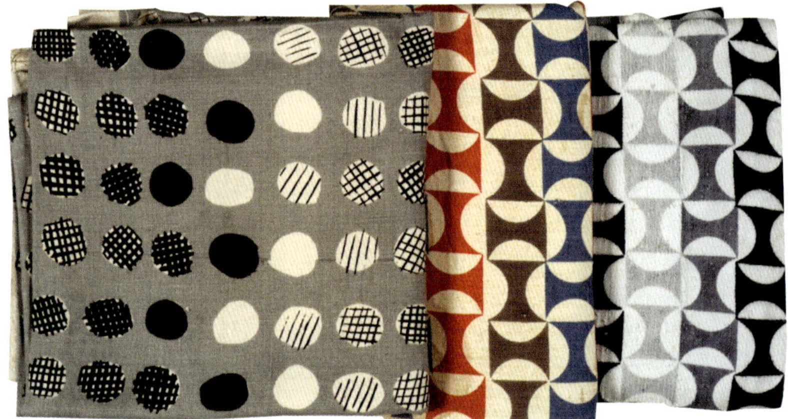

2.48

Gyorgy and Juliet Kepes, textile designs
for Laverne Originals. © The Estate of
Gyorgy Kepes.

physiognomic perception), Hilla Rebay—then director of the Museum of Non-Objective Painting, later known as the Solomon R. Guggenheim Museum—initially said no. She could not permit Kepes to use the images because she did not understand what he was doing. She writes:

> I hope you will understand that it is very difficult for me to have our paintings used in connection with exhibitions and writings which are totally opposed to that for which we stand…. [T]o think that [Kandinsky's work] would be taken for a land-scape or a wallpaper pattern, or any other out-of-the-frame-going decoration would do tremendous harm to what we believe in…. I don't know what you mean by visual design, and hesitate to consent to the use of the Kandinsky reproduction in an undertaking which may be contrary to our beliefs and principles.[118]

As Rebay saw it, Kepes's project was a gross distortion of Kandinsky's art, one that would reduce Kandinsky's painting to the mundane and familiar—to mere visual culture, something as common as the view out the window or onto the wall. She believed in the sanctity of art, in art as a noble defense against a proliferating culture of images—and Kepes threatened to dismantle these hierarchies ("the <u>new landscape</u>, as you call it, … is a very dangerous title," she warns). The universalizing imperatives behind visual design, the way Kepes was willing to study any image at all, would denigrate fine art, turning it into, as Rebay writes derisively, "a pattern or a landscape or a wallpaper or decoration," maybe even a "tapestry," all of which "becomes banal with a second look." Rebay's concerns were fair enough: Kepes's program did equate Kandinsky's paintings with banal images from daily life—all those dizzying patterns. Even more to the point, Kepes had also recently designed, with his wife Juliet, actual wallpaper and textile patterns sold by Laverne Originals for home decoration (fig. 2.48).

But Rebay also made a more substantive critique, claiming that Kepes threatened not only to legitimate a wildly expanding visual culture, but also to legitimate something worse: science and technology. In an early, unsent draft of her letter, she explicitly ties Kepes's institutional affiliation to such a concern. "Of course, I quite realize that belonging to the Institute of Technology [MIT], in the School of Architecture and Planning, art is not what you intend to propagate." She assumed that Kepes intended to propagate the rational and logical worldview that she understood as characteristic of the Institute. "I find it most advisable to suggest that you leave Kandinsky to us who represent art and creativeness."[119] She failed to grasp the purpose of Kepes's project as the cultivation of this very same creativeness.

In the end, this draft was left unsent. Kepes offered guarantees, and Rebay acquiesced, offering the necessary permission, but her criticism was an early warning of the backlash Kepes would face in subsequent decades.

Her remarks also suggest the extent to which visual design actually did promote MIT's core research interests in science and technology even while it contained a subtly subversive critique of those interests. A report titled *Art Education for Scientist and Engineer*, authored by the Committee for the Study of the Visual Arts at MIT—Kepes served as a consultant to the faculty panel in 1952—argued for visual design as a source for "intuitive" rather than "logical" forms of knowledge (fig. 2.49). The report implies that these intuitive ways of seeing and thinking could counter and correct MIT's skewed academic culture; visual design and the arts more broadly would "develop the capacity of the technician to undertake responsibility for the forms that his technical training creates," as if the militaristic applications of MIT's research could be allayed by creative pursuits that would magically generate "responsibility."[120] Kepes really did believe that intuitive forms of knowledge might serve a reparative function. But the way this program was officially sanctioned in a formal report by a special committee—and with Kepes's blessing—also indicates its compensatory role, at least when viewed from the administration's perspective; even if the program originated as a subtly subversive intervention, it could still be embraced by powers that be as justification for the destructive forms of research that the Institute would continue to vigorously support.

Rebay's remarks also anticipated another striking episode. Just weeks later, at the 1952 Walter Gropius Symposium—an honorary event for the Bauhaus founder held at the American Academy of Arts and Sciences in Boston, dedicated to the theme "The Future of Design"—Gropius, Kepes, and the architects Serge Chermayeff, Pietro Belluschi, and Charles H. Burchard discussed the continuing value of the Bauhaus idea. The speakers all affirmed their commitment to modernism as put forth decades prior at the Bauhaus. Kepes promoted visual design as a pedagogical project that would allow one to "understand how to tie things together visually," a simple summation of the universalizing ambitions in his *New Landscape* projects.[121] During the question-and-answer session that followed, however, an unnamed audience member asked whether this program was sufficient to the need, especially in a world where science and technology—in the form of the nuclear bomb—posed a mortal threat:

> Isn't there something tragic about the fate of the … artists and architects…. Aren't they fiddling, really, while Rome is burning, meaning while the scientists, physicists and other militarists are preparing the third world war? Doctor Gropius mentioned destruction of dreams. The third world war is being prepared now. Isn't this a tragedy we all have to face and we have to encourage our students to face, to think through, to feel it through?

Gropius said very little in response: "I, for one, refuse to think of the third world war. I don't want to do it." Kepes said nothing at all.[122]

To me, Kepes's silence indicates not his failure to consider the role of visual design in the context of the Cold War—this chapter demonstrates his intentional formulation of a program that used visual culture as a means of intervention against military power at MIT—but an awareness, faint as it was, that his project could not achieve such ambitions.[123] There were blind spots in visual design; the project suffered from a certain myopia. Kepes must have sensed these limitations, understanding that his purely visual solution would not actually solve the complex problem of militarism on MIT's campus. "Fiddling" with patterns and puzzles would only ever be a clever diversion, an elaborate distraction from the Institute's real business of designing weapons and planning wars. In this way, Kepes's project was utopian in both a positive and a negative sense. He idealistically believed in the transformative power of vision, but he also overlooked the ways in which the arts could not counter the sciences without also compromising their own integrity—without becoming part and parcel of the very system Kepes had hoped to change. This doubled perspective of Kepes's project, one that sees visual design as progressive and regressive, is fully consistent with his dialectical worldview; perhaps it is also the truest picture that we can draw.

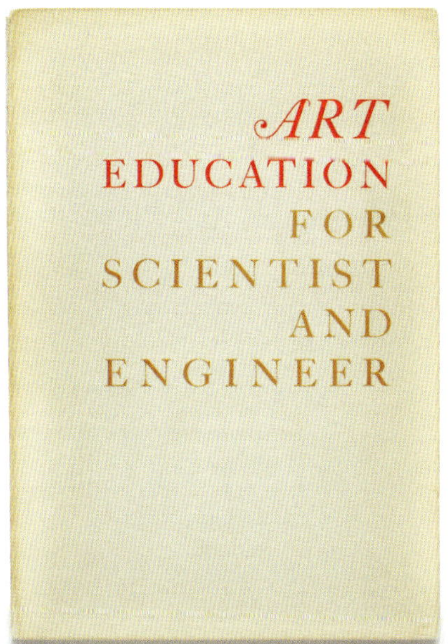

2.49

Art Education for Scientist and Engineer [The Report of the Committee for the Study of the Visual Arts of the Massachusetts Institute of Technology, 1952–1954]. Cambridge, MA: MIT, 1957. Courtesy MIT Museum, Cambridge, MA.

chapter

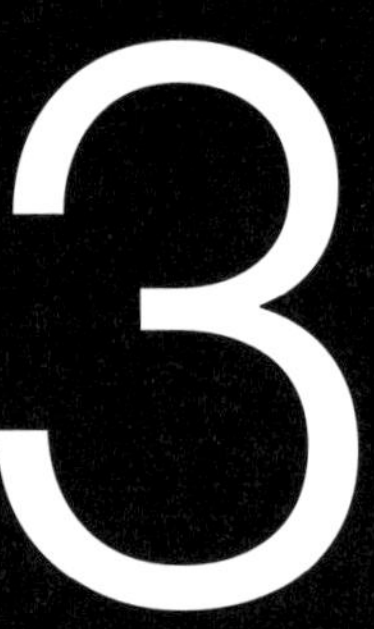

3

"To read what was never written." Such reading is the most ancient: reading before all languages, from the entrails, the stars, or dances. Later the mediating link of a new kind of reading, of runes and hieroglyphs, came into use. It seems fair to suppose that these were the stages by which the mimetic gift, which was once the foundation of occult practices, gained admittance to writing and language. In this way language may be seen as the highest level of mimetic behavior and the most complete archive of nonsensuous similarity: a medium into which the earlier powers of mimetic production and comprehension have passed without residue, to the point where they have liquidated those of magic.

Walter Benjamin, "On the Mimetic Faculty," 1933[1]

Visual Knowledge

The first image in *Module, Proportion, Symmetry, Rhythm*, a volume in Gyorgy Kepes's *Vision + Value* series, depicts a fantastical device called The Engine (fig. 3.1): a wooden frame twenty feet square holds many tiny cubes, each one pasted with paper covered in an unusual script spelling out every word the world has ever known, in all possible declensions and variations. With a simple turn of the mechanical cranks, the cubes clatter and clank, shifting to create a new arrangement. The magical device could generate infinite knowledge through these random combinations of visual signs—algorithmic permutations of a code without end. The Engine serves as an illustration for an essay by MIT physicist Philip Morrison, who pulls the image and its description from the pages of Jonathan Swift's 1726 novel *Gulliver's Travels*. In the novel, Swift describes Captain Gulliver's encounter with the machine at the Grand Academy of Lagado, a mythical center of absurd scholarly pursuits. Students at the Academy transcribed the traces of meaning The Engine produced with the utmost seriousness; "the professor showed me," writes Swift, "several volumes in large folio, already collected, of broken sentences, which he intended to piece together, and out of those rich materials, to give the world a complete body of all arts and sciences."[2]

Morrison's reference is ironic. Swift's description was satirical, lampooning not only the mechanical contrivances used to calculate mathematical equations but also the futility of encompassing all knowledge in a series of books; the novel uses the experiments conducted at the Academy to mock enlightenment reason and its pretense at progress. But in his essay for Kepes's book, Morrison instead argues that the "ambitious professor of Lagado was logically without flaw," for the random configurations of information actually anticipate the calculations of the modern computer.[3] Progress was not a foolish pretense. Morrison creates a lineage of similar devices to the present, from Joseph Marie Jacquard's famous loom, which operated by means of coded punch cards, and Charles Babbage's differential engine, an early mechanical calculator, to Herman Hollerith's tabulating machines. He ends with the IBM System/360 mainframe, which by the 1960s had already begun transforming knowledge into an informational blur of digital code, an endless permutation of ones and zeros. "More than that," Morrison continues, "the extension of his [the professor's] method has convinced us today that, just as all our prose might indeed appear from his mindless but modular frame, so our pictures, fabrics, devices, formulae—all knowledge—share the modular nature." All knowledge "can be counted and listed" and through its "combination" in visual and textual forms, a "richness beyond human grasp contained in the interacting multiplicity" is revealed.[4]

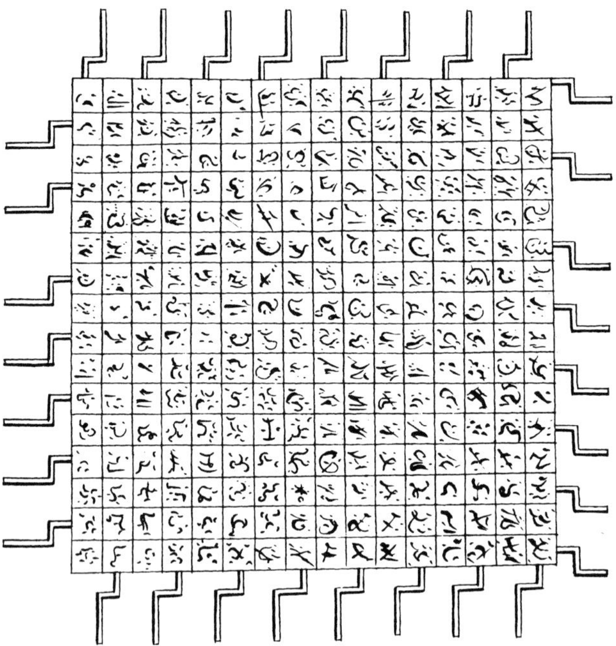

3.1

The Engine, from Jonathan Swift, *Gulliver's Travels*, 1726, reproduced in *Module, Proportion, Symmetry, Rhythm*, edited by Gyorgy Kepes. New York: George Braziller, 1966. Courtesy George Braziller.

Morrison's reference to the collections of visual signs is doubly ironic, however; not only does Morrison embrace with sincerity what Swift offers as satire, he does not seem to realize—and perhaps neither does Kepes—that Swift's commentary could very well apply to the series of which Morrison's contribution was a part, a series that also attempted to piece together an encyclopedia spanning all arts and sciences. In *Gulliver's Travels*, Swift quotes the professor, who describes "how laborious the usual method is" of generating knowledge—by thinking—while The Engine, by contrast, allowed even "the most ignorant person" to "write books in philosophy, poetry, politics, laws, mathematics, and theology, without the least assistance from genius or study."[5] Such an ambition could well describe Kepes's *Vision + Value* series, which attempted to master all such disciplines.

Indeed, the very same volume of the series also includes a picture essay assembled by Kepes demonstrating a similar method. Titled "Modular Ideas in Science and in Art: Visual Documents," the spread opens with images that look not unlike the unusual script generated by The Engine, with arrangements of black and white depicting the magnified crystal of a virus and the core memory from an IBM computer, as photographed by Ansel Adams (fig. 3.2). On subsequent pages, Kepes provides seemingly random combinations of visual signs: the thirteenth-century mosaic tiles from the floor of the church of San Giovanni Evangelista in Ravenna, Leonardo da Vinci's drawings from the *Divine Proportione* of Luca Pacioli, polyhedrons from the *Perspectiva Corporum Regularium* of Wenzel Jamnitzer, sunflower seeds, a textile, a snowflake, a flower, and on and on. Like the volumes of broken sentences and visual nonsense, the images are a certain gibberish, a compendium of similarity and difference across meaningless modular black and white elements, all so seductive in their detailed perfection. Kepes is the modern-day professor of Lagado, whose bizarre technological contrivance is not a proto-computer, but a series of inscrutable encyclopedias. The series is intended to create a new form of knowledge, a comprehensive, all-encompassing, all-embracing "visual knowledge."[6]

Creative Altruism

Published by George Braziller in two sets of three volumes in 1965 and 1966, and followed by a slightly smaller final book in 1972, the *Vision + Value* series (figs. 3.3–3.9) explored a range of open-ended visual themes—visual pedagogy proper was the focus of the series' centerpiece (*Education of Vision*), but other volumes examined more abstract aesthetic principles, like communication (*Sign, Image, Symbol*), organization (*Structure in Art and in Science*), movement (*The Nature and Art of Motion*), the standard unit (*Module, Proportion, Symmetry,*

3.2

Crystal of virus, magnification: 122,700 X compared to magnetic core storage element of an IBM computer in a photograph by Ansel Adams, from *Module, Proportion, Symmetry, Rhythm*, edited by Gyorgy Kepes. New York: George Braziller, 1966. Courtesy George Braziller.

3.3

Book cover to *Education of Vision*, edited by Gyorgy Kepes. New York: George Braziller, 1965. Cover designed by Gyorgy Kepes. Courtesy George Braziller.

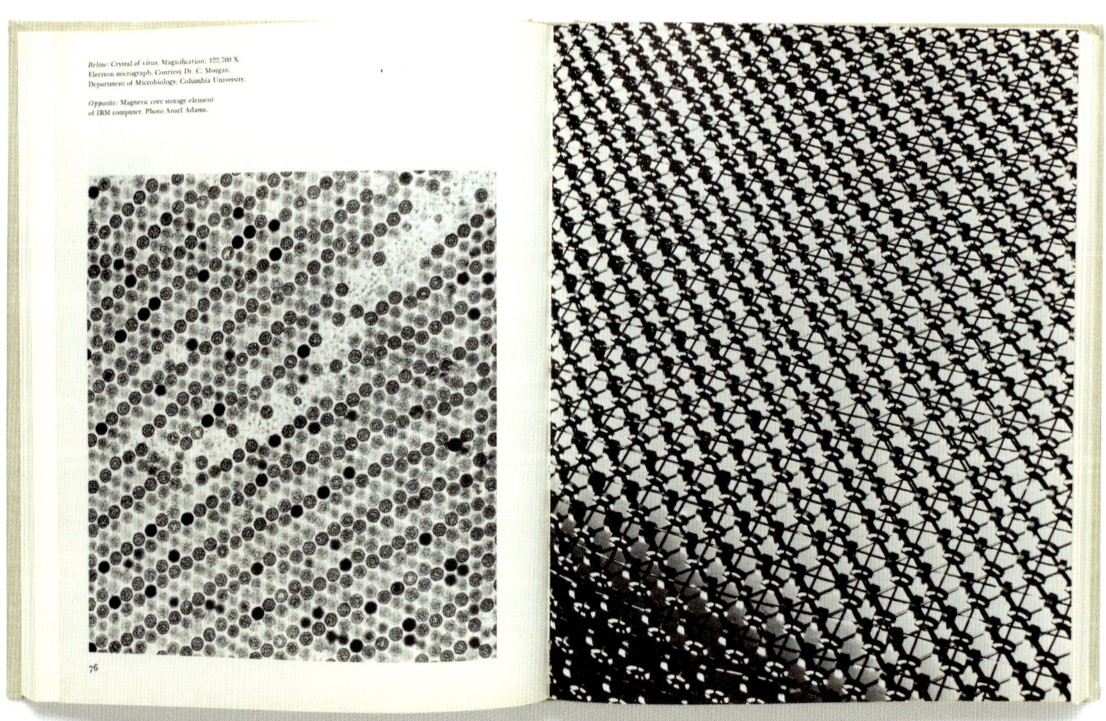

Education of Vision

edited by Gyorgy Kepes

Rudolf Arnheim   Mirko Basaldella   Julian Beinart   Will Burtin

Anton Ehrenzweig   William J. J. Gordon   Bartlett H. Hayes, Jr.

Gerald Holton   Johannes Itten   Tomás Maldonado

Wolfgang Metzger   Robert Preusser

Paul Rand   Robert Jay Wolff

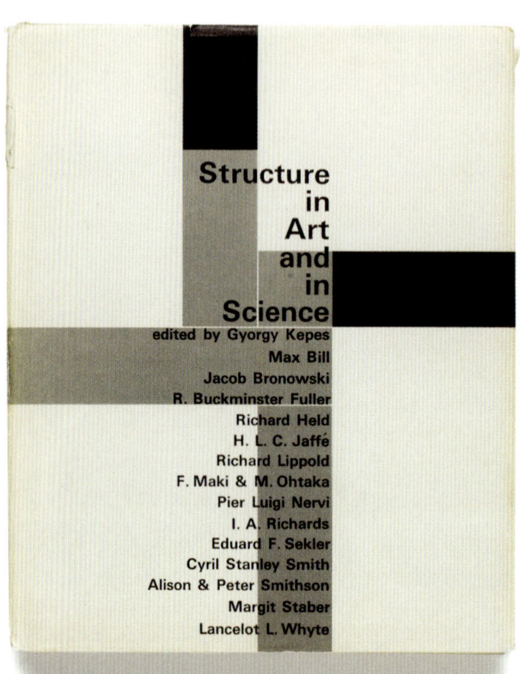

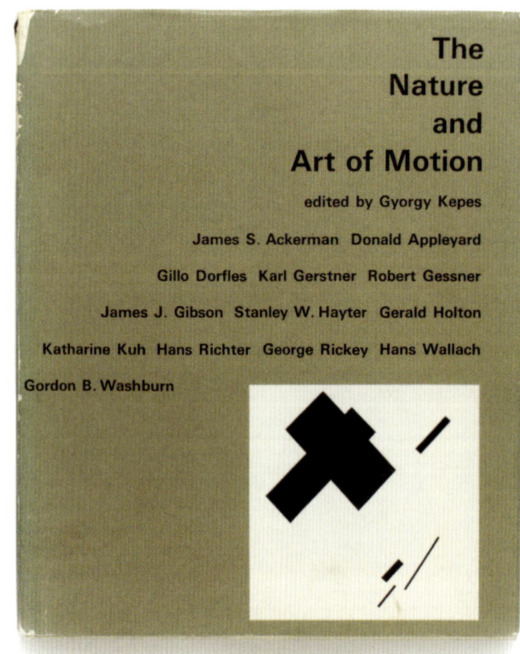

3.4

Book cover to *Structure in Art and in Science*, edited by Gyorgy Kepes. New York: George Braziller, 1965. Cover designed by Gyorgy Kepes. Courtesy George Braziller.

3.5

Book cover to *The Nature and Art of Motion*, edited by Gyorgy Kepes. New York: George Braziller, 1965. Cover designed by Gyorgy Kepes. Courtesy George Braziller.

3.6

Book cover to *Sign, Image, Symbol*, edited by Gyorgy Kepes. New York: George Braziller, 1966. Cover designed by Gyorgy Kepes. Courtesy George Braziller.

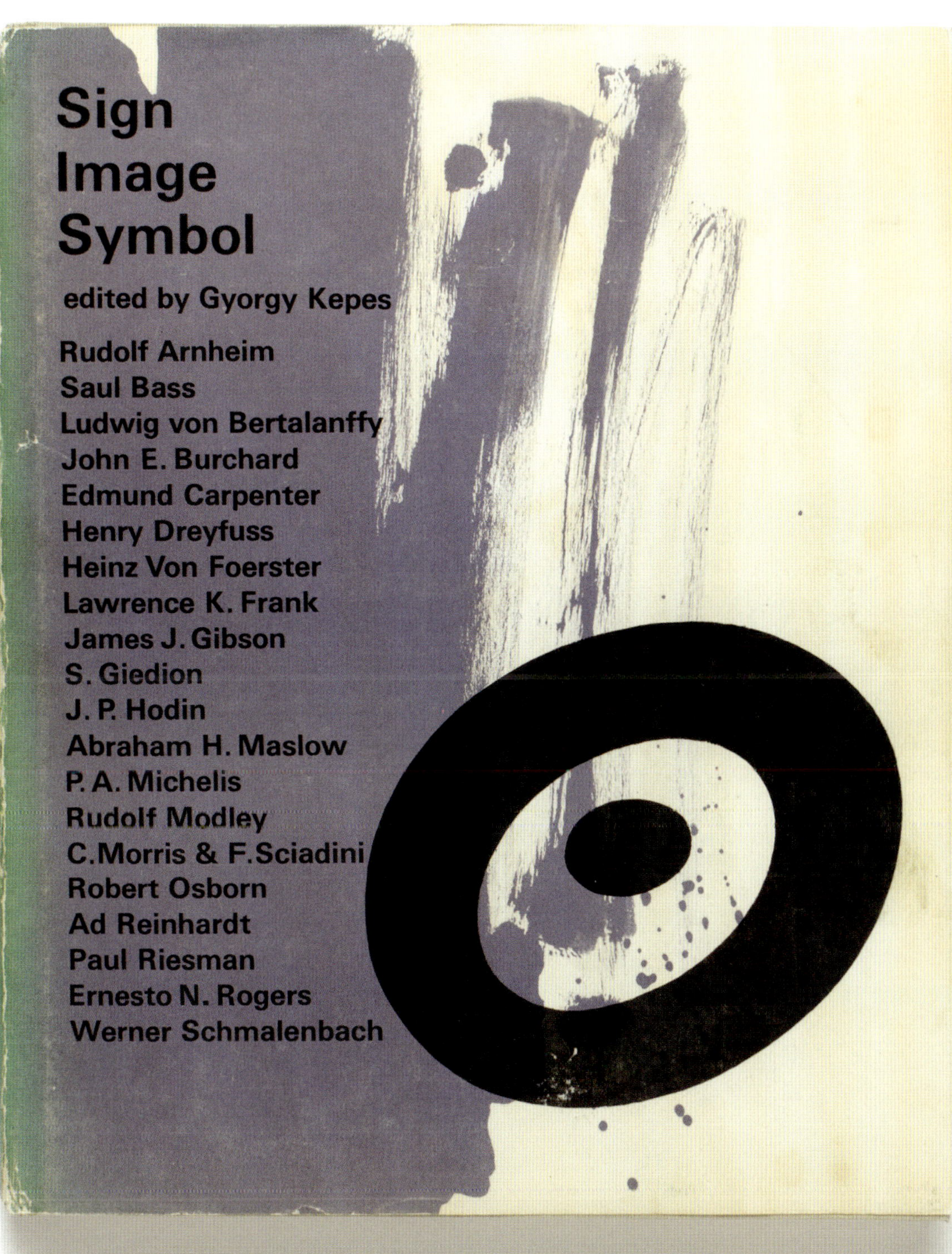

Sign
Image
Symbol
edited by Gyorgy Kepes
Rudolf Arnheim
Saul Bass
Ludwig von Bertalanffy
John E. Burchard
Edmund Carpenter
Henry Dreyfuss
Heinz Von Foerster
Lawrence K. Frank
James J. Gibson
S. Giedion
J. P. Hodin
Abraham H. Maslow
P. A. Michelis
Rudolf Modley
C. Morris & F. Sciadini
Robert Osborn
Ad Reinhardt
Paul Riesman
Ernesto N. Rogers
Werner Schmalenbach

Module
Proportion
Symmetry
Rhythm

edited by Gyorgy Kepes

Lawrence B. Anderson
Rudolf Arnheim
John Cage
Ezra D. Ehrenkrantz
Anthony Hill
Ernö Lendvai
Arthur L. Loeb
Richard P. Lohse
Francois Molnar
Philip Morrison
Stanislaw Ulam
C. H. Waddington

3.7

Book cover to *Module, Proportion, Symmetry, Rhythm*, edited by Gyorgy Kepes. New York: George Braziller, 1966. Cover designed by Gyorgy Kepes. Courtesy George Braziller.

3.8

Book cover to *The Man-Made Object*, edited by Gyorgy Kepes. New York: George Braziller, 1966. Cover designed by Gyorgy Kepes. Courtesy George Braziller.

3.9

Book cover to *Arts of the Environment*, edited by Gyorgy Kepes. New York: George Braziller, 1972. Cover designed by Gyorgy Kepes. Courtesy George Braziller.

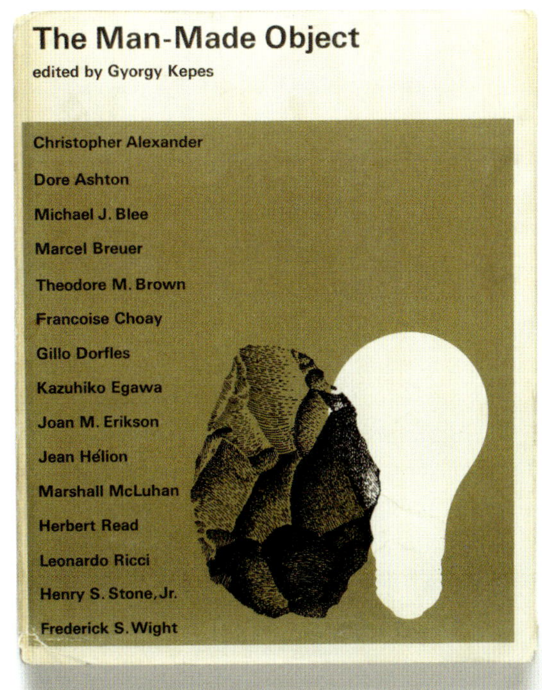

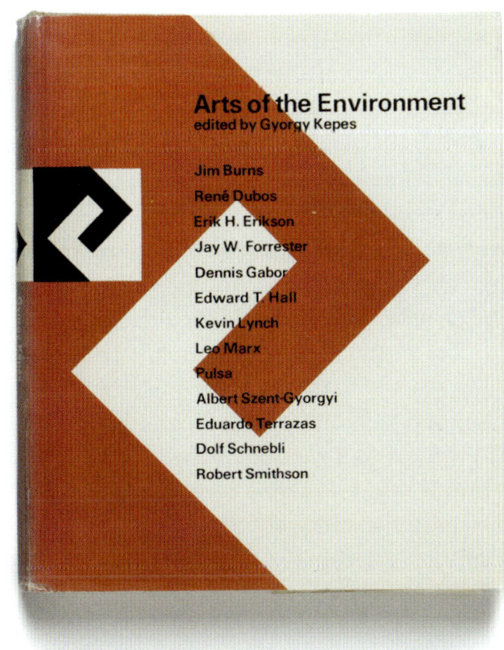

Rhythm), the designed artifact (*The Man-Made Object*), and the natural and human-made environment (*Arts of the Environment*).[7] Kepes had plans for an additional volume, to be titled *Arts of Participation*.[8]

The series has notable precedents. Kepes repeats the general aims of earlier design books—namely those by László Moholy-Nagy and originally published as part of the Bauhaus books, like *Painting, Photography, Film*, or Moholy's posthumous *Vision in Motion*, as well as Kepes's own *Language of Vision*, which was a New Bauhaus iteration of the form. Such precedents trained readers through pedagogical arrangements of images and text, often organized around particular design topics.[9] But in *Vision + Value*, Kepes also expands this program in new ways, beyond the familiar aims of Bauhaus and New Bauhaus design pedagogy, by including a very wide range of authors.

The series had other, more immediate, precedents as well. It builds on the visual design program Kepes first theorized and practiced as part of his *New Landscape* projects. In fact, the contract for *Vision + Value*, negotiated with publisher George Braziller in 1961, lists the series' tentative title as "A Revision of Vision"—the very same tentative title Kepes had first given to *The New Landscape in Art and Science*.[10] Kepes had also begun inviting speakers from far-ranging fields to his visual design courses around 1951, when he assembled his art-and-science exhibition. He "conducted seminars at the Massachusetts Institute of Technology for fifteen years," he explains, with the hope that "students might gain confidence that they could combine and reinforce their knowledge with the knowledge of other fields."[11] These seminars included visits by speakers ranging from "Weisner to Wiener," from electrical engineer Jerome Wiesner to cyberneticist Norbert Wiener, both members of the MIT faculty.[12] They eventually developed into a collaborative issue of the journal *Daedalus*, the interdisciplinary publication published by the American Academy of Arts and Sciences, edited by Kepes and devoted to "our visual culture."[13]

But perhaps the more relevant, if also more obscure, precedent for the series is in the origin for the word "value" in its title. Value refers to the study of human values that became an academic topic in the 1950s, specifically to a 1957 conference organized by sociologist Pitirim A. Sorokin at MIT and titled "The First Scientific Conference on New Knowledge in Human Values." Sorokin assembled the event through the Research Society for Creative Altruism, a group he founded with Harvard and MIT intellectuals including Kepes.

A preface to a volume of the conference's papers edited by psychologist Abraham H. Maslow, later a contributor to *Vision + Value*, explains the mission of the conference through the original letter of invitation circulated to invited participants—a letter Kepes signed. According to the invitation, the conference would "form a report to the public by objective, deeply concerned, hopeful scholars in the broad field of human values—moral, spiritual, aesthetic, economic—and their application to the affairs and institutions of mankind."[14] The conference was also highly interdisciplinary, representing a "cross-section of academic disciplines," all united by a "common framework" through the study of values.[15] It included Jacob Bronowski and Ludwig von Bertalanffy, both later participants in *Vision + Value*.

The preface, authored by Maslow, explains that "the ultimate disease of our time is valuelessness," a cultural condition that is "more crucially dangerous than ever before in history" and threatens human civilization with the "real possibility of annihilation" through a future world war.[16] The treatment of valuelessness was the cultivation of values: "The cure for this disease is obvious. We need a validated, usable system of human values, values that we can believe in and devote ourselves to because they are true." Sorokin's foreword, which follows Maslow's preface and was authored as a general statement on behalf of the Society, explains creative altruism as a means of preventing a "crisis" that "could develop into an Apocalyptic war capable of terminating mankind's creative history," a reality actually already proven "by the Second World War, by a multitude of hot and cold wars, and by a legion of bloody revolutions and disorders."[17] The study of values was a means of "fostering the spirit of altruism in man," especially through "the development of *new knowledge* and *new methods* for its application."[18]

The "Human Values" conference is long-forgotten; its intellectual impact was marginal at best, and the Society for Creative Altruism effectively disbanded shortly thereafter. Kepes's brief text included in the volume of conference proceedings is a condensed summary of observations pulled from *The New Landscape in Art and Science*—it is not particularly compelling.[19] The phrase "creative altruism" does not appear in any of Kepes's writings. It was not one he used in his research notes. But the mission of the Society nevertheless illuminates the motivations behind Kepes's *Vision + Value*. Creative altruism was apolitical politics—it opposed revolution and rebellion as a viable means of changing the world, and instead embraced values, common understanding. It was idealistic; it would prevent conflict, not inspire it.

While Kepes did not refer to creative altruism as such, his notes do contain a wealth of related phrases. One note compares "art for art sake" with "art for love sake."[20] Another note contrasts "I.Q. = logic" against "E.Q. = logic of heart."[21] He made lists of altruistic values, with emphasis on what he calls "peace research":

sensibilities
values
love
emotional literacy
E.Q.
ethical literacy
moral, esthetic literacy
peace research[22]

There is a certain naiveté in these notations—they are too earnest, too eager. But they also indicate the authenticity of Kepes's beliefs, the sincerity of his faith in creativity as a form of altruistic world-making.

This mission was also premised directly against the work of other artists, most of whom Kepes rejected as self-serving, the opposite of altruistic. Kepes's introductory comments in the *Vision + Value* series are explicit in framing the books as a polemic directed against others, specifically those who Kepes believed had "renounced the public forum and recoiled to the innermost privacy of unsharable singular moments of existence," particularly through "rebellious gesture"—an allusion to the gestural abstraction of the abstract expressionists, who were his main target. He makes this argument by way of Harold Rosenberg: "'The big moment came,' as an articulate spokesman of this group has put it, 'when it was decided to pain ... just to paint.' The gesture on the canvas was a gesture of liberation from value—political, aesthetic, moral."[23] Such a "liberation from value" was precisely the problem for Kepes: what was the purpose of art if it was not serving human values, if it was purposely abandoning them?

Kepes's notes from the period are littered with bitter, personally aggrieved descriptions of this lack of values. One note attacks the commercial gallery system in general terms—"manipulated market," "merchants of second hand novelties," "rear guard of avant guard [*sic*]"—and targets action painting in particular as an expression of distorted values:

St. Vitus dance
fretting
convulsive motion – combat with oneself
mutilating in wild fervor
scorched earth policy
brutal ferocity
pent up emotions[24]

Kepes understands the erratic bodily movements characteristic of the neurological disorder Sydenham's Chorea, or "St. Vitus dance," as a result of a cultural illness, of valuelessness. It also evoked Pollock's action painting, which he understood as physically violent but also mentally unstable: "The intellectual emotional bleeding to death of Pollock[,] De Kooning[,] Kline," he writes.[25] These artists were "reading their own guts" or "inspecting emotional entrails," a disturbing metaphor of grisly evisceration, of both self-obsession and self-destruction.[26]

But Kepes also rejected other postwar trends. As he writes in one note on the merging of art and consumer culture: "parallel with action painting the hero worship of the banal … pop art—orgy of philistinism—banal—orgy of the banal."[27] He rejected pop as juvenile, as a "childish sticking tongue out, frustrated child esthetic."[28] He even rejected artistic trends to which he ought to have been more sympathetically inclined, on account of their scientific and technological proclivities. In the introduction to *The Nature of Art and Motion*, he denounces kinetic art ("watchwork-like toy machines") and op art ("amusing, well-groomed eye teasers)."[29] I read Kepes's comments as personal resentment over his marginal position in the art world; his own paintings shared more than a passing likeness with abstract expressionism, and yet he garnered only cursory attention from the art-critical establishment, often no more than brief blurbs in the back of magazines.

But I also read these comments as a statement in support of art as altruism. Kepes positioned his visual design project as separate from whims, fads, and fashions; visual design did not require the art world at all, it could take place in the technocratic realm of the research laboratory. "By and large, the art world has become the scene of a popularity contest manipulated by appraisers and impresarios who are blind to the fundamental public role of the artistic image," he argues.[30] His *Vision + Value* anthologies would be publicly oriented, socially engaged, more relevant, more transformative—directed with intention to making creativity an altruistic pursuit.

Super-Topology

What makes *Vision + Value* most unusual, however, and what distinguishes the books from their precedents is their startling reach; nearly one hundred experts participated in the anthologies. The cast of characters that Kepes assembled includes artists, architects, and designers, from Robert Smithson to Marcel Breuer and Paul Rand—all figures who, although they represent different creative cultures, make sense in such a series. But Kepes was ecumenical. He also secured contributions from across the natural and social sciences, from physicists, psychologists, biologists, anthropologists, chemists, linguists, mathematicians, and specialists with more obscure fields of study, like metallurgy or musicology—and it is this surprising range of disciplines that prompts the question every reader asks upon opening a *Vision + Value* volume: What actually relates the book's strange contents? While each volume is framed around a particular principle, the essays and images can only be described as divergent. Some are qualitative, others quantitative; some scientific, others artistic; some intuitive or literary, others exacting in their technical precision. Does this assemblage have any cohesive meaning at all?

What unites the books' contents is the peculiar methodology Kepes used in organizing it all, one he hoped to teach readers by way of demonstration. The participants in his project employed myriad iterations of "systems discourse"—the universalizing interdisciplinary intellectual languages that exploded across the academy at midcentury. The discourse of systems encompassed both the cybernetics of Norbert Wiener and the General Systems Theory of Ludwig von Bertalanffy (Bertalanffy's writing appears in *Sign, Image, Symbol*, but so does the writing of Heinz von Foerster and Lawrence K. Frank, both of whom participated in the Macy Conferences that established cybernetics).[31] The series borrows particular methodological precepts from both, like the use of common terms ("homeostasis" and "feedback") across otherwise isolated fields—although here such terms are not only linguistic, but also visual, the common ways of seeing that might be mirrored across disciplines. Like all systems thinkers, Kepes applied this approach to topics that were not previously understood through systems discourse—that, after all, was the very purpose of interdisciplinary intellectual languages, which allowed specialists in different fields to understand one another through shared ideas and images. Kepes subsumed distinct intellectual projects into his books. Some of these were philosophically aligned to systems, like Gestalt psychology, which shares a universalist orientation and which both cybernetics and systems theory often evoked. Others were more idiosyncratic. For example, Kepes includes an essay on "graphic symbols for world-wide communication" compiled by Rudolf Modley, a former assistant to Vienna Circle philosopher and pictographic pioneer Otto Neurath, who invented the visual language Isotype; the symbols reflect the global aspirations of all systems thinking, here employed for literal communication across the globe—even though Modley's symbols have no specific connection to cybernetics or systems theory.[32] The industrial designer Henry Dreyfuss illustrates another text in Kepes's volume with a sign for engaging mechanical gears, one recognizable to people of all tongues (fig. 3.10). Dreyfuss was addressing a particular graphic design problem, but his diagram represents the methodology of Kepes's program as a whole, which intended to transcend the limits of language, specifically disciplinary language, and thus the limits of cultural, social, and political boundaries—it aimed to facilitate transparent exchanges that would unite worlds.

In this way, systems speak defined the series in terms of the explicit systems-related specializations of some of its authors, and the general methodology that created new relationships between them. But to speak of systems was also, for Kepes, a more profound aesthetic gesture. More than that: it was an ethical ideal, a moral imperative, an altruistic statement of belief that cooperation, collaboration, and a seamless exchange of knowledge could solve global issues. Systems were reparative, restorative, a way of holding together a world that felt as if it was falling apart, spinning too fast, expanding ever outward.

The series' book jacket copy offers the clearest statement of this mission. The text explains this ambition as creative altruism, a "search for values." It describes "a crisis in communication" caused by an explosion of information and the ease with which it is exchanged—a condition that has only become far worse today. Unprecedented connectivity creates a very disconnected culture, one suffering from "the fragmentation of experience and the dispersion of knowledge into many self-contained disciplines, each with its own ever growing, increasingly private language."[33] The result is a fractured worldview—one the books would repair. "The aim of this series is to stimulate the circulation of ideas, to find channels of communication that interconnect various disciplines and offer us a sense of structure in our twentieth-century world."[34]

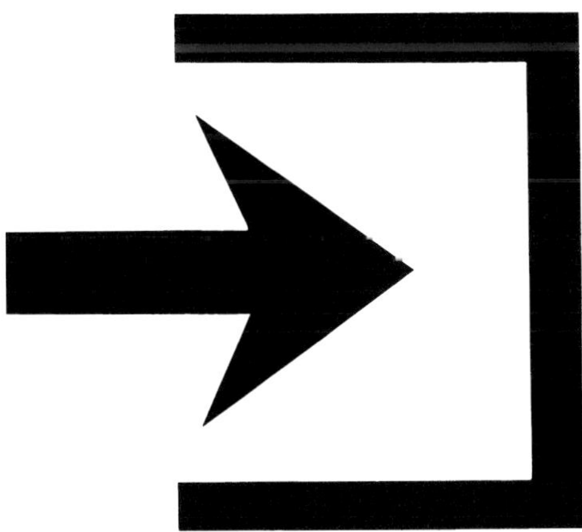

3.10

Henry Dreyfuss, symbol representing the term "engage" with linguistic signifiers, from *Sign, Image, Symbol*, edited by Gyorgy Kepes. New York: George Braziller, 1966. Courtesy George Braziller.

Kepes thought of this crisis as a scale crisis—a problem of being out-of-scale with a rapidly expanding network society and the resulting information overload. From his research notes:

> out of scaleness
> scale conflicts
> the individual is <u>buckling</u> under the compressive loading of knowledge and
> demands of life. Need bridges – character depends upon span
> pile + beam bridge
> arch bridge
> suspension bridge[35]

The metaphor of building bridges recurs throughout the books. "Using our extraordinary vision, we can fashion such bridges," he writes in one introduction.[36] Another research note similarly describes a type of bridgelike "topology," a vast structural system that might subtend all this information:

> we need a <u>super topology</u>
> knowledge of relationships
> independent of the size etc
> connecting art – science[37]

Kepes understood the problem of scientific and technological change through a range of metaphors. Another note describes "art as genetic coding," a means of programming a "model of interdependence" by creating, through vision, various "roads, paths, edges, nodal points, [and] channels" that relate unrelated ideas; the goal was to "establish networks, interfaces, communication."[38] He describes the difference between "horizontal educators," whose teaching focuses on a wide "diffusion of knowledge," and "vertical educators," whose teaching focuses on "deepening" knowledge within areas of expertise—the problem was how to find a joint linking the horizontal and vertical dimensions.[39] Sometimes the metaphor was framed around bodily processes like illness or injury. In a note labeled "Braziller intro," he describes "limited sectors of knowledge," and the resulting "crisis of sensibilities," in terms of physical ailments: as a "blood clot" that might be treated by a transfusion of ideas—"need blood circulation of knowledge."[40] Elsewhere the metaphor was less about the body and more about the mind: the atomization of ideas and impossibility of gaining a total vantage was understood as the cause of mental instability, as neurosis or psychosis: "out of scale sickness—motion sickness—strains + stresses" causing "uncertainty, insecurity, fear."[41]

Kepes hoped to build his "super-topology" through approaches he more specifically called interthinking and interseeing. He explains these ideas through the work of George Gaylord Simpson, a paleontologist and evolutionary biologist. Simpson, writes Kepes, "has commented that, as organic evolution was brought about by interbreeding, so our further cultural evolution today will come about through broadscale 'interthinking.'"[42] The reference derives from the Simpson's 1949 book *The Meaning of Evolution: A Study of the History of Life and Its Significance for Man*, which describes how interbreeding between geographically proximate biological species accelerates evolution, just as interthinking between intellectually proximate research domains might accelerate cultural evolution.[43] In another introduction from another volume, Kepes extends this idea still further, bringing it into the realm of vision. Each book "suggests some seminal patterns of 'interseeing' needed for the living process of growing understanding."[44] In borrowing the term "interthinking" from Simpson, and then transforming it into "interseeing," Kepes demonstrates the very concept behind it. Ideas and images might be read through and against each other, producing new ideas and new images. Such intellectual miscegenation would create new knowledge; to think between ideas, to see between images, might then advance human culture. Kepes's research notes show him contemplating the concept: "cross fertilization → interthinking," he writes, and "evolutionary process → cultural and psycho social evolution."[45]

The Education of Vision

Kepes created a sophisticated theory for *Vision + Value*—but could this attempt to build bridges, to create a super-topology, really be achieved in something as ordinary as a book? How would it all work? To begin, each volume followed a basic formula. The first three, from 1965, include introductions authored by Kepes that explain the aim of the series and the focus of the particular installment. Kepes did not write introductions for the three volumes from 1966—a fact that suggests some doubts about the series (and a point to which I will return at the end of this chapter). Each book also contains a uniquely Kepes art form: a picture essay. Some of these are presented as compilations of source images, as in "Structure: Visual Documents." Others are more thematically organized. Kepes cataloged images in *Sign, Image, Symbol* under various classifications: he provides examples of visual writing systems; maps, models, and data visualizations; images indicating movement; expressive images; and personal symbols. A spread of "codified signs and symbols" (fig. 3.11) demonstrates the approach: the spread's two pages encompass an accumulation of graphic forms, including numerals, chemical symbols, the cardinal directions used with a ship's azimuth compass, the musical notion for John Cage's score *Cartridge Music*, and a visual language Tomás Maldonado created for electronic data-processing machines.

These image archives, even when categorized under thematic headings, are merely given as "documents," without further explanation. There are no instructions for how to read each picture essay. Some seem to be simple accumulations, but others create surprising relationships through visual likeness, like the spreads in "Object Forms and Functions: Contrasts and Analogies," one of which compares a shaman's copper rattle, created by the Chilkat tribe—a Tlingit group indigenous to Alaska—to an image from an advertisement for a Zenith-brand television set (fig. 3.12). Kepes lets us speculate on meaning: the rattle is activated by the shaman shaking it, summoning the being represented in its form; the TV, a midcentury equivalent to the ritual object, magically comes to life with a simple turn of a knob. In the same essay, a nineteenth-century stove used to heat a railroad station appears next to a Frigidaire-brand two-oven electric range (fig. 3.13). Rather than construct any specific argument through these morphological, and pseudomorphological, relations, the comparison is left open-ended, unexplained. It is the viewer's job to make meaning.

Actually, there are instructions, of a sort: the textual essays that comprise most of the books' contents. These do not address particular images in specific picture essays, but they offer general ideas about how to see across time, place, and object category. *Education of Vision* serves as a training manual or primer for this task; it teaches methods readers might use to study the series as a whole. The opening essay in the book, by psychologist Rudolf Arnheim, names the essential approach: "visual thinking." Arnheim introduces the topic with an anecdote about telling time; he describes how two people, asked to determine what time it will be in a half hour, go about the task by different methods; the first calculates time "intellectually," by adding thirty minutes to the current hour, whereas the second imagines the circular face of a clock and shifts the minute hand ahead by a semicircle, thus thinking "visually" (fig. 3.14).[46] As the examples demonstrate, making vision a means of knowing is commonplace. "Visual thinking is constantly used by everybody."[47] But Arnheim rejects the notion that this ubiquity renders visual thinking mundane; "perceptual thought processes are as exacting and inventive and require as much intelligence as the handling of intellectual concepts."[48]

The essays that follow provide a Bauhaus pedigree for visual thinking. Johannes Itten describes his foundational course; Maldonado explains the industrial design education at the Hochschule für Gestaltung in Ulm—one of many successors to the Bauhaus; and Robert Preusser and Robert Jay Wolff share their teaching at MIT and Brooklyn College, respectively, where they both followed the model of Bauhaus pedagogy (Preusser was a student at the New Bauhaus, where Wolff was a founding instructor). Kepes's own visual design courses are illustrated with abundant student examples.

"Codified Signs and Symbols," from *Sign, Image, Symbol*, edited by Gyorgy Kepes. New York: George Braziller, 1966. Courtesy George Braziller.

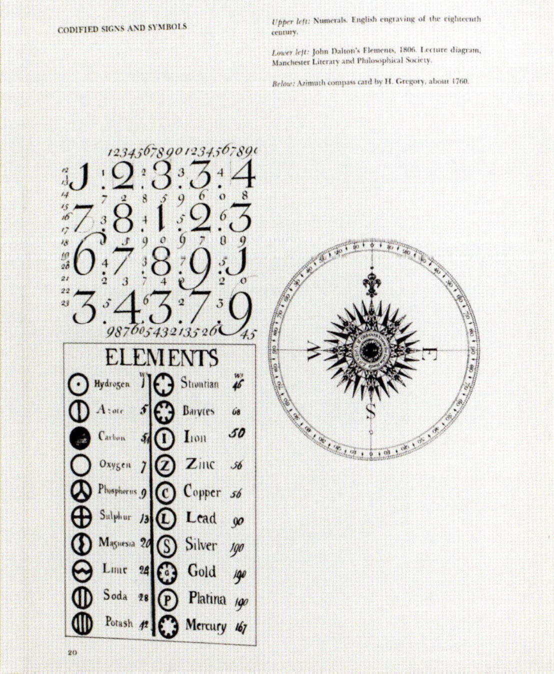

3.12 and 3.13

"Object Forms and Functions: Contrasts and Analogies," from *The Man-Made Object*, edited by Gyorgy Kepes. New York: George Braziller, 1966. Courtesy George Braziller.

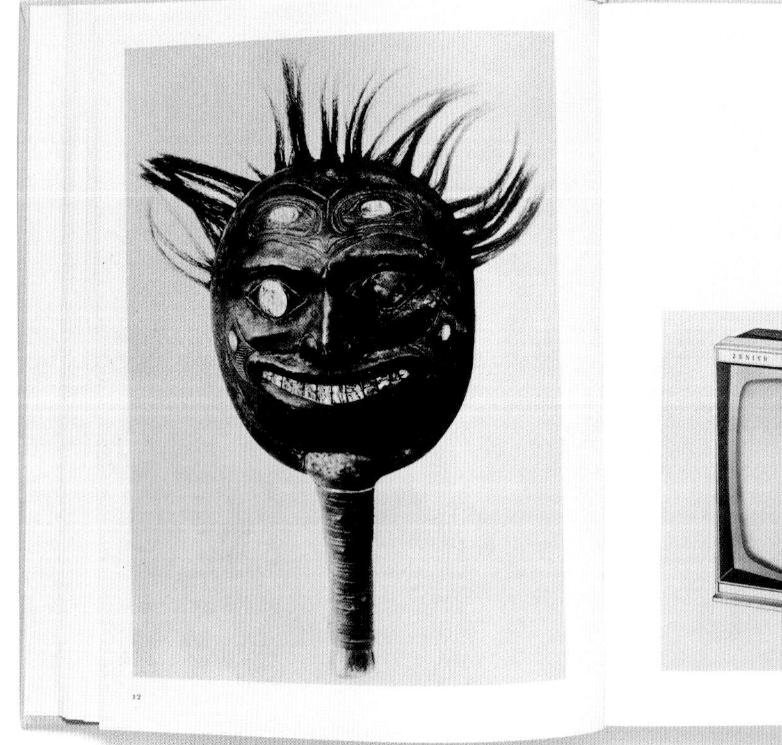

Shaman's copper rattle, from Alaska. Peabody Museum, Harvard University.

Portable TV set. Photo courtesy Zenith.

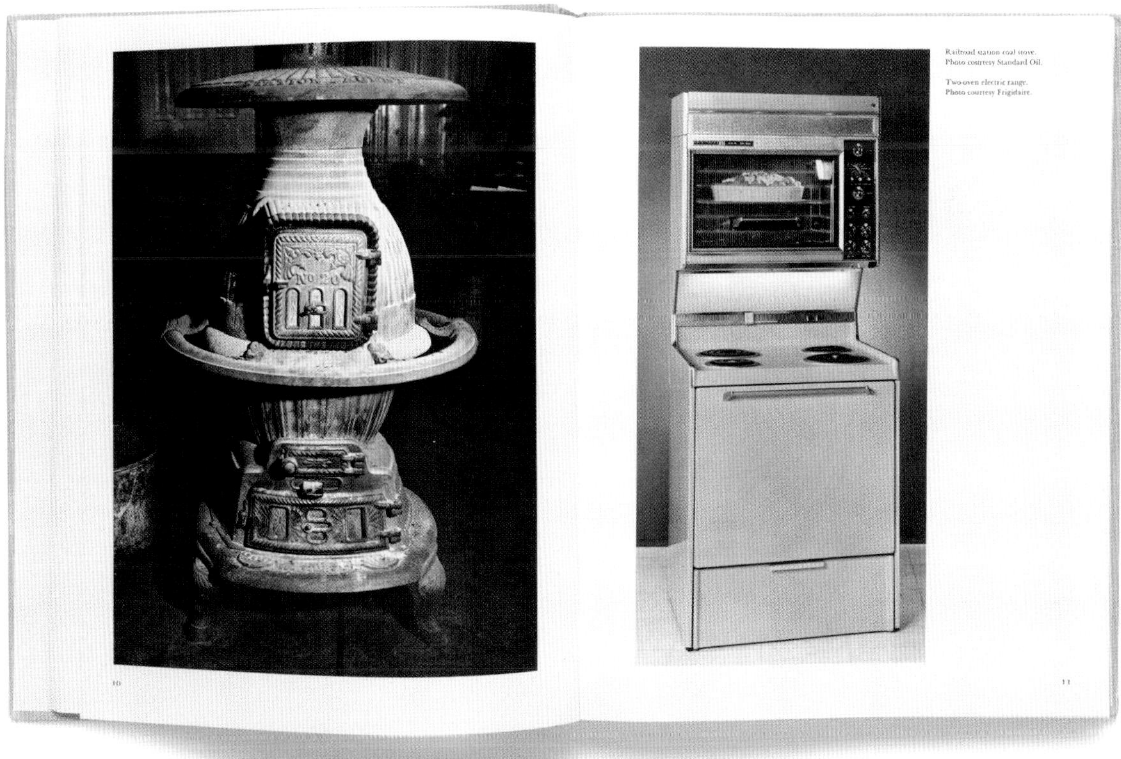

Railroad station coal stove.
Photo courtesy Standard Oil.

Two-oven electric range.
Photo courtesy Frigidaire.

But Kepes also extends visual thinking beyond the Bauhaus and New Bauhaus, beyond the arts. The expansion of visual thinking to fields not necessarily considered to be visual was a goal of many of the series' authors. Heinz von Foerster models the interactions between the organism and the environment in a diagram included in his contribution to one volume, which implicitly suggests the role of vision in mediating interactions (fig. 3.15). Another model by François Molnar, the theorist associated with the Groupe de recherche d'art visuel, or GRAV, places the eye in such a system (fig. 3.16). In looping circuits, the diagram suggests that a stimulus perceived by the eye—it could be anything at all—might change the mind of the individual, and thus society at large; and that, in turn, society simultaneously structures the individual's mind, the mind structures how we see, and how we see structures the visual stimuli we create. The resulting system is "so complex," "so vast," that it cannot be mapped in its particulars.[49]

These diagrams demonstrate an expansion of the Bauhaus and New Bauhaus program from art and design as commonly understood to a wider environment, a wider society, to the world at large. The authors in Kepes's books offer many strange tools with which to think visually in this way. Von Foerster discusses a series of puzzles that demonstrate the "breakdowns of the perceptive apparatus" (fig. 3.17), as represented, for example, when one studies an optical illusion like the "triple-pronged fork with only two branches," a figure that is impossible to see in its totality but nonetheless looks complete in its details.[50] The mind's eye short-circuits, seeing both simultaneously, even though neither exists. A 1535 anamorphic woodcut by Erhard Schön creates another ambiguous image: at first, it appears to capture a meandering, if distorted, landscape filled with rivers and roads; viewed from a grazing side angle, portraits of Emperors Charles V, Ferdinand I, Frances I, and Pope Paul III appear—all four are even identified with textual labels. A set of meaningless shapes—a star and a scribbled mark—are given the nonsensical names "Ooboo" and "Iratzky."

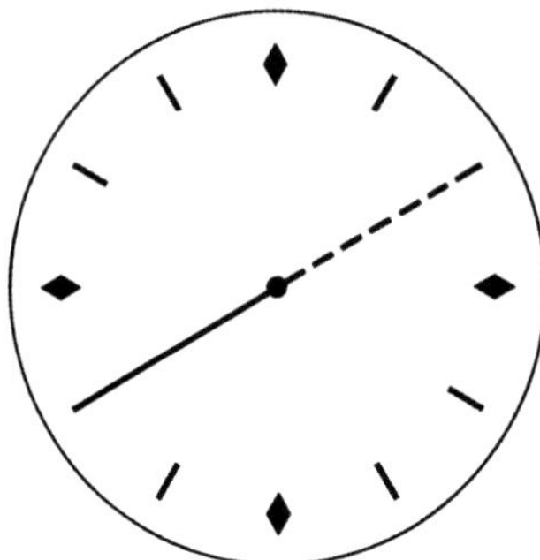

3.14

Diagram from Rudolf Arnheim, "Visual Thinking," from *Education of Vision*, edited by Gyorgy Kepes. New York: George Braziller, 1965. Courtesy George Braziller.

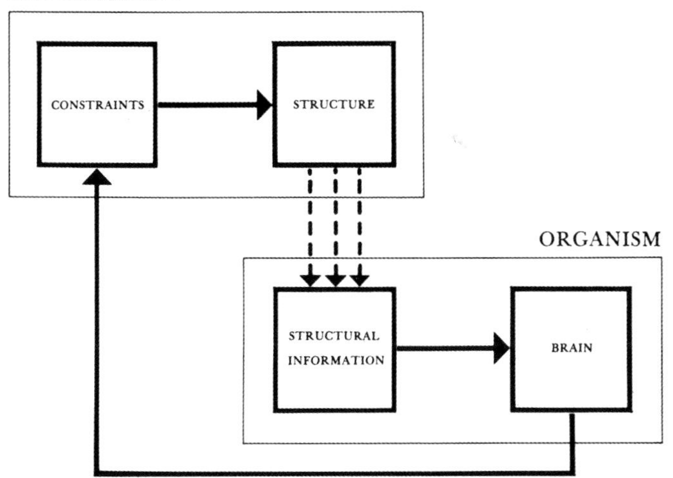

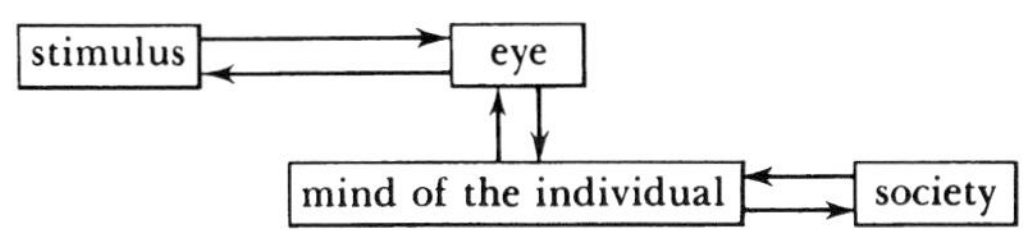

3.15

Diagram from Heinz von Foerster, "From Stimulus to Symbol: The Economy of Biological Computation," from *Sign, Image, Symbol*, edited by Gyorgy Kepes. New York: George Braziller, 1966. Courtesy George Braziller.

3.16

Diagram from François Molnar, "The Unit and the Whole: Fundamental Problem of the Plastic Arts," from *Module, Proportion, Symmetry, Rhythm*, edited by Gyorgy Kepes. New York: George Braziller, 1966. Courtesy George Braziller.

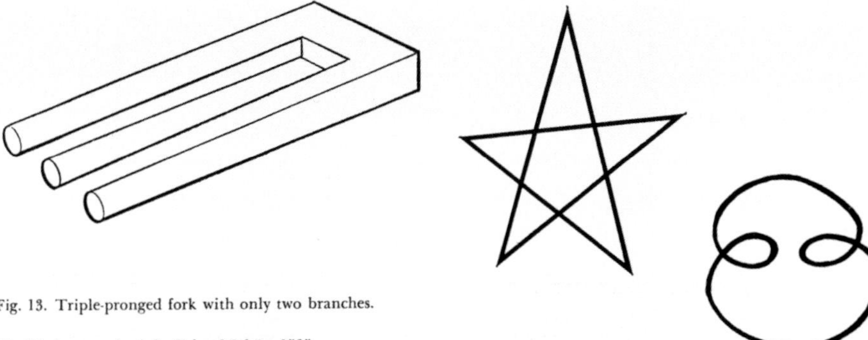

Fig. 13. Triple-pronged fork with only two branches.

Fig. 14. Anamorphosis by Erhard Schön, 1535.

Fig. 15. "Ooboo" and "Itratzky."

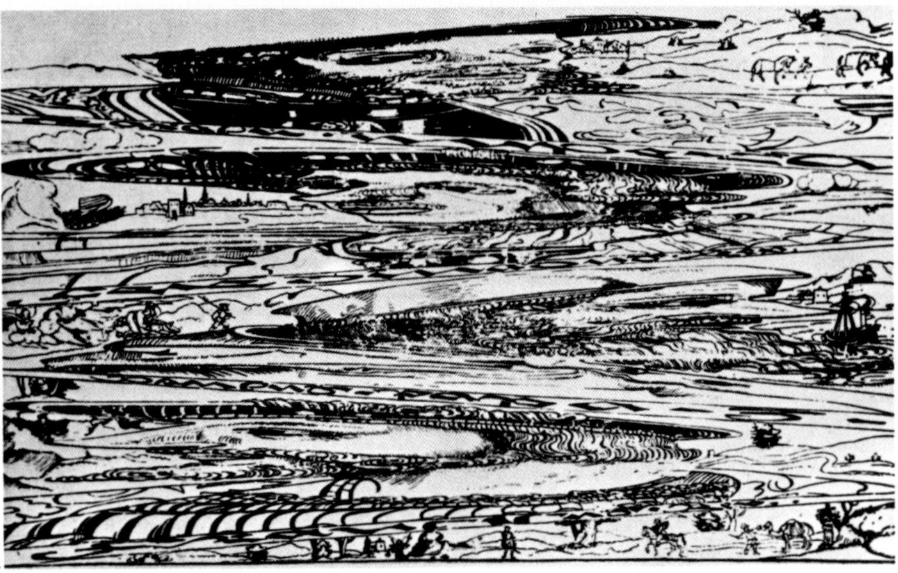

While word and image are random, the mind nonetheless creates an indelible link between verbal association and visual cue. All of these perceptual tricks demonstrate visual knowledge as phenomenological—something experienced, not just thought. Or, as von Foerster puts it, the exercises reflect the integration of the senses; the ear and the eye, the eye and the mind, all work as one: "there is 'together-knowledge,' there is *con-scientia*."[51]

Similar tricks are integrated in essays throughout the series. In a survey of kinetic art, George Rickey demonstrates movement as an optical phenomenon by including, within the flow of his text, a lattice of equally spaced lines and special viewing instructions: "stretch a thread diagonally just above this black and white striped panel here below and move it slowly up and down over the lines" (fig. 3.18).[52] Doing so produces a moiré shift. Such demonstrations indicate that the eye is a powerful source of visual knowledge.

3.17

Page from Heinz von Foerster, "From Stimulus to Symbol: The Economy of Biological Computation," from *Sign, Image, Symbol*, edited by Gyorgy Kepes. New York: George Braziller, 1966. Courtesy George Braziller.

3.18

Moiré demonstration from George Rickey, "The Morphology of Movement: A Study of Kinetic Art," from *The Nature and Art of Motion*, edited by Gyorgy Kepes. New York: George Braziller, 1965. Courtesy George Braziller.

Interthinking and Interseeing

The *Vision + Value* series was premised on teaching readers not only how to think visually, but also how to think visually between unrelated fields and across separate disciplines—to interthink and intersee. Such a process, outlined by Kepes in the vaguest of terms in the series' introductions, is not explained with any specificity; readers are left to discover how to interthink and intersee on their own.

The series encourages interthinking and interseeing even within single essays and single images; unique readings, or rather misreadings, facilitate the creation of this interdisciplinary visual knowledge. In his contribution to one of the *Vision + Value* volumes, the metallurgist Cyril Stanley Smith uses a photograph of soap bubbles to illustrate a common concept in crystallography: the "grain boundaries" that form between molecules in a crystal—the bubbles provide a visual model for a phenomenon that is otherwise invisible (fig. 3.19).[53] The two-dimensional "lines of disorder" in the soap film simulate the "planes of disorder" that form between differently aligned areas in the three-dimensional crystal.[54] The grain boundary is paradoxical, a "source of both strength and weakness."[55] It indicates disorganization within a structure but also provides an infrastructure holding the whole together, creating cohesion. Disorder creates a higher order; a line of weakness that actually strengthens the aggregate form. No doubt Kepes was captivated by the image due to its appearance—it is exquisite in its fine detail, an imperfect array of perfect bubbles—but it also serves as an intriguing metaphor for his project as a whole, and for any interdisciplinary endeavor. Separate disciplines, art and science, also form "grain boundaries." Paging through Kepes's volumes, these boundaries create confusion and disorder. But perhaps they also provide a hidden strength, a higher order, a framework for relating images and ideas. This metaphorical reading of the grain boundary is neither one Kepes imposed upon the image, nor merely my own interpretation; Smith encouraged it. "Do not these simple structures of crystals and the simpler ones of bubbles graphically illustrate some important features of the world and our appreciation of it, aesthetically as well as intellectually?"[56] Smith saw the world in a soap bubble, and asks his readers to do the same.

3.19

Image of grain boundaries in soap bubbles, from Cyril Stanley Smith, "Structure, Substructure, Superstructure," from *Structure in Art and in Science*, edited by Gyorgy Kepes. New York: George Braziller, 1965. Courtesy George Braziller.

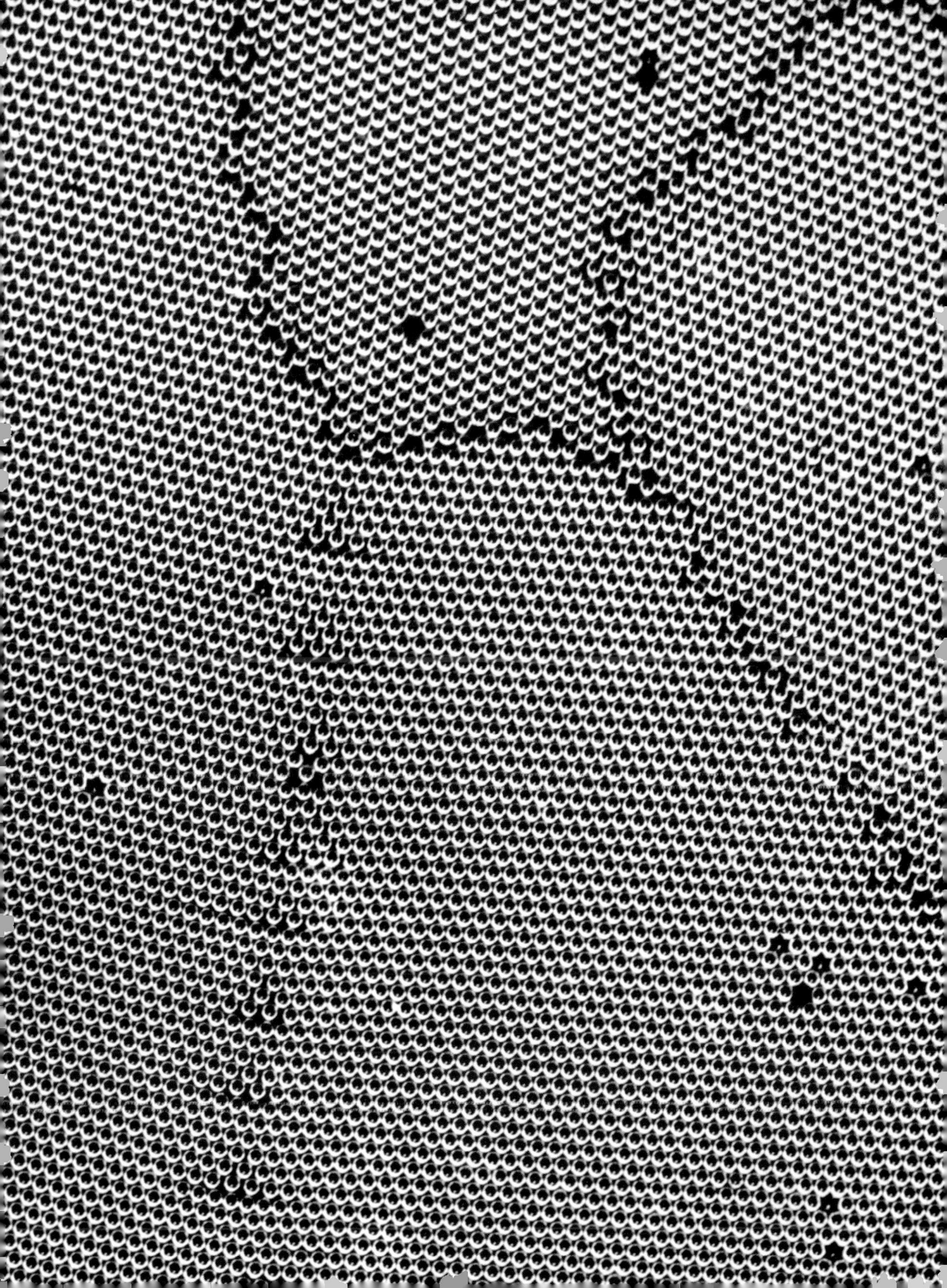

3.20

Copper-silicon alloy before and after cold-rolling, from Cyril Stanley Smith, "Structure, Substructure, Superstructure," from *Structure in Art and in Science*, edited by Gyorgy Kepes. New York: George Braziller, 1965. Courtesy George Braziller.

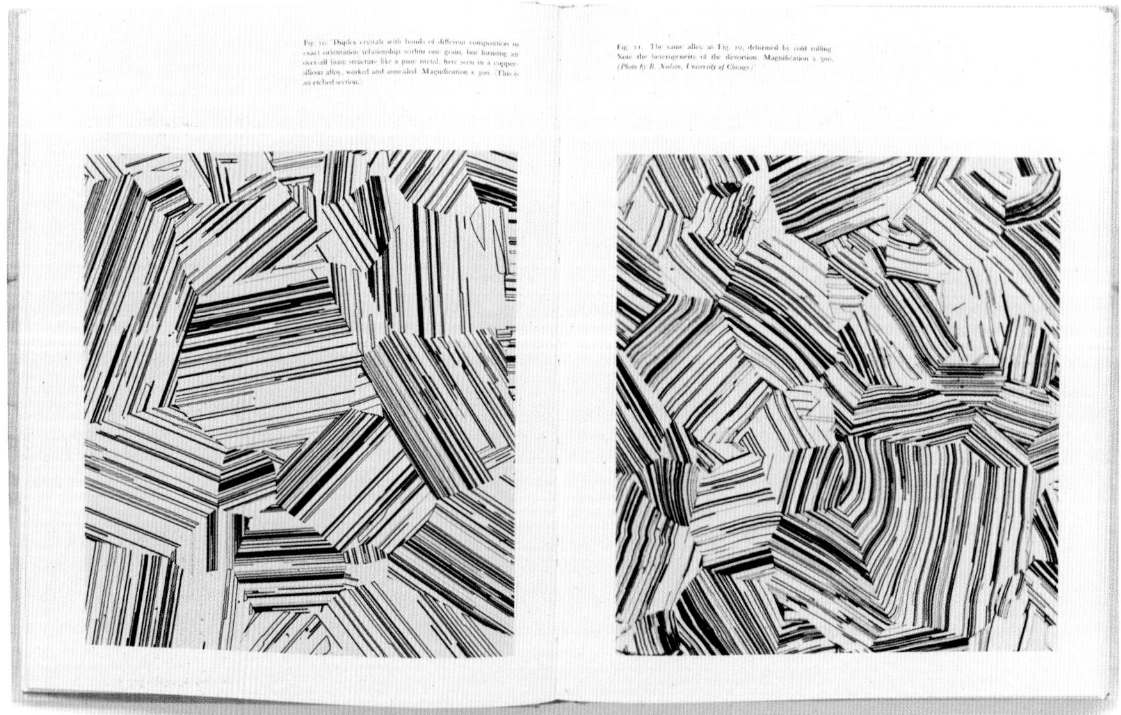

Another spread of images in Smith's essay similarly encourages this type of interthinking and interseeing. It depicts two photomicrographs of the same copper-silicon alloy under three hundred times magnification (fig. 3.20); the image on the right, however, is altered by cold-rolling, resulting in the deformation of actual grain boundaries within the metal. I do not think Kepes was interested in the specifics of metallurgy, in copper-silicon alloys or the cold-rolling process. He would have read these images against the "grain boundary" of their discipline. His fascination would have focused on the peculiar similarity yet distinct difference between the images, a montage effect created by their juxtaposition.

Creating the super-topology connecting all knowledge that Kepes imagined possible required the use of these comparative methods across multiple essays and multiple images. Let us attempt to read one *Vision + Value* volume this way. In *Module, Proportion, Symmetry, Rhythm*, mathematician Stanislaw Ulam describes two-dimensional patterns created by mathematical formulae (fig. 3.21). First, a square is extended in a series, with each new square adjacent to only one square of the previous generation. The result is a simple cruciform pattern. Modifying the formula produces variations in the resulting patterns; by adding new squares that do not touch any other existing square, even ones in future generations, a modified cruciform pattern appears, this one in the style of a Maltese cross (fig. 3.22). When extended without end, these patterns create gridlike extensions in planar space. Computer processing allowed Ulam to model such patterns for a variety of shapes—squares, triangles, hexagons (fig. 3.23). The mathematical rules become increasingly complicated, and Ulam's essay is quickly tangled in equations.

But the mathematics behind the equations is less important than the generative process Ulam illustrates a process we might interthink and intersee in a new way. A later text in the volume, by painter Richard P. Lohse, adopts similar approaches. One of Lohse's concrete paintings (fig. 3.24) shows a grid composed of four groups, each formed by "four equal right angles and four equal squares"; while the color combinations within each group are different, the pattern of their arrangement is the same, only rotated ninety degrees in each turn.[57]

Ulam's patterns and Lohse's paintings are very similar, but is there more than a superficial relationship between these texts? It is safe to assume that Ulam did not know the work of Lohse, and that Lohse had never heard of Ulam. But lines of influence are not the point; Kepes's intention was to show an interrelation between supposedly divergent fields of knowledge.

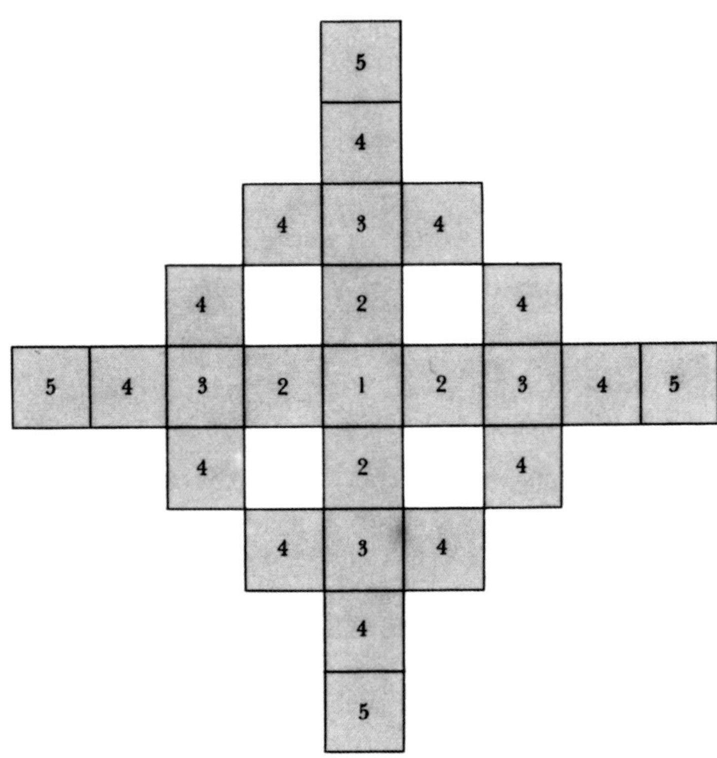

3.21

Diagram from Stanislaw Ulam, "Patterns of Growth of Figures:
Mathematical Aspects," from *Module, Proportion, Symmetry,
Rhythm*, edited by Gyorgy Kepes. New York: George Braziller, 1966.
Originally published as figure 1 (p. 216) in Stanislaw Ulam, "On some
mathematical problems connected with patterns of growth figures,"
in *Mathematical Problems in the Biological Sciences*, edited by R.
E. Bellman, Proceedings of Symposia in Applied Mathematics, XIV.
American Mathematical Society (1962), 215–224. © 1962 American
Mathematical Society.

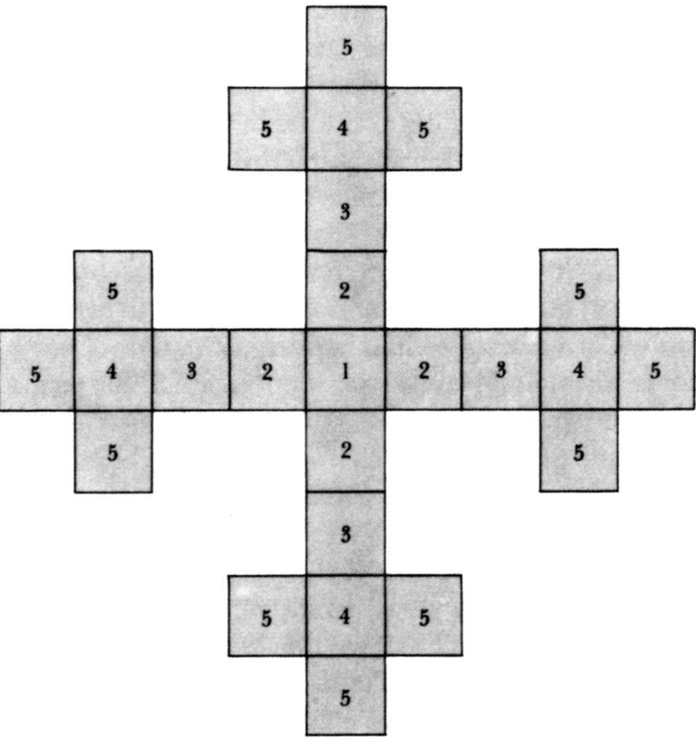

3.22

Diagram from Stanislaw Ulam, "Patterns of Growth of Figures:
Mathematical Aspects," from *Module, Proportion, Symmetry,
Rhythm*, edited by Gyorgy Kepes. New York: George Braziller, 1966.
Originally published as figure 2 (p. 217) in Stanislaw Ulam, "On some
mathematical problems connected with patterns of growth figures,"
in *Mathematical Problems in the Biological Sciences*, edited by R.
E. Bellman, Proceedings of Symposia in Applied Mathematics, XIV.
American Mathematical Society (1962), 215–224. © 1962 American
Mathematical Society.

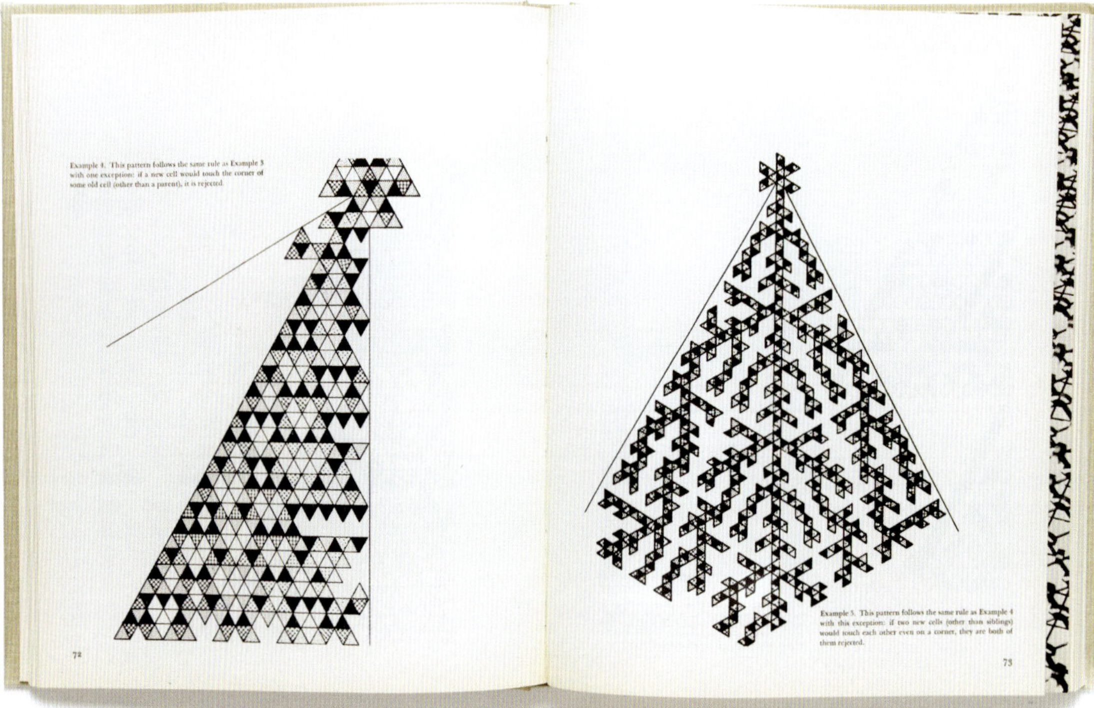

3.23

Diagrams from Stanislaw Ulam, "Patterns of Growth of Figures: Mathematical Aspects," from *Module, Proportion, Symmetry, Rhythm*, edited by Gyorgy Kepes. New York: George Braziller, 1966. Originally published as examples 4 and 5 (pp. 223–224) in Stanislaw Ulam, "On some mathematical problems connected with patterns of growth figures," in *Mathematical Problems in the Biological Sciences*, edited by R. E. Bellman, Proceedings of Symposia in Applied Mathematics, XIV. American Mathematical Society (1962), 215–224. © 1962 American Mathematical Society.

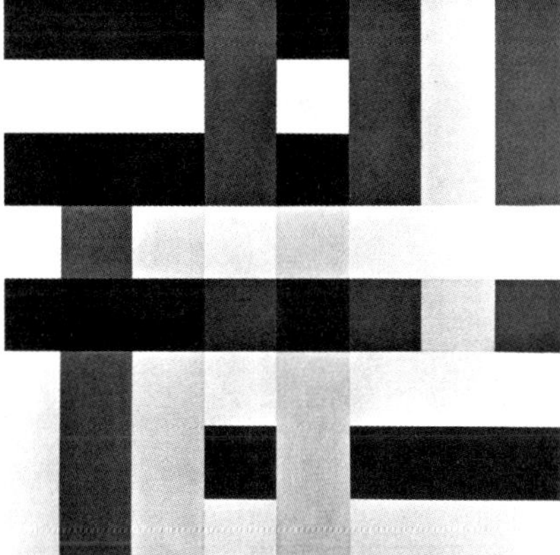

3.24

Richard P. Lohse, *Four Similar Groups*, 1965, reproduced in Lohse, "Standard, Series, Module: New Problems and Tasks of Painting," from *Module, Proportion, Symmetry, Rhythm*, edited by Gyorgy Kepes. New York: George Braziller, 1966. Courtesy George Braziller.

In the interest of creating these sites for interthinking and interseeing, a super-topology, Kepes's volumes could even encompass essays from authors who did not agree with his mission; John Cage, for one, used his contribution for *Module, Proportion, Symmetry, Rhythm* to argue against all four of the concepts embedded in the volume's title. In response to Kepes's invitation to participate, Cage explained that he "wasn't interested in any of those things."[58]

Intent on making his contribution a protest, Cage prepared a text based not on proportion or symmetry at all but on aleatory procedures. He used his 1960 work *Cartridge Music*, which employed a series of cutout shapes and transparent plastic sheets, as a device with which to compose a fragmentary narrative text. The result was a strange collage: "Take as an example of rhythm anything which seems irrelevant," reads one line; "Symmetry. Pure Symmetry. Doesn't exist," reads another.[59]

But Kepes seems not to have noticed, or to have cared, that Cage's text functioned to undermine the book's ostensible theme; he even placed Cage's remarks immediately following an essay by musicologist Ernö Lendvai that analyzes the use of the golden section—that pure proportional concept so abhorrent to Cage—in the music of Béla Bartók. Robert Smithson would also exchange angry letters with Kepes in the late 1960s, and his embrace of disorder rather than order was antithetical to Kepes's program; Kepes happily accepted Smithson's contribution to *Arts of the Environment*.

These disagreements were not just ones of methodology; they were disagreements of worldview. But such disagreements were to the point of the *Vision + Value* series. Kepes's concept of visual knowledge was extremely flexible, able to expand as needed to subsume all perspectives. His encyclopedias could include anything at all, even forms of knowledge that seemed to militate against the very logic of encyclopedic knowledge; interthinking and interseeing allowed for dissension between particular contents. The volumes are free from a certain dogma; they do not advance any particular agenda (aside from the value of vision), or, more subtly, they advance an agenda of totality, of super-topology.

These planes of disorder and grain boundaries recur throughout the series. In *Structure in Art and in Science*, the concrete painter Max Bill ends an essay with the declaration that "the special form of a work of art grows out of the general structure"; just pages later, the sculptor Richard Lippold opens an essay with an inverse declaration: "Structure is illusion."[60] *Sign, Image, Symbol* contains other conflicting contributions. A text by Ad Reinhardt uses repeating incantations to assert the purity of art:

> **The beginning in art is not the beginning.**
> **Creation in art is not creation.**
> **Nature in art is not nature.**

Art in nature is not nature.
The nature of art is not nature.
Life in art is not life.[61]

Reinhardt's manifesto, written in the hermetic style of a minimalist polemic, argued for purifying art by purging all that is not art—it was part of a series of essays that came to be known as Reinhardt's "art-as-art dogma." But the text is followed by an exuberant, if rambling, statement from the illustrator Robert Osborn that asserts much the opposite: "We require new means to make our sensations visible and, more important, *felt*. Precisely that! To show how things FEEL!"[62] By way of example, Osborn includes a series of cartoons, titled "The Hangover," illustrating the effects of excessive alcohol consumption. One spread depicts a figure suffering after a night of drinking; he is reduced to a pile of limp hair, hanging by a thread; his head becomes blocky, stepped and stacked; it inflates with air, or fills with cotton; and his mind is finally split open, the brain exposed (fig. 3.25). The images are amusing, demonstrating expressive exaggeration through caricature. They are also jarring after Reinhardt's ascetic pronouncements.

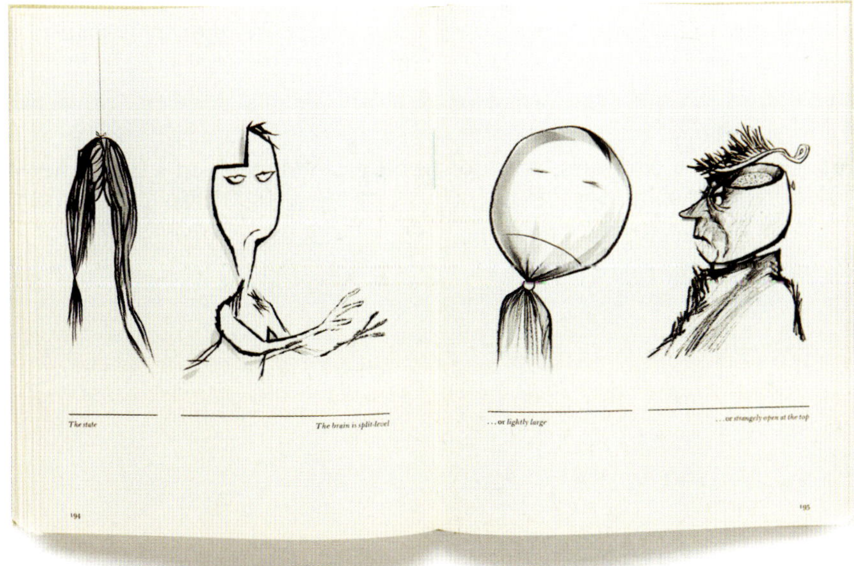

3.25

Two-page spread from Robert Osborn, "The Hangover," from *Sign, Image, Symbol*, edited by Gyorgy Kepes. New York: George Braziller, 1966. Courtesy George Braziller.

These tensions are themselves allegorized within the series by Lippold, who punctuates his text, otherwise illustrated rather conventionally with photographs of his art, with a series of full-page spreads. One presents two unprinted blank white pages (fig. 3.26). Another presents two fully saturated black pages (fig. 3.27). Between these two, Lippold includes a third spread (fig. 3.28) in a fine mesh of black and white (the image is actually a detail of a diagram he produced for one of his sculptural works). "Into the space of the white pages have filtered energetic particles of black, or one can say that the space of the black pages has fissured, and whiteness has isolated the particles of black, like a gigantic magnification." The result is, as Lippold explains, a synthesis of opposites; "both something *and* nothing; it is black *and* white, it is empty *and* full."[63]

Vision + Value imagined knowledge itself as aesthetic object; organizing it required a certain artistry. Kepes refers to such a goal throughout his introductions to the series. He cites Tommaso Campanella's utopian treatise *City of the Sun*, first published in 1623, which describes a mythical city whose walls serve as the painted pages of a vast encyclopedia, filled with mathematical figures and equations and hundreds of illustrations of precious stones, minerals, plants, animals, instruments, and technologies: a compendium of visual knowledge accessible to all who see. As Kepes explains, the city contains "a universal knowledge depicted on central walls in powerful images."[64] The implication was that the *Vision + Value* volumes would re-create this visual knowledge, projecting it onto book pages. In the introduction to *The Nature and Art of Motion*, he distills this ambition as the establishment "of a true *universitas*—a living fabric of the best knowledge of a given time." The word "universitas" is Latin for totality. "Today, more than ever, we have the opportunity to found such a *universitas*."[65]

3.26–3.28

Spreads from Richard Lippold, "Illusion as Structure," from *Structure in Art and in Science*, edited by Gyorgy Kepes. New York: George Braziller, 1965. Courtesy George Braziller.

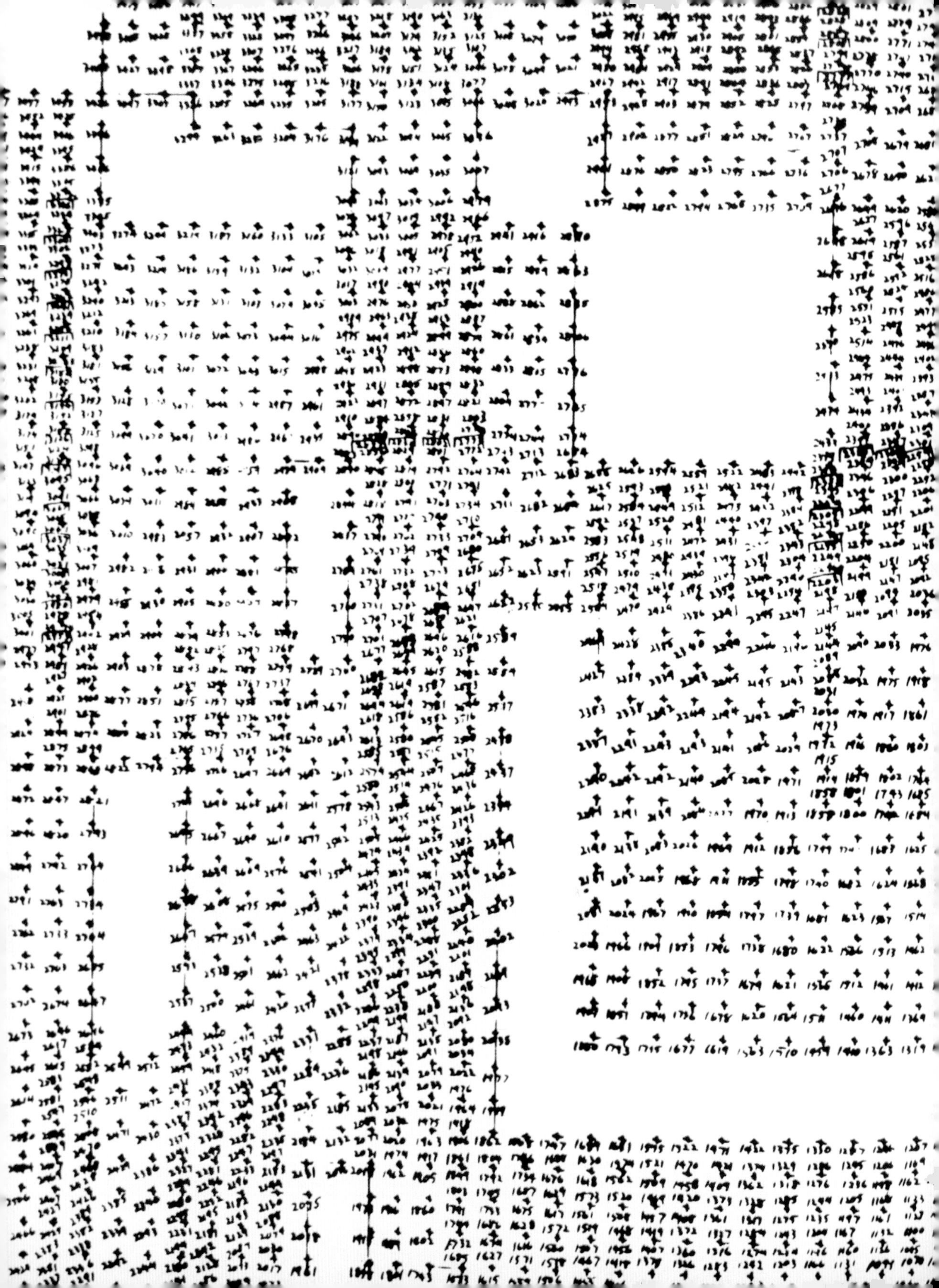

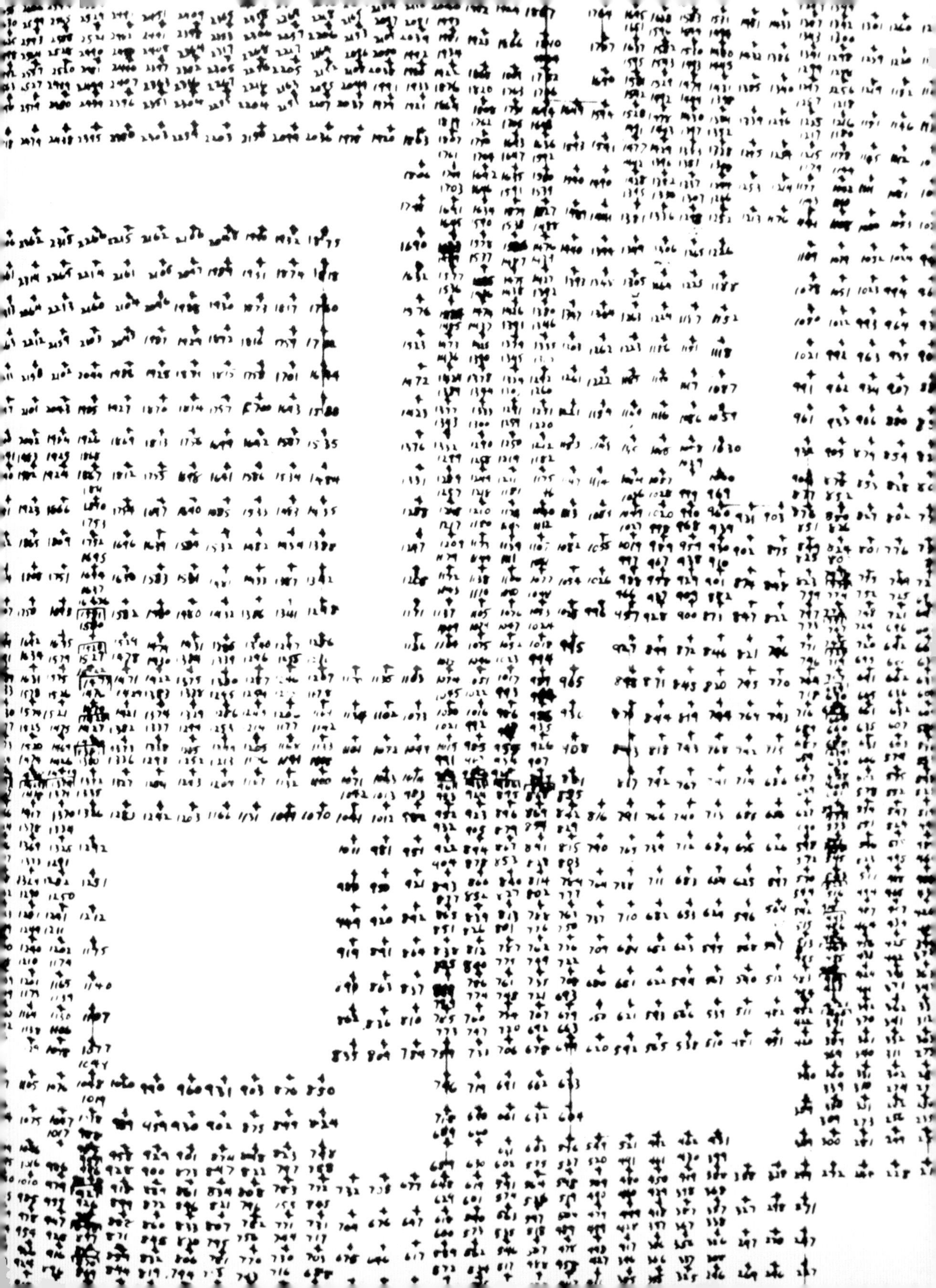

Strategies and Tactics

What does Kepes's super-topology, his interseeing and interthinking, actually accomplish? Allow me to be cynical. These methods were self-serving; they allowed Kepes to align himself with experts and elites from a wide range of fields. They allowed him to connect visual design to intellectual contexts that previously had no connection to the arts, thereby securing validation for his program through the power of other discourses. The methods achieve a form of social networking. Geoffrey Bowker refers to this process as "legitimacy exchange"—the exchange of validation from one field to another through discursive translation, a process in which the discourse of one field is made legible to another.[66] The appeal of systems thinking was its ability to create the appearance of meaningful relationships were none existed. It transferred authority between individuals, symbiotically benefiting both.

Kepes's cover designs make this mission clear: each cover includes a simple visual emblem overlaid with the names of contributors, a who's who of systems thinkers. The names are the only real contents that matter. This strategic and tactical goal was purposely self-aggrandizing, even if it was rhetorically altruistic.

The books used these methods to exchange legitimacy between surprising figures; many of the contributors to Kepes's volumes were also entangled with military power. Heinz von Foerster's essay was sponsored by the Air Force Office of Scientific Research.[67] Stanislaw Ulam's mathematical patterns were calculated on "electronic computing machines at the Los Alamos Scientific Laboratory."[68] The biographical sketch on Ulam at the end of Kepes's volume explains that his "basic ideas, along with those of Edward Teller, led to the development of the thermonuclear weapon."[69] This entanglement did not include only experts in the natural sciences. Those in the social sciences also applied their specialized skills to warfare. Abraham H. Maslow, who wrote so eloquently on human values, had also written a pioneering study of personality types predisposed to authoritarianism—a study with military applications during the Second World War and again in the Cold War.[70]

Knowledge has origins, it is used and abused, it is created for a purpose. What meaningful values—what human values, to refer to the origins of *Vision + Value* in creative altruism, in "love"—do Ulam's computerized arrangements of squares and triangles even convey, even if we ignore his other, more nefarious, interests? The production of visual knowledge about seemingly neutral themes like "proportion" and "structure" may not be so neutral after all.

And yet we need not be completely cynical. The strategic and tactical function of the books as a means of exchanging legitimacy also had a possibly humanistic, maybe even altruistic, end. Kepes was, after all, fully aware of all these connections to militarism; he makes them clear in his research notes: "offensive + defensive military potential directly related with the scientific + technological progress."[71] Another note goes further, describing a dystopian future in which such progress (he specifically cites "nuclear energy, computer[s], system[s] thinking") would produce the domination of the masses: "with further extreme development of scientific technology … coupled with dictatorial organisations [*sic*], supervision of big brother (totalitarian[ism]) the resources of the world could be manipulated, populations controlled."[72] These connections to militarism are actually intrinsic to the project's goal. Kepes hoped to shape and shift science and technology through his super-topology; his exchange of legitimacy was a form of discourse translation but also a form of discourse transformation, a means of creating a "revision of vision" by transforming war research into a form of "peace research."

A Total Picture

The *Vision + Value* series garnered very mixed reviews. Some were positive, lauding the series as a unique intellectual achievement. But even the positive reviews commented on the impenetrability of the volumes and their failure to create a total picture; the contents were too disparate, the links between images and ideas left unexplained. Others were less forgiving, calling out its legitimacy exchange as a means of achieving prestige and power without creating anything of value at all; the series was a vanity project. Even George Gaylord Simpson was skeptical about Kepes's project; he declined to participate in the series because he did not understand what Kepes was trying to do. He did not understand "interthinking," even though Kepes cites him as origin of the idea. ("I am however somewhat puzzled," Simpson writes. "I am quite skeptical as to whether concepts and canons developed in one field will have a really useful application in the other.… I must decline your invitation as inappropriate for me.")[73]

The most incisive critique of the books, however, was articulated by none other than Sibyl Moholy-Nagy, Moholy's widow (following Moholy's death, Moholy-Nagy established herself as an important architectural historian and critic).[74] Moholy-Nagy had commented on Kepes's work before. In a review of *The New Landscape in Art and Science*, she enthusiastically celebrates "the wonder with which a truly artistic eye perceives and enjoys the richness of forms and patterns." But Moholy-Nagy also raised the question of ideology:

> **Much as this reviewer admires the pictorial message of this book, its ideological message arouses a most emphatic protest. No "corrective measures" in the world will and should bridge the gap between art and science.**[75]

Moholy-Nagy attacked the way Kepes made meaning out of nothing at all, out of purely incidental surface effects—all those pseudomorphological allusions that fill *The New Landscape*.[76]

Moholy-Nagy's negative review was only the opening salvo of an argument with Kepes that lasted years. In an astonishing sequence of letters from the 1960s, she expands her critique of the "ideological message" embedded in *The New Landscape in Art and Science* into a devastating critique of *Vision + Value*, and, by extension, of any project that attempted to relate art and science in the context of the Cold War.[77] The letters respond directly to her receipt of the first three volumes of the series; Kepes had sent her complimentary copies as a collegial—and perhaps competitive—gesture. It is worth quoting from Moholy-Nagy's letter at length:

> I have been at a total loss how to react to your shipment of the 3 volumes of the Vision and Value Series, and I still don't know how to find a diplomatically correct answer....
>
> The overall idea behind the books as outlined in the introduction to <u>Education of Vision</u> is so absurdly contradictory to the actual content that I would have stopped rather then and there. What you say in this introduction is a mere paraphrasing of what Moholy said in his introduction to <u>Vision in Motion</u> and what you said in <u>Language of Vision</u>. Not that I accuse you of any plagiarism. These ideas developed logically in the 1920's and '30's, and all I am saying is that you or anybody else are harping on the same string. This is beautifully demonstrated in the fact that most of your authors are the ancient oldtimers [*sic*] from yesterday, starting with [Johannes] Itten, and that what they have to say does not vary from anything they have said over the last 30 years or more.
>
> The trouble is that this clashes stridently with your vaunted claim at "the re-integration of all aspects of our life through 20th century knowledge and power," repeated over and over in such phrases as "the new parameter of 20th century knowledge[,]" "immense vistas of science which we have failed to utilize or to share wisely," "construct a truly 20th century environment" etc etc etc. [*sic*]
>
> Now, all of this eulogizing of this century stands side by side with your highly justified denouncements of the devastations this self same century has wrought on human environment and human vision. If one would take you seriously, it would boil down to a process by which science and technology offered untold riches of social, personal and material happiness to mankind, and mankind – through failure of vision – failed to accept and use these riches. You then call for the old old [*sic*] moral panacea of personal salvation, man-to-man morality, the inner light and everything else that has been mixed up with art, design and architecture since Ruskin started the pernicious confusion of morality and art.[78]

This remarkable letter responded in part to personal grievances—Moholy-Nagy felt, as she puts it, "aggravated by personal resentment."[79] She was blindsided; not only had Kepes failed to stay in touch with her, he failed to ask her to contribute to his series, despite their shared interests.[80] This oversight reminded her of Kepes's abrupt departure from the School of Design many years ago.[81]

But her "personal resentment" was also professional, related to what she perceived as Kepes's unacknowledged appropriation of Moholy's ideas. While she claims not to accuse Kepes of plagiarism ("Not that I accuse you of any plagiarism"—the remark reveals a clever use of apophasis, a rhetorical device that allows Moholy-Nagy to make the accusation through insinuation while simultaneously withdrawing it), she does suggest that his project is, at a minimum, derivative, a paradigm repetition of the Bauhaus, but one without any meaningful change in its program in response to the altered status of science and technology postwar. "It seems to me that you have copied a very old pamphlet, and then enlisted the support of every oldtimer [*sic*] who, rightly in his own day, was an honest insurgent."[82] This criticism of Kepes as the tragic epigone of Moholy, condemned to repeat the master's work, was perhaps the most enduring critique Kepes faced across his career; Moholy-Nagy had made this very same comment behind Kepes's back nearly twenty years earlier.[83] But now she extends the argument further, critiquing Kepes's program not only as derivative, but also as anachronistic. The *Vision + Value* series fetishized the visual as a means of cultivating values at a time when visual culture, from newspapers to magazines and especially television, had already displaced the written word and saturated contemporary life with a blur of images. "Aren't you tilting against the <u>long long vanished</u> enemies of Visual Education: books, words, figures, that exercised the progressive teachers from Comenius to Moholy? The prime characteristic of our age is a total denunciation of words and a mania for PICTURES. Words have become completely devaluated. People neither read nor write, they LOOK," she explains.[84] In this way, Kepes's project seemed not only late on arrival in its art-historical lineage but also reactionary in its content, fighting battles for the education of vision at a moment when visual thinking was ubiquitous, the standard means of navigating a society of images; educating people to read and write, not to look, was the real challenge of the time. Teaching students to think, not to see, should have been Kepes's priority.

But Moholy-Nagy's criticism extends still further, beyond the personal and professional, back to the ideological. For her, Kepes's project was academic in the worst way: too bookish, too pedantic, too removed from the actual experience of science and technology, especially at a moment when both were intertwined with military power. *Vision + Value* was an escapist fantasy, a retreat into the esoteric and arcane, into "book puzzles," as she derisively calls them, into the depths of the library, into useless visual knowledge, old old knowledge, from long long ago.[85] She continues:

> The most telling sentence in your whole introduction is: "We cannot renounce the new scientific efforts and technological achievements of the 20th century because they were bought by human distress."
>
> Even if one gives you the benefit of semantic doubt and assumes that you meant <u>with</u> human distress instead of <u>by</u> human distress, the involuntary joke is considerable. BY human distress would mean what actually did take place: that this cursed century made mankind so desperate that they were and are willing to buy all the fake progress offered by science and technology…. "Bought WITH human distress" which I hope you meant might at best indicate that science and technology have sacrificed all human values for a fiercely romantic chase into nothingness: moon shots; computerized education, love life, poetry etc.[86]

The romanticism of Kepes's project is its redeeming quality—Kepes's fascination with the three-prong fork with two branches and the grain boundaries between soap bubbles is what makes his project interesting. To Moholy-Nagy, such fascination was a retreat, a fall into the abyss, a plunge into obsessions with irrelevant visual facts. The semantic distinction between scientific and technological advancement "bought by" human distress and "bought with" human distress was intended to indicate that Kepes actually supported an advancement in knowledge linked to human suffering—"an almost impenetrable chaos of regressive progress," as she puts it, in a very Marcusean turn of phrase.[87] Moreover, Kepes's actual position on science and technology was nonsensical. Unlike "the impressively imbecile doggedness of Buckminster-Fuller or [Konrad] Wachsmann or [Christopher] Alexander"—the technocratic architects of the era (and all participants in Kepes's projects; both Fuller and Alexander wrote for *Vision + Value*, and an iconic image by Wachsmann of a structural system appears in *The New Landscape in Art and Science*)—Kepes was of two minds.[88] He was both in awe and in fear of science and technology, and his attempt to reconcile these positions with visual tricks appeared to Moholy-Nagy as plainly ridiculous. Whereas other postwar figures embraced science and technology without reservation, or rejected both with conviction, Kepes seemed ambivalent, and Moholy-Nagy pressures this equivocation as akin to a more damning moral ambivalence about the role of the education of vision at institutions like MIT. Her letter crescendos to a final critique that brings us back to that crucial, inescapable, unavoidable issue of instrumentality that looms over all of Kepes's work:

> But how can you reconcile this kotow [*sic*] before the 20th century plunge into self-annihilation, and then fill 3 books with all the antiquated totally personal, totally "prima donna" observations on <u>CREATIVITY</u>. Christ, if only there were a way of putting an embargo on this word which you bandy around as any other EDUCATOR. If anything, this is an age of analysis and tabulation. Is this why we have to rub the old lamp from the Third Avenue Antique Store to conjure up the geni [*sic*] of "Creative art"?

> But even if it made sense to worship MIT while denouncing what its graduates have done to society, isn't the whole premise of your books completely obsolete? … Does it not occur to you that the computer boys of MIT are playing a game of words and figures that could kill every last instinct for sense values if it were to conquer the territories it so naively invades now? …
>
> I am sure the books will sell like hotcakes because teachers and designers are so shellshocked by the accusation of "renouncing the new scientific efforts and technological achievements" that the compromise between verbalization and ancient ancient [*sic*] visual solutions you offer will be a welcome straw. No matter how many new puzzles you publish, Gyuri, they will never make a total picture that is a perspective in the future. What you need more than anything else is to get out of MIT and Cambridge as fast as you can to save your soul and your talents and your integrity. You have no idea how silly and 19th century deterministic their programs look from the outside. –
>
> There can be no "integration." We can only be saved by drawing new boundaries.…
>
> Best regards and never mind. I did my best to meet your challenge purely for old times sake [*sic*], although I could ill afford the time. But 33 years is a long time and I still hope for your disenchantment with the Bauhaus program.[89]

Moholy-Nagy's letter must have stung. I can only imagine that Kepes was hurt.[90] I do not think it is a coincidence that he failed to write introductions for the next three *Vision + Value* volumes, published the following year. Moholy-Nagy tested his faith. He faltered, his belief wavered.

But his response to her was still surprisingly measured and muted, reflecting his characteristically mild manner and an overpowering desire to avoid all conflict: "I appreciate the time you took to comment on my books." "I cannot help but respect your frankness." "The only thing I regret is that some unintended carelessness on my side may have caused some personal resentment."[91] He refuses to engage Moholy-Nagy in a line-by-line confrontation. "I will not try to defend the books or get into verbal duels."[92] But he did affirm a continued commitment to the Bauhaus idea—even some three decades after the Bauhaus had closed, even in the dramatically changed circumstances of the Cold War:

> And indeed, you ended your letter by saying that "33 years is a long time," and you hope for my "disenchantment with the Bauhaus program." I don't exactly know what you mean by "Bauhaus program" – the program of Gropius? Klee? Kandinsky? Moholy? Albers? As I understand it, the Bauhaus stood for the effort to juxtapose centripetal forces against the centrifugal forces that ran rampant early in this century. In this sense, I am for the Bauhaus program, though with the understanding that conditions have changed, problems are different, and therefore the techniques of collaboration have to be re-thought and restructured. Please don't hold my having beliefs against me (or against anyone), because I feel that belief fused with knowledge is our only hope of getting out of this sophisticated mess we have gotten into. Once more, inspite [*sic*] of everything, I <u>would</u> like to see you again very soon. Many thanks for your letter.[93]

Visual fundamentals

motion / futurist approach

"Knowledge" in their (futurist) mentality language, was the subjective response to an inward or "centripetal" tendency of the object which draws the parts together to constitute a whole. "Apparition" was the manifestation of the diffusive or "centrifugal" tendency which causes the object to resolve itself into emanations whose nature is determined by the energy of surrounding objects." Looking back at futurism

R. Trillo Clough

the dynamic figure + ground relationship mostly in terms of magnetic electric energies attraction – repulsion – tension. Their goal was a fusion of object + milieu their terms object milieu corresponding to space time.

3.29

Gyorgy Kepes, untitled and undated drawing.
Gyorgy Kepes papers (M1796). Dept. of
Special Collections and University Archives,
Stanford Libraries, Stanford, Calif. © The Estate
of Gyorgy Kepes.

For Kepes, "centripetal" force—the desperate search for a center anchoring a world of chaos, a vital center (to call forth the 1949 book by Arthur M. Schlesinger, Jr.)—was the core of the "Bauhaus program" and thus the core of his own mission, the program of Kepes.[94] His approach was constructive; he hoped to build new ways of seeing, not to break old ones. He explains: "in spite of all contrary signs, some of the twists in our life can be straightened out by pulling constructive efforts together"—rather than pulling them further apart.[95] This formulation—balancing outward forces with inward ones, turning inward toward a center—recalls the emblem embossed on the cover of his *New Landscape* book, composed of two bands from an oscilloscope circling together toward a stable midpoint (fig. 2.11). It also recalls drawings throughout Kepes's papers that fixate obsessively and compulsively on finding a visual center (fig. 3.29). In a notation, he writes: "not centrifugal—acquisitive, but centripetal—integrative. Looking for connectedness."[96] Another note drafted in preparation of the *Vision + Value* volumes: "today centrifugal force tearing apart, growth uncontrolled, scary.... Need centripetal forces—new meeting points, joints, anchors, structural pattern."[97]

But things do fall apart; the center cannot hold; mere anarchy is loosed upon the world. No joints and anchors, no new meeting points, no super-topology can prevent the accelerating spiral of knowledge from expanding, extending, tearing ever outward. Allow me to fast-forward, previewing the context we will more fully investigate in the fifth chapter of this study. By 1968, the Cold War had turned hot. Students and faculty on campuses across the country had started connecting the dots, tracing the outlines of a new constellation, one that Kepes, despite his agile eye, refused to see: the military-industrial-aesthetic complex.

In February of that year, weeks after the Tet Offensive and weeks before the My Lai massacre—events that made plain the utter terror of the war in Vietnam—Kepes sent Moholy-Nagy another letter. It was an invitation to a symposium honoring the establishment of his Center for Advanced Visual Studies (CAVS), the research institute he founded to "integrate"—and I use here a term Moholy-Nagy mocks—to "integrate" the arts and sciences through collaboration. The Center was the culmination of Kepes's career—the "university of vision" and "research lab" he had imagined decades prior, or the *universitas* he had more recently put forth in *Vision + Value*. The very name of Kepes's new institution—it was a "Center," after all—announces a concern for centripetal as opposed to centrifugal force: for stability, not chaos. After receiving Kepes's invitation, Moholy-Nagy confirmed her attendance but also reiterated her reluctance to endorse Kepes's project in any way. This time she was blunt, explaining her animus in concrete terms. The war had entrapped Kepes.[98] The futility of the education of vision was now obvious; it served, at best, as a compensatory gesture intended to balance a university that poured most of its resources into weapons and war. It offered humanistic gilding for an otherwise inhumane institution. It provided justification, and thus accommodation, for MIT's military investments. Moholy-Nagy writes:

> I shall come to Cambridge on Thursday and Friday to attend the sessions. I do this out of an honest hope to understand perhaps by personal confrontation what escapes me so completely in written statements – this whole homogenization of art and technology, or rather of designed environment and scientific experimentation. That vehement protest I sent you some years ago (1965?) when I received the first 3 volumes of your Braziller series would hardly be written by me today – not because I KNOW differently but because the dehumanization and impending annihilation of American civilization seems to furnish proof for my life thesis that is beyond argument. The staggering guilt of Vietnam lies ultimately with the ideologists of technological progress, the way the ultimate guilt of Hitlerism lay with Nietzsche and the ideologists of the Germanic super race. – Science and Technology live by permission of the military establishment. This is the way I feel now; I do hope that the various offerings at the symposia will teach me differently – that the miracle will occur that reconciles what MIT is doing for the war effort with your vision of creativity, that "the reintegration of all aspects of our life with the new parameter of 20th century knowledge and power" – as you wrote – will be free of compromise. In any case – I shall come to listen and to learn – hopefully.[99]

Kepes responded with forced pleasure: "I am delighted. . . . I wish we could make such an impressively convincing presentation that you would reconsider your own frame of reference."[100] But his reply would be the last in what was already a very long exchange of letters. The correspondence ends. We can only assume that Kepes's "vision of creativity" ultimately failed to persuade; that the "frame of reference" did not shift; that no new miracle occurred.[101]

Hitler's Revenge

This long epistolary excursus is but one private exchange in what soon became a very public debate over the legacy of the Bauhaus. Moholy-Nagy's letters to Kepes anticipate an essay she penned that year for *Art in America*.[102] Titled "Hitler's Revenge" and dated 1968, the article condemned Marcel Breuer's plans for a skyscraper designed to rise over the Beaux-Arts Grand Central Terminal in Manhattan (after a legal battle that went before the Supreme Court, the skyscraper remained unbuilt). The article suggests that Germany's architectural influence was tainted by a latent totalitarianism, poisoned by a few corrupt individuals who would betray the modernist cause with megalomaniacal, and thus essentially fascistic, architectural distortions. They would offer Hitler's revenge on the United States.[103]

Moholy-Nagy directs her critique not only against Breuer and his tower, but also against Walter Gropius and Mies van der Rohe—all of whom she names at the outset of the essay, and all of whom, along with Moholy and Moholy-Nagy herself, brought the Bauhaus idea to the United States. But Moholy-Nagy also implicates Kepes in this circle. She describes how "Harvard, M.I.T. and the Illinois Institute of Technology established through their European design teachers a totally new curriculum" that indoctrinated American students in the Bauhaus program. By citing Kepes's institution of higher learning, and not only those occupied by Gropius and Mies, she aligns him with the problem.[104]

Moholy-Nagy's provocation, however, was her argument—wild with literary flourish—that the Bauhaus was doubly flawed, that it was corrupt because of its fascism but also its capitalism; she connects Breuer's building plans for Grand Central to the money of corporate elites (a "profit dictatorship") while also suggesting that the building would glorify scientific and technological power.[105] She describes a dystopian future in which "the emerging generation of mega-structure functionalists will want to honor their ancestor by using his masterpiece"—she means the as yet unbuilt Breuer tower looming menacingly atop Grand Central Terminal—"as foundation for a High Technology Center of Computerized Existence."[106] This description of a Center functions as another veiled reference to Kepes, recalling the Center for Advanced Visual Studies that she had recently visited, and the religious allusions bring to mind her barbed remark to Kepes that he was too busy "worshiping MIT" to notice MIT's real role in the Cold War (the scene she sets is also ironic considering that Kepes's notes include his own private musings about dystopian futures). With biting wit, she continues her projection of a postapocalyptic world, one that would be destroyed by science and technology, as epitomized by the hydrogen bomb: "Above that the ape men, returning after the hydrogenic holocaust, might want to worship the divine slabs salvaged from the set of *2001*. And in the zenith of heaven will float the dazzling satellite of a Gold Medal, 'highest award of architectural excellence,' which falls automatically, like an oxygen mask, from the Parnassus of the American Institute of Architects whenever hardening of conceptual arteries and gross office income have reached genius level."[107]

The essay shocked other members of the Bauhaus and New Bauhaus diaspora. Gropius was "appalled by Sibyl's impertinence," as he wrote in a letter to Robert Jay Wolff.[108] In his reply to Gropius, Wolff even refers to Moholy-Nagy as "Moholy's greatest mistake."[109] In an inscription added to his letter *ex post facto*, intended for discovery in posterity, Wolff describes the exchange as a response to "a disrespectful article by Sybil [*sic*] in which she referred to Marcel Breuer's project for a building over Grand Central Station in New York as reminiscent of some of Hitler's public building schemes"—though this is an inaccurate description of her essay, which does not refer to Nazi building plans as such, but rather to a more essential dictatorial control over space, history, and human life that she claims Breuer, Gropius, and Mies—and Kepes—promoted.[110]

The tensions disclosed in these letters are personal and professional, the result of long-held resentments, but they also reveal how the legacy of the Bauhaus became a site for a more significant ideological battle. Moholy-Nagy defies expectations that she be an obedient defender of her husband's legacy and an observant Bauhaus apologist; she savages that legacy, or at least what she sees as its perversion at the hands of Moholy's colleagues. To her, the Bauhaus idea, through its alignment with science and technology, and the alignment of science and technology with military power, had become debased. The relentlessness of her attack reminds us that collaboration is easily compromised, that every interdisciplinary exchange involves troubling concessions.

•••

Visionaries are dangerous. They see too much, or maybe too little—not what they are supposed to see—and for this they are often fated to suffer torture and abuse. Their apparitions inspire intense scorn. Consider Tommaso Campanella, the late-Renaissance thinker whom Kepes so admired for his spectacular dream of a universal visual knowledge projected on the walls of the utopia he imagines in *City of the Sun*: Campanella was denounced for heresy during the Inquisition and thrown into jail for nearly thirty years; he was tortured on the rack because of his heterodox views, and because of his embrace of esoteric traditions like astrology and magic. Campanella spent these years of imprisonment furiously writing in an attempt to assemble a complete encyclopedia spanning all fields of knowledge. Or, for that matter, consider Kepes: he is like the medieval mystic who sees both too much and too little. He, too, spent his years at MIT assembling an encyclopedia spanning all visual knowledge. He, too, referred to esoteric traditions—esoteric, at least, by the conventions of the art world in the 1960s, as suggested by the unexpected quotations from the likes of Campanella, the writings of Ludwig von Bertalanffy and Heinz von Foerster, the images of soap bubbles and anamorphic woodcuts. He also was interested in unusual methods, methods with a certain magic, like interthinking and interseeing. He saw the world in a way that often inspired confusion, if not outright condemnation. His exchange of letters with Moholy-Nagy suggests to me the type of abuse faced by those who embrace a dangerous vision, who see with excessive imagination. The sincerity of Kepes's conviction is the greatest virtue of his project, just as it is the most obvious flaw. We must remain skeptical of these convictions and the institutional agendas they supported—but we must also consider the possible value visual study offered.

The exchange of letters takes me back to a single spread from Kepes's *New Landscape in Art and Science*, one that surprised me when I first encountered it among the intricate patterns and puzzles (fig. 3.30). It includes a detail of Christ's crown of thorns from Mathias Grünewald's *Tauberbischofsheim Altarpiece*, created in the early sixteenth century. Kepes presents the detail alongside famous prints by Leonardo da Vinci and Albrecht Dürer that depict elaborate labyrinths created by a knotted thread.

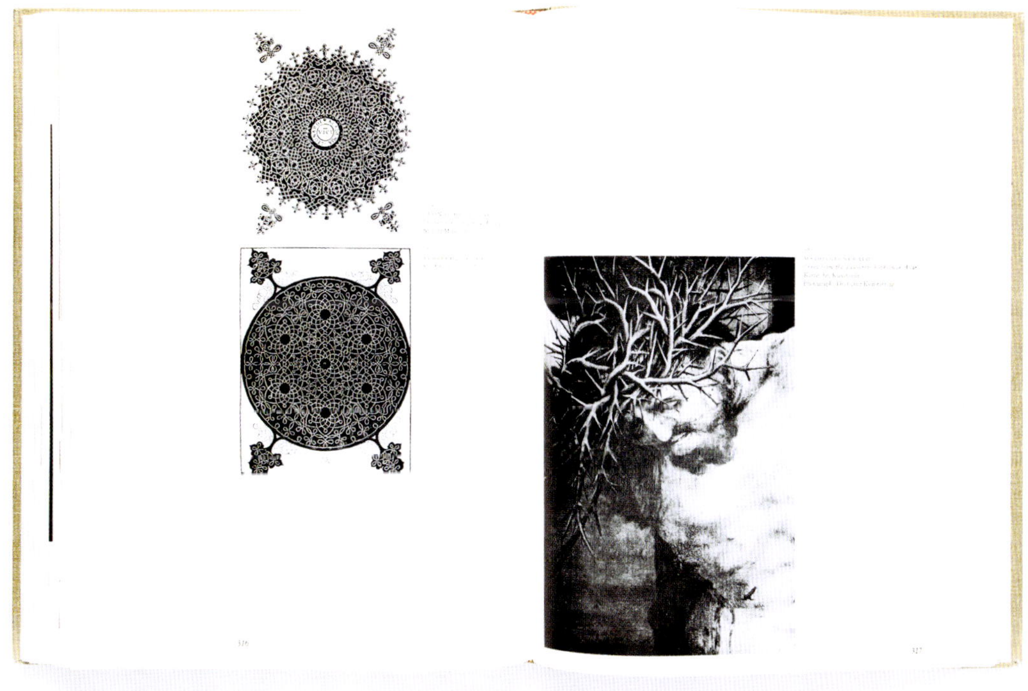

3.30

Two-page spread from Gyorgy Kepes,
The New Landscape in Art and Science.
Chicago: Paul Theobald, 1956. © The
Estate of Gyorgy Kepes.

The pseudomorphosis, Kepes's favorite aesthetic tool, and one he uses across *The New Landscape* and the *Vision + Value* volumes, here permits him to include among his scientific images what would otherwise have been an out-of-place—embarrassing, really—reference to religious iconography. Such themes are embodied in the image of Christ but also in Leonardo's and Dürer's prints, which abstractly render celestial maps of God's universe as an endless sequence, a Great Chain of Being.[111] Kepes's visual quotation also refers us back to an image of total abjection. The full panel from Grünewald's altarpiece depicts Christ's crucifixion: gangrenous flesh pockmarked by thorns, feet grossly deformed, fingers distorted, ribcage distended—but for believers, this visceral depiction of Christ's earthly ruination, seen in in the dark of night, serves to reveal what remains unseen, a heavenly light. In his accompanying text in *The New Landscape*, Kepes describes this opposition of light and dark—one that stands in for all the other oppositions, for art and for science. He refers to Grünewald's altarpieces (specifically the Isenheim altarpiece) as "the most despairing of all Crucifixions, the most triumphant of all Resurrections."[112]

I am encouraged by these religious references—by the explicit ones leveled disparagingly by Moholy-Nagy against Kepes, and the hidden ones that Kepes himself buried in his books—to see Gyorgy Kepes as a martyred figure. Part of what makes him compelling is his willingness to endure all this abuse. He was principled, but not in the way many felt he should have been. While other artists would have taken the easier path of denouncing MIT's technocratic culture, of condemning the military-industrial-aesthetic complex, Kepes instead remained part of that culture, even while he resented what it represented. Ultimately, he was lost in a labyrinth; he was entrapped, entangled in the knots, unable to straighten out all the twists. He was caught. He hoped—foolishly, naively, maybe admirably—to change MIT through vision and value alone. For his faith, and for all the compromise it required, he withstood an onslaught of vitriol. It did not alter his devotion to a knowledge that only vision can provide.

Darkness into Light

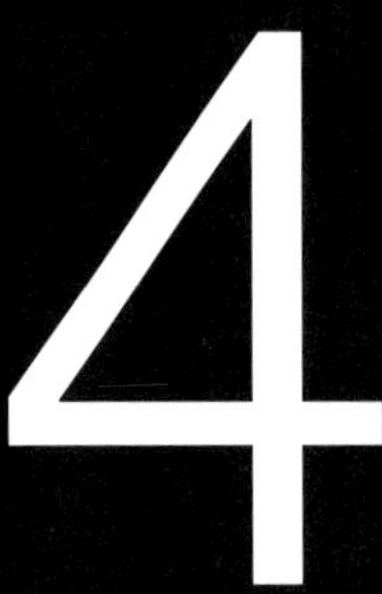

Fiat Lux

In a two-page typed fragment from the manuscripts Gyorgy Kepes prepared for what he called "The Light Book," the sprawling research project on light that preoccupied him for decades but which remained unfinished and ultimately unpublished, Kepes describes the most magnificent and most terrifying work of art ever created: the explosion of the nuclear bomb. Upon detonation, the bomb used fission to split atoms (in the case of the atomic bomb) or fusion to join them (in the case of the hydrogen or "thermonuclear" bomb, so named for the immense thermal energy required to start the process), thus initiating a chain reaction that created, in Kepes's words, "gigantic destructive luminous wonders." These spectacles of light far surpassed conventional forms of aesthetic experience in their brilliance. The noiseless flash from a nuclear blast—first there was light, then heat and radiation—was blinding. It would dazzle retinas even as it destroyed worlds. Unlike the "fire storms created by incendiary bombs at the last war," those "that may be unleashed in a future thermonuclear war" would rival nature in their power of illumination. Such fires "could flare up into a conflagration so tremendous in scale that it may compete with a meteorological event," generating its own wind and weather: the clouds of radioactive particles that would rain down in a radius of hundreds of miles from ground zero. A future world war fought with nuclear weapons would create hundreds of these "apocalyptic sunsets," each more spectacular than the actual setting sun.[2]

For Kepes, as for many others, the bomb was the quintessential symbol of midcentury science and technology, one that attained near-religious significance.[3] But its meanings were highly conflicted. It was a discordant image—an icon both sacred and profane, to be worshiped out of respect but also contempt. The bomb symbolized scientific thought and technological rationality, the pinnacle on the arc of progress; the bomb symbolized thoughtless barbarism and brutality, a regressive, out-of-control technics.[4] It was the greatest human accomplishment, an instrument of peace, but only because it could annihilate the greatest number of humans—millions, maybe billions—in one horrible instant. In these ways, the bomb perfectly epitomized the contradictions that characterize the halting movement of the dialectic of enlightenment, a point Kepes makes clear in a concise inscription in blue ink at the top margin of his typescript: "each step has Janus face."[5]

This esoteric reference to Janus was fitting; the Roman god with two faces might be something like the bomb's patron saint. Janus looked both forward and backward, and therefore presided over transitions, beginnings and ends, and both war and peace—just as the bomb presided over the Cold War and its duplicitous state of conflict. Kepes expands on the Janus-faced quality of science and technology in his text: "Our run-away technology has both constructive and destructive aspects. Our potent new illumination tools have their still more potent destructive by-products." Despite the creative possibilities that science and technology might offer the arts, there was still this threat; "parallel with our amazing affluence of brightness," there was the "menacing face of our light-creating technical know-how."[6]

I open with this example, one of many in the hundreds of typed pages that comprise Kepes's Light Book manuscripts, to demonstrate how Kepes motivated the seemingly elusive quality of light, charging it with meaning and significance. Kepes's consideration of "a future thermonuclear war" as an aesthetic event worthy of artistic consideration dramatizes the very real stakes of his project: light is poetry, sure, but it is also politics. It is ideological. It functions as a side effect or by-product of science and technology, a consequence in both a literal and figurative sense; new technologies like the bomb actually emit light as artificial illumination, while the sciences metaphorically evoke the concept as a representation of knowledge—a resonance evident in common notions like "the light of reason" or the grand project of "enlightenment." Light therefore also manifests—renders visible—the profound anxieties over science and technology that marked the Cold War. Light reflects and refracts these cultural conflicts as a uniquely visual phenomenon. By studying light, then, even the artist might participate in such debates.[7]

Kepes illuminates these points in his typescript. "Light, the begetter of vision, could become the blinder of vision," he writes, suggesting that the bomb's ideological ambiguity, its "constructive and destructive aspects," are given visual form in the light it releases. As evidence he cites an example from the Pacific Proving Grounds, where the United States military conducted extensive nuclear testing from 1946 to 1962. "Experiments during the atomic bomb tests in the Pacific above Johnston Island in 1958 showed that burns to the retina could occur as far away as 350 statute-miles."[8] Kepes refers to the detonation, on 1 August 1958, of the Teak shot, one of a series of tests undertaken that spring and summer as part of Operation Hardtack. The trial was one of the more disturbing nuclear weapons effects tests of the era. A Redstone rocket sent a 3.8-megaton thermonuclear device nearly fifty miles into the atmosphere above Johnston Island—essentially to the edge of outer space—where it was then discharged. The purpose of the Hardtack Teak test was to study the effects of high-altitude atmospheric explosions, which might have defensive or offensive roles in a nuclear war.[9]

A secondary purpose of Hardtack Teak was to study the bomb's light, and it was this goal that interested Kepes. When the device went off, minutes before midnight, the dark of night was erased, as if it were midday—an effect intensified by a frightening malfunction that accidentally caused the device to detonate directly over Johnston, at the correct altitude but the wrong latitude, rather than at its planned target six miles south of the island. It was as if the sun had fallen to the earth.

Caged rabbits onboard naval vessels and in airborne aircraft were intentionally exposed to the flash, their eyes forced open in the direction of the blast by a metal apparatus. Permanent burns to the rabbits' retinas due to light exposure occurred as far away as approximately 350 nautical miles from hypocenter.[10] Kepes's characterization was correct: Hardtack Teak proved that light was not only a "begetter of vision" but also a "blinder of vision" (his original choice of words was more accurate, if overblown; he called light a "murderer" but later crossed it out). Kepes's remarks also call to mind John Hersey's description, in his 1946 account *Hiroshima*, of the photographic quality of the nuclear blast, which recorded the physical world in reverse, searing permanent shadows of structures and figures into the ground as if the earth were light-sensitive photographic paper.[11] Kepes's two-page typescript was subtitled "Tools," suggesting how the bomb might be understood as an artistic tool, like the paintbrush or chisel, but one that shaped not pigment or stone but atomic structure, all in one flash of light.

For Kepes, then, negative and positive were one and the same, a dialectical unity. The light of the bomb symbolized the sheer terror of science and technology, their capacity to annihilate, but also their unrealized aesthetic potential. Kepes therefore concludes his dark text, concerned as it is with nuclear holocaust, on a hopeful note: "But in spite of all these fears of apocalyptic events—fire balls that may carry tons of vaporized material from the ground upward into the air, carrying fission products imbedded in particles of grit—light could serve its great role in human life." If wielded for artistic ends, new forms of light—perhaps even the light from a thermonuclear device—might offer a form of redemption, an absolution from the original sin of Hiroshima. As if to emphasize these quasi-religious, quasi-spiritual intentions, Kepes titled a related typescript from his Light Book manuscripts "Fiat Lux," the Latinate version of God's commandment in Genesis 1:3: "Let there be light."[12] The dissonance between Kepes's declaration and the precarious state of the world is to the point. Kepes is optimistic in the face of consummate pessimism. He sees light even in darkness. He puts it plainly in a notation penciled between typed lines: "light has immense promise."[13]

This chapter explores Kepes's idea of redemption through light as put forth in an important but inscrutable project: the Light Book. The project exists today as textual and visual fragments—a work not unlike Walter Benjamin's Arcades Project in its scale and scope, but also in its confusion.[14] Nonetheless, the Light Book is still the most elaborate aspect of a larger discourse on light that motivated many of Kepes's projects. This chapter attempts a speculative reconstruction of selected portions of the Light Book, but also considers how this discourse developed across Kepes's career. In a set of preparatory notes for a lecture, he explains: "The central theme … in all my work is to light in all its actual + potential creative use. Convinced that light as creative tool has the most significant contribution to make." For Kepes, light is "the key to [the] visible world."[15] As he states in another set of lecture notes, he is "obsessed by light."[16] Such a discourse informed his photography, painting, pedagogy, curatorial endeavors, and the environmental works that he referred to as "mobile kinetic light murals."

Kepes diagrams this discourse in an image created for the June 1946 cover of *Interiors and Industrial Design* magazine (fig. 4.1), a trade publication for architects and designers. He casts a spectrum of colored beams across the page. The rays of the primary colors—red, yellow, and blue—intersect to create their complements: orange, green, and purple. In the center, where all three beams perfectly meet, they produce a pure white light. Kepes marks this prism with a single keyhole. A matching key, collaged above, is not merely an iconographic illustration, for it is itself created in light and by light. Kepes produced it as a photogram, exposing a key resting directly on light-sensitive photographic paper to bright light, thereby leaving an impression in reverse on the developed paper. Kepes diagrammed the technique in his papers, with a drawing of a key and a note to himself: "find graphic symbols of single object … like shadow letters."[17] The key thus symbolically interlocks with the drawn keyhole even as it literally links light and shadow.

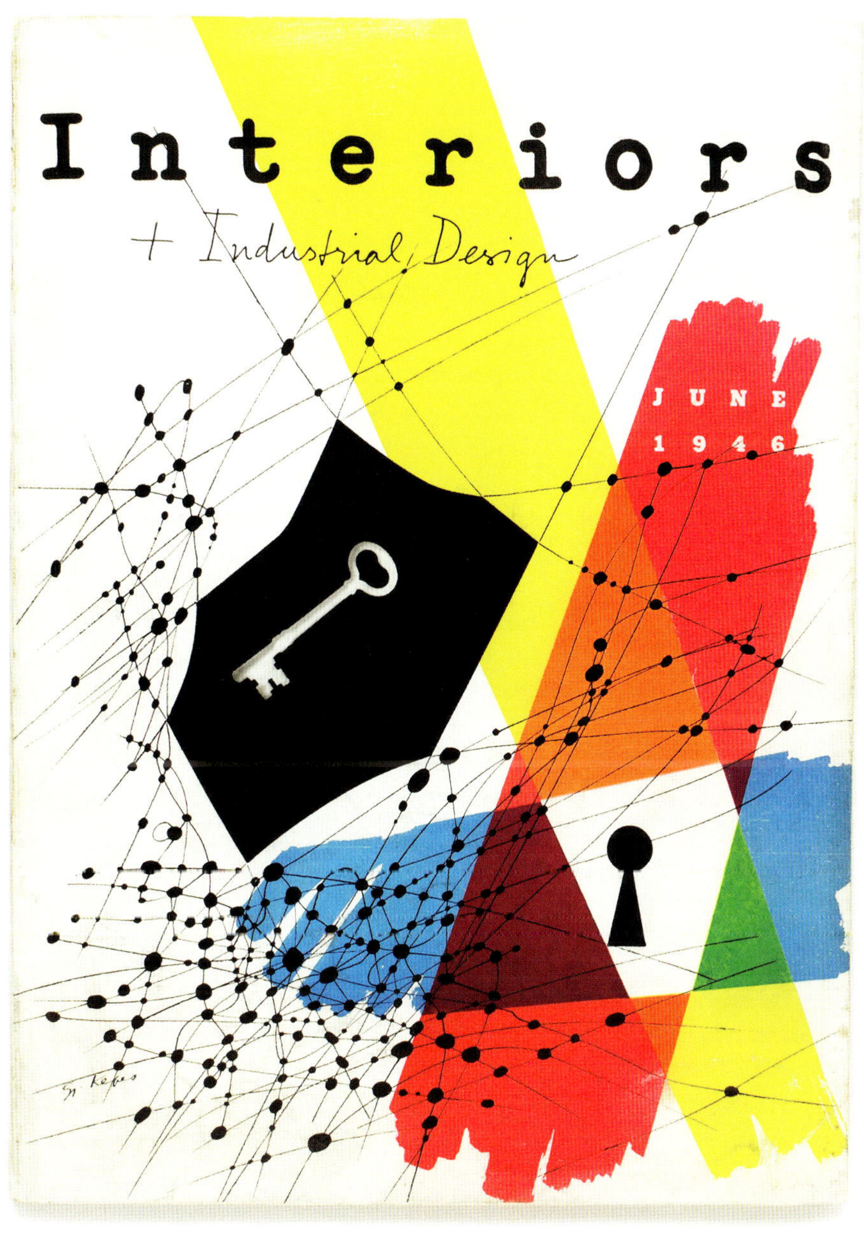

4.1

Interiors and Industrial Design, June 1946.
Cover designed by Gyorgy Kepes.
© The Estate of Gyorgy Kepes.

Kepes elaborates the stakes of this discourse in one version of a possible introduction to his Light Book (he would write many different versions):

> In spite of our immense range of artificial light sources, from the mad, incinerating glare of the exploding hydrogen bomb to the faintest glow of the numbers on a wrist watch, in spite of the ease with which we shape and direct it, our wealth of light has a quality of emptiness…. The new richness of light, a wonder of the twentieth century, lacks the true meaning of light, its inner sheen. In washing away the clear boundary between night and day, we have washed away the inner sheen that made light, in its history, a potent symbolic force. It is typical of our manner of existence that for the warm living play of firelight on the hearth we have substituted the bluish, greenish television screen with its stream of meaningless images. In making artificial illumination bright and ample, lighting engineers have lost the wonderful, ever-changing live quality of natural light and frozen it into stability, thus draining the vitality that sustains our vision.[18]

Kepes's comments are directed not toward the lack of light so much as the lack of virtue in light—despite the ability to produce it at will. "What are our contemporary values of light, the unifying symbols that bring richness and clarity to our lives on the new level of our incredible light technology? It would be more in order to ask where are they?"[19] These comments resonate with the roughly contemporaneous remarks of philosopher Gaston Bachelard, who declares, in reference to the rise of artificial illumination in the twentieth century: "We have entered an age of administered light."[20] Can the artist change a culture of administered light, light managed and controlled, light purged of significance? In a revised version of the text I quote above, Kepes distills this problem into a single question: "Given the vastness of our resources, how can the substance be restored to our experience of light?"[21] This chapter explores Kepes's attempts to restore such value.[22]

Allegories of Light

In all of these projects, Kepes presents light allegorically, as a form of purification. Using science and technology to generate light might transform science and technology, making both more humane. This ambition is clear in Kepes's meditation on "a future thermonuclear war"; by framing a scientific and technological event, one that is terrifying, as nonetheless artistic, Kepes gains allegorical purchase over it.

Light has always been allegorical, of course. Kepes's handwritten notes compiled for the Light Book document the resonances of light from time immemorial. In one notation, Kepes recounts creation stories and origin myths, like the Babylonian legend of Marduk, the god of light, who defeats darkness as embodied by the monster Tiamat. Kepes recalls the Persian mystic Zoroaster, who preached a religion that worshiped Ahura Mazda, the hypostasis of light, as the inverse of Ahriman, the symbol of destruction. He ponders the Greek myth of Phaethon, the son of the solar deity Helios; Helios drove the chariot of the sun across the vault of the sky every day, but when Helios gave Phaethon the reins to the fire-breathing horses, Phaethon struggled and lost control, crashing into the earth. Zeus struck Phaethon dead with a thunderbolt "to save the cosmos from conflagration," as Kepes puts it. Kepes framed these origins myths in terms of the present, writing "hydrogen bomb" and "A bomb" below.[23] Light was allegorical in the past, but also the present.

Kepes was not alone in reading current events through allegories of light. President Kennedy used such a metaphor when he addressed the nation in 1963 to report the signing of the Partial Test Ban Treaty, the international agreement that ended aboveground nuclear testing, like that of Operation Hardtack at the Pacific Proving Grounds: he called the treaty "a shaft of light cut into the darkness" in a television and radio broadcast.[24] In an inverse reference, President Johnson's campaign during the 1964 presidential race aired a famous television advertisement known as "Daisy." The ad juxtaposed a young girl pulling petals from a flower with the countdown of a nuclear bomb blast. After the flash of light, reflected in the girl's eye (an apt reference), President Johnson gravely intones: "These are the stakes. To make a world in which all of God's children can live, or to go into the dark. We must either love each other, or we must die."[25]

The most resonant reference to light was its repeated invocation in relation to the Vietnam War. Perhaps Kepes was familiar with the aerial bombardment campaigns called "Operation Flaming Dart," "Operation Rolling Thunder," and "Operation Arc Light"—the code names given by the US military to the missions aimed at destroying targets in North Vietnam, Laos, and Cambodia in the 1960s. The names of all three evoke the incendiary displays of aerial warfare; the name of the last—a reference to the band of light that crackled between two filaments in the arc lamp, a lighting technology reserved for grand public spaces—conjures the types of effects that fascinated Kepes.

As early as 1953, French General Henri Navarre described the situation in Southeast Asia with an optimistic metaphor: there was, he reassured the public, "a light at the end of the tunnel." The phrase then became an all-too-common refrain, most infamously uttered by General Westmoreland on the eve of the Tet Offensive. Soon Johnson was saying it, and, as the war continued, so was Nixon. Walter Cronkite declared on television, after the fall of Saigon in 1975: "In Vietnam we finally have reached the end of the tunnel—and there is no light." A political cartoon from the early 1970s caricatures this darkness; Nixon is trapped in the tunnel, with a train labeled "Vietnam" hurtling toward him (fig. 4.2).

Kepes's collection of handwritten research notes makes these types of association clear. Under a quotation from the art historian Henri Focillon's *The Life of Forms in Art*, specifically a reference to "light as form" in religious architecture ("Light itself is form, since its rays, streaming forth at predetermined points, are compressed, attenuated or stretched …"), Kepes writes: "The flashlight of the bomber—the search lights scanning the sky seeking the enemy plane, the flares of antiaircraft—the mushroom of A bomb, the black out + the pulseout."[26] These forms of light were the most awe-inspiring of the early Cold War, more awe-inspiring than the light streaming through stained-glass windows in medieval churches.

In Kepes's collection of typed quotations, also compiled for his Light Book, more than a few also speak to nuclear fear. One page bears lines from a 1905 notebook belonging to the Russian composer Alexander Nikolayevich Scriabin: "I am fire enveloping the universe, / Reducing it to chaos. / I am the blind play of powers released. / I am creation dormant, Intellect quenched."[27] A second page quotes lines of verse written by another Russian, the Symbolist poet Alexander Blok: "Blacker and Blacker shall grow the terrible world,

The Light At The End Of The Tunnel . . .

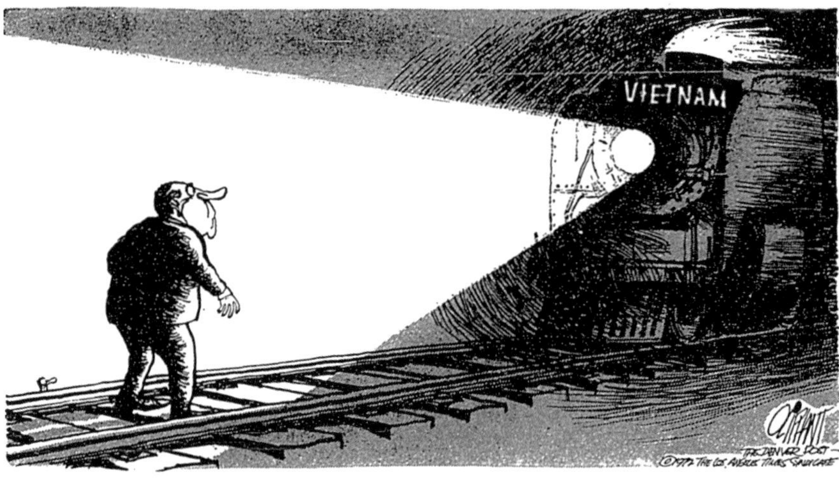

4.2

Pat Oliphant, "The Light at the End of the Tunnel …," 1972. Oliphant © Andrews McMeel Syndication. Reproduced with permission. All rights reserved.

Madder and madder the reeling dance of the planets."[28] This carefully selected line of verse suggest how Kepes used words from the past to understand the present, to imagine the Cold War not as a geopolitical conflict between superpowers, but as a more profound war to preserve humankind against the threat of science and technology.

Kepes also composed his own lines registering such sentiments on a handwritten page labeled "light scale XX century background"—more preparatory material for his Light Book. The notation's frantic, frenetic script, in a wildly looping scrawl, telegraphs a manic sense of urgency. Its broken lines fall at a clipped staccato, like the rapid-fire imagery of a cut-up collage or beat poem:

> playing the gigantic toys of destruction, death making, accidental chance to play
> reckless arrogant intellectuals who toy precariously
> megalomaniacal whims
> control + communication
> theory of games in warfare, rationality of chance in survival
> push button war
> earth orbiting satellites
> launching pads for nuclear missiles
> new scale of reification of death + life
> new scale of war, strategist who invented the words "overkill" [and] "megadeath"
> computer scale of calculation
> ever spiraling scale of destructive force …
> the eclipse of values[29]

The fragment is a stark impression of Cold War terror. Kepes provides a complete inventory of the technologies and techniques for nuclear war. He refers to computers, missiles, and satellites, but also to cybernetics—referenced by "control and communication," the subtitle of Norbert Wiener's book on the subject—and to the game theory of John von Neumann and Oskar Morgenstern.[30] He draws his language from the militaristic lexicon of the defense strategists who built the Cold War while ensconced in think tanks, military bureaucracies, and university laboratories. He cites terms popularized by Herman Kahn, the infamous thinker of the unthinkable from the RAND Corporation: *megadeath*, the death of a million people, especially as a unit of measure for estimating the effects of nuclear warfare; *overkill*, destruction in excess of strategic requirements, especially the capacity to kill and destroy many times over in nuclear warfare. Kahn made both terms common parlance in notorious texts like *On Thermonuclear War* of 1960 (he also lectured at MIT in this period).[31]

But what makes the fragment striking is that Kepes felt this terror both personally and professionally. He had his own intimate involvements with the "megadeath intellectuals," as they were then called.[32] More than a few, like

Wiener, were at MIT. Consider James R. Killian, Jr., the President of the Institute from 1948 to 1959; Killian also served the President of the United States as Eisenhower's first special assistant for science and technology following the launch of Sputnik in 1957. In this position, he chaired the President's Science Advisory Committee or PSAC, a panel that crafted defense policy. As chair of PSAC, Killian oversaw Operation Hardtack, the series of nuclear tests that so fascinated and disturbed Kepes.[33] This panel also included Jerome Wiesner, a future MIT president. Kepes was cordial with both men; both owned his paintings—canvases that captured light in luminous pigment (Kepes provided them as gifts, a means of ingratiating himself with MIT's administration).[34] Killian and especially Wiesner used their influence to advocate for disarmament, but the point is that Kepes became part of a culture that he also fulminated against in his private Light Book musings. This unusual position—part of, but also against—perhaps made him uniquely able to offer this allegorical redemption, to guide the Institute with a different light.

Kepes titled this particular note "light scale XX century background." The inscription indicates that these fears were the ground upon which Kepes built his Light Book, an attempt to describe a new scale of experience in the twentieth century that would frame a new aesthetics of light. But could light actually address this exploded scale? Could light meaningfully respond to social, cultural, and political debates about science and technology and the destruction they had wrought?

Light Start

Light is at the core of Kepes's pedagogy, as fundamental to his thought as visual design. Consider a spread he produced, in collaboration with Juliet Kepes and Thomas McNulty, for a regular feature called "Interiors to Come," again for *Interiors and Industrial Design*, the same magazine that earlier published his graphic cover. As the editorial copy explains, the purpose of the feature was to showcase "ideas from a designer's notebook"—not practical proposals but thought pieces, provocations.[35]

Kepes used the feature to imagine a laboratory devoted to the study of light. He based his design on *The New Atlantis*, Sir Francis Bacon's 1627 fable about a technocratic utopia. The center of Bacon's imaginary society is Salomon's House, a great academy devoted to the arts and sciences. Bacon describes the academy as both the "lantern" and "eye" of his imaginary world; it both projected and reflected the light of reason. Salomon's House is further divided into distinct research laboratories, each filled with instruments for the experimental study of natural phenomena, all for the betterment of humankind. As Bacon explains, one of these laboratories is a "perspective-house" devoted to "demonstrations of all lights" and "demonstrations of shadows." Kepes's *Interiors* article quotes Bacon's original description of this laboratory:

> We represent also all multiplications of light, which we carry to great distance, and make so sharp as to discern small points and lines; … We find also divers means yet unknown to you, of producing of light originally from divers bodies.… We make artificial rain-bows, halos, and circles about light. We represent also all manner of reflexions, refractions, and multiplications of visible beams.[36]

Kepes illustrates this "perspective-house" as what he calls a "perspective tower," an abstract impression of fractured planes. On the next page, he depicts the vista from this tower as a gridlike spatial expanse punctuated by the looping curves of a Lissajous figure—described here as a "floating spiral" composed of intersecting beams of light (fig. 4.3). And on the pages following, he creates a kaleidoscopic spectacle, with undulating bands that ripple across the pages (fig. 4.4). Kepes collages text and image, including his own photographs and those of students and colleagues like Carlotta Corpron, Dorothy Pelzer, and Thomas McNulty; these explain particular experiments that might occur in his fictional laboratory: investigations into color perception, visual pattern, and the reflections of light through prisms and lenses. The spread is a fantasy of science as aesthetic subject and technology as aesthetic tool. Elsewhere, he also made renderings of this perspective tower as a tower of light (fig. 4.5).

Kepes's perspective tower was not entirely imaginary. In an article from 1947, one Kepes likely read, the philosopher Milton C. Nahm argues that Bacon's academy exists in the present. Nahm relates the apparatuses in Bacon's perspective-house to "telephones, radios, aeroplanes, automobiles, railroads, telegraphs, radar"—all devices that expand and compress the experience of light and space across distances.[37] Kepes similarly mimicked Bacon's fabled perspective-house through his Light and Color Workshop at the New Bauhaus in Chicago and his Light and Color courses at MIT; he appropriated basic science and technology—lessons using mirrors to reflect light or lenses to refract it—and put them to new aesthetic purposes.

A single file folder from Kepes's papers labeled "Light Start 1937" contains his earliest attempts at theorizing this project.[38] A few loose pages are covered with sketches and notations in both Hungarian and English. The folder's date indicates that Kepes drafted these notes the same year that he left London for Chicago in order to teach at the New Bauhaus. He may have drafted them during the journey across the Atlantic. (In a later interview, he recalls: "My preparation for this teaching was the five days on the *Queen Mary*, … where I put down a curriculum in writing.")[39] Matching Kepes's drawings to sources cited in bibliographies he later compiled for his Light Book allows us to reconstruct this curriculum.[40]

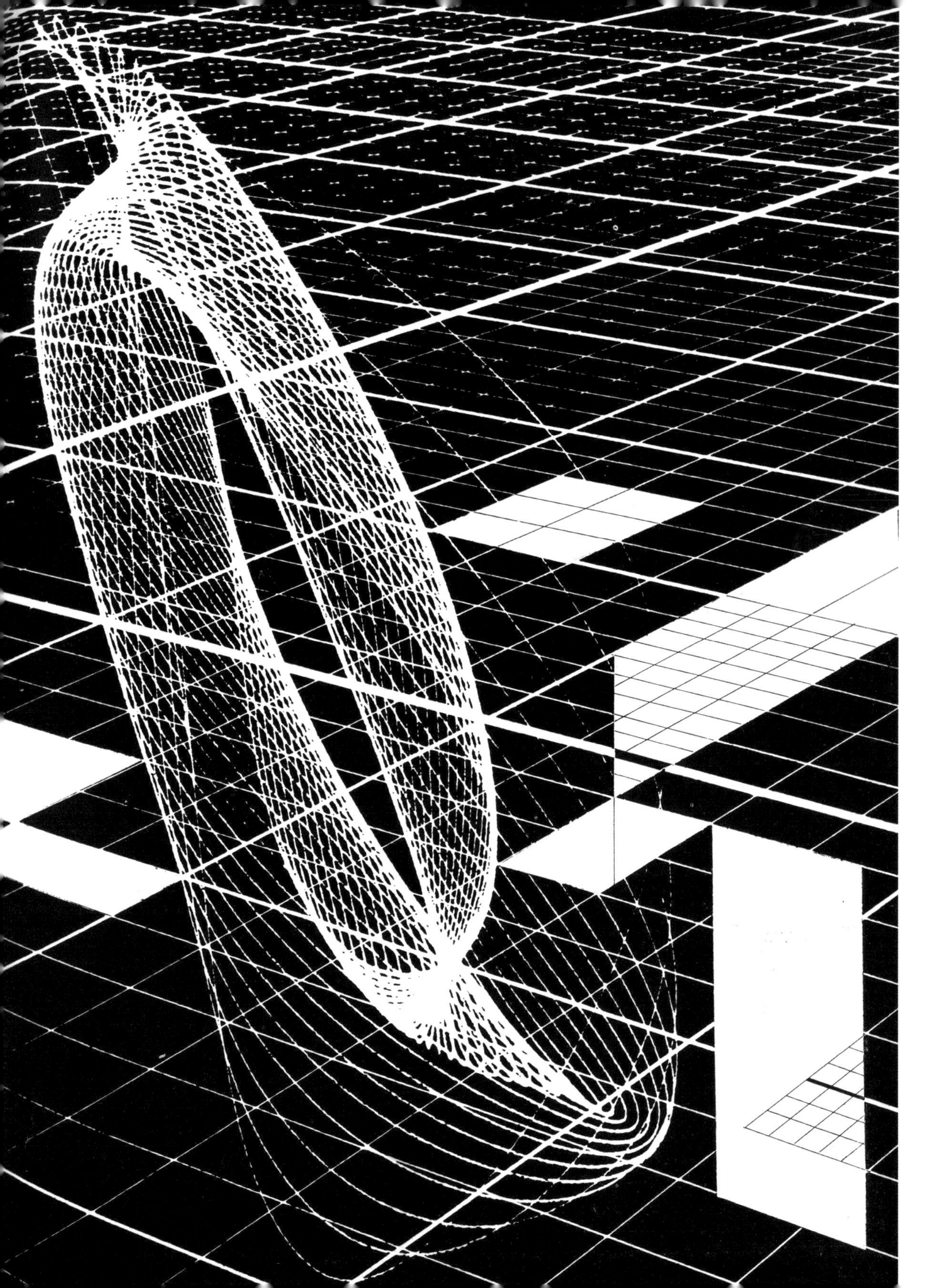

4.3

"View from Perspective Tower," from "Kepes
Looks at Light; Sees Spots," *Interiors
and Industrial Design* 110, no. 6 (January 1951).
© The Estate of Gyorgy Kepes.

4.4 (following pages)

Gyorgy Kepes, "Kepes Looks at Light;
Sees Spots," *Interiors and Industrial Design*
110, no. 6 (January 1951). © The Estate of
Gyorgy Kepes.

Chapter 4

Below: Wall surfaces in Kepes' project would demonstrate effects of light on different textures. This one was evolved by Dorothy Pelzer to coordinate acoustic requirements with purely decorative patterns.

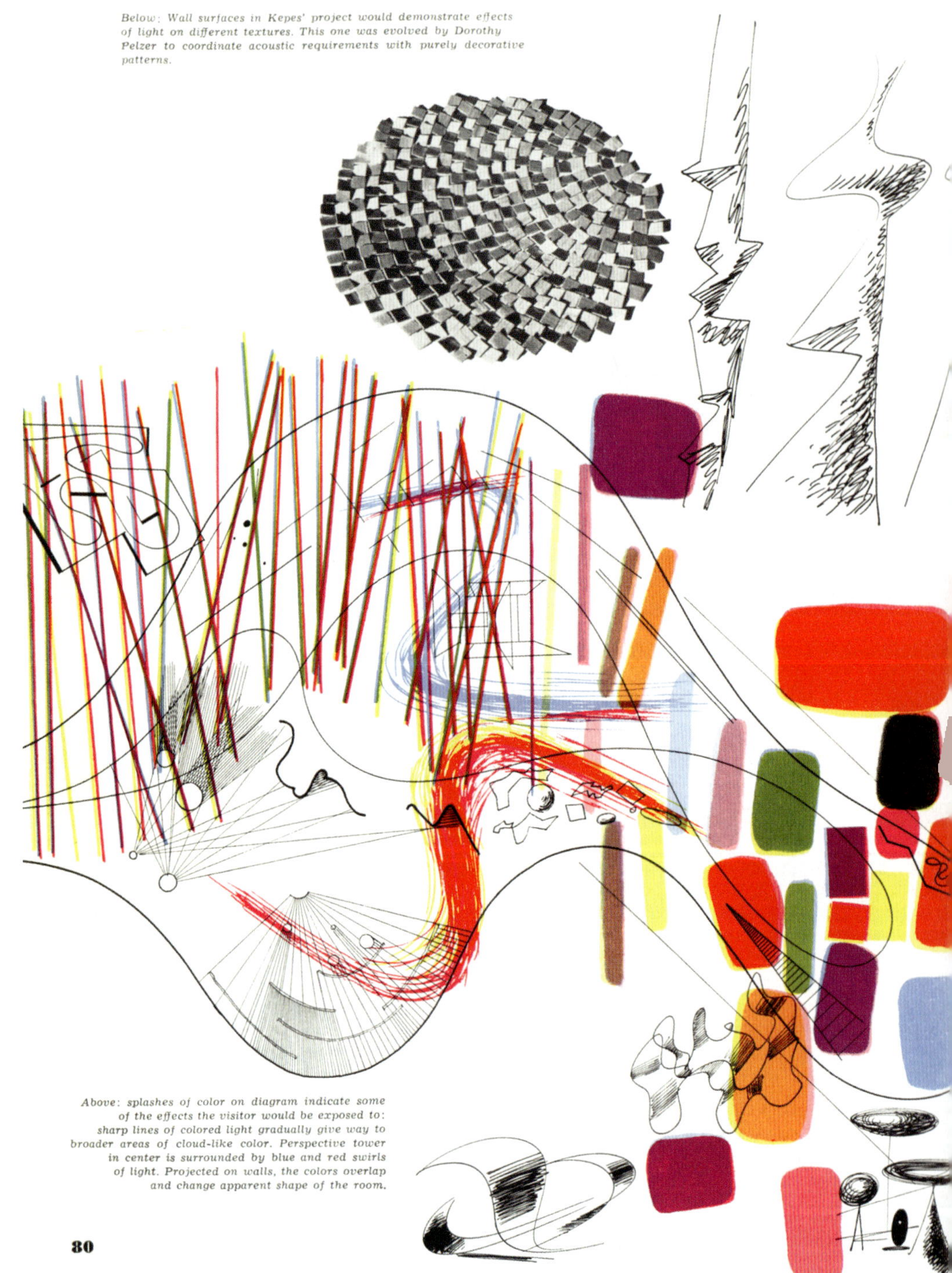

Above: splashes of color on diagram indicate some of the effects the visitor would be exposed to: sharp lines of colored light gradually give way to broader areas of cloud-like color. Perspective tower in center is surrounded by blue and red swirls of light. Projected on walls, the colors overlap and change apparent shape of the room.

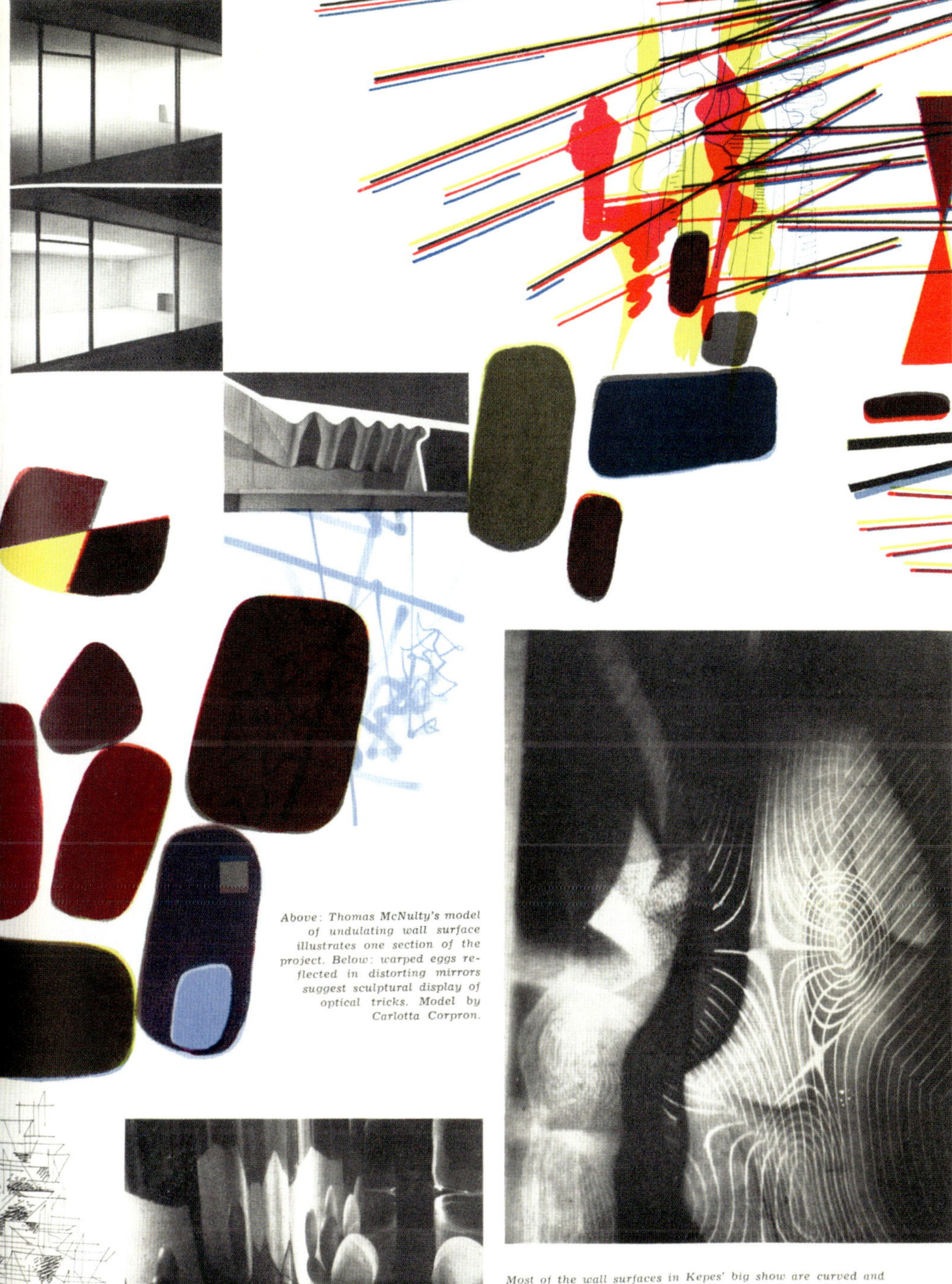

Above: Thomas McNulty's model of undulating wall surface illustrates one section of the project. Below: warped eggs reflected in distorting mirrors suggest sculptural display of optical tricks. Model by Carlotta Corpron.

Most of the wall surfaces in Kepes' big show are curved and angled. Photograph above illustrates patterns made by lights in different positions projecting overlapping beams onto the same surface.

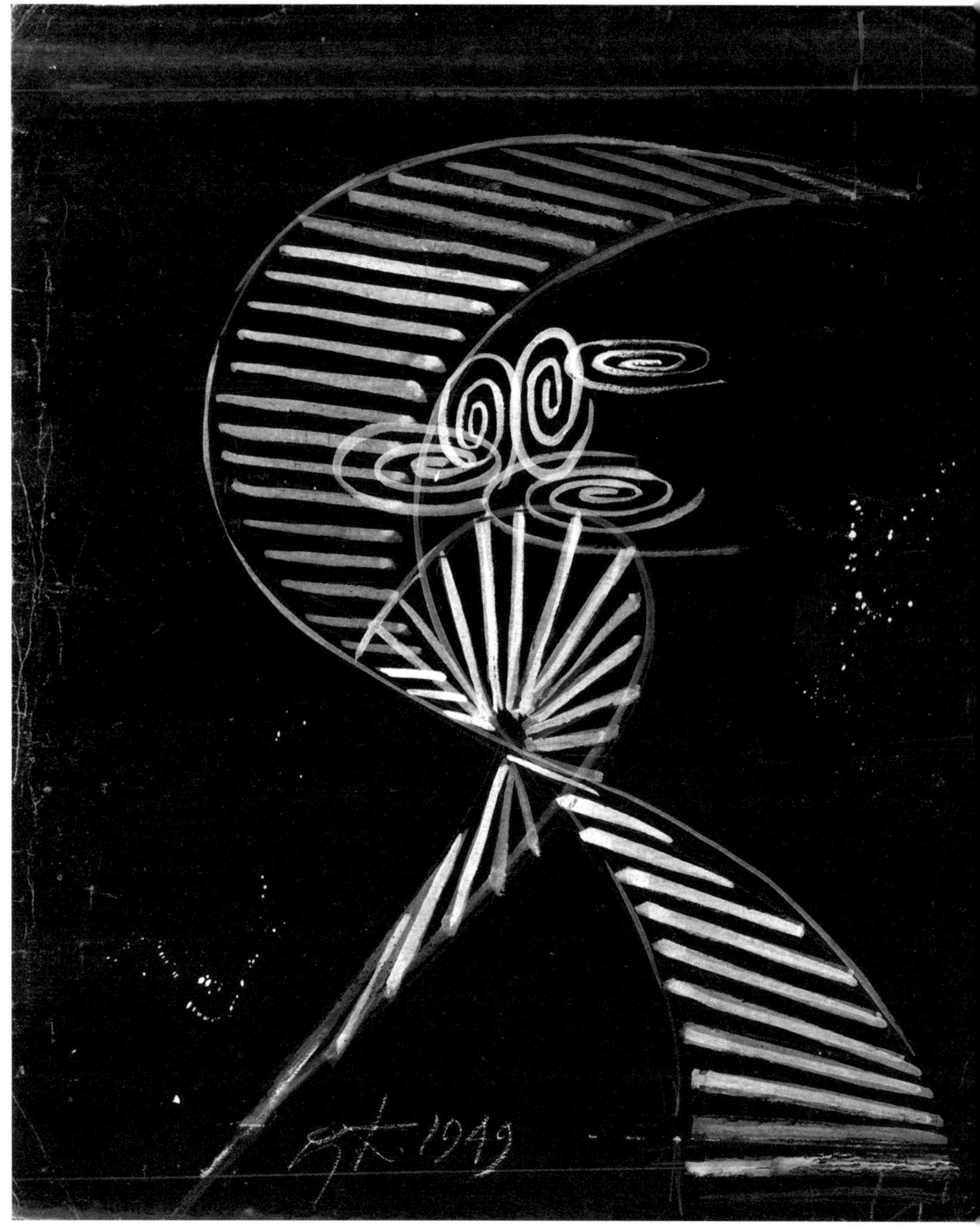

4.5

Gyorgy Kepes, Light Tower design, 1949.
Gyorgy Kepes papers (M1796). Dept. of
Special Collections and University Archives,
Stanford Libraries, Stanford, Calif.
© The Estate of Gyorgy Kepes.

One page from "Light Start 1937" indicates that Kepes consulted physicist William Bragg's 1933 introduction to optics, *The Universe of Light*, which use basic experiments to demonstrate the foundational principles of light (fig. 4.6). A diagram depicts the simulation of light waves through a tank of water; ripples bounce off the tank's wall, just as light rays reflect from a mirrored surface. Another depicts light emitted from a lantern and reflected from a single mirror; still others show the same principle using two or three mirrors positioned at right angles, causing entering and returning rays to align. Kepes diligently copied Bragg's diagrams into his notes (fig. 4.7). He also extends them, imagining mirrors reflecting a single beam of light to create what Kepes calls a "mirror space."

Kepes also consulted the work of Matthew Luckiesh, a physicist who ran General Electric's Lighting Research Laboratory. In a 1916 volume titled *Light and Shade and Their Applications*, Luckiesh explores the "art and science of lighting."[41] One set of diagrams shows the characteristics of reflecting and transmitting materials (fig. 4.8). Again, Kepes diligently copied these diagrams, labeling various surfaces—"white blotting paper," "aluminum," "semigloss paper," "plate glass," "dense opal glass"—and drawing the resulting pattern of light produced on each (fig. 4.9).

These comparisons not only establish an origin for Kepes's light experiments, they also index an approach that defined all of his engagement with the topic: Kepes created an aesthetics of light by simulating the science of light. He imitated an empirical approach, reading technical literature and reiterating its experiments. But Kepes's pedagogy of light also exceeded the limits of empiricism. He simulated science not only to mimic its rational investigation but also to reveal its irrationality, its surreal quality. He aimed to transform the quantitative method into something qualitative through an aesthetic alchemy, a process that might expose light's more mysterious qualities.

4.6 (following pages)

Diagrams from William Bragg, *The Universe of Light*. London: G. Bell and Sons, 1933. Courtesy of the Royal Institution.

4.7 (following pages)

Gyorgy Kepes, drawings from a folder labeled "Light Start 1937." Gyorgy Kepes papers (M1796). Dept. of Special Collections and University Archives, Stanford Libraries, Stanford, Calif. © The Estate of Gyorgy Kepes.

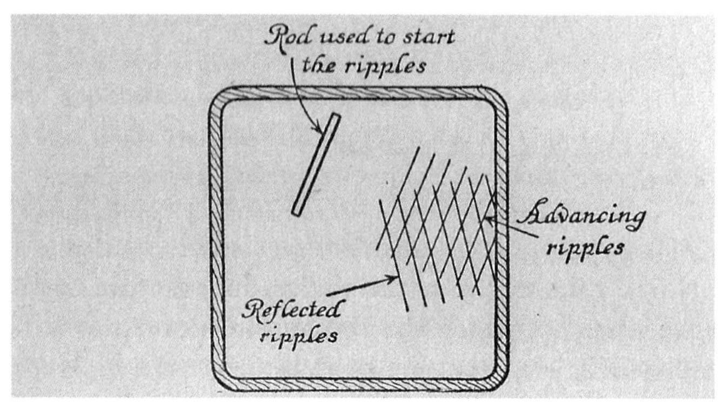
Rod used to start
the ripples
Advancing
ripples
Reflected
ripples

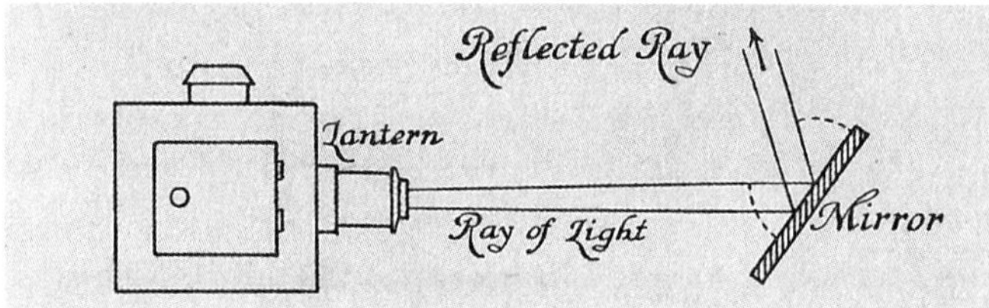
Reflected Ray
Lantern
Ray of Light
Mirror

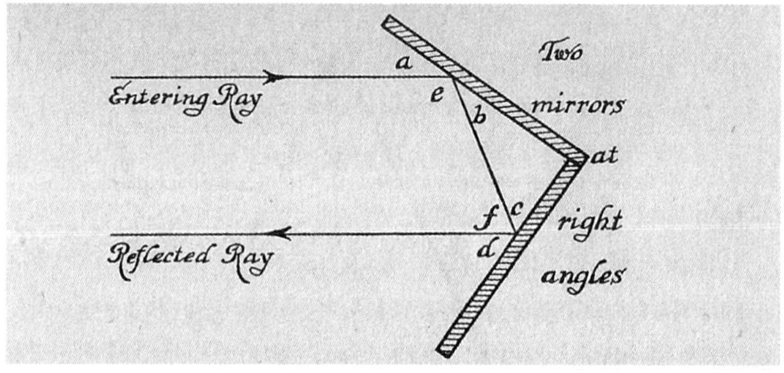
Two
mirrors
at
right
angles
Entering Ray
a
e
b
f
c
d
Reflected Ray

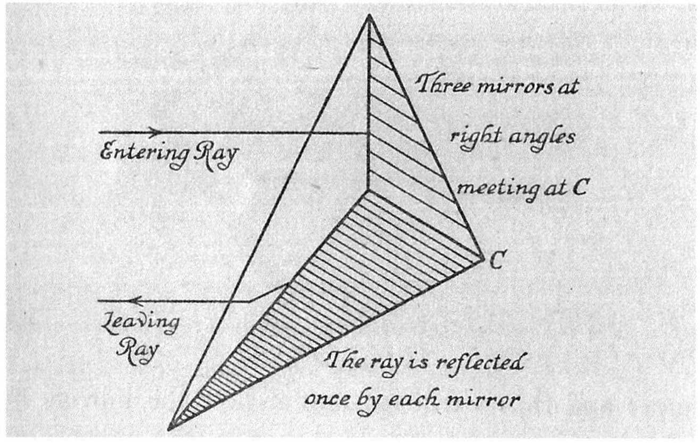
Three mirrors at
right angles
meeting at C
Entering Ray
C
Leaving
Ray
The ray is reflected
once by each mirror

(fény) reflection

a térerős és visszaverődés szöge ugyanaz

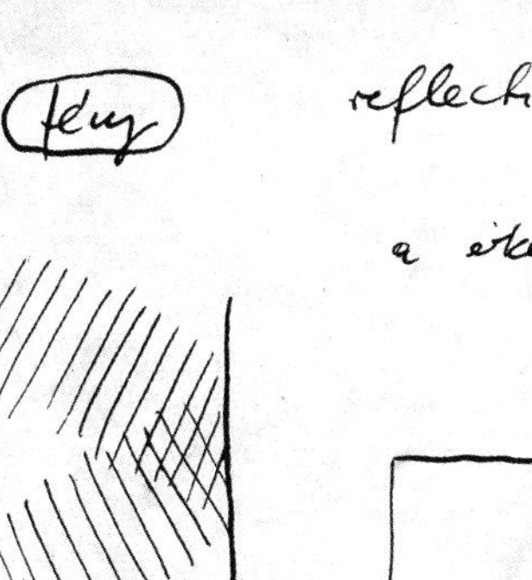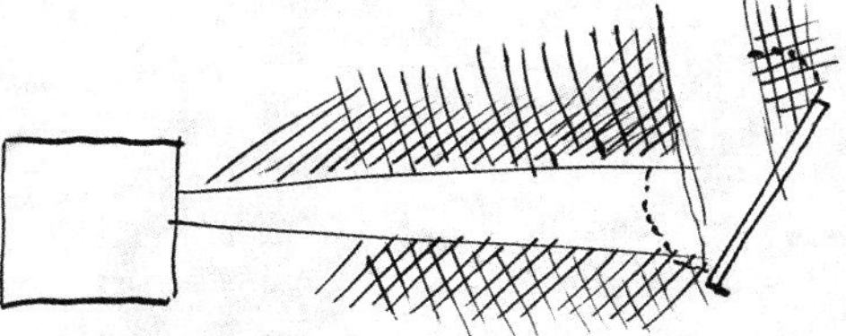

Kísérletek tükörrel, fénykupposzni a fénycsövét

mirror spell

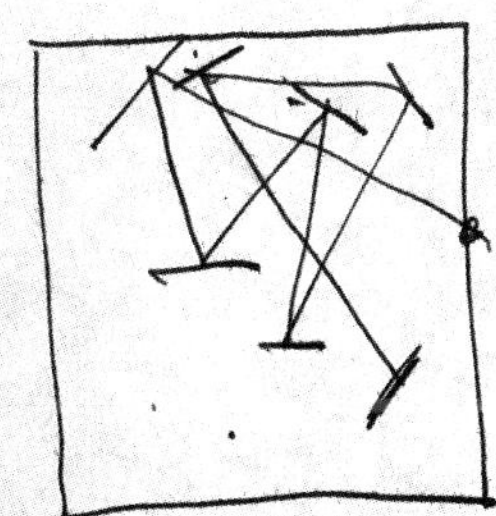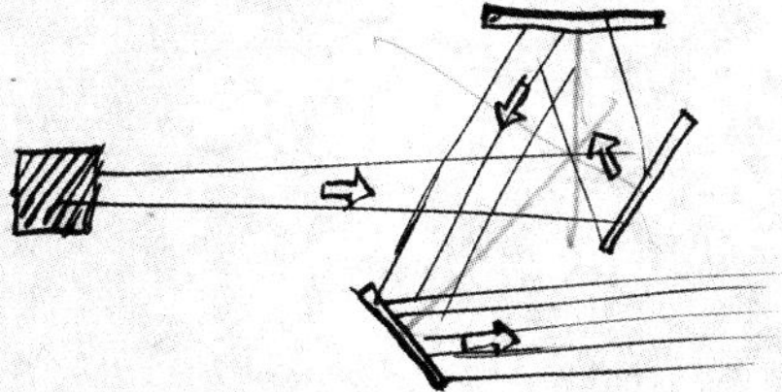

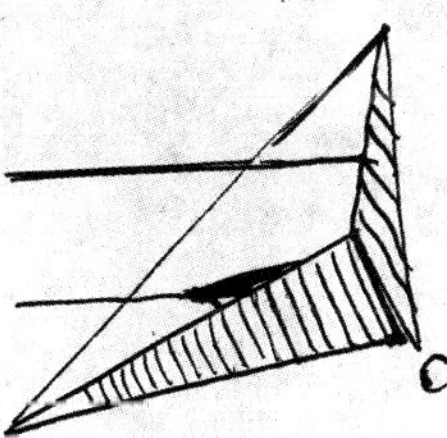

3 tükör derékszögbe
találkozik c pontban.

a fény ugyanabban az
irányban megy vissza
amelyikben jött

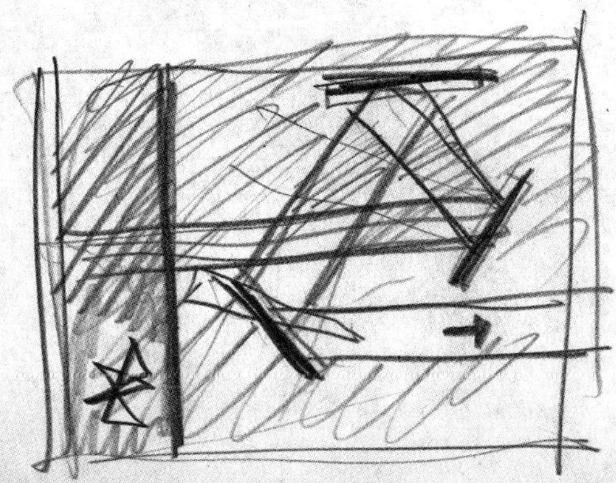

Chapter 4

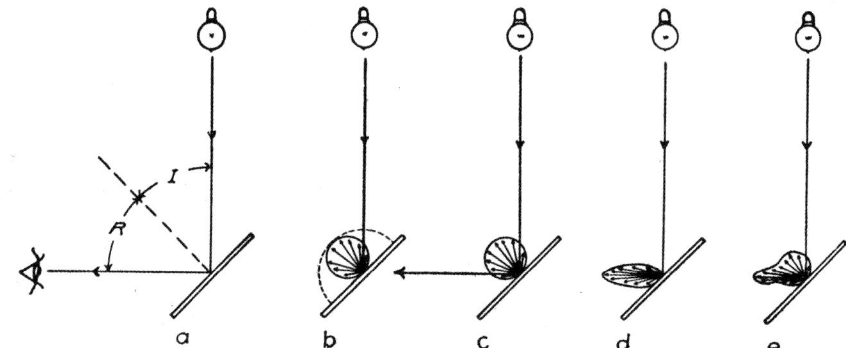

Fig. 1 — Characteristics of reflecting media.

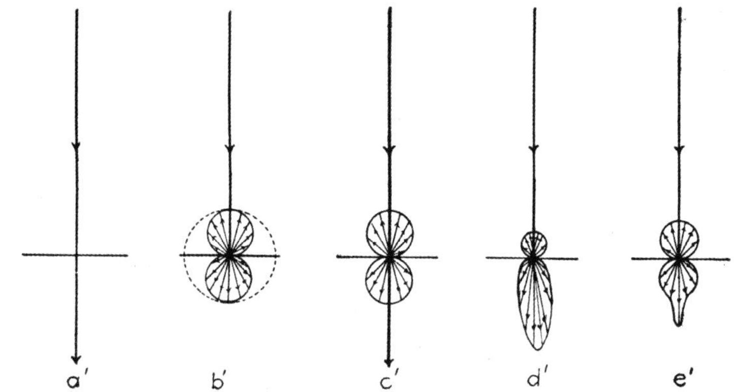

Fig. 2 — Characteristics of transmitting media.

4.8

Diagrams from Matthew Luckiesh, *Light and Shade and Their Applications*. New York: D. Van Nostrand, 1916.

4.9

Gyorgy Kepes, drawings from a folder labeled "Light Start 1937." Gyorgy Kepes papers (M1796). Dept. of Special Collections and University Archives, Stanford Libraries, Stanford, Calif. © The Estate of Gyorgy Kepes.

light   optical laws of reflecting surfaces

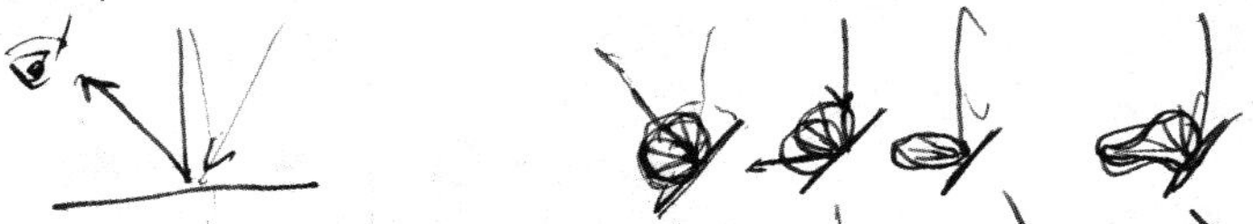

mirror   white blotting   glass +   aluminium   semiglossy
                paper   blotting p.   paper

                                                      minute
                                                      mirrors oriented
                                                      at random

transmitting surfaces

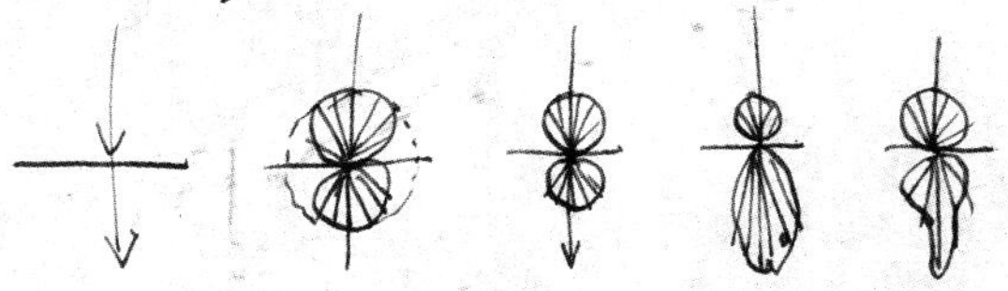

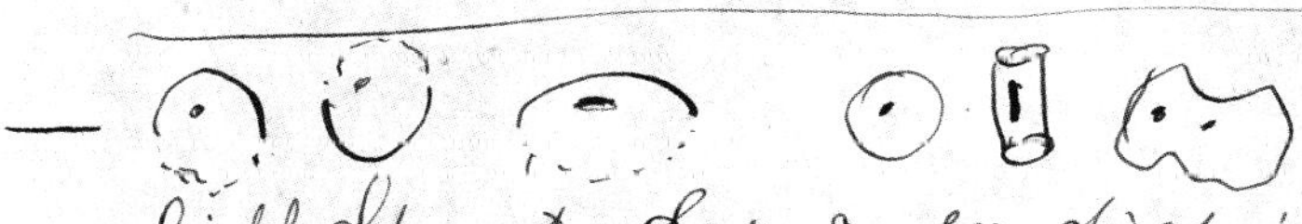

plate glass   dense opal   milky   sand   sand blasted
               glass   glass   blasted   clear glass
                                         opalglass

highlights. As shape, number, direction
depends upon light source, distance, number
quality, shape of, angle of incidence of linear light
spot, flood, window, indirect light &
depends on reflecting quality metallic rough etc
                                    glazed
glossy — varnished black pigment
matte   powdered "   "
glazed
polished | metallic
                   textile
                   velvet
                   powder

Such an approach is evident in the name he gave to the cameraless images he began creating in 1937, concurrently with the development of his light pedagogy. Kepes did not call them "photograms" (meaning light-drawing or light-writing), the term popularized by Moholy, and he avoided rival terms for the same technique from modernist precedents, like Man Ray's "rayograph" or Christian Schad's "schadograph." Instead, Kepes called these works "photogenics," a word meaning *light-producing* that indicates his interest in propagating light rather than merely recording it—for Kepes, the drawn or written images that resulted from his experiments were secondary to this act of producing light.[42] Kepes borrows the anachronistic term "photogenic" from William Henry Fox Talbot, who used it for his cameraless contact prints. This terminology allowed Kepes to trace his lineage to a figure who was as much a scientist as an artist, while also laying claim over the nineteenth century—a chronology of significance for Kepes's philosophy, as we shall see.

Kepes's appropriation and use of technical equipment to create these photogenics was also a form of aesthetic alchemy, a process of transformation; he turned the darkroom into a laboratory, albeit an antiquated one. Consider an unusual apparatus from physicist John Tyndall's 1875 volume *Six Lectures on Light* (fig. 4.10).[43] Tyndall devised a means of projecting an enlarged image of a magnetic field onto a screen through a device that incorporated a lamp, a mirror, a prism, a set of magnets, and a handful of iron filings. The apparatus allowed Tyndall to render visible a typically invisible natural phenomenon: magnetism. Kepes followed the essential principle behind Tyndall's apparatus by placing a magnet and iron filings either directly on photographic paper or on a sheet of glass resting above it, and subjecting these materials to light. The magnetic field appears in the photogenic image (fig. 4.11).

Physiologist Hamilton Hartridge's 1949 book *Colours and How We See Them* illustrates similar devices (though published after Kepes's period of intensive photographic investigation, Hartridge's book presents classic optical experiments first popularized in the nineteenth century).[44] Hartridge constructed an "apparatus for projecting an artificial rainbow" and an "apparatus for projecting an artificial sunset"; both used light and water, poured into either a beaker or a glass vessel, to create optical effects (fig. 4.12). In the first case, the spectrum of a rainbow is captured on a flat surface near the beaker. In the second, an orange or yellow light is cast upon a screen. Kepes did not re-create these apparatuses in a specific way, but he did produce photogenics by projecting light through liquids poured and pooled on glass plates and acetate films. His technique derived most obviously from the antiquated *cliché-verre* method, a nineteenth-century photographic process that involved painted glass negatives.[45] Kepes often used mixtures of pigment and casein in these works; pressed between glass sheets, the mixtures created the inkblot impressions of decalcomania (fig. 4.13). But the origins of Kepes's experiments were not only in the history of art; they were also in the history of science, they were also approximations of Hartridge's "artificial rainbows" and "artificial sunsets."

Kepes also studied a device invented by physicist Charles Wheatstone in 1827.[46] Wheatstone, borrowing the name of physicist David Brewster's kaleidoscope, called his apparatus a "phonic kaleidoscope" or kaleidophone. It consisted of multiple thin metal rods mounted on a circular wooden base; a mirrored bead was affixed to the end of each rod (fig. 4.14). Striking the rods would set them into rapid motion, vibrating the reflective bead. Illuminated by bright light, the movement of the reflective bead created the curves of the Lissajous figure; the persistence of vision caused these visual vibrations to appear as cohesive patterns. Wheatstone recorded the common variations of these patterns in a set of illustrations that accompanied a report on his invention (fig. 4.15). Kepes simulated the kaleidophone's effects through a teaching tool he called the "virtual volume." A reflective metal wire was twisted and bent, then rapidly spun and photographed; the persistence of vision caused the spinning wire to similarly appear as cohesive patterns (fig. 4.16). The virtual volume is a standard Bauhaus pedagogical exercise, but it is significant that Kepes traced the technique back to Wheatstone's scientific experiments (Kepes saved multiple diagrams of Wheatstone's kaleidophone in his collection of Light Book photographs).

Perhaps the most famous of Kepes's optical tools was the light box. The use of the device is typically attributed to his student Nathan Lerner, who purportedly first used it as a photographic instrument in Kepes's Light and Color Workshop in 1937, as I discussed in the first chapter of this study (it was applied to Kepes's camouflage research).[47] Kepes, however, maintained that the use of the light box was first his idea—a repurposing of devices, like the camera obscura, that earlier aided Renaissance painters. "The ideas of light box experiments came from Tintoretto that I learned in my Art history course at the Art Academy of Budapest in about 1924."[48] Regardless, and despite its possible art historical pedigree, the instrument was also a scientific tool. In *The Universe of Light*, Bragg illustrates two such boxes used for the study of optical phenomena (fig. 4.17). A beam of light entering a hole at one end of a closed box exits a hole at the other end without leaving an impression; with no interfering material to scatter the light, the inside of the box remains dark.

FIG. 26.

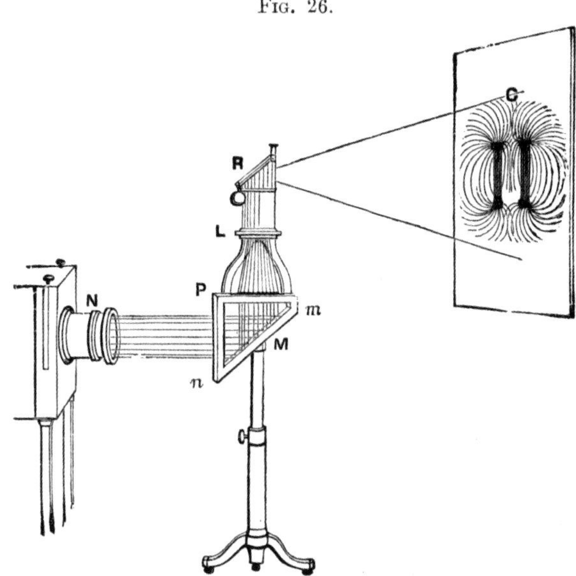

N is the nozzle of the lamp; M a plane mirror, reflecting the beam upwards. At
P the magnets and iron filings are placed; L is a lens which forms an image of
the magnets and filings; and R is a totally-reflecting prism, which casts the image
G upon the screen.

4.10

Diagram from John Tyndall, *Six Lectures
on Light: Delivered in America in 1872–1873*.
London: Longmans, Green, 1875.

4.11

Gyorgy Kepes, *Magnetic Fields*, 1938.
Gelatin silver print, 8 3/8 by 6 1/2 inches.
© The Estate of Gyorgy Kepes.

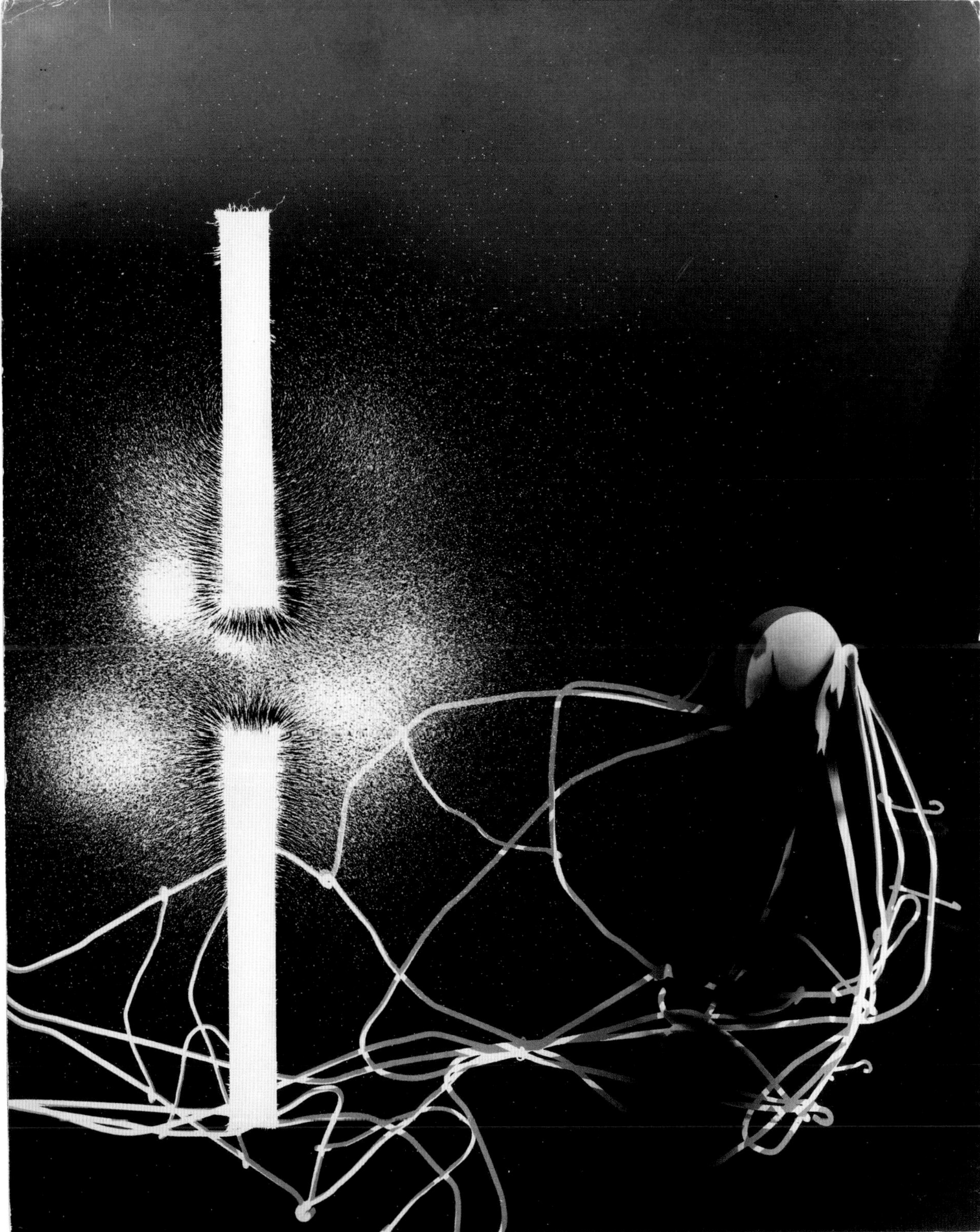

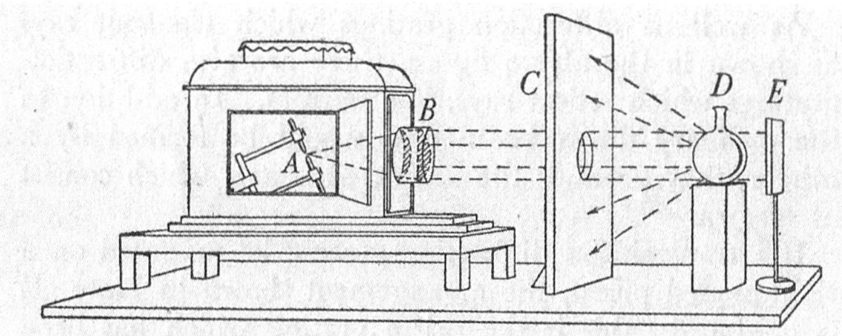

Fig. 18.   Apparatus for projecting an artificial rainbow

Light from the electric arc lamp A, having been rendered parallel by the condensing lenses B, passes through the hole in the screen C to fall on a spherical flask filled with water D.   The rainbow is produced on the right-hand side of the screen C.   Unwanted light rays are cut off by the small screen E.

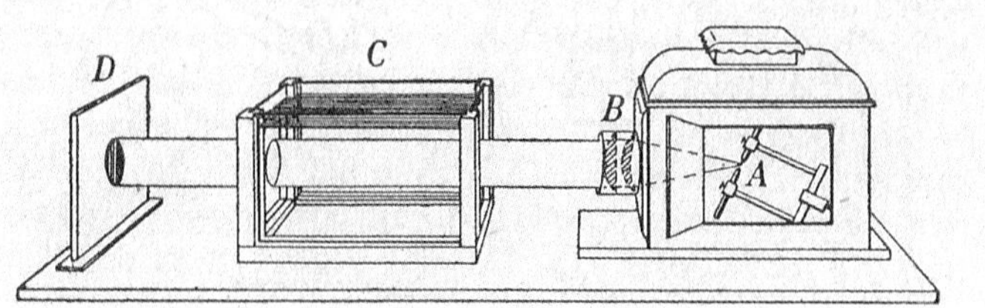

Fig. 23.   Apparatus for projecting an artificial sunset

Light from the electric arc lamp A, having been made parallel by means of the condensing lenses B, passes through the trough of water C to reach the screen D.   When the water contains very fine particles in suspension some of the blue rays are deflected sideways so that the light which falls on the screen is deficient in these rays and appears yellow or orange in colour.

4.12

"Apparatus for projecting an artificial rainbow" and "Apparatus for projecting an artificial sunset." Diagrams from Hamilton Hartridge, *Colours and How We See Them*. London: Bell, 1949.

4.13

Gyorgy Kepes, *Abstraction—Surface Tension #2*, c. 1940. Gelatin silver print, 14 x 11⅛ inches. Courtesy the Museum of Modern Art, New York. © The Estate of Gyorgy Kepes.

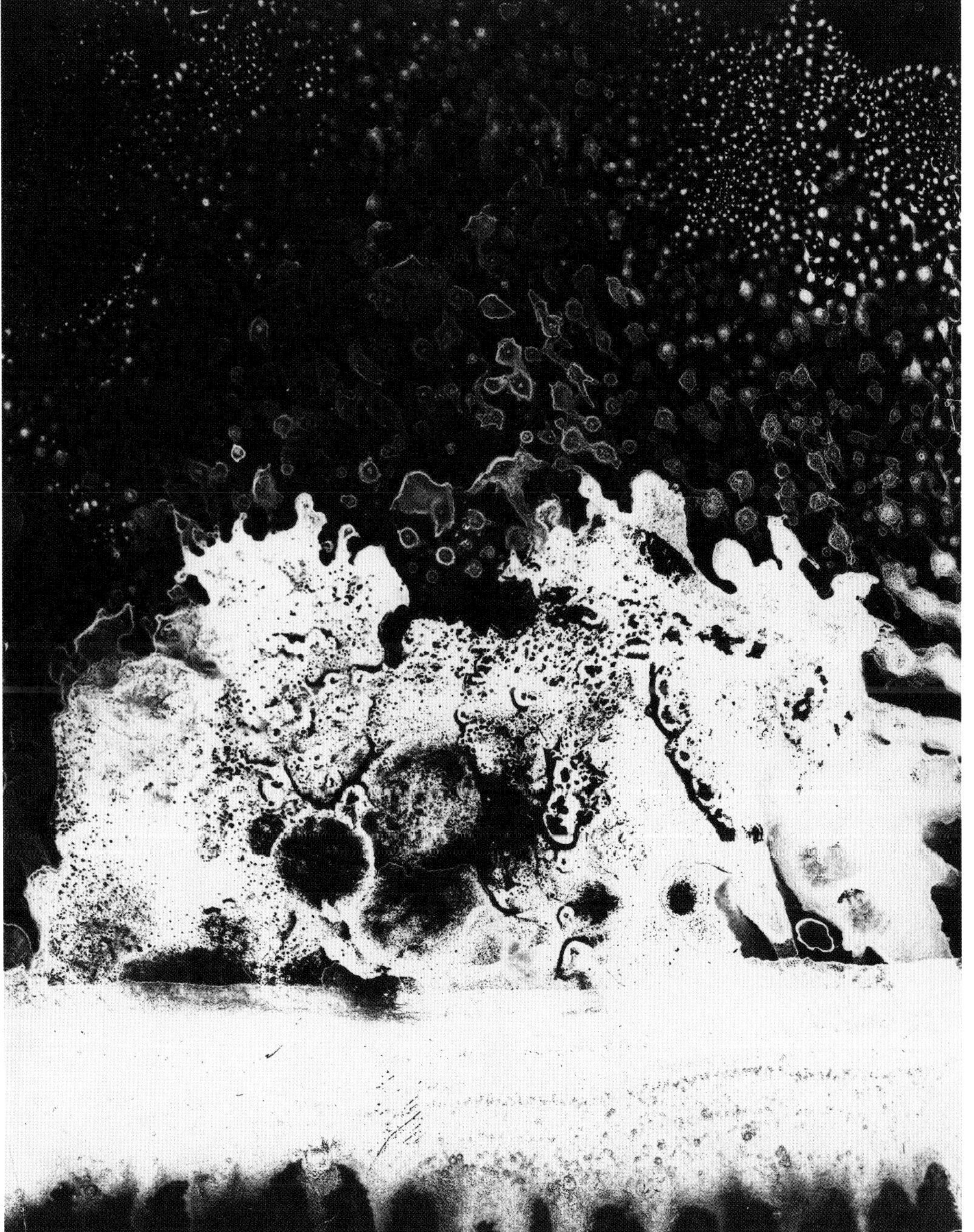

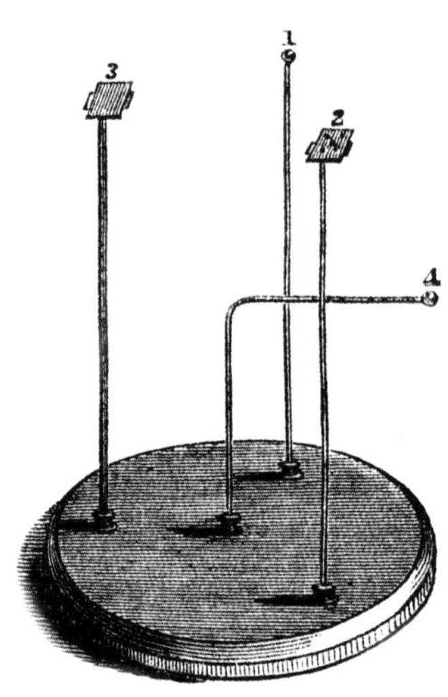

4.14

Diagram of the kaleidophone. From Sir Charles Wheatstone, *The Scientific Papers of Sir Charles Wheatstone*. London: Taylor and Francis, 1879.

4.15

Kaleidophone images. From Sir Charles Wheatstone, *The Scientific Papers of Sir Charles Wheatstone*. London: Taylor and Francis, 1879.

4.16

Theodore M. Brown, virtual volume, 1952. Produced in an MIT Visual Design course taught by Gyorgy Kepes. Gyorgy Kepes papers (M1796). Dept. of Special Collections and University Archives, Stanford Libraries, Stanford, Calif. © The Estate of Gyorgy Kepes.

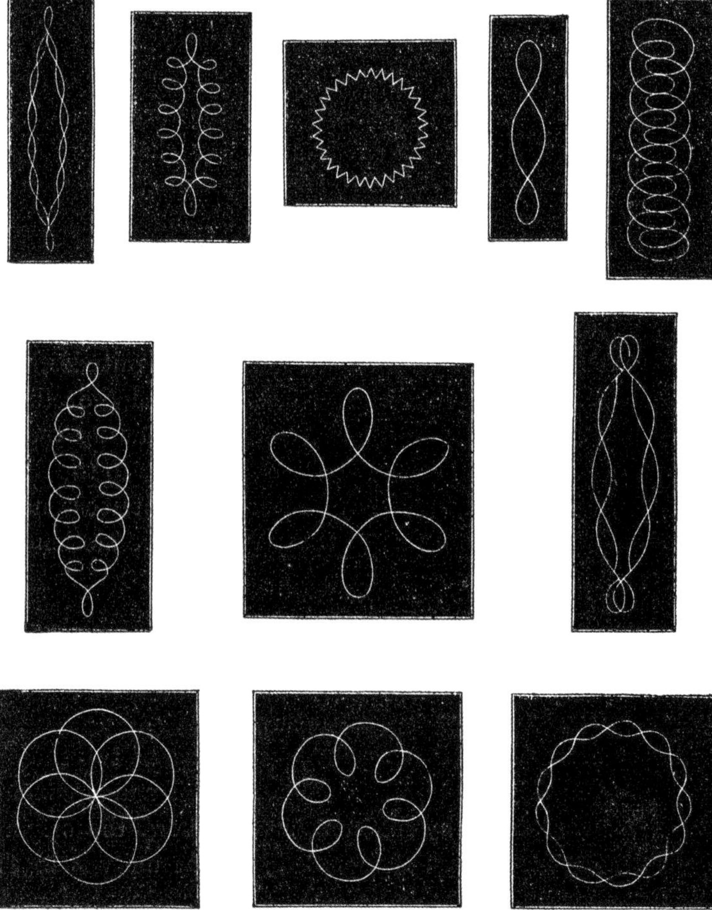

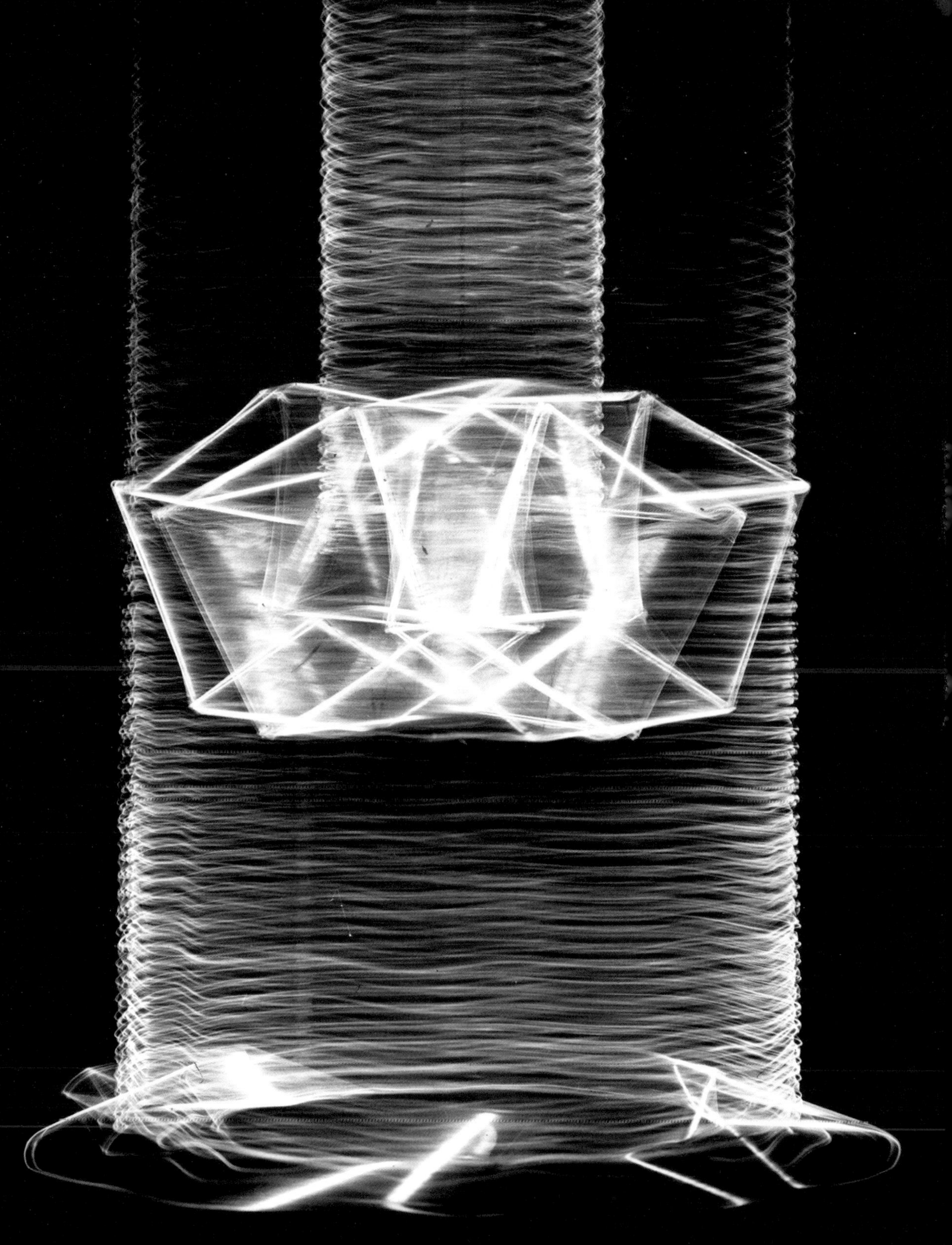

Fig. 9.   Light passes through dust-free air without appreciable scattering.

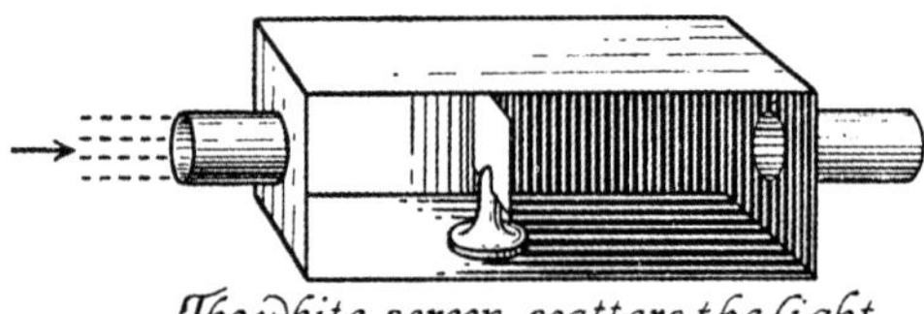

Fig. 10.   A piece of white paper intercepts the light and scatters it, becoming a brilliant secondary source of light.

4.17

Diagrams from William Bragg, *The Universe of Light*. London: G. Bell and Sons, 1933. Courtesy of the Royal Institution.

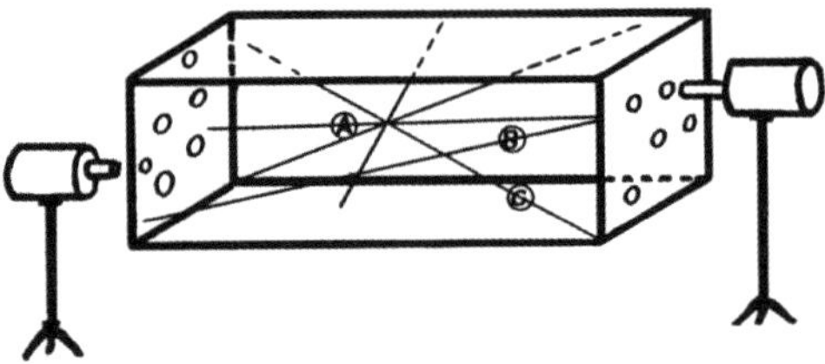

4.18

Nathan Lerner, diagram for a light box. From László Moholy-Nagy, *Vision in Motion*. Chicago: Paul Theobald, 1947. © The Estate of Nathan Lerner.

4.19

Nathan Lerner, *Light Experiment*, 1937. Gelatin silver print, 16 by 20 inches. © The Estate of Nathan Lerner.

By placing objects, like a small screen, within, the light scatters, illuminating the box's interior. In his "Light Start" notes, Kepes illustrates the technique; so, too, does Lerner in a diagram published in Moholy's *Vision in Motion* (fig. 4.18). By placing a camera at one hole, it was possible to capture an image of the box's illuminated interior—an image of light purified from outside interference. A photograph created by Lerner under Kepes's supervision demonstrates the results (fig. 4.19).

All of these experimental tools—the light box, Wheatstone's kaleidophone, Tyndall's magnetic projector, Hartridge's devices for "artificial rainbows" and "artificial sunsets"—demonstrate how Kepes transformed the ostensibly empirical method of science and its basic laboratory technologies, wielding them now for more metaphysical explorations.

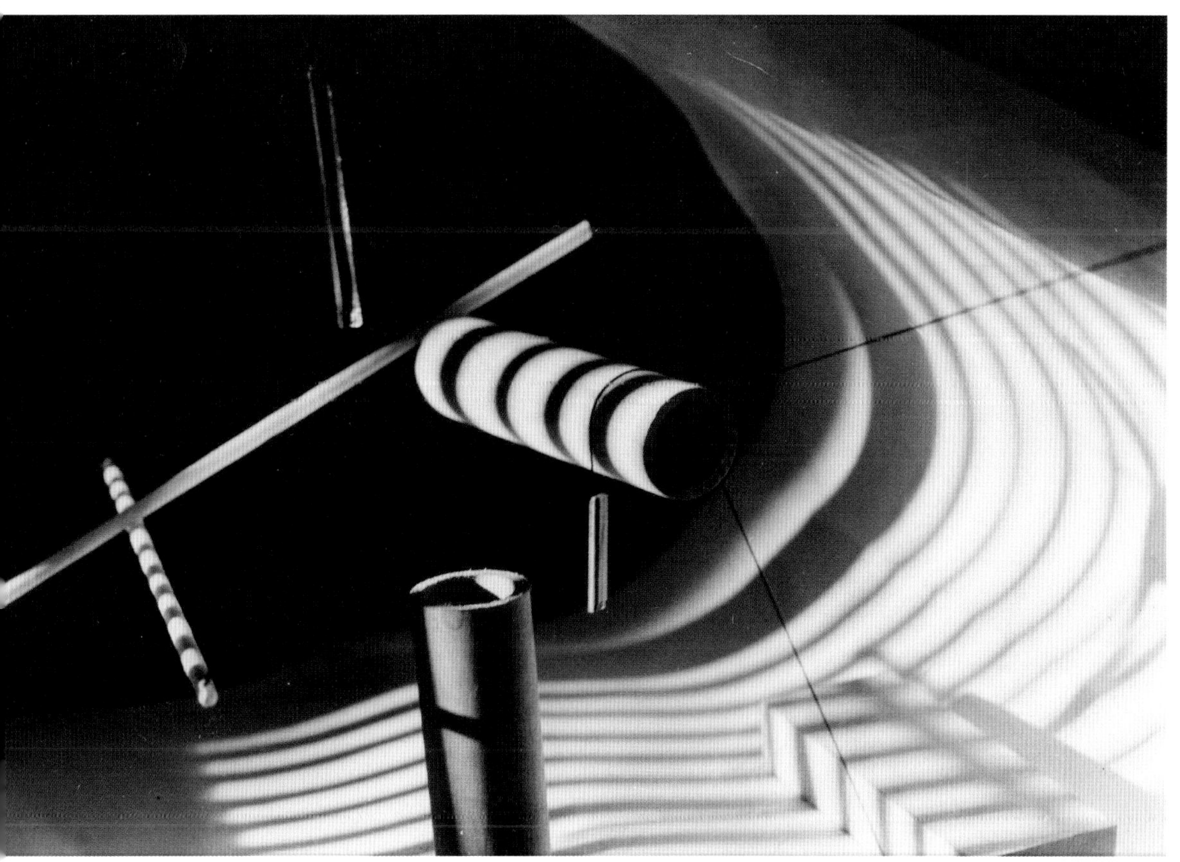

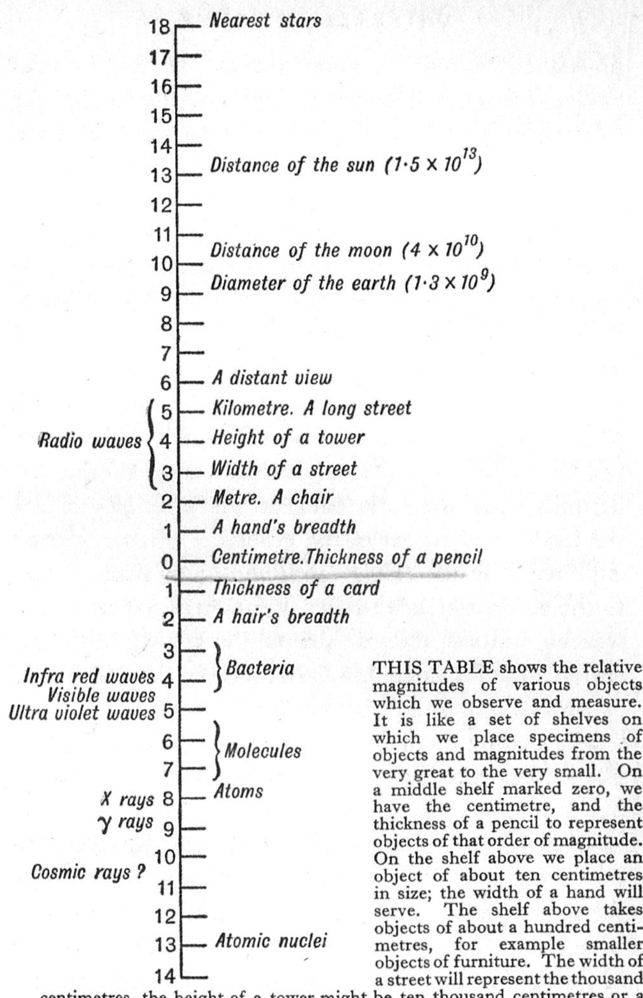

THIS TABLE shows the relative magnitudes of various objects which we observe and measure. It is like a set of shelves on which we place specimens of objects and magnitudes from the very great to the very small. On a middle shelf marked zero, we have the centimetre, and the thickness of a pencil to represent objects of that order of magnitude. On the shelf above we place an object of about ten centimetres in size; the width of a hand will serve. The shelf above takes objects of about a hundred centimetres, for example smaller objects of furniture. The width of a street will represent the thousand centimetres, the height of a tower might be ten thousand centimetres or a hundred metres, and so on. Below the zero shelf comes first a shelf holding something of the order of a millimetre in thickness, as a card; then the hair's breadth on the next shelf and so on. Bacteria are at various heights on the third and fourth shelves down; molecules on the sixth and seventh, atoms nearly down to the eighth. On the other side of the vertical line the various wave-lengths are shown in the same way. Distances are sometimes given in figures. The sun's distance is fifteen million million centimetres or in symbols 1.5×10^{13}. This goes therefore on the thirteenth shelf up.

4.20

Table from William Bragg, *The Universe of Light*. London: G. Bell and Sons, 1933. Courtesy of the Royal Institution.

But Kepes was not only interested in creating images, despite this period of sustained photographic investigation. He soon sought to expand the scope and scale of such engagements. Ideas about light's expanded scale were already present in the scientific literature Kepes consulted as early as 1937. *The Universe of Light*, for example, embraces such expansion in its very title. The book was not only practical but also poetic, eliding the scientific with the sublime:

> **Light brings us the news of the Universe. Coming to us from the sun and the stars it tells us of their existence, their positions, their movements, their constitutions and many other matters of interest. Coming to us from objects that surround us more nearly it enables us to see our way about the world: we enjoy the forms and colours that it reveals to us, we use it in the exchange of information and thought.**[49]

Bragg illustrated this concept with a chart (fig. 4.20) that climbs vertically both up and down from zero—representing a single centimeter—by powers of ten (the chart anticipates the canonical film *Powers of Ten* created by Ray and Charles Eames decades later). At the top of this light scale, and thus furthest from the single centimeter, are the nearest stars, followed in descending order by the distance of the sun and moon from the earth, and, rapidly falling, the height of a tower, the width of a street, the size of a bacterium, and the speck of an atomic nucleus. For Bragg, the purpose of this chart was not to illustrate the great chain of being (these links call to mind the images Kepes compiled in his *New Landscape* projects) but to map it all onto a related system of light, with "light" conceived broadly as the full electromagnetic spectrum and the varying scale of its wave phenomena (specifically the amplitude of waves). He indicates as much at the left of his chart, where he shows how light spans the same great magnitudes, from large radio waves to visible light waves and, rapidly falling, to the infinitesimal waves of cosmic rays. Light, for Bragg, was a universal system, a way to conceptualize the world through interrelations of scale. For his Light Book, Kepes would later prepare quotations from Bragg's volume that emphasize this sublime quality. Light, writes Bragg (and quotes Kepes), "is akin to the matter of which all things animate and inanimate are made. The universe is its sphere of action. We do it no more justice when we speak of the Universe of Light."[50]

In this way, Kepes began imagining light as an expansive phenomenon, one that exceeds the confines of any single image. Light could be exploded to the scale of the total environment.

Kepes compiled extensive lists of his observations on disparate and diffuse light phenomena, on this "universe of light." Many of his observations are banal: "water in glass on table," he writes on one page. Elsewhere he notes the reflections in "polished brass buttons" and a "water puddle." He made another notation about the "mirrored lights in shop windows" and the "pulsating brightness on the pavement."[51] But still other observations are rather nuanced, suggesting how Kepes saw light everywhere. Another list: "embroidery of light on the skyline," "windy snow in the beam of car headlights," "the moving shadow play produced by bypassing motor car light—projected on the wall (through curtain)," "raindrops running down window pane or rain on screen," "bubbling water," and "diamond, brilliant," a reference to the faceted lapidary cut of the gemstone and the wealth of light it scatters.[52]

These observations grew increasingly elaborate as Kepes compiled more and more lists (this activity was one of the core components of his Light Book research; he made hundreds of such pages). One records the dynamic effects of constantly moving automobile lights, the luminous traffic of the street: "bypassing bus—moving Xmas tree," "the red threads of moving backlights of the cars" and the "oncoming glaring headlights," all "woven into [a] mobile fabric."[53] Another describes various "lightspace sequences" in which the experience of light modulates over time, as in the transition from light to dark and then light again while driving a car through a highway tunnel. Kepes categorized his observations, ordering them into themes and assigning them examples from both lived experience and the history of art: reflections (a chrome car bumper or the metallic sheen of a Brancusi sculpture), figure-ground relationships (the dark silhouette of the city against the sky or a Franz Kline canvas), natural and artificial light sources (the sun and stars or the paintings of Rembrandt), screens and transparent surfaces (light filtering through a tree canopy or a Japanese shoji screen).[54] One example from his notes indicates the vastness of light experience: "man's most embracing cosmic frame of reference," notes Kepes, is the "majestic firmament," the light created by the "curtain of sparks[,] stars[,] moon[,] sunshine."[55] Kepes summarizes this expanse with his own aphorism: "The whole visible world, natural and man-made, is a light world. Its heights and depths, its great outlines and intimate details, are mapped by light."[56]

In addition to studying and analyzing the universe of light, categorizing its endless variety, Kepes began to theorize the possible relationships between its constituent parts. This was also a major component of his light research. He described the experience of light as a "triadic" relationship linking eye, object, and light source; he diagramed it as a simple triangle (fig. 4.21).[57] But, as in all of Kepes's projects, this system grew increasingly elaborate. Introducing variables like the motion of the eye or a multiplicity of objects and light sources creates an infinitely more complicated optical situation.

Another highly impressionistic drawing from his papers imagines a "mobile eye," "mobile light," and "mobile object" at the top of the page, indicating that the "triadic" relationship was in flux (fig. 4.22). At the center, multiple reflecting surfaces—perhaps mirrors—orbit a central light source. The eye of the observer is depicted at left; it is a mobile eye that encircles this apparatus, as if running on a track. Kepes labels the diagram "total situation" and "system." An accompanying note describes the eye as a computer that processes a field of visual relationships; one change in this system alters all of the system's constituent parts. The text suggests the allegorical connotations Kepes gave these relationships. In large script: "interdependence community."[58] Kepes's simulation of scientific processes, his appropriation of technological devices, his exhaustive lists of observations, all offer a romantic vision of light as a unifying force, a way to build community and create interdependence.

Kepes's environmental ambitions were not just imaginary, however; he also explored the possibility of realizing these light systems through designed environments. Kepes first explored the possibilities of using light as an environmental medium in a series of commissions for the Sylvania Electric Company, a Boston-based manufacturer of light bulbs, vacuum tubes, semiconductors, transistors, and components for computers. Sylvania hired Kepes as a consultant—Kepes recalls his role as a "kind of light 'tutor'"—who could advise the company on applications for its lighting technologies.[59] This 1947 collaboration presaged the relationships he developed at his Center in the late 1960s. The collaboration also produced one of his first significant writings about light, a lengthy memorandum on "Color Illumination" which Kepes prepared that year. In his report, Kepes argues that scientific and technological studies of light, like those conducted in the research-and-development laboratories at Sylvania, were inadequate, for they "leave out of consideration certain important human needs," especially "aesthetic needs."[60] Kepes proposed manipulating light to address such needs, though his proposals were hardly realistic or even realizable. In the report, he refers to his ideas as "doubtless Utopian." Years later he recalls them as a series of "overcourageous [*sic*] projections for the future."[61]

Kepes begins his "Color Illumination" memorandum by arguing that human vision is inherently dynamic, and that light must be as well. "Seeing is a mental act," he explains, one that is not only quantitative and physiological—based on the electronic processes of the eyeball and brain—but also qualitative and psychological, built upon subjective experience. Light has "emotional connotations" and resonates with "inherited historical experiences, cultural forms, habits, mores, and symbols"—valences purged by static and sterile artificial

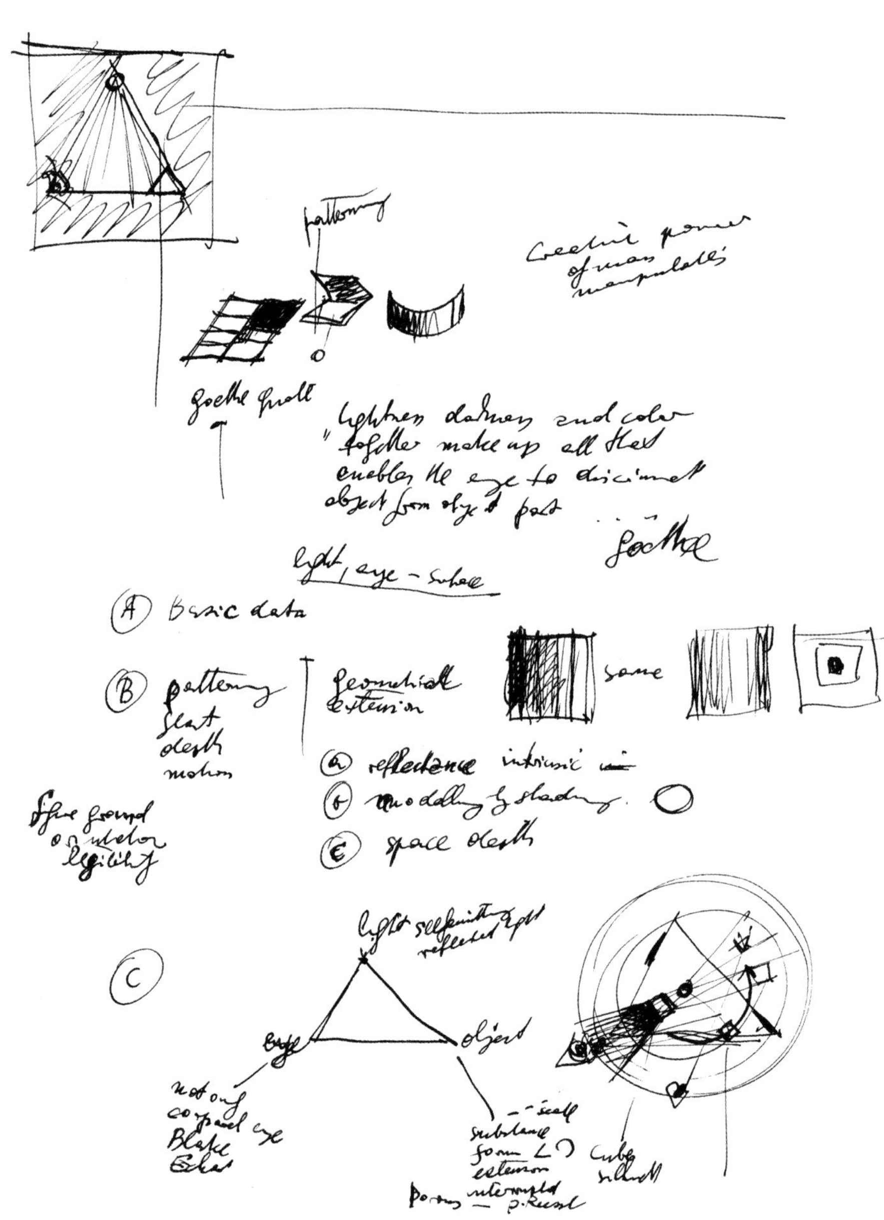

4.21 and 4.22

Gyorgy Kepes, untitled and undated drawings.
Gyorgy Kepes papers (M1796). Dept. of
Special Collections and University Archives,
Stanford Libraries, Stanford, Calif.
© The Estate of Gyorgy Kepes.

spatial reality? — tactile — (pottery, anatomy, olympic games
optical projective transformation (series)
luminous substance — interdependent
field

infinite **series**

mobile eye
mobile light
mobile object/surface) .

view   scale of magnification,
angle + distance
vertical scale

micro

surface → form
form → surface texture
(aerial)

strength of light/darkness
scale of
magnification

opaque — transparent

aerial

system   not object thing directed

triadic

eye as computer
reads the relationship

interdependence
community

total situation

unified system
field relationship
any change in
any factor — implies
change .

polarity complementarity
is needed: constant/change
fixed/closed — open/transient

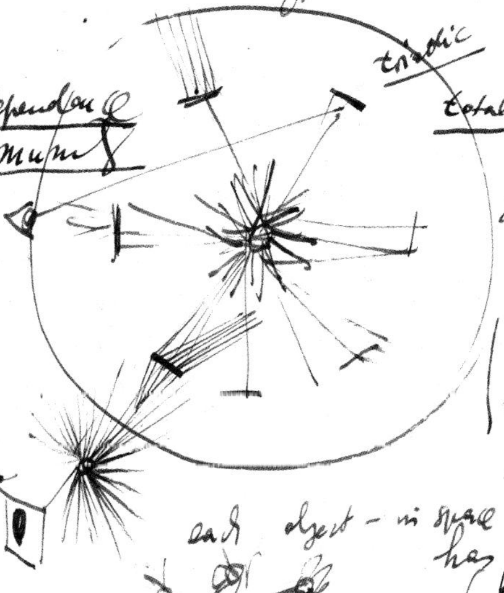

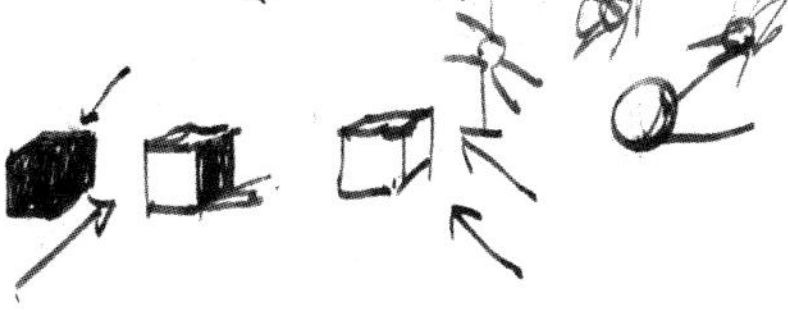

belongs to
has a (illumination family)
each object — in space   has many faces
(projective family)

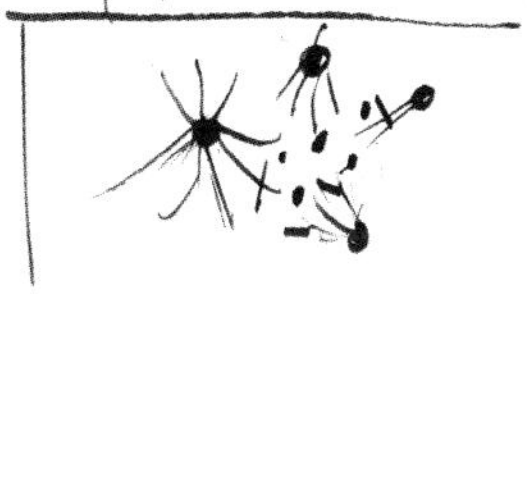

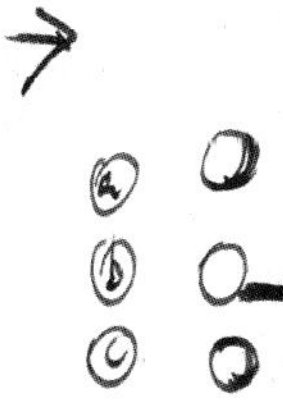

illumination. In order to recover these subjective qualities, Kepes suggests displays that would employ "luminous change" activated by constantly altering the location, strength, color, and direction of lights. The "visual sense is fed by renewed excitation and dulled by permanent excitation," he explains. "The eye needs, as much as any other organ, a sort of continuous gymnastic to retain its elasticity." The goal was to generate these random optical experiences through a series of fluid, ever-changing, ever-shifting patterns of light.[62]

The actual means to accomplish this goal, however, are vague. "We have to rethink the entire issue of light sources, bulbs, and fixtures," Kepes declares, instead offering a conception of light as an "architectonic element" or system in space that combines light sources, reflecting and refracting media, and the observer in a triadic relationship. He envisions the use of luminous "wall panels, screens, and other structural elements." He suggests transforming familiar lamps by changing their shape, coating them "with color filters or color shades," and using them as "flexible tools." Together, these devices and approaches would create an immersive "luminous environment." Such methods would transcend the limitations of the work of art as a discrete two-dimensional or three-dimensional object—a photograph, a painting, a sculpture—but also the work of architecture as a discrete physical space. They would liberate light from scientific and technological apparatuses. Kepes christened these works "projected mobile color murals" (he later refers to them as "projected kinetic light forms").[63]

Nothing came of Kepes's "Color Illumination" memo, however, no doubt because his plans were so impressionistic. But five years later, in 1952, Sylvania again consulted Kepes, this time about actually developing "mobile art forms fashioned of sign tubing; and including fluorescent, incandescent and possibly panelescent" lamps.[64] Kepes accepted this second commission, but his design was no less vague. A now lost drawing submitted to Sylvania depicts his projected kinetic light form (the drawing was photographed with black and white reversed so that the drawn elements of the design appear as if illuminating a dark space) (fig. 4.23). Filaments, bulbs, lit tapes, and an array of diodes fill a thinly outlined spatial expanse. The drawing is less a technical rendering than an aesthetic provocation, not unlike his fantastical "perspective tower." The light box is exploded to an environmental scale; it creates the "total situation" Kepes dreamed about in his drawings.

In a description written years later, he more fully outlines how this projected kinetic light form would function:

> A delicate opaque mobile was to be positioned between a screen made up of photoelectric cells and a strong light source aimed at the screen. A lamp of high thermal character was to be placed under the mobile so that the random air current generated by the heat would keep the mobile in continuous random movement.

4.23

Gyorgy Kepes, Sylvania design drawing, 1954.
Photograph by William Alphin of Sylvania
Electric Company. Original lost. Gyorgy Kepes
papers (M1796). Dept. of Special Collections
and University Archives, Stanford Libraries,
Stanford, Calif. © The Estate of Gyorgy Kepes.

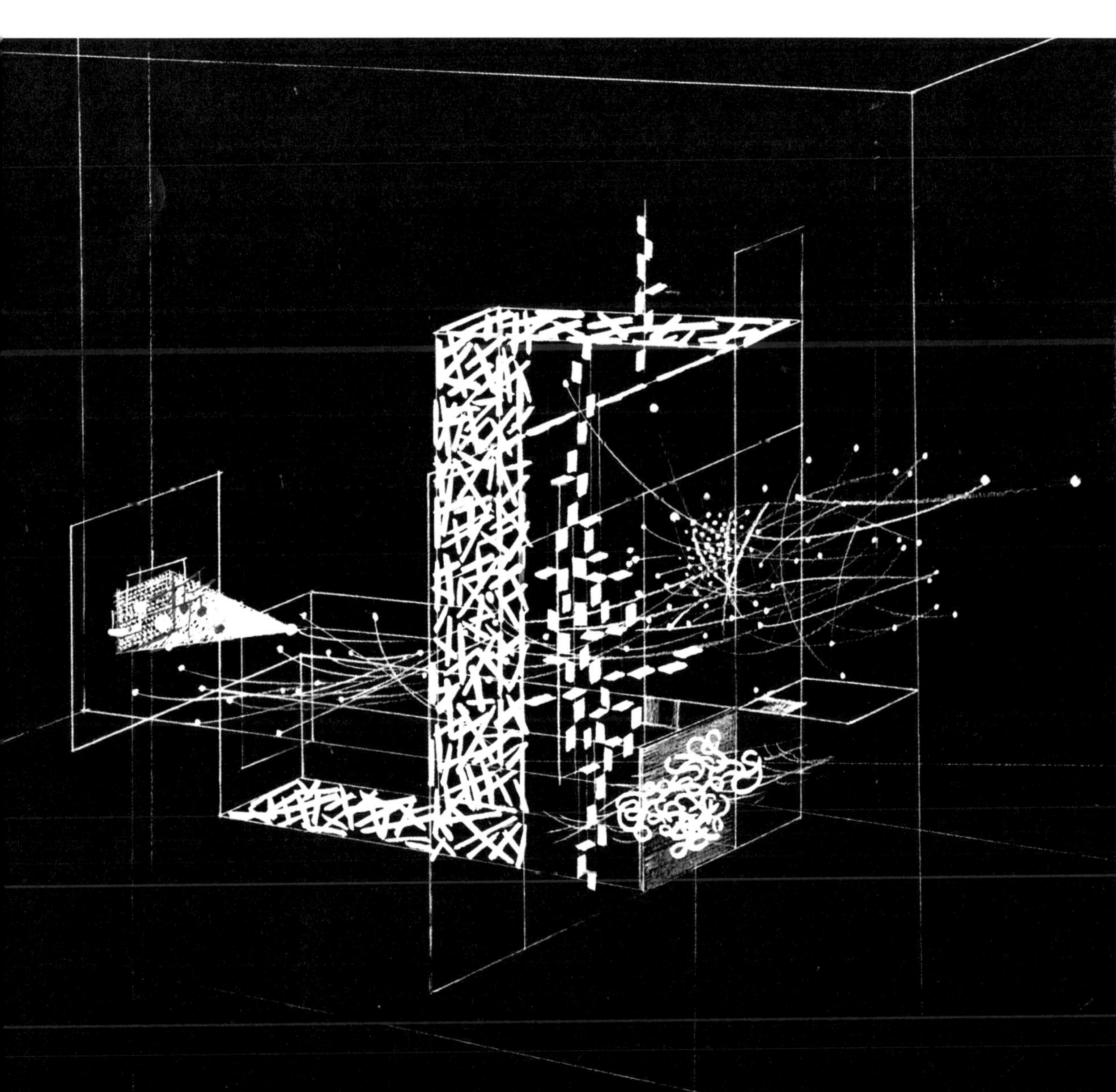

Random changes in the movement of the mobile, and, thus, in the shadows it cast, would, in turn, activate every compartment of the screen of photoelectric cells. Since each of these cells was connected with light sources of specific character placed in different positions within the space, the created illumination was to appear and disappear, racing and jumping from wall to wall and from corner to corner. A further increase of variety and diversity of the effects was to be achieved on the one hand by dimmer, relays, and complex random circuitry, on the other hand by various light-reflecting or light-refracting devices – flat and curved metallic surfaces with different textures, mirrors, lenses, and prisms – each of them able to redirect, multiply, or filter light in some particular way. These explosive kinetic lights were to be contained within a legible architectonic setting or spatial volume. The basic intention was to create a luminous envelope that had no stationary dimension but still revealed a persistent space pattern as it fluctuated, pulsated, opened, and closed.[65]

Kepes proposed a spectacle of light in constant flux (he uses the word "random" four times in this paragraph), with an array of lamps, photocells, electronic circuitry, dimmers and relays, and a series of mirrors, lenses, and prisms, all reflecting and refracting light. Sylvania had no idea what Kepes was proposing, or how they might realize it; after viewing the preparatory sketch and Kepes's budget estimate of $11,500, they declined to proceed.[66]

But the project is still significant. It reveals Kepes's shift of emphasis from the creation of images to the creation of environments. It also continues the method of his earlier photographic investigations through the simulation of scientific processes and appropriation of technological devices. He began to understand this approach as a form of aesthetic transformation; in a much a later note, prepared as part of his Light Book material, Kepes addresses this process, specifically how technologies used by the military could be redirected and repurposed—"channelized," he calls it—toward new ends. "Today electronic calculating devices trace movements of distant rockets [and] release and guide defensive weapons—techniques developed there could be channelized in controlling complex dynamic illumination patterns." These dynamic illumination patterns would use military power to provide, as Kepes describes it, the "reverse" or "counterpart" of military power, a more human, more humane vision. Parenthetically, he connects these ambitions back to his initial plans for Sylvania: "see my Sylvania project."[67]

And indeed, Sylvania's technology at the time of Kepes's collaboration was associated with warfare; in addition to light bulbs for the consumer market, the company also manufactured electronics for the military. Sylvania actively pursued research-and-development contracts; they opened a classified laboratory devoted to defense research in Silicon Valley in the 1950s.[68]

But Kepes's intervention was not just a matter of applying technologies to new uses; he was also, in the process, transforming the visual qualities they

create. He aimed to challenge the rationality and logical control of scientifi-
cally and technologically mediated experience by turning his light forms into
something that was, he believed, more alive. Cold, static, artificial light would
become warm, kinetic, and natural. Closed systems would become open.
He describes his intended effect as "randomness in sequence"—an ordered
disorder that would produce constant variation in the strength and form of
light, and would invert the "mechanical repeat pattern," the "standardization,"
inherent in Sylvania's equipment. He considered this transformation an act
of "rebelling" against "fixed stationary forms."[69] The goal was to create an
experience that was consistent but shifting, stable but changing—or "invariant
in transformation," as Kepes described it. Elsewhere, he elaborates randomness as a "road to escape contrived, controlled, outworn gestalt." Randomness
could "reveal hidden deeper aspect [sic]," changing the "technical," "mechanical," and "logical" attributes of these light forms into a "glimpse of richness."[70]

Kepes's ideas derived in part from his own study of then emerging
research on what we now call "new media." He read an essay by Ralph K. Potter, a Bell Telephone Laboratories acoustical engineer, which put forth an early
argument for the use of electronic equipment in the visual arts.[71] Titled "New
Scientific Tools for the Arts," Potter's 1951 article claimed that the artist's traditional tools were obsolete. Painting was still practiced with paintbrushes, but
new tools could invigorate it; Potter lists ideas ranging from "ways to paint
electrically on television type screens" to "'electronic brains' to act as robot
assistants in searches for interesting visible and audible designs" (both ideas
would be realized in subsequent decades through, for example, video art; they
suggest areas of interest at Kepes's CAVS). Potter offers new tools for old arts.

But he also makes a more significant proposal, suggesting that technology need not be applied to old artistic disciplines but could create entirely new
ones: new tools for *new* arts. The "arts of complex tools remain to be discovered," he notes. By way of example, he offers a brief history of the search for a
machine that could synthesize sight and sound through colored light displays,
referring to mathematician Louis Bertrand Castel's 1720 "ocular harpsichord,"
a keyboard-like instrument that could project light through stained-glass
panes; and the artist Bainbridge Bishop's 1873 "Color-Organ," which used
pedals and keys to control gas lights illuminating a white screen. Bishop was
interested in "painting music," as he called it.[72] The painter Alexander Wallace
Rimington designed another version of the apparatus; so, too, did Thomas
Wilfred with his Clavilux.

Kepes was fascinated by the potential of these technologies, and by the
way Potter's examples brought together the "two cultures" (Castel was a

scientist, for example, and Bishop was an artist). Evoking Potter's historical account, Kepes describes a "palette of light technology" that includes materials that could be used as new tools for new arts: candles; matches; incandescent bulbs; tungsten lights; arc lamps; stroboscopes; and tubes filled with argon, neon, sodium, or mercury. He added to his palette more unusual lighting sources: chemiluminescence (light generated by chemical reaction); triboluminescence (the creation of sparks from friction); and phosphorescence (the release of light without combustion).[73]

Kepes also engaged new media in ways that depart from Potter's examples; he challenged technological and scientific newness by paradoxically employing this "palette of lighting technology" for atavistic, even folkloric uses. He embraced the new in order to discover the old; the futuristic in order to find the ancient. Kepes frequently described his light forms as "tapestries" in light, and his notes indicate a special fascination with the textiles of the Paracas civilization, an ancient pre-Incan people from present-day Peru who innovated new forms of weaving. Their textiles are "incomparable in chromatic richness," with "triumphant harmonious chords of colors alternating in rhythmical sequence," notes Kepes.[74] He was also fascinated with sixteenth-century Incan textiles, which he described as carpets "woven by light." He was especially intrigued by the way these textiles were both organized and accidental, or invariant in transformation; changing one's angle while viewing them seemed to modulate the fabric itself, generating an "oscillation of chromatic richness."[75]

Kepes's Sylvania design was never built, only imagined. But some of his other commissions were constructed. In 1950, he designed a mural for the flagship Radio Shack store in downtown Boston that animated a depiction of a radio receiver, as illustrated in a technical diagram of its electronic components (fig. 4.24). Alternating neon and black-light tubes flash against a porcelain enamel and metal surface on the storefront's façade. A schematic radio at the center of the mural translates the AM waves of a radio signal, at left, into output waves or audio, at right. The diagram depicts a transformation of the inaudible into the audible. But for Kepes, the mural diagrammed the transformation of radio waves not only into sound but also into light—the radio waves were, after all, actually inscribed in neon tubes. His preferred photographic documentation of the piece, taken by his assistant at MIT, Nishan Bichajian, rendered it not as an architectural mural but as a pulse of optical signals: pure light (fig. 4.25). In his notes, he further transforms the technical and scientific logic of radio signal modulation into visual metaphor. Kepes provides another impressionistic drawing (he writes "crude diagram" at the bottom of

Gyorgy Kepes, *Kinetic Outdoor Light Mural*,
1950, for Radio Shack, Boston, Massachusetts.
Photo: Nishan Bichajian. © The Estate of
Gyorgy Kepes.

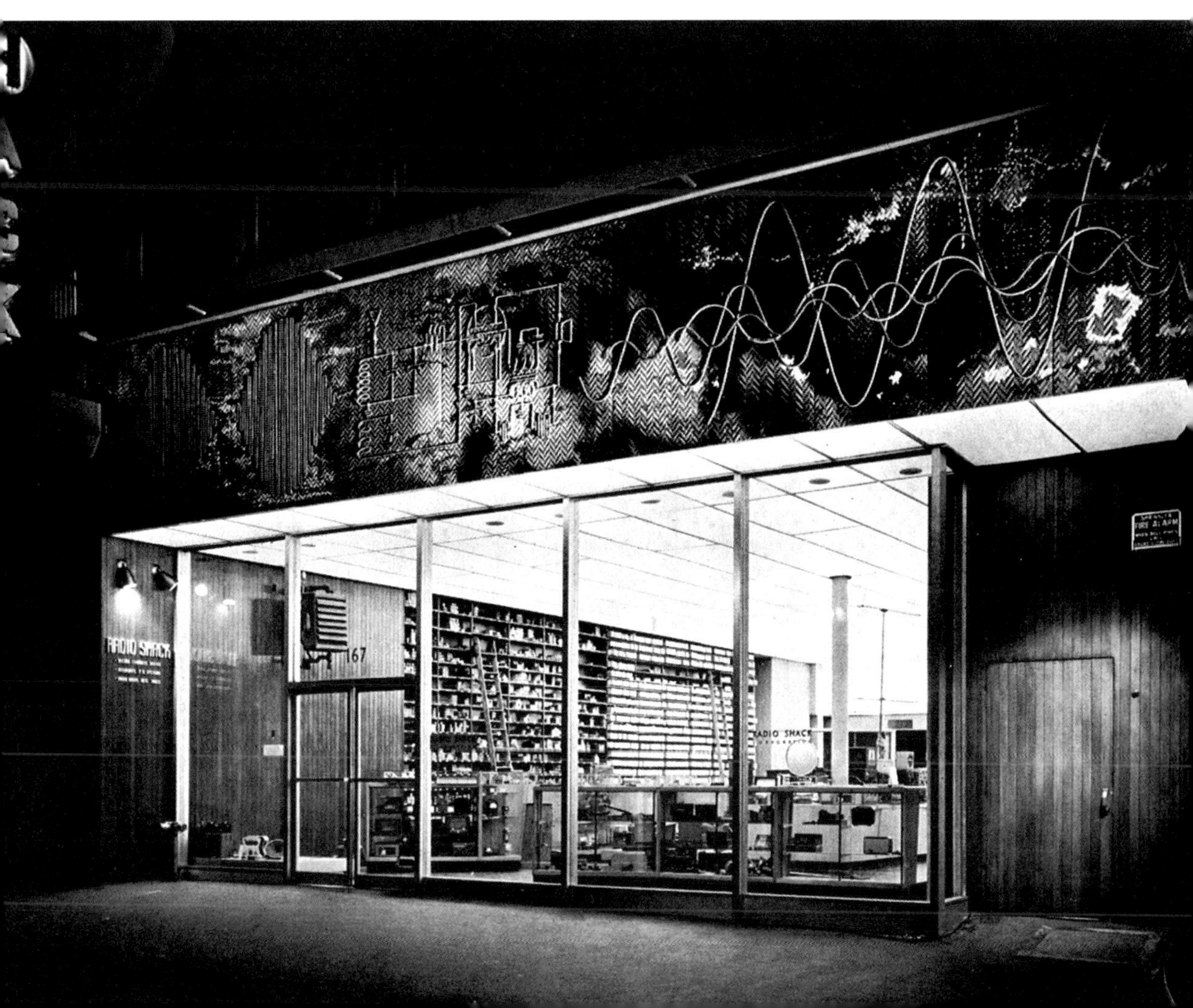

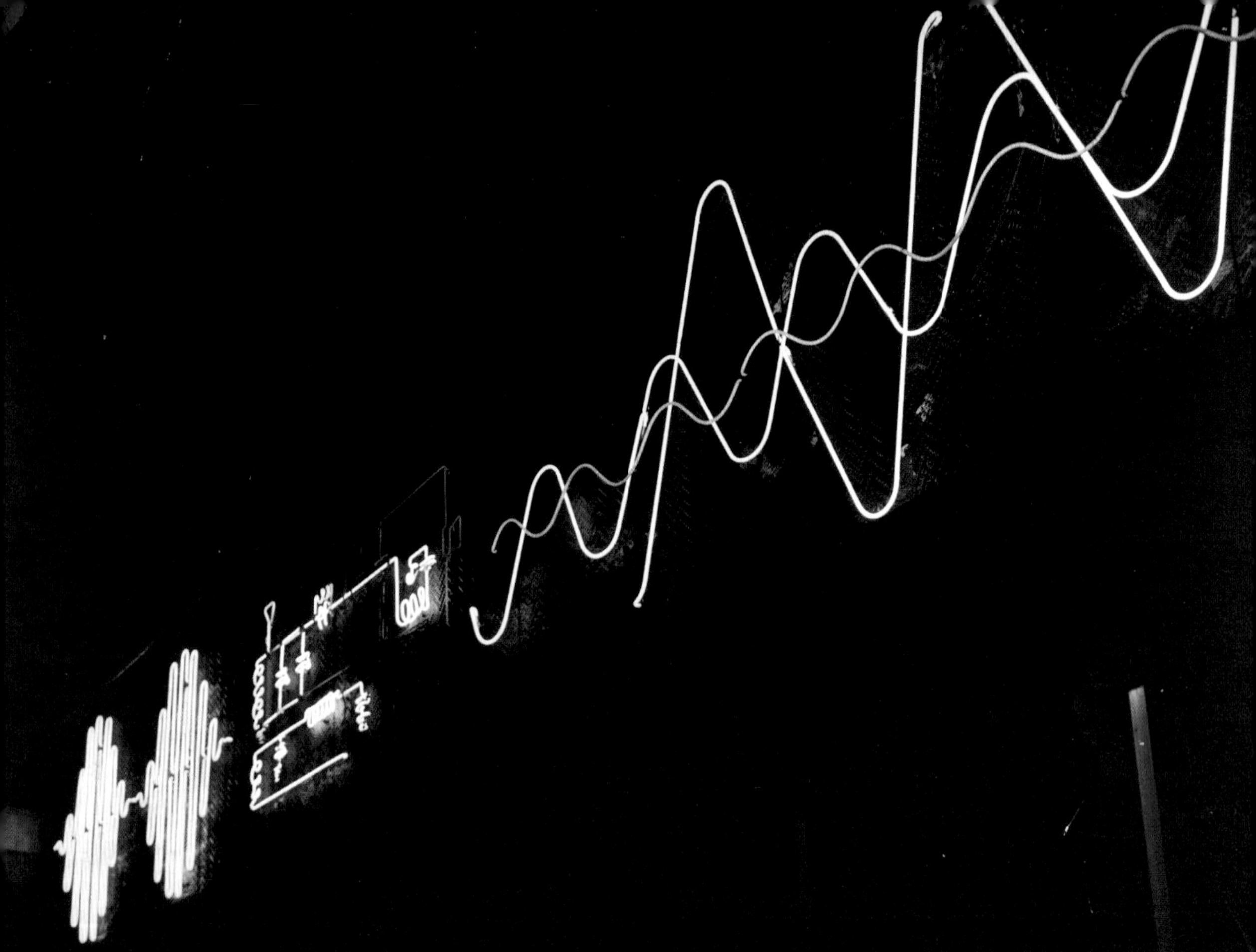

4.25

Gyorgy Kepes, *Kinetic Outdoor Light Mural*, 1950, for Radio Shack, Boston, Massachusetts. Photo: Nishan Bichajian. © The Estate of Gyorgy Kepes.

4.26

Gyorgy Kepes, untitled and undated drawing. Gyorgy Kepes papers (M1796). Dept. of Special Collections and University Archives, Stanford Libraries, Stanford, Calif. © The Estate of Gyorgy Kepes.

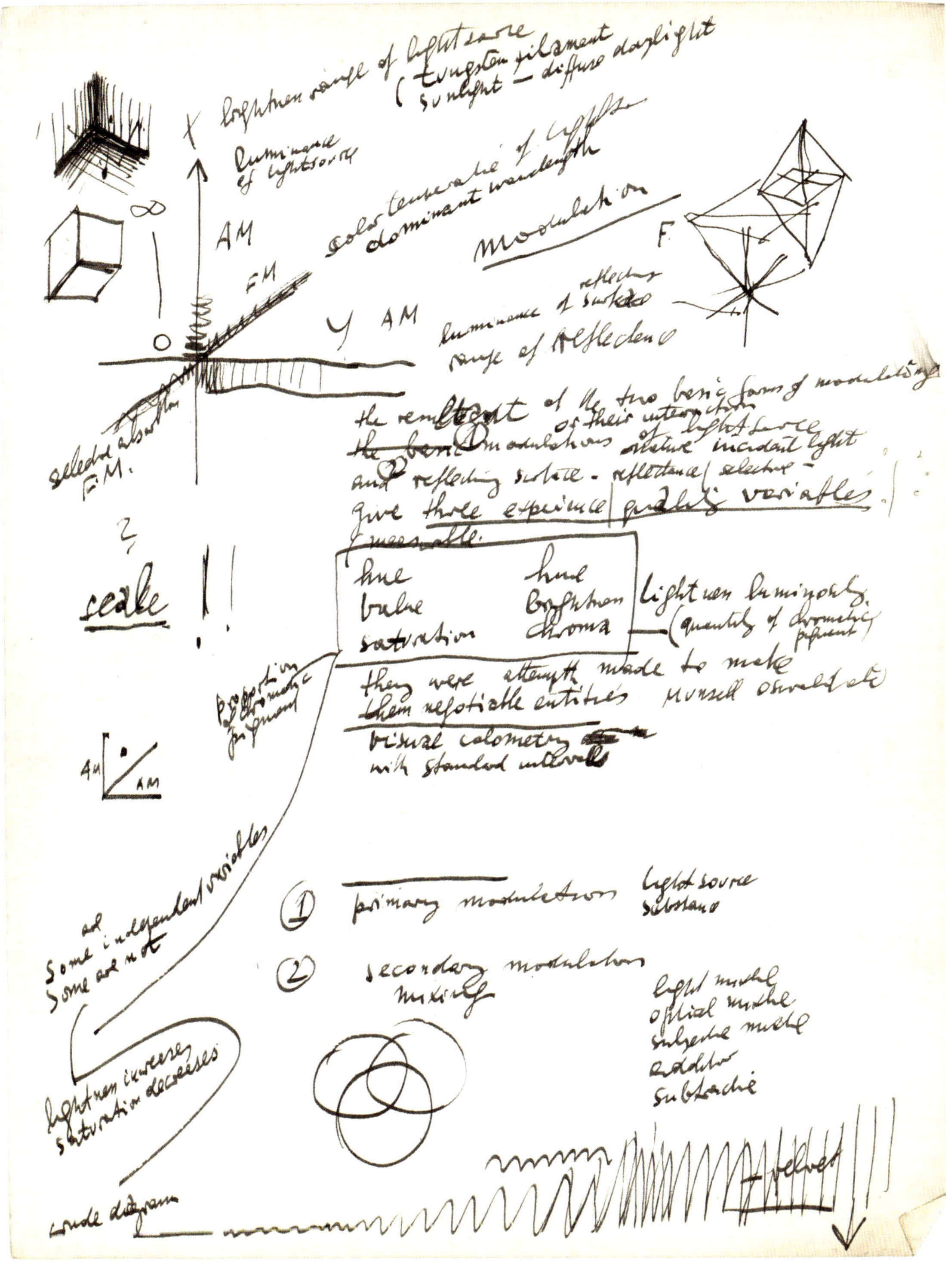

lightness range of lightsource
(tungsten filament
Sunlight — diffuse daylight

luminance
of lightsource

X

colour temperature of lightsource
dominant wavelength

AM

FM

Modulation   F

Y AM

luminance of surface
range of reflectance

selective absorption
F.M.

the resultant of the two basic forms of modulating
the beam of modulations of their interaction
and reflecting surface — reflectance (selective —
give three experience[?] practical[?] variables
of measurable.

scale ! !

| hue | hue | lightness luminosity
| value | brightness | (quantity of chromatic
| saturation | chroma | pigment present

they were attempts made to make
them negotiable entities   Munsell ostwald etc

visual colometry
with standard intervals

proportion
of chromatic
pigment

AM

some independent variables
some are not

lightness increase
saturation decreases

① primary modulation   light source
                             substance

② secondary modulation   light mixture
   mixing   optical mixture
                              subjective mixture
                              additive
                              subtractive

crude diagram

the page) (fig. 4.26). At the top left, he creates a graph labeled AM (amplitude modulation) on the x and y axes, and FM (frequency modulation) on the three-dimensional z axis; he then attempts to diagram light through the technical language of signals. According to Kepes, AM might represent the "luminescence of light source" and the "luminescence of reflective surface." FM might represent the "color temperature of lights." The graph is nonsensical—radio waves and light waves are not identical—but it suggests how Kepes merged the scientific with the aesthetic in an attempt to make sense of the infinite dimensions of light, how it might be expanded by shifting variables to create seemingly endless variations. The graph is another fantasy.

Kepes's "mobile kinetic light form" for the KLM ticket office on Fifth Avenue in Manhattan, which I first introduced in relation to the aerial view, also employed the extended palette of light technology, an array of electronic equipment, to invert the mechanical and standardized attributes of that equipment (fig. 4.27). Kepes also preferred photographic documentation of the work that pictured it not as a static architectural mural but as an animated luminous expanse; Bichajian photographed it from an oblique angle to emphasize this quality (fig. 1.5).

Kepes believed all of these projects were too limited, too controlled, too contained, still too static. "Kinetic light art on the scale of the whole environment has no actual, full-size physical examples. In this sense it does not yet exist."[76] This belief in unrealized potential was perhaps the clearest initial impetus for his Light Book, which attempted to articulate these utopian aspirations.

The Light Book

This survey of Kepes's experiments in light serves as an introduction to the Light Book, but we have actually already entered the project; its accumulated contents have animated my discussion of Kepes's photography and mobile kinetic light forms, and the origins of his ideas about light's transformative and redemptive potential.

The Light Book consumed Kepes for years, especially during intensive research and writing through the 1960s. But it remains an unknown undertaking. Never finished and never published, despite Kepes's sustained efforts to complete it, the Light Book would never see the light of day. Aside from the primary material Kepes left behind, there is almost no evidence of the project in the historical record; where it has been acknowledged, it is the project's incomplete state that is its most noteworthy attribute.[77] Otto Piene notes in 1973 that Kepes "has been working for a long time on the definitive book on light art, a *magnum opus* whose scale currently forbids it to get published."[78]

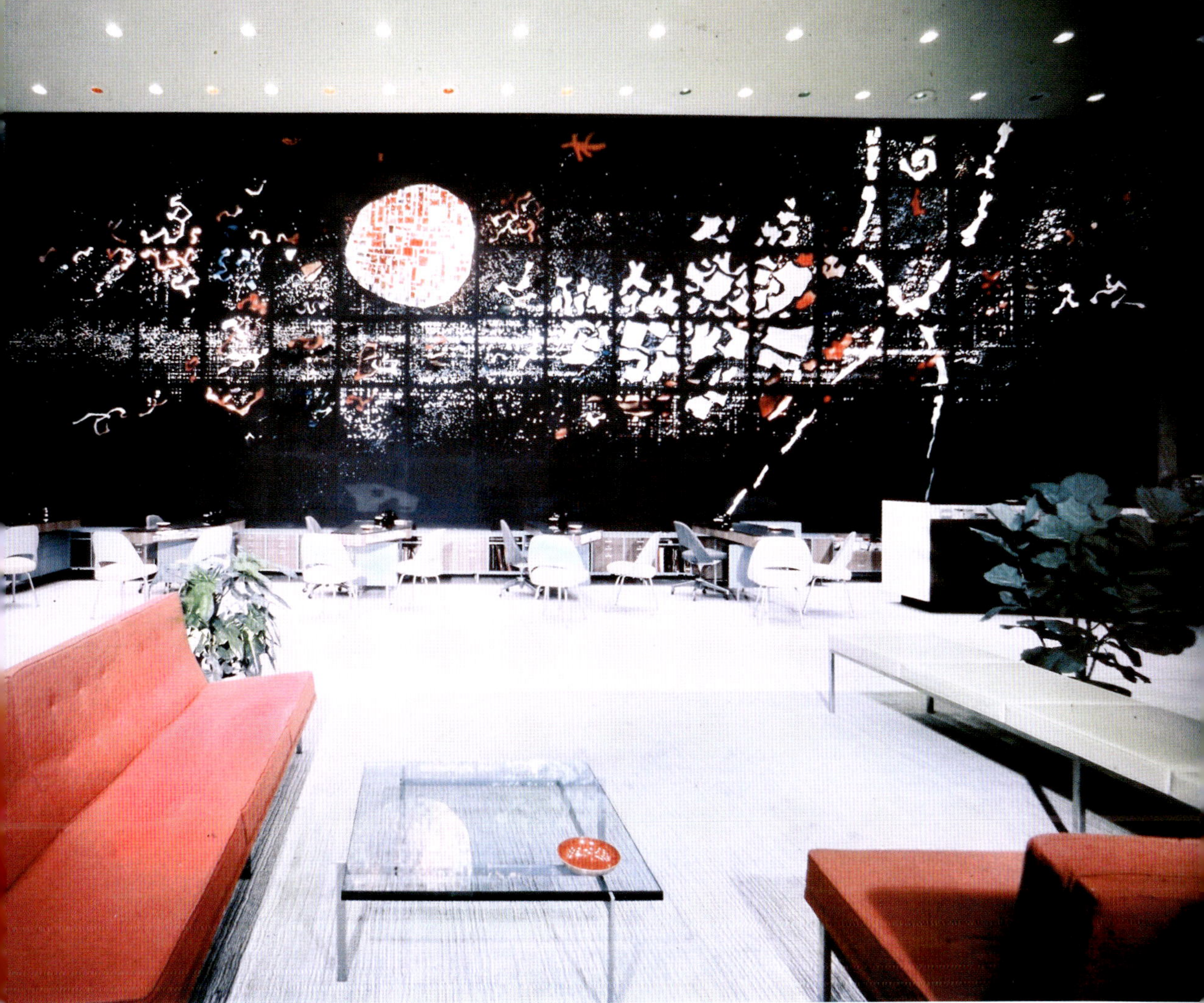

4.27

Gyorgy Kepes, *Programmed Light Mural*,
KLM ticket office, New York, 1960.
Gyorgy Kepes papers (M1796). Dept. of
Special Collections and University
Archives, Stanford Libraries, Stanford, Calif.
© The Estate of Gyorgy Kepes.

When Kepes retired from MIT the following year, a handful of news articles about his departure similarly explained the first task of his emeritus career as, at long last, the completion of the Light Book—an endeavor that never succeeded.[79] The Light Book is the magnum opus that never was.

It is therefore inaccurate to describe Kepes's Light Book as a cohesive project or a coherent work, let alone a proper "book." There is no single finished final manuscript; there is no book at all. The Light Book exists instead as an archive of textual and visual fragments, originally bundled in manila file folders and stuffed into old Kodak boxes (these were an appropriate, if improvised, repository, for they once housed Kodak-brand photographic paper coated in a light-sensitive emulsion).[80] Within this archive are hundreds of pages of typescripts, thousands of black-and-white photographs, and countless stacks of handwritten preparatory notes filled with drawings and diagrams, all composed and compiled over decades.

To sift through this accumulation—to study its images and texts, and to attempt to decipher a relation between the two—is to discover an unusual artistic and literary artifact. The material had a shifting, porous relationship to Kepes's other projects. He began the Light Book as a single manuscript, accompanied by its own illustrations, around 1960.[81] But the project slowly subsumed other texts and images associated with tangential projects, like his Sylvania design and his "Color Illumination" memorandum, resulting in a palimpsest.[82] The Light Book would take one form, only to later be buried under new layers. The project is therefore opaque, hard to see, but also prismatic in its associations. Light, after all, is allusive; it accommodates a range of meanings. The project thus forms a *mise en abyme* of texts and images, all reflecting and refracting their thematic center. It is as if the Light Book is not only *about* light but also modeled *on* light, attempting to reflect the topic not just in subject matter but also in form and structure. The failure to publish the book may be a natural result of this form, an inevitable consequence of Kepes's aesthetic ambitions.

The Light Book's inception can be formally dated to 1959, when Kepes won a Guggenheim Fellowship for a study on "The Creative Use of Light."[83] At the time, he conceived of the project as a book-length study of light in art, architecture, and design. It would have surveyed "light as a creative medium" in contemporary work from the 1950s through 1970s, especially work employing scientific and technological means to propagate light. "This book is the prolegomenon of an emergent art, the creative management of light," Kepes declares in one version of the work's introduction, suggesting how he considered himself avatar—and his book sacred text—of an artistic movement waiting to be christened.[84]

Kepes had a broad understanding of light as medium; such works might employ electronic components, from blinking bulbs to computer diodes; they might utilize conspicuously optical materials like refractive acrylic or reflective metal; they might create kinetic effects through perceptual illusion. This medium-specific ambition is evident in Kepes's potential titles for the project; while he referred to it as "The Light Book," one page lists other ideas: "Light art and Light architecture," "Light in art to Light Art"—the capital "A" in "Light Art" emphasizing the significance of new medium-specific movement—or perhaps simply "Light-Art."[85] It is also evident in the images he collected for possible inclusion. Through contacts like Howard Wise Gallery and Galerie Denise René, the journals *Signals* and *Leonardo*, and the critic and curator Jasia Reichardt, Kepes collected reproductions of Julio Le Parc's plastic constructions, Günther Uecker's textured surfaces, Lucio Fontana's punctured paintings, and the works of many others—Dan Flavin, Nicolas Schöffer, Gyula Kosice, Richard Anuszkiewicz, and more.[86]

And yet, the categorical requirements quickly became untenable; despite Kepes's calls for "light as a creative medium," his project began encompassing works that explicitly defied the restrictions of medium specificity. There was no medium, only light. Kepes shifted the definition of the work of art from the closed logic of a discrete object to the open logic of media, multimedia, intermedia, and mixed media, to the wider universe of light. Additional possible titles for his book project emphasized this broad scope: he would call it "Language of Light," conveying his universal aspirations, or "Nature and Art of Light," suggesting a continuum across the sciences and the arts.

But the Light Book also shifted from image to environment. The book's central topic continually expanded on the premise that *all* visual experience is essentially composed of and created in light. Kepes moved out from familiar aesthetic topics toward less familiar ones, shifting the project's temporal range into history (eighteenth-century optical toys, Greek mythology, creation stories) and its disciplinary scope across fields (holograms, lasers, fiber optics). The photographs proliferated: the textured ceiling of a mosque, light filtering through a thatch roof, a department store lit up with neon, grand chandeliers, a fireworks display.

Kepes attempted to maintain order over it all. He filed images in taxonomic categories under familiar art-historical terminologies ("architecture," "sculptures," "art using mixed mediums," "paintings on canvas"). Other classifications were more abstract, recalling the visual effects of light ("form modeling," "surface modeling," "projected lights") or Kepes's peculiar interests ("scientific photos," "computer/light displays," "microbiology," "outer space").[87] Yet the images are not arranged within these categories, and the groups themselves are not related to one another. There is no key.

This ever-expanding scope is mirrored in the organization of the book's textual elements. A handwritten "light outline" suggests seven original chapters: "role of light as creative tool," "19th century," "20th century," "light modulation," "cognitive," "esthetic," and "symbolic." But a later typed table of contents—the only formal contents Kepes set to paper—suggests an altogether different book; the historical interest in the nineteenth and twentieth centuries is suppressed, existing only obliquely in a single chapter concerning "ancient light technology (metaphysics of light)"; the other chapters are gone. Kepes instead imagined this version as a compendium rather than a focused analysis. In his original application for a Guggenheim Fellowship, these encyclopedic ambitions are clear; he proposed a study of light as "a creative medium" but also "a systematized knowledge."[88] His ambition was to explore "light itself as a field for the creative imagination, not merely as an adjunct of architecture and planning," by "combining research in areas unassociated at present":

> At present there are only fragments of understanding divided among laboratories, workshops, drafting rooms, studios, darkrooms, and theaters. Our knowledge is splintered among the many fields for which light has major importance. This knowledge can be pooled and we can sift its bulk, find the interconnections and common denominators, select and sort and combine until a hundred tangled currents of light use become one broad, straight stream.[89]

It is impossible to know how Kepes may have assembled the visual elements in such a project. But he likely would have compared and contrasted photographs in the manner of his other books, creating pseudomorphological constellations of images across carefully articulated page layouts. Pulling randomly from his *Light Book* archive: one photograph depicts light streaming through the dome of Hagia Sophia. Another depicts a similar but different image of luminous splendor—the Esso Standard Oil refinery in Baton Rouge, Louisiana. The refinery's cooling towers gleam like a modern cathedral; the images contrast light past and light present (figs. 4.28–4.29).

Or, pulling another from the *Light Book* archive: a bizarre photograph of a nuclear bomb blast captured immediately after explosion shows a grotesque, deformed fireball suspended in the sky (fig. 4.30).[90] The Atomic Energy Commission contracted the production of these images to Edgerton, Germeshausen and Grier, Incorporated (later known as EG&G), a defense firm founded by MIT engineer Harold Edgerton with two of his graduate students. EG&G used a "rapatronic" (meaning "Rapid Action Electronic") camera—a device that employed the same technology that triggered the bomb's ignition—to document the blasts at intervals of mere milliseconds. The photograph allows one to see the otherwise unseeable; the fireball looks like a speck of dust, but also a distant galaxy.

I imagine that Kepes intended to contrast this image, a frightening statement of light as a "blinder of vision," with countervailing images of light as a "begetter of vision." Many photographs in his collection mimic the bomb blast: a mandala-like psychedelic burst created by the pioneer of computer graphics, John Whitney; an image of concentric circles blurred under a distorting lens by Karl Gerstner; a sphere of changing light forms by Julio Le Parc (figs. 4.31–4.33). The black-and-white image renders all of them analogous. In a letter to Edgerton requesting the photograph for possible publication, Kepes reassures the engineer that there would be "no political overtone in the visual commentary."[91] But I imagine Kepes instead creating a visual commentary that is very much as political as it is aesthetic. By forming a visual continuum across the arts, one that would relate to but also rival, even surpass, an image of science, Kepes would provide a highly abstract critique of Edgerton's image and what it represents. He would have used its disturbing quality, its darkness, as a way to reveal light.

4.28 (following pages)

Hagia Sophia, from Kepes's Light Book photographs. Gyorgy Kepes papers (M1796). Dept. of Special Collections and University Archives, Stanford Libraries, Stanford, Calif.

4.29 (following pages)

Esso Standard Oil refinery, Baton Rouge, Louisiana, from Kepes's Light Book photographs. Gyorgy Kepes papers (M1796). Dept. of Special Collections and University Archives, Stanford Libraries, Stanford, Calif.

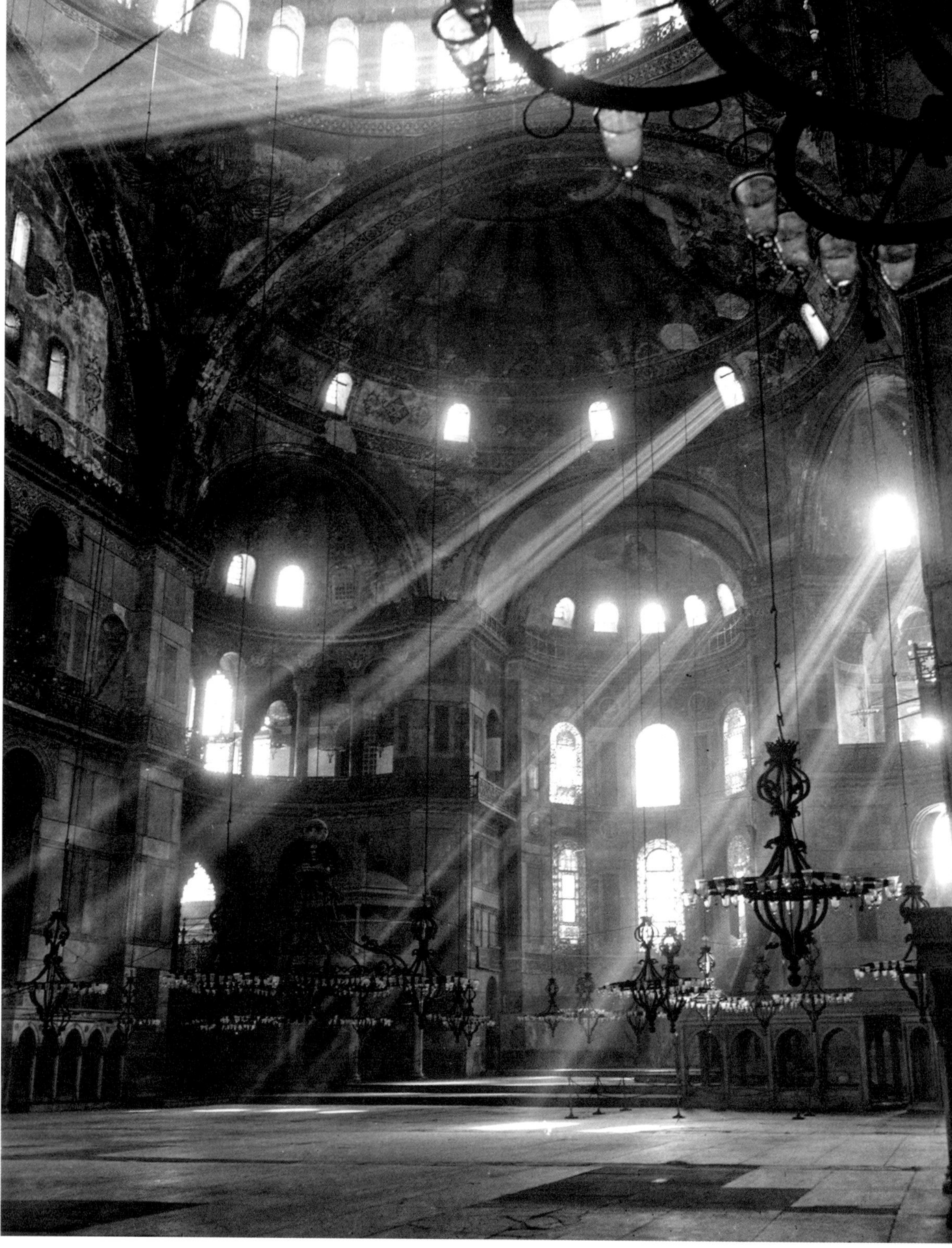

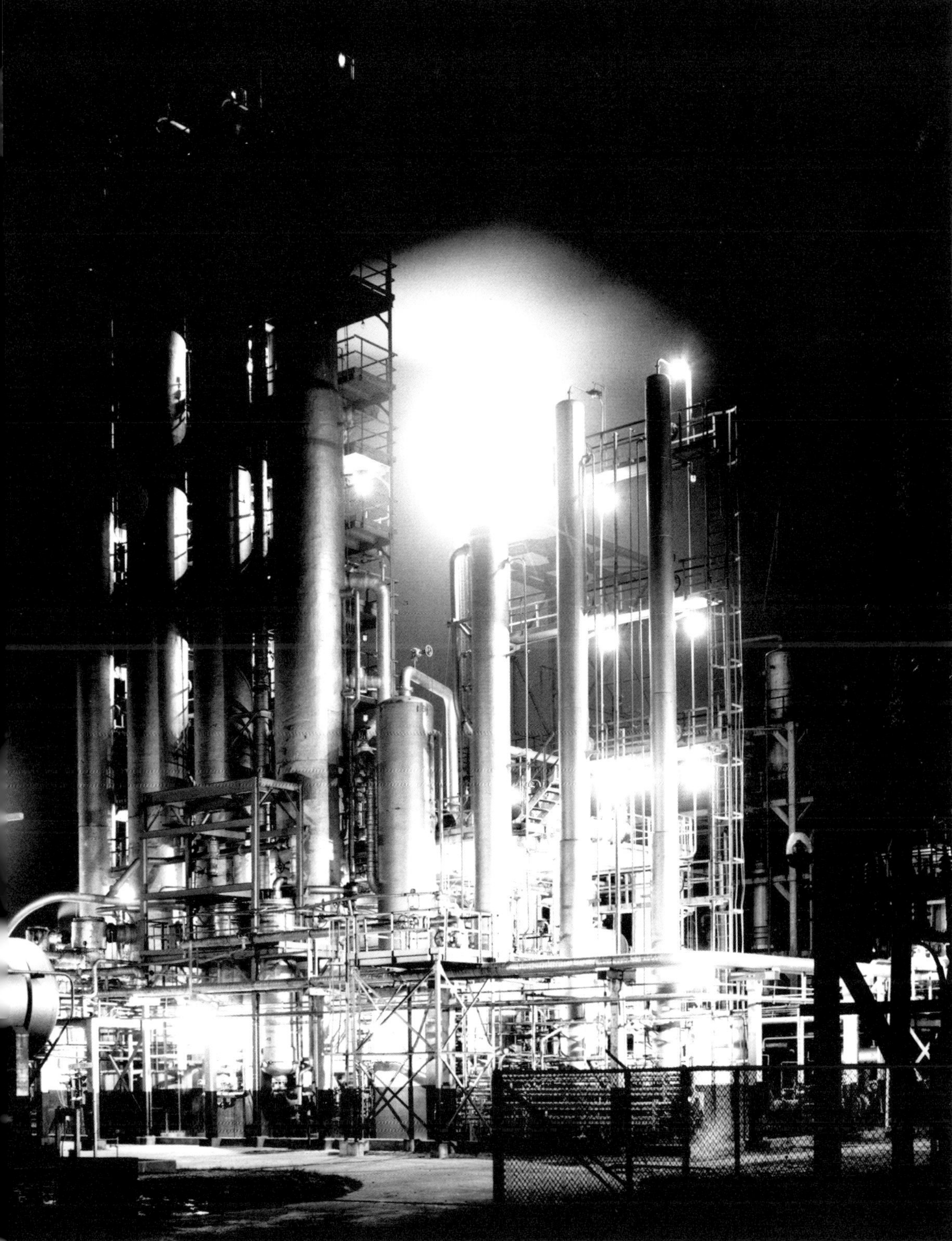

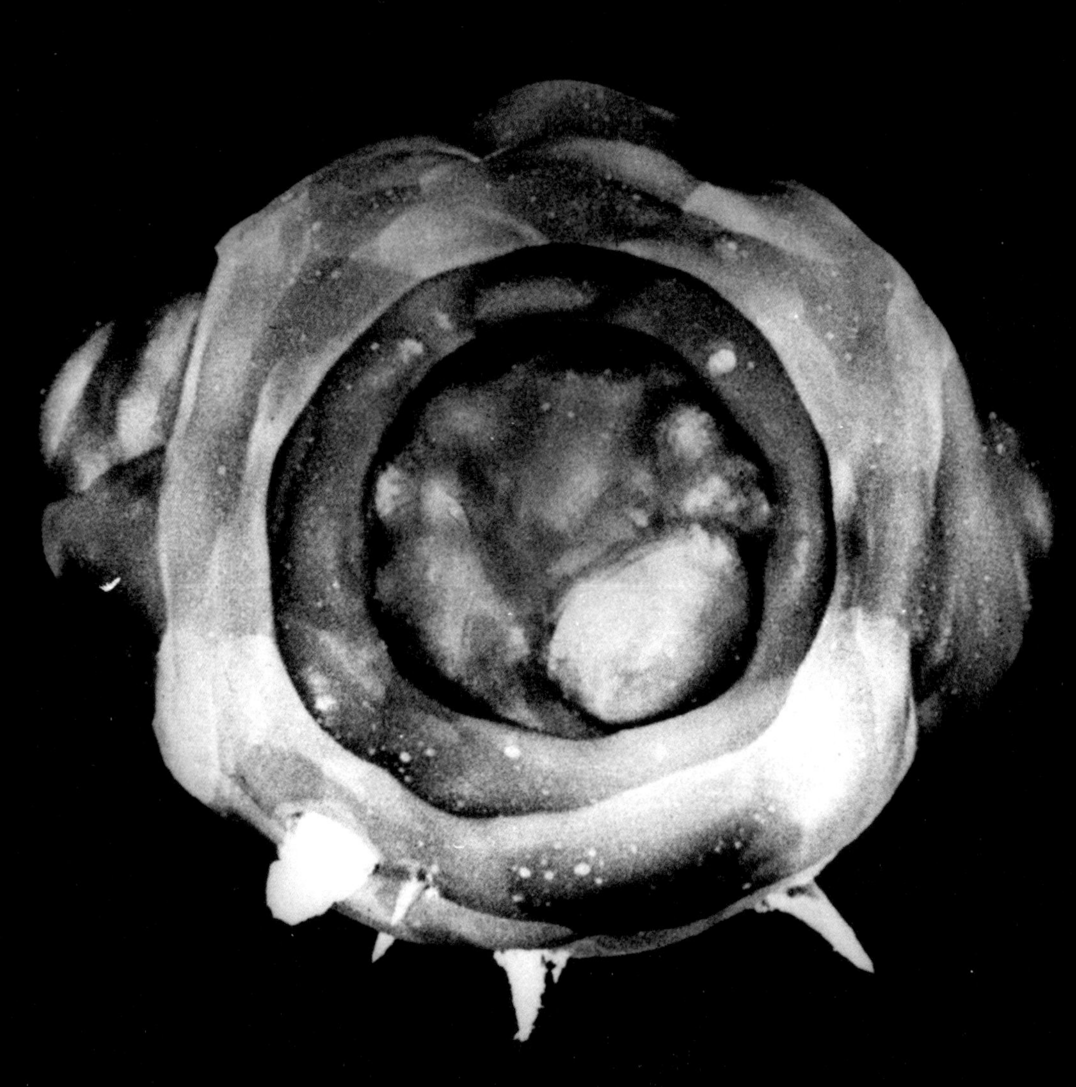

4.30

Nuclear bomb blast, from Kepes's Light Book
photographs. Gyorgy Kepes papers (M1796).
Dept. of Special Collections and University
Archives, Stanford Libraries, Stanford, Calif.

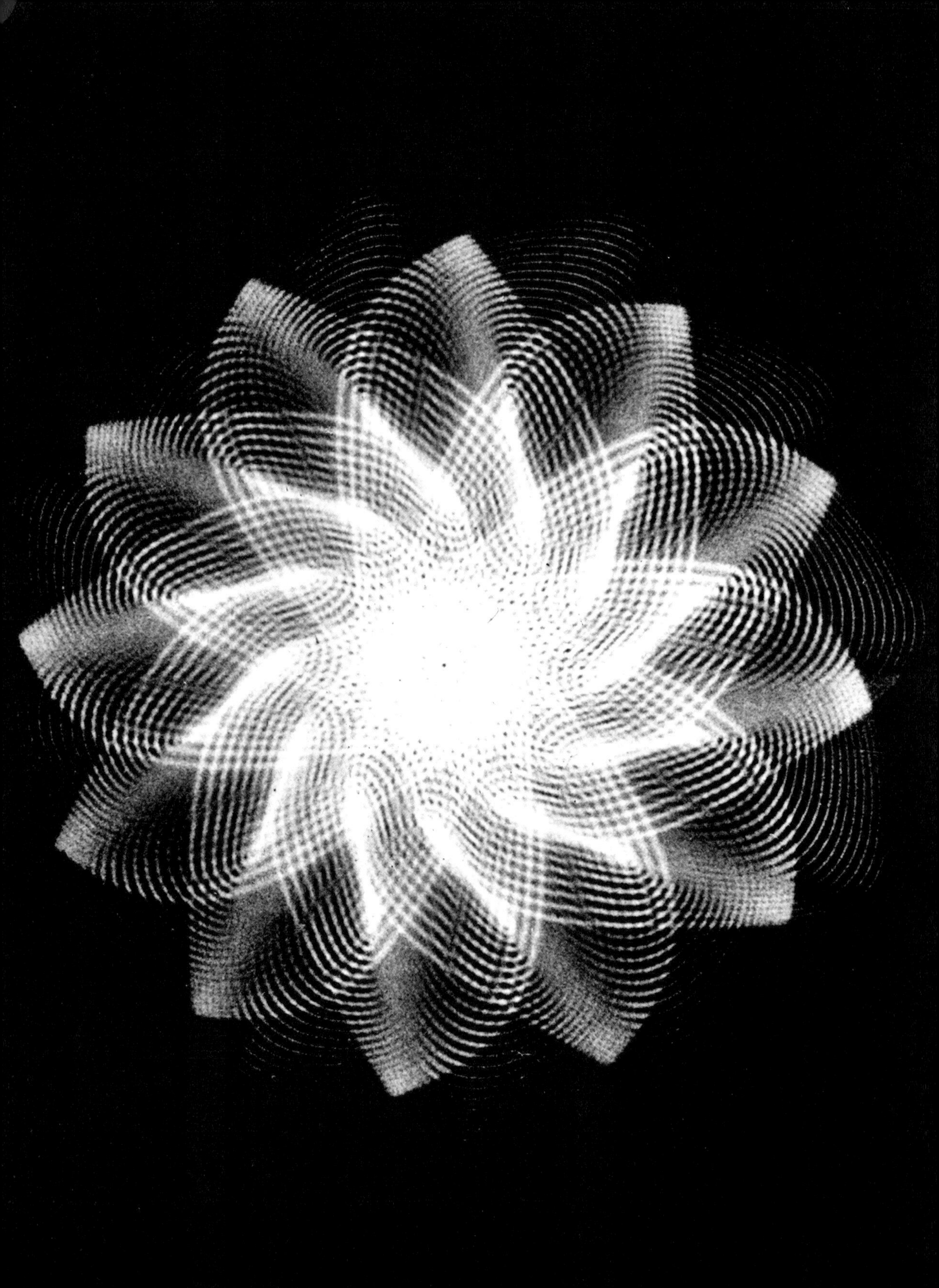

4.31

John Whitney, computer graphic, c. 1965.
From Kepes's Light Book photographs. Gyorgy
Kepes papers (M1796). Dept. of Special
Collections and University Archives, Stanford
Libraries, Stanford, Calif.

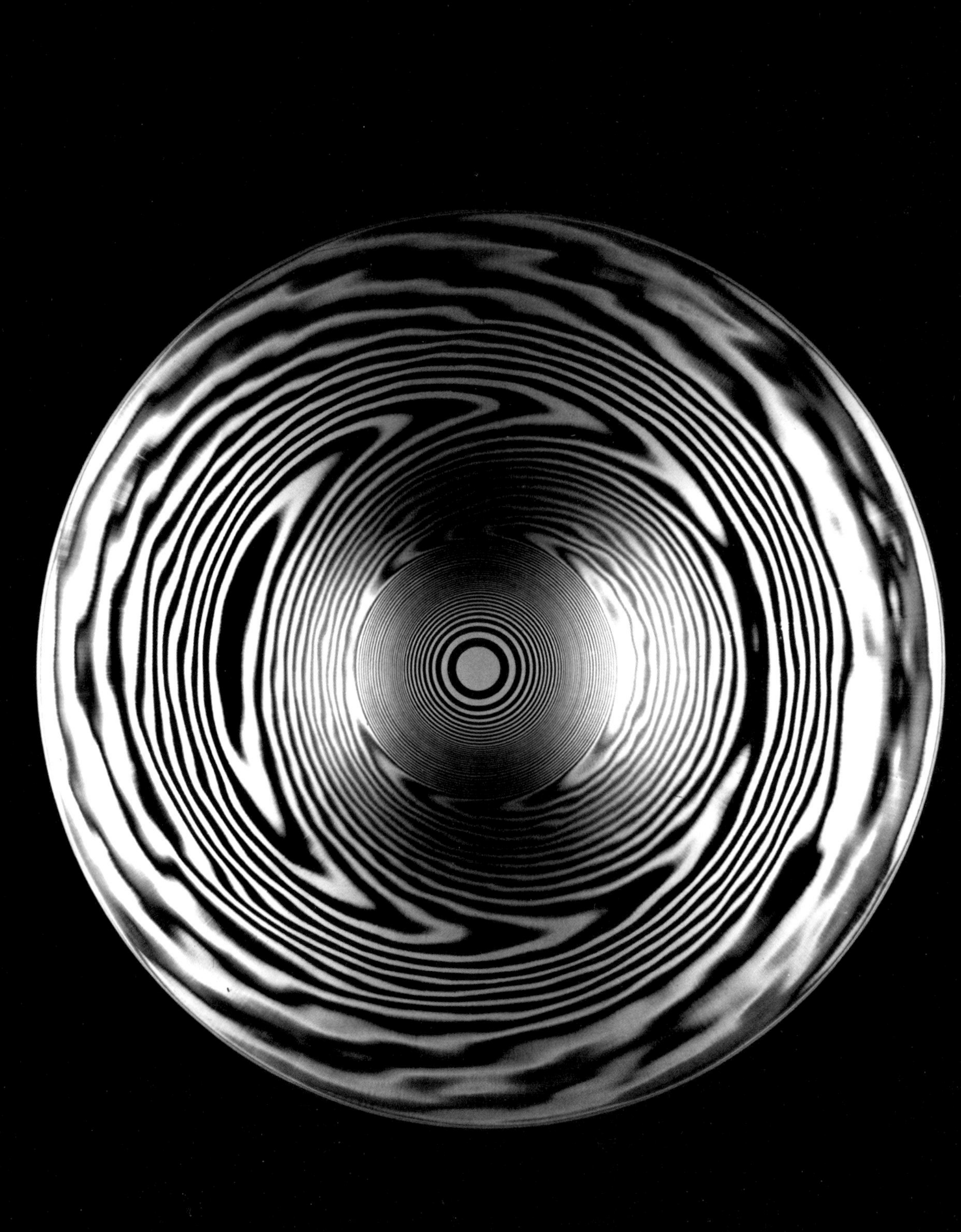

Karl Gerstner, *Lens Picture No. 15*, 1964.
Concentric circles viewed through Plexiglas
lens. Gyorgy Kepes papers (M1796).
Dept. of Special Collections and University
Archives, Stanford Libraries, Stanford, Calif.

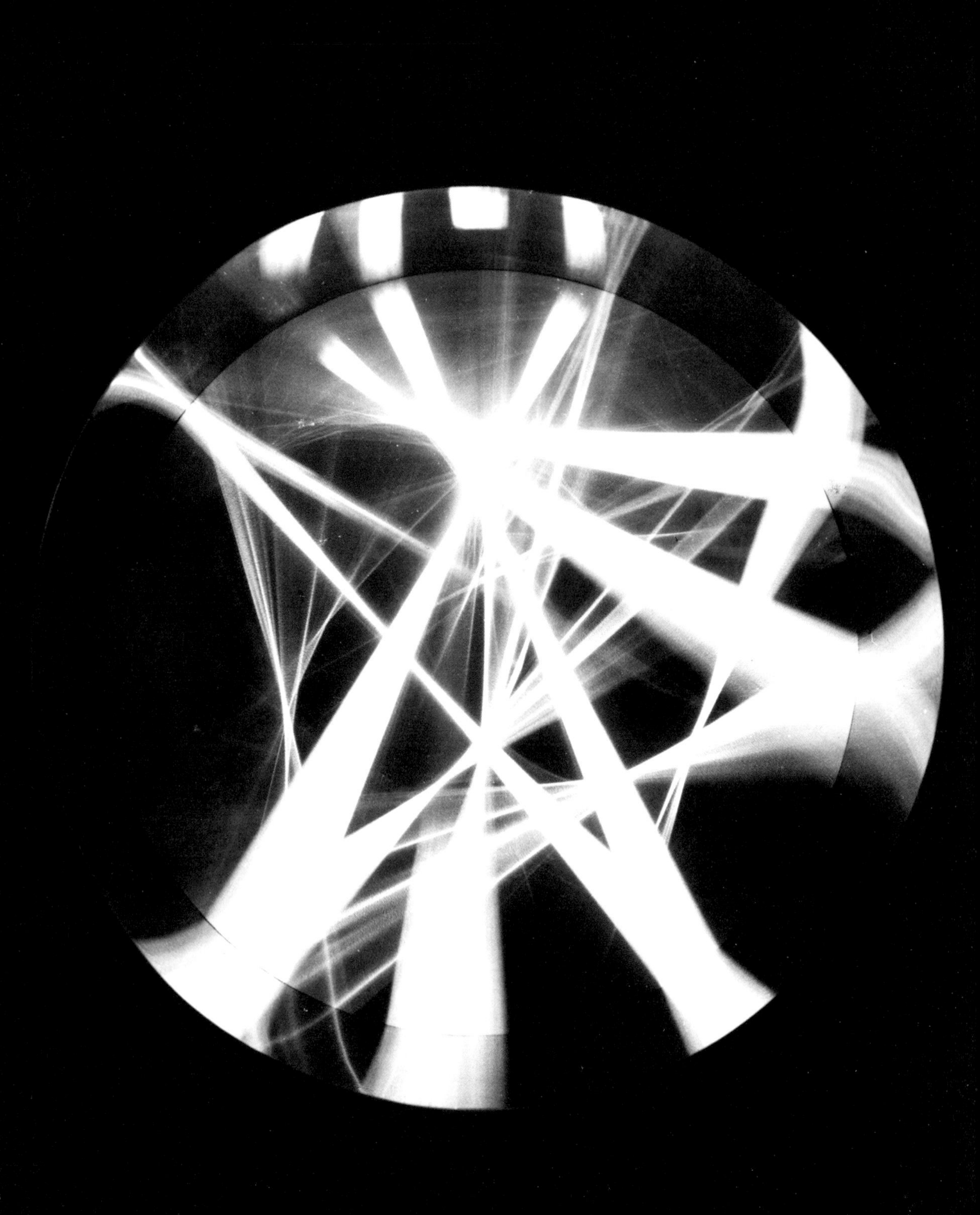

4.33

Julio Le Parc, *Continuel Lumière Cylindre*, kinetic
light design, 1962. Gyorgy Kepes papers (M1796).
Dept. of Special Collections and University Archives,
Stanford Libraries, Stanford, Calif. © 2018 Artists
Rights Society (ARS), New York / ADAGP, Paris

Old and New Romanticism

All of these themes are more fully developed in completed portions of Kepes's Light Book texts, especially a series of short typescripts—the start of Light Book chapters—labeled "The Nineteenth Century" and "The Twentieth Century." In both, Kepes frames his allegorical reading of light with historical specificity, comparing the meanings of light in the nineteenth century to its meanings in the present.

Kepes first became interested in the nineteenth century in the mid-1930s, while exiled in London, where he encountered works by the English romantic painters John Constable and J. M. W. Turner, the "painter of light":

> I was utterly overwhelmed by some of the later watercolors of Turner. I tried to understand the significance of the fact that Turner, who created such tremendous crescendos of light … grew up in England just when the first industrial slums overwhelmed certain areas of the country. I concluded that it could not be an accident that Turner and Constable, who worshipped light, movement, space and color, emerged as creative artists at the time when there was the first real destruction of the quality and clarity of the chromatic world around them.… Because they felt that this quality of life was being destroyed, they tried to replace it with dream images that contained everything missing from reality.[92]

Kepes began studying Turner and Constable, even obtaining a copy of Turner's papers on microfilm from the British Museum. He read the romantic poets, especially William Blake, William Wordsworth, Percy Bysshe Shelley, and Lord Byron. He consulted the writings of the American transcendentalists Henry David Thoreau and Ralph Waldo Emerson. John Ruskin was one of Kepes's favorite thinkers; so too was William Morris, the socialist leader of the Arts and Crafts movement.

Kepes wanted to understand how the cultural figures of the romantic movement reacted against the values of the Enlightenment—reason, logic, and order—by embracing countervailing values: the irrational, the illogical, the sublime. Kepes used their critique of science and technology as a template for his own, bringing the romantic cause from the nineteenth century into the twentieth. He explicitly embraced such a label; a 1960 issue of *Art in America* included a detail of Kepes's painting *Garden of Light* on its cover (fig. 4.34). The theme of the issue was "Old and New Romanticism"; it included an essay by curator Sam Hunter on romantic responses to modernity. A text by Kepes on his mobile kinetic light forms is included within the magazine.[93]

In his "Nineteenth Century" typescript, Kepes situates romanticism as a response to the literal darkness of the period: the pollution generated by the Industrial Revolution. He opens with a reference to "dark satanic mills," the factories described by Blake in the preface to *Milton*: "With the spread

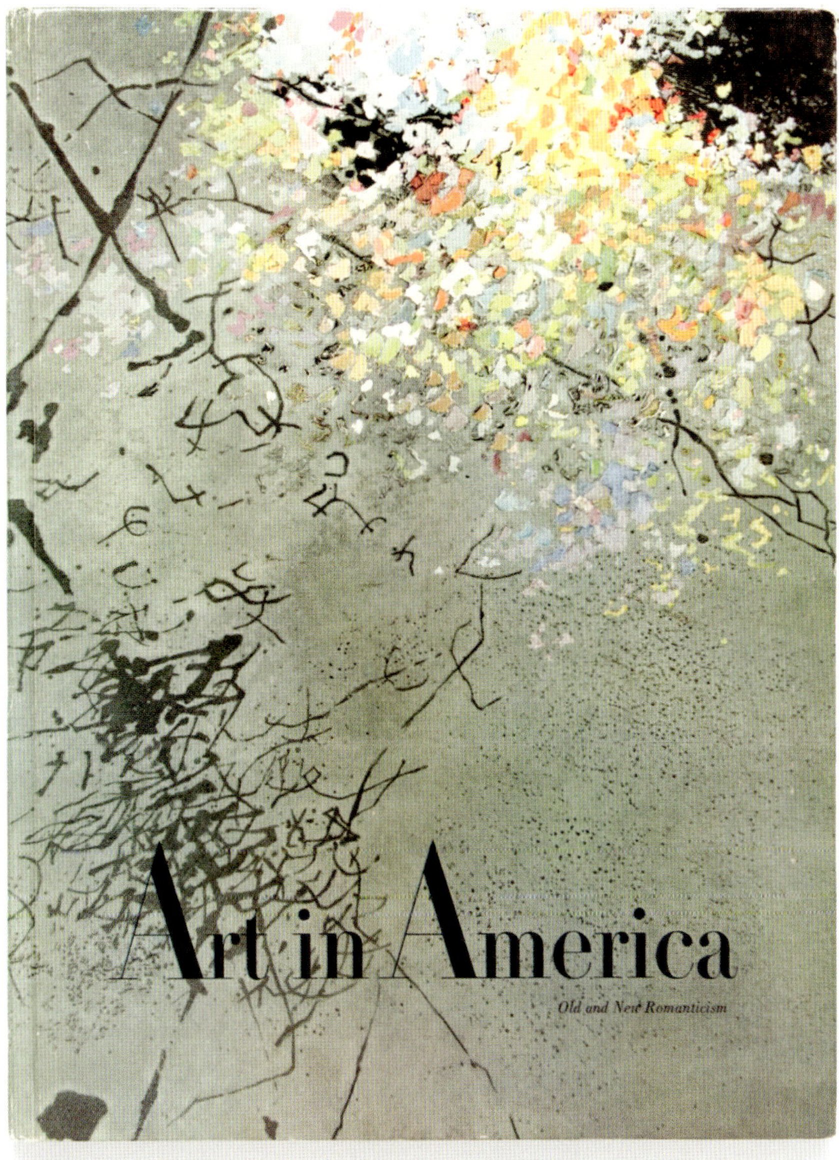

4.34

Cover of *Art in America* 48, no. 4, 1960. Cover depicts a detail of Gyorgy Kepes, *Garden of Light*, oil on canvas, 1959. Originally published by *Art in America*, 1960. Copyright © Art Media Holdings, LLC, New York. Reproduced by permission. © The Estate of Gyorgy Kepes.

of Blake's 'satanic mills,' cities were bathed in smoke and dirt," writes Kepes. "Darkness, crowding, and filth" covered the landscape. He cites the window tax—a tax on light, assessed on the number of windows at a given property.[94] Cramped, dingy tenements "devoured space, killed privacy, and slaughtered light and color. Squalid, unsightly, overcrowded streets were characteristic of the early mercantile and industrial slums."[95]

Not only literal darkness but also the metaphorical darkness of industrial capitalism defined this bleak age. Kepes refers to factory labor, mass production, and the temporal control of the clock and the railroad. He was especially concerned about the "alienation and isolation of the most sensitive and creative members of society—the artists." He explains their response: "When industrialization wrought havoc with the physical environment, stealing light and color from the first industrial cities, the painters of England tried to reclaim their lost paradise, the luminous richness of the visible world."[96] As Kepes understood it, artists preserved utopia in dream images while historical forces destroyed reality. In their "adoration of light and color," artists freed light from material form and physical boundaries. There was a "liberation of light … from the restriction of defining surfaces, from the limitation of describing objects." For the romantics, "light for its own sake and color for its own sake became a kind of limitless fluid wonder, an infinite extension of brightness." While earthly existence was "chopped into narrow, gray, and dirty slabs," painters projected an alternative: "the radiant splendor of blazing light." As the world began to fade, the "painters did the replenishing in an imaginary plane." They responded to darkness "with the marvelous human inner eye."[97] Kepes described this gesture as "dissent by imagination."[98]

In notes related to these typescripts, Kepes created long lists of isolated vignettes: "squalor, hunger," "pickpocket," "Dickens London," "gin as escape," "degradation of individual," "fake life," "man in disguise."[99] Another list captures the Marxist dimension of this darkness: "monotony of motion, repetitive work, reduction of initiative, hypnotic effects of machine rhythm, fatigue, long work hours, … separated from the means of production."[100] Kepes also formed pages of notes contrasting dark and light. One side of a page lists more of bleak qualities: "16 hours day work," "child labor," "industrial slums." The other side of the same sheet lists Kepes's notes on Constable's oil studies of clouds: "sky clouds, breezy freshness of morning, placid movements of clouds, rainbows upon the stormy sky, burst of sunshine." Kepes even included an inscription left by Constable on the reverse of one of Constable's own studies: "very bright and fresh grey clouds running fast over a yellow bed, about halfway in the sky," writes Constable (and, in the act of copying the passage, Kepes).[101]

Kepes understood these remarks as an inversion of darkness, its transformation into light. He literalized this transformation by using the front and back of this single sheet of paper.

In turning darkness into light, either real or imagined, Kepes's project again took on a political charge, if a highly abstract one. Kepes pulled many of his ideas from the nineteenth-century radical tradition. One quotation from the "Nineteenth Century" typescript considers the window tax: "What is the good of living in an empire where the sun never sets if you also live in a court where it never rises?" In his notes, Kepes attributes the quotation to one of the "radical speakers of 19 C. Victorian England."[102] It comes specifically from Will Crooks, a socialist politician. His papers also include scattered references to light from the words of Karl Marx. Kepes mined the Marxist art historian Francis D. Klingender's 1943 volume *Marxism and Modern Art: An Approach to Social Realism* for primary sources, often pulling pages out of the volume and underlining passages.[103] He highlighted a quotation from an 1856 speech in which Marx employed the metaphor of light: "Even the pure light of science seems unable to shine but on the dark background of ignorance." In the passage, Marx explains that the promise of science and technology, its "pure light," is always in contrast with lived experience.

Kepes's typescript also examined actual works of art.[104] An 1872 engraving by French printmaker Gustave Doré titled *Over London by Rail* depicts dense tenement housing, as seen through a brick railway arch at an elevated perspective (fig. 4.35). Rows of chimneys, laundry lines, and claustrophobic courtyards continue into the distance, where a train billowing smoke passes overhead, nearly obliterating the sky. Against this image of industrial gloom, Kepes compared the works of Turner and Constable: "light pours into the picture and creates and animates space in endless chromatic play. Sunrise, sunset, and the pageantry of the seasons—screened out of existence in Birmingham and Manchester by filters of soot and smoke—are brought back in their deepest poetic significance." From his notes: "the XIX century pilgrims of industrialized wasteland were building the images of luminous splendour [*sic*]."[105]

In his second typescript, "The Twentieth Century," Kepes connects these themes to the present. "The advent of the twentieth century," he writes, "did not resolve the conflicts of the nineteenth century; it exacerbated them."[106] In his notes, he often referred to this contemporary darkness as the "mole life," a "fallout shelter future." He had in mind a series of drawings produced by Henry Moore of civilians taking shelter in the London Underground during the Second World War ("gloom, H. Moore subways ghastly shelter," writes Kepes).[107] In one of Moore's drawings from the series (fig. 4.36), reclined

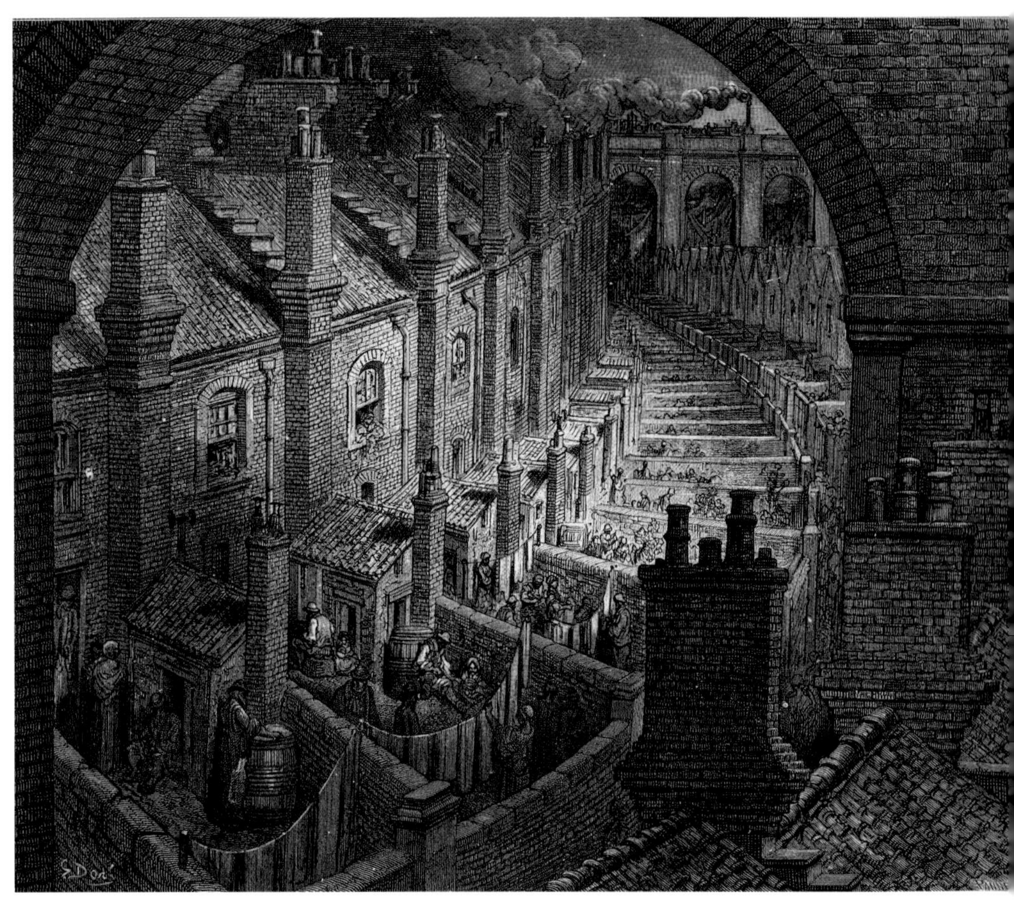

4.35

Gustave Doré, *Over London by Rail*, engraving from *London: A Pilgrimage*, 1872. Courtesy the British Library.

4.36

Henry Moore, *Tube Shelter Perspective*, 1941. Graphite, ink, wax, and watercolor on paper.

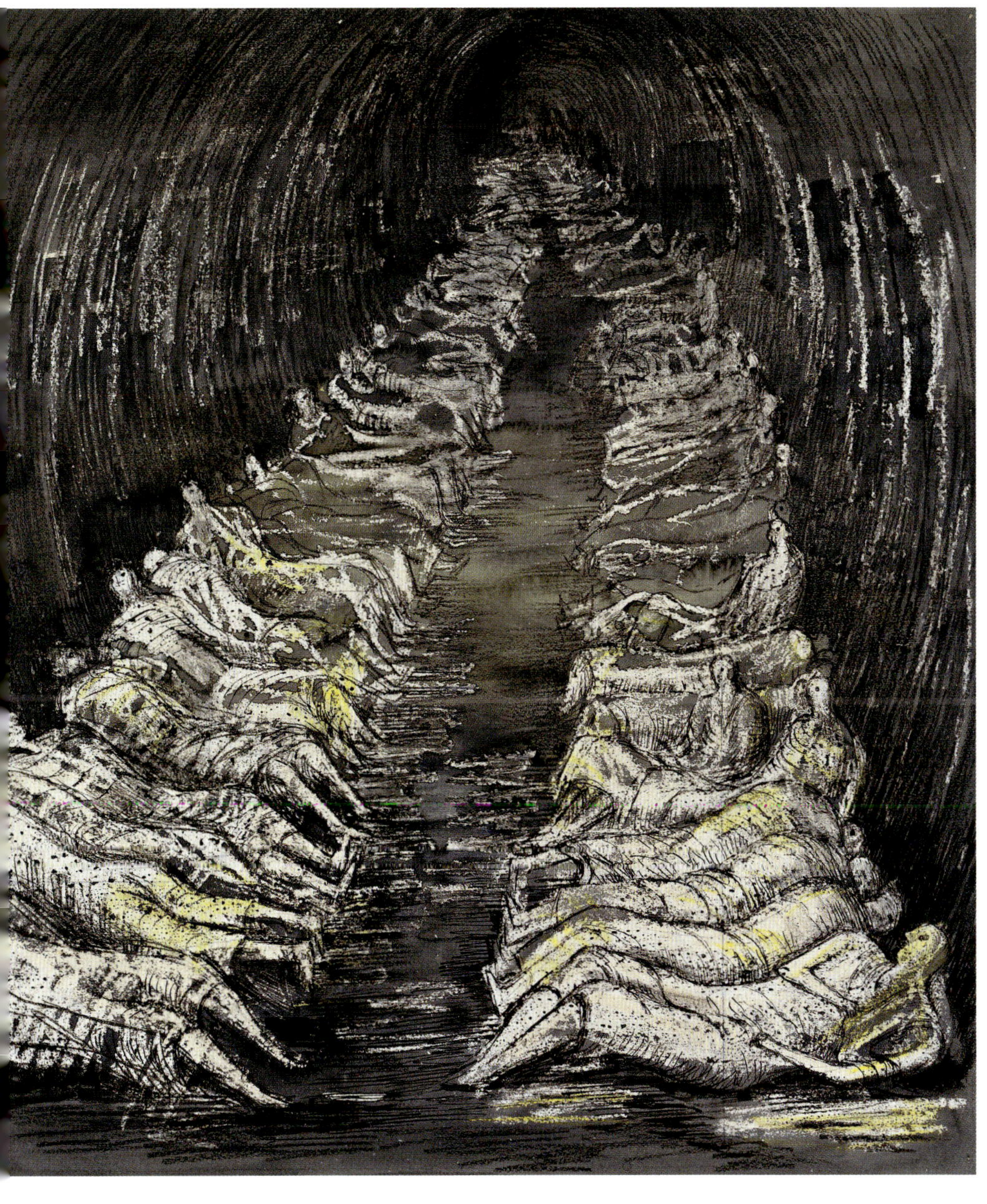

bodies extend into the distance in a dark tunnel. Kepes relates this allusion back to the trenches of the First World War but also forward to the Cold War; he was thinking of the bunkers built for protection from radioactive fallout.

The fallout Kepes feared was literal—the "vaporized material" and "particles of grit" that would drift thousands of miles from hypocenter. One pamphlet from 1959—the year Kepes prepared his Guggenheim Fellowship application—illustrates how a bomb blast over Washington, DC, would not only obliterate the nation's capital but also send a dark cloud raining radioactive debris all the way up the northeast corridor (fig. 4.37). But the fallout Kepes feared was also figurative—he also refers to the delayed effects of advanced science and technology as cultural fallout. He connected these themes to the inverse search for light: "in spite of mole life of shelter buried underground, light is the higher key of body + mind + the dark tunnel we may dig ourself [*sic*] in makes us aware of broader clearer sky. The deeper we may bury ourself [*sic*] underground the deeper the longing."[108]

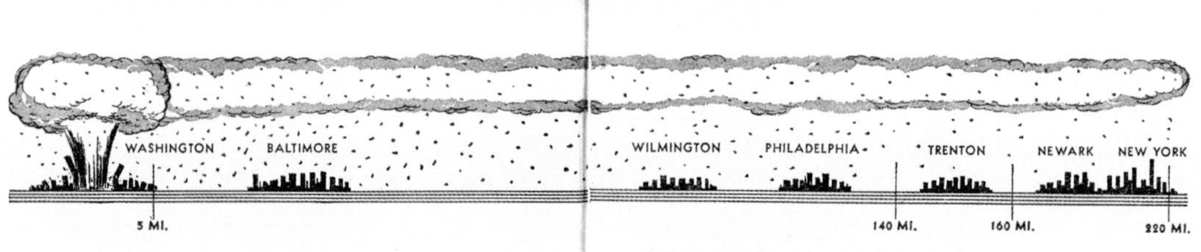

Fallout area

It is impossible to know in advance how large or where the area of dangerous fallout would be. During the 1954 tests in the Pacific, the fallout that showered Japanese fishermen was radioactive enough to dangerously contaminate an area extending downwind for 220 miles and varying in width up to 40 miles. This cigar-shaped area would have been large enough to reach from Washington, D. C., to New York City, including the cities of Baltimore, Wilmington, Philadelphia, and Trenton.

When does fallout start?

The larger radioactive particles from a nuclear explosion start to fall out shortly after the bomb is fired. Coarse, heavy particles fall faster and settle closer to the point of the explosion.

10

11

4.37

Fallout diagram from "What you should know about radioactive fallout," Office of Civil and Defense Mobilization pamphlet, 1959.

Kepes thus adapted his argument about light and dark into an abstracted critique of science and technology. He liked to recite a story from the journals of Ralph Waldo Emerson in which Emerson recounts visiting Nauset Light, a lighthouse on the Cape Cod coast. The lighthouse-keeper told Emerson that there had once been opposition to constructing the tower because "it would injure the wrecking business," a reference to those who profited from shipwrecks along the coast. Kepes applied the anecdote to the present:

> **Such is our situation today. A highly articulate group is involved in the wrecking business; its members are afraid of light, the use of light, and the meaning of light. Their creative range is a spectrum of destruction and despair. Their desolate images have taken the foreground of the creative horizon. Nevertheless, the great new tool of twentieth-century technology will have its future.[109]**

For this "wrecking business," Kepes had in mind the art world, but also the world of science—the "megadeath intellectuals" against whom he had railed in his private musings. It was his hope to build light towers, to imagine a new light art that would counter and correct the uses and abuses of science and technology.

Dark Ages

The light of reason … falls, even on the happiest day, on its irresolvable contradiction: the calamity which reason alone cannot avert.

Max Horkheimer and Theodor W. Adorno, *Dialectic of Enlightenment*[110]

In the end there is always darkness. Kepes's Light Book, like many of his projects, remained unfinished and unrealized; it exists only on the plane of the imagination, it exists only in dream images. When a publisher inquired in 1962 about the possibility of putting the Light Book into print, Kepes declined the offer: "It would be wonderful to be able to say that the book is finished … but I seem to get derailed again and again."[111] He repeated the same sentiment a full decade later. To another publisher in 1972: "I was proud and touched by your confident desire to publish my light book; but, as I told you, I have a deeply ingrained fear of making commitments with 'unfinished symphonies.'"[112]

One problem for Kepes was that light is limitless. It is universal, but it is this universal quality that also makes it ephemeral, seemingly everywhere and nowhere at once. It is allusive, accommodating a range of meanings, but it is also elusive, immaterial, hard to grasp. But the other problem was that light could never accomplish the task Kepes set for it to accomplish. Light could not transform science and technology; it could not really redeem or absolve.

4.38

Installation view of *Light as a Creative Medium*, Carpenter Center for the Visual Arts, Harvard University, 1965. Courtesy Center for Advanced Visual Studies Special Collection, MIT Program in Art, Culture and Technology. © Massachusetts Institute of Technology.

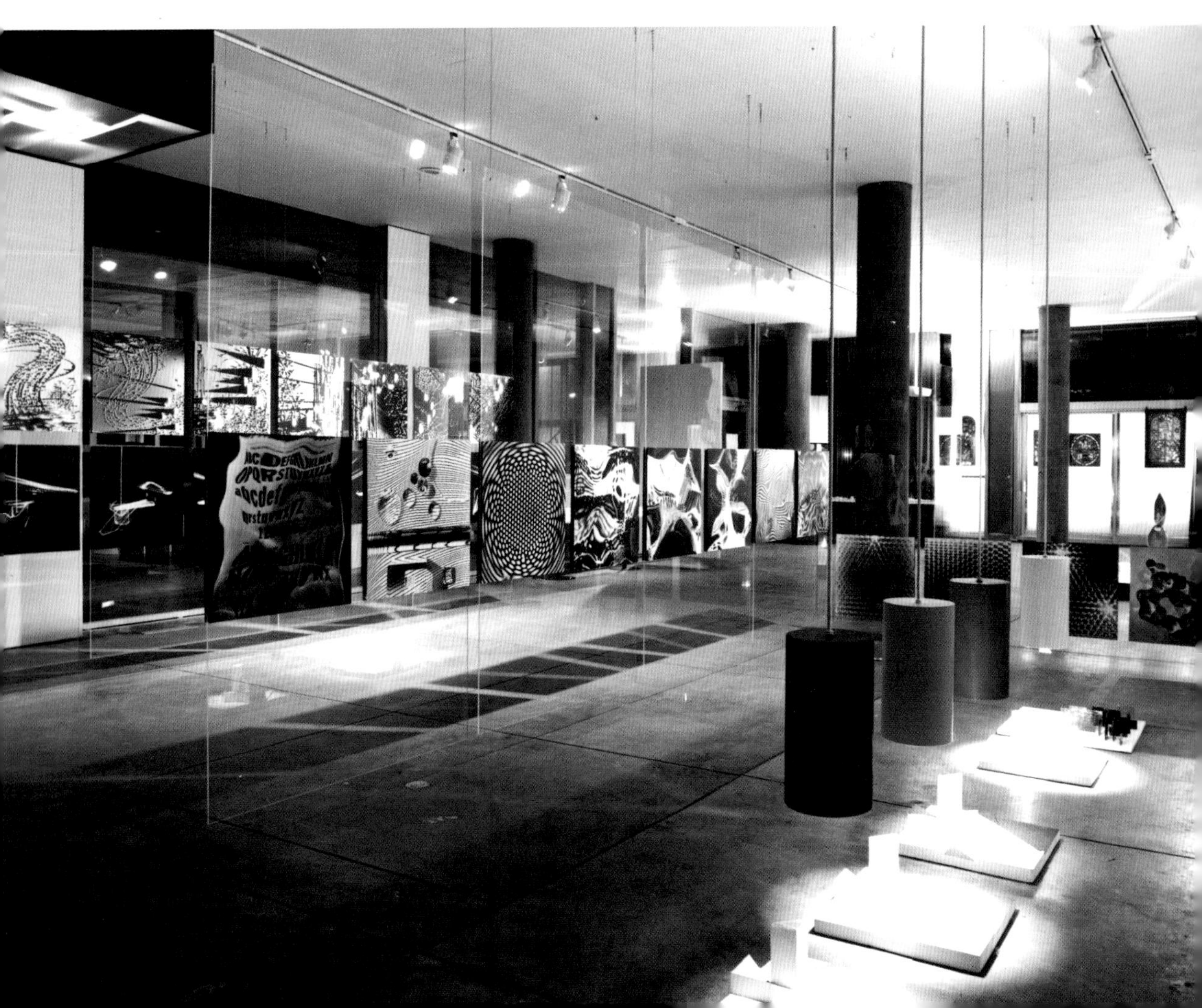

Kepes was aware of these limitations. In a lecture likely given on the occasion of an exhibition he curated at Harvard University's Carpenter Center for the Visual Arts in 1965 titled *Light as a Creative Medium* (fig. 4.38)—a show that resulted from his Light Book research—Kepes summarized his grand narrative. With slides of Doré's prints and Turner's paintings illuminating the lecture hall, projected in beams of light, Kepes described the Industrial Revolution: the "spatial world was chopped up into small ugly cubby holes where space was not anymore space, where light was not anymore light, where colour was not anymore colour." He explained how artists responded by creating "wonderful symphonies, where light, colour, and movement became one." He declared the relevance of his allegory to the bleak view of midcentury: "light is missing and has to be replaced."

But Kepes ended his talk with doubt: "I have to repeat, I am a partisan, I am one-sided, I may see much more in things than is there, but my only duty is just to tell what I think and feel about it and it's up to you to judge whether it is just a caricature of a hope, or a really meaningful hope."[113] He imbued light with more meaning than it could sustain.

Of all his many collected light quotations, Kepes was especially fond of an aphorism from Meister Eckhart, the medieval Christian mystic. Writing in the depths of the so-called Dark Ages, Eckhart states: "One really finds light best in the darkness. Thus when a man suffers and knows discomfort, he is nearest to the light."[114] Kepes responded to the dark conditions of his moment by retreating into this allegory.

The Military-Industrial-Aesthetic Complex

chapter

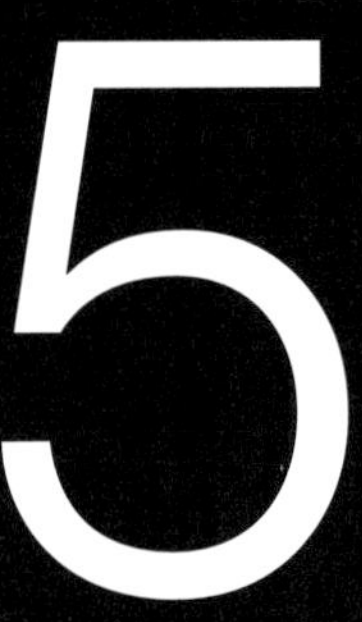

5

Toward a Human Community

In 1970, Gyorgy Kepes's Center for Advanced Visual Studies (CAVS) at MIT organized its first collaborative exhibition. As you enter a dimly lit gallery, its walls painted black, works within suddenly come alive (figs. 5.1–5.3). Sound triggers waves of motion in a forest of vibrating metal rods. Illuminated by pulsing stroboscopic light, the oscillations mold gentle sculptural forms in the air, like reeds in the wind. Polarized plastic on the ground erupts in a riot of color, each footstep recorded in vivid pooling patterns. Enormous artificial flowers billow and bend as they inflate and deflate with air. Video screens flash, lasers ricochet, and a crystal-coated Mylar wall exudes a radioactive glow. The materials on view are cold and mechanical—motors, metal, vinyl, and plastic—but also highly futuristic: phosphor powders, fluorescent liquids, luminescent tapes, even a telecopier machine. Although the exhibition is technological to the extreme, it is still incessantly organic in its metaphors.

The show, presented first under the title *Exploration* at MIT's Hayden Gallery and then in expanded form as *Explorations* at the National Collection of Fine Arts (now known as the Smithsonian American Art Museum) in Washington, DC, was the first project fully realized by Kepes's Center and thus the culmination of Kepes's ambitions for his new, think-tank-like art-and-science research institute.[2] It debuted the Center's inaugural fellows—the artists Jack Burnham, Ted Kraynik, Otto Piene, Harold Tovish, Wen-Ying Tsai, Stan Van-DerBeek, and Takis Vassilakis—alongside many others, including Newton Harrison and Les Levine.[3]

But Kepes rejected the standard dictates of contemporary art and its display. Not only was the white cube painted black, but there were no wall labels identifying individual artists; specific works were instead integrated into an immersive group experience. In this way, Kepes hoped to transform

5.1

Installation view of *Exploration*, Hayden Gallery, MIT, 1970. Photo: Nishan Bichajian. Courtesy Center for Advanced Visual Studies Special Collection, MIT Program in Art, Culture and Technology. © Massachusetts Institute of Technology.

5.2

Wen-Ying Tsai, *Cybernetic Sculpture System*, 1969. Stainless steel, aluminum, electric motor, stroboscopic light, electronic audio feedback control system. 48 x 24 x 17 inches. © Tsai Art and Science Foundation.

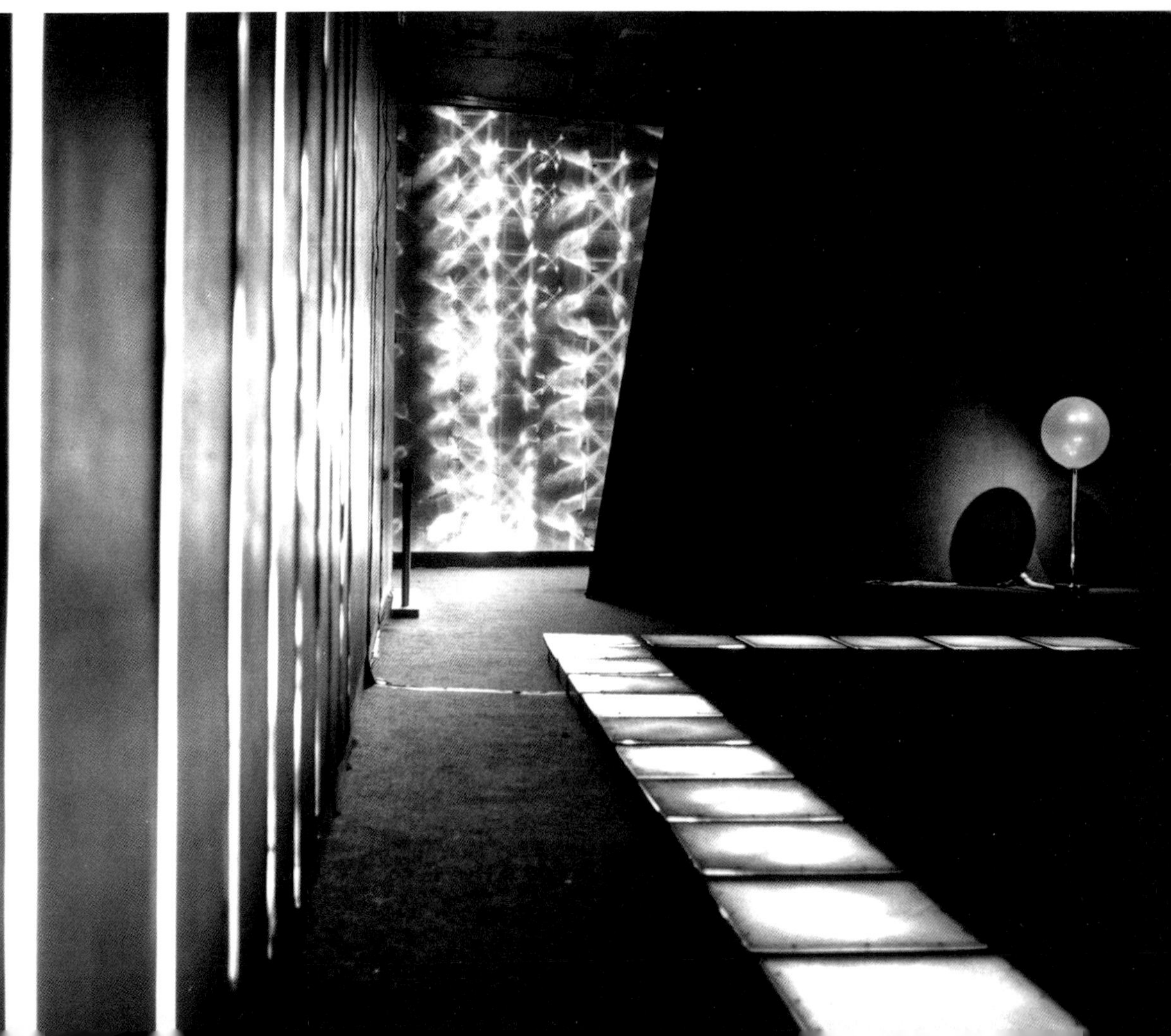

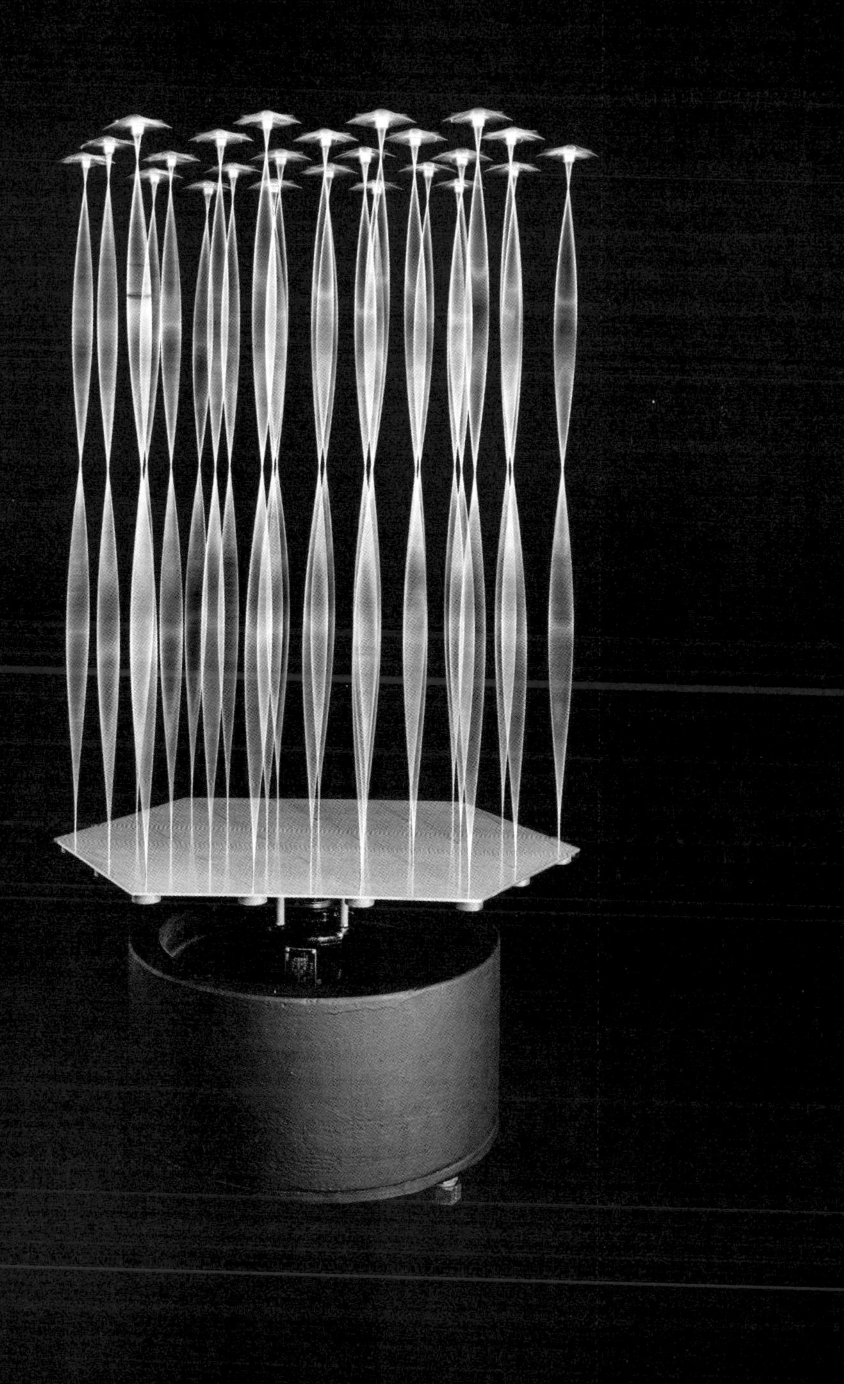

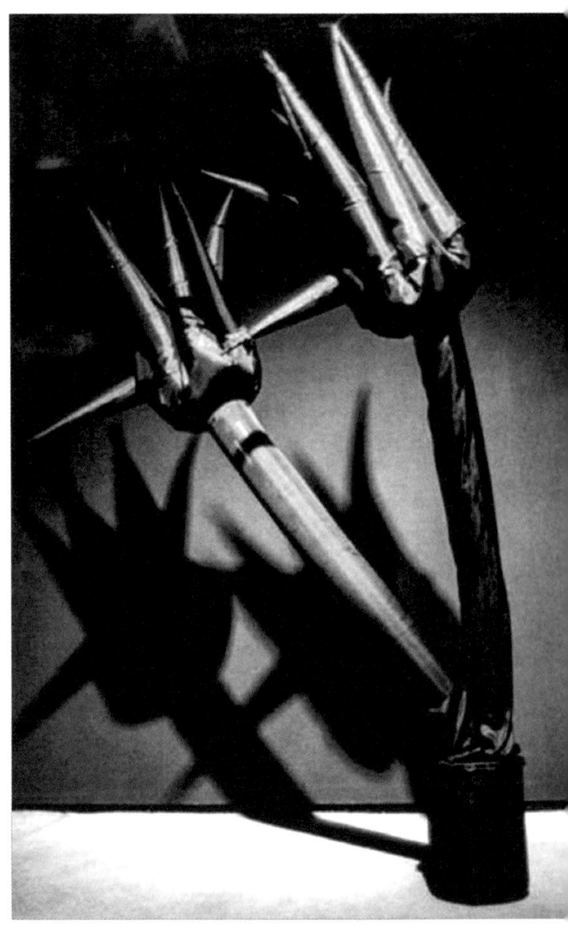

5.3

Otto Piene, *Fleurs du Mal*, 1969. © The Estate
of Otto Piene.

the show's lifeless technical media, all the electronic equipment and clunky machines, into something natural, to turn rote technology into an environmental ecology. The show's phenomenological technics—interactive sculptures, responsive materials, and ephemeral sensations of flashing light and frenzied motion—enveloped viewers in a sensory-rich simulacrum of a natural landscape. It was, explains Kepes, like a "little forest," an "ecological garden," an "environment which is alive because of hidden forces." Moreover, this new landscape echoed the "synergetic system" of an ideal society; it reflected, he says, "the dream of a human community"—a complex communications network that would connect all things. Kepes believed that the aesthetics of collaboration would stimulate, or at least simulate, collaboration on a higher plane, not just between the objects on view but also between the groups and institutions that trace ever outward from the work of art. This "interacting quality," he explains, would produce an experience of profound interdependence; it would be deeply "humane."[4]

Kepes's conception of the show's curatorial mission is clear in some of the objects he included. Many evoked natural phenomena by simulating them through technology; Takis's *Signals*, blinking lights on long vertical poles, looked like "lily pads vibrating in the water," explains Kepes.[5] Lila Katzen's *Liquid Environment* created a luminous cave viewers could enter; walking into the fluorescent tunnel was "like being in the center of an opal" (fig. 5.4).[6] Kepes's own work (fig. 5.5), a walkway designed in conjunction with the engineer William Wainwright and titled *Photoelastic Walk*, used plastic sheets and polarizing screens to produce a captivating optical illusion. Physical pressure against the walkway caused the refractive index of the plastic to change, revealing brightly colored lines—"rainbow colors"—that evoked the rippling effects of water in sunlight.[7] (Kepes recalls a memory from the Hungarian countryside: "As a child I loved to wade through water and watch the rhythms my movements made. That has something to do with this floor.")[8] These works all conveyed Kepes's larger ambition—the creation of a peaceful global connectivity—through formal experiences of interactivity. Communication and collaboration were thematized as aesthetic content.

In his exhibition catalog essay, a manifesto of sorts titled "Toward Civic Art," Kepes explained his theory that these natural experiences, produced though they were through scientific and technological means, might return us to a purer past—and thus guide us to a better future. Kepes evoked the quasi-spiritual, quasi-mystical language of systems. Long ago, "life was everywhere, in men, beasts, plants, stones, and water," he writes, conjuring the infinite circle of life.

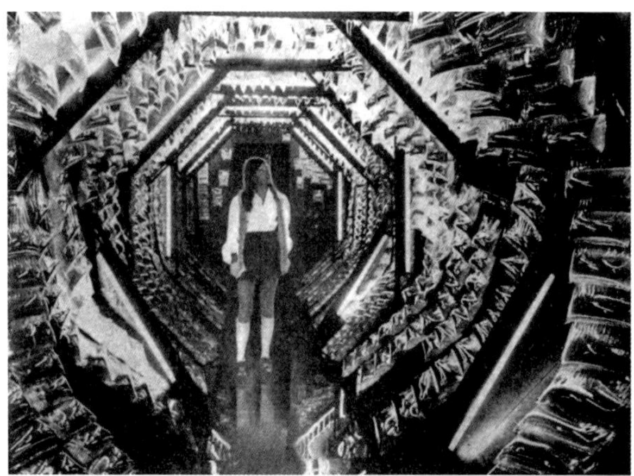

5.4

Lila Katzen, *Liquid Environment*, 1970.
© 2018 Philip Katzen and Lila Katzen
Living Trust / Licensed by VAGA at Artists
Rights Society (ARS), NY.

5.5

Gyorgy Kepes with William Wainwright,
Photoelastic Walk, 1969 (detail). Photo:
Nishan Bichajian. Courtesy Center for
Advanced Visual Studies Special Collection,
MIT Program in Art, Culture and Technology.
© Massachusetts Institute of Technology
and the Estate of Gyorgy Kepes.

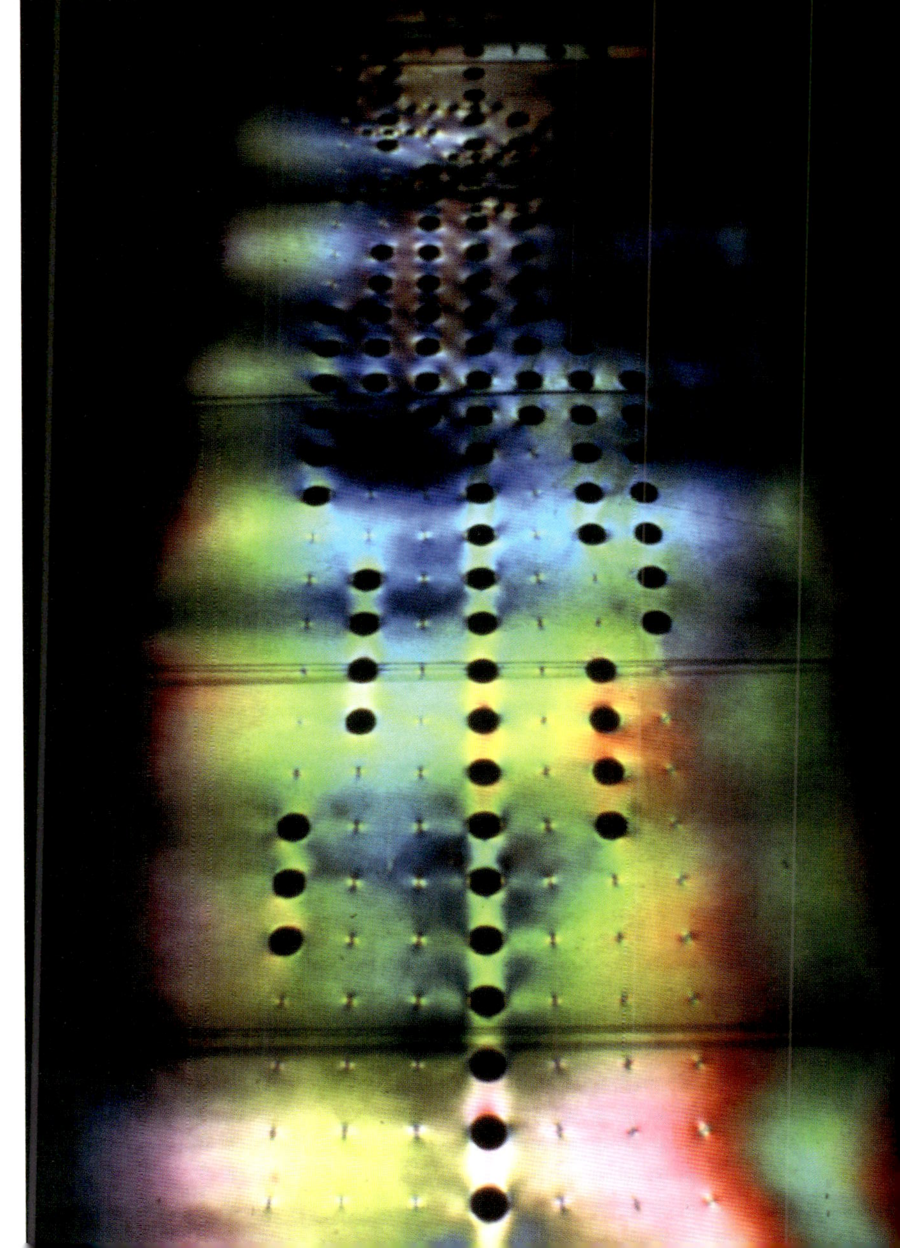

> The pearly iridescence of sea shells, the sparkling of a crystal, the phosphorescent glow of the sea at night, and the sunlight caught in droplights above a waterfall are all signs of an embracing, living thing…. Everything is permeated by life. Everything seems in contact, interacting, interliving.[9]

By mimicking these effects, the works in *Explorations* created "a dwelling for the human spirit not unlike the dimly remembered Garden of Eden," a paradise free from original sin—a time before science and technology.[10]

The public was enthralled. *Explorations* was a blockbuster. At MIT's Hayden Gallery, the show had "the largest attendance in the history of MIT exhibitions."[11] Same at the National Collection of Fine Arts: "According to the Smithsonian staff," Kepes later boasted to John E. Burchard, then MIT's Dean Emeritus of Humanities and Social Studies, "it has had an almost magical effect resulting in an attendance almost four times as many as at any previous exhibition."[12]

But the critics unanimously hated the show. It garnered unusually negative reviews (Kepes called it "acid criticism")[13] after arriving in Washington, the most biting of which came from Lawrence Alloway, who saw the show as an obscene aestheticization of science and technology, and thus also an obscene perversion of art:

> The sum of these works is a frivolous and gross fantasy of technology…. Many of the technologically derived works can be placed and seem to belong in a spectrum of fantasy that runs from horror movies through comic books to science fiction…. It seems clear, after "Explorations" and all the other shows of this kind, that the junction of art and technology does not result in cultural lodestones, but in an art of mostly trivial effects. You feel more a part of the 20th century by making a phone call from the Metroliner on the way home from Washington.[14]

Alloway targeted not only Kepes's curatorial vision, but also the works he had included. VanDerBeek's grandly titled *Panels for the Walls of the World*, a large mural created by transmitting images via fax machine from MIT to Washington, hardly impressed Alloway: "I can only conclude that the technical expertise here is, finally, pointless." Neither did Otto Piene's *Fleurs du Mal*, a series of inflated polyethylene sculptural forms which looked ridiculous alongside the gadgets and gimmickry; it was like "mutilated cacti in a mad scientist's underground laboratory." Alloway understood the show's ideological program, which he considered a mindless embrace of science and technology rather than their humane transformation, as an extension of the Bauhaus—and he further related the Bauhaus, as Sibyl Moholy-Nagy had before, to fascism.

The mass media echoed these comments: the show was "a pretentious and tacky collection of entertaining scientific toys," said the *Washington Post*, and, according to the *New York Times*, "little more than laboratory demonstrations."[15] *Newsweek* agreed: Kepes "has failed."[16] Emily Wasserman of *Artforum* also denounced Explorations as an outright "failure."[17] Elaborating on Alloway's commentary, she saw the show as a lot of "very silly stuff," with the "exotic effects of the new media" degraded to nothing more than "new forms of entertainment." But Wasserman also critiqued a more fundamental problem: the aesthetic results that Kepes produced could never fulfill the goals he set out in his soaring rhetoric. Kepes could not capture the infinite circle of life in a mere exhibition. *Explorations* "failed to end up looking as significant as its possibilities of cooperative design, vision, and communication suggested."[18]

These critiques were consistent with the reception of other popular but polarizing art-and-science and art-and-technology ventures from the period, most notably the Art and Technology Program at the Los Angeles County Museum of Art (LACMA) and the cooperative work supported by Experiments in Art and Technology, or E.A.T.[19] As Anne Collins Goodyear has demonstrated, the negative reception of these projects reflected the associations attached to science and technology as instruments of war, especially as public opinion shifted decisively against the war in Vietnam in 1969; they were understood as an aestheticization not only of science and technology but also, by extension, of military power.[20]

But *Explorations* was unique among these projects, for it was entangled in debates that exceed the iconographic resonance of new media circa 1969. Kepes opens his statement "Toward Civic Art" by situating art in a very wide context (a "civic scale"); he believed works of art could not be contained in galleries and museums and would instead "outgrow the exhibition format and find their real territory in the dynamic life situation outside." To take seriously the argument that art exists in a system and network, that everything is connected to everything else in one infinite circle ("Everything seems in contact, interacting, interliving …"), is to spiral out of the art world and into the far more conflicted terrain beyond.[21]

This chapter explores that terrain. Kepes first organized the exhibition in 1969 as the United States section for the tenth São Paulo Biennial; in this official capacity, the show was to be a national representation of artistic culture from the United States, specially packaged for an international forum. But the show never happened. After nine of the twenty-three artists Kepes had invited to participate publicly withdrew their participation in protest that spring, Kepes was forced to cancel the original exhibition. A collaborative undertaking cannot exist without its collaborators.[22]

The explicit target of the artists' protest was Brazil's reigning military government, which came to power in a 1964 *coup d'état* that replaced the democratically elected left-wing government of João Goulart and his Partido

Trabalhista Brasileiro (Brazilian Labor Party) with successive right-wing dictators backed by the Brazilian military. The United States embraced regime change as a means of containing the spread of Communism across Latin America, but also as a means of shaping economic policy to the benefit of US corporate interests—as a way to open new markets. The Johnson administration even planned to secure the takeover's success through a secret military campaign called "Operation Brother Sam." Designed by the Pentagon, the White House, and the Ambassador to Brazil and cold warrior Lincoln Gordon, the plan would have provided logistical support for the Brazilian military should the country have fallen into civil war as a result of the coup. "Operation Brother Sam" remained a contingency—Goulart was overthrown with little violence, and the United States was able to formally recognize the new government immediately—but declassified documents have since revealed the extent to which the Johnson administration was willing to secure Goulart's fall from power. A naval fleet that included an aircraft carrier, six destroyers, and four petroleum tankers was deployed to Brazil, with shipments of arms, ammunition, oil, gas, and jet fuel at the ready.[23]

The artists who withdrew reacted not against these clandestine circumstances—they were rumored but not yet known—but against the repressive policies used by one of the leaders the coup installed: President Artur da Costa e Silva, a former Army Marshal and Minister of War who, with the Brazilian military's backing, consolidated power in 1966 as Brazil's de facto dictator. He used this position to silence political opponents, including artists, with impunity. Knowledge of his authoritarian rule had become widespread by the late 1960s. The artists were also reacting more generally against the way in which events in Brazil mirrored events in Vietnam.

The debate over São Paulo was thus international in scope, a matter of art as global geopolitics—a form of diplomatic statecraft that Kepes could not master. "Toward Civic Art" argues that artists might recover the "long lost role of cultural leadership," but Kepes, now forced onto the world stage, was hardly prepared to provide such leadership.[24]

But the debate over São Paulo was also domestic; Kepes's home institution was engulfed in a related set of controversies. MIT, as I have argued throughout this study, was the archetypal "Cold War university," enmeshed in US military's research agenda.[25] As the Cold War turned hot, students and faculty began to recognize and reject the insidious relationships between government, military, industry, and the academy. One booklet distributed by students in Berkeley, California, identifies the emergence of a frightening new formation, mapped as a pentagon of power: WAR INCORPORATED (fig. 5.6).[26]

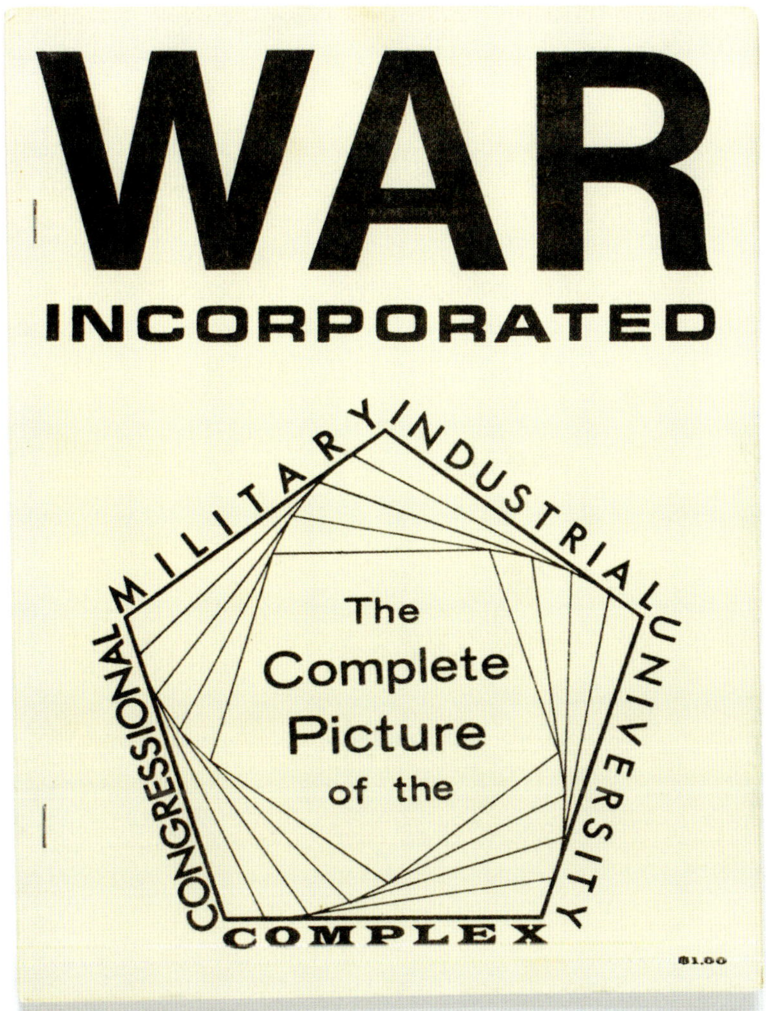

5.6

Cover of Janet Brown et al., "War Incorporated:
The Complete Picture of the Congressional-
Military-Industrial-University Complex," booklet
published by the Student Research Facility,
Berkeley, c. 1970.

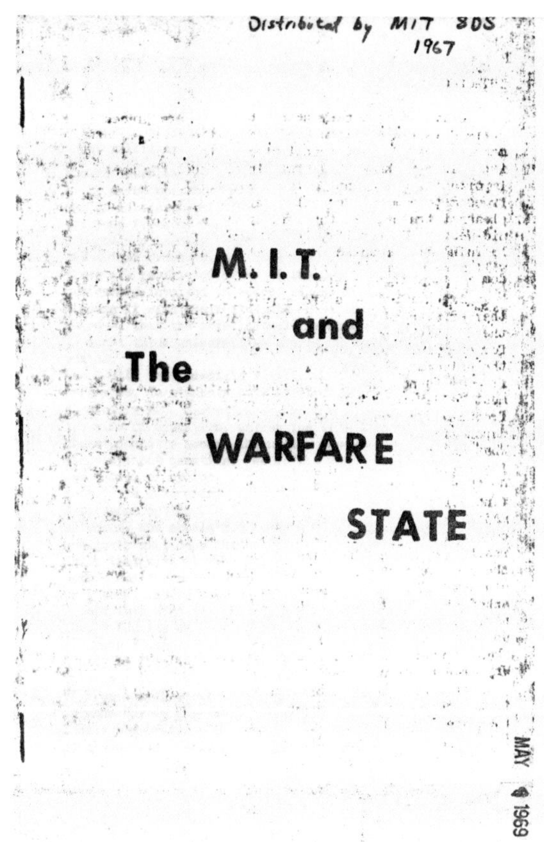

5.7

Cover of "M.I.T. and the Warfare State,"
pamphlet published by the Massachusetts
Institute of Technology chapter of Students
for a Democratic Society, 1967. Courtesy
MIT Museum, Cambridge, MA.

This protest was slow to emerge at MIT—students at the Institute were not especially political. Kepes referred to his own students as "kind of goal-oriented" and "very little experimental"—somewhat euphemistic language meant to signify their great technical minds but lack of humanistic focus.[27] But a 1967 pamphlet published by the MIT chapter of Students for a Democratic Society (SDS), titled "M.I.T. and the Warfare State," set the campus antiwar movement into motion (fig. 5.7). "MIT has indirectly become an instrument for waging modern warfare," the pamphlet reads, noting how contracts secured through the Department of Defense supported classified research at the Institute that had direct applications in Vietnam. The pamphlet emphasized how MIT's relationships with military power generated ethical questions that students and faculty had avoided for far too long:

> **MIT's complicity in the war in Vietnam raises the issue of institutional and individual morality and responsibility. Institutions are made of men, and men must bear responsibility for their actions or non-actions. The MEN of the MIT Corporation must bear the responsibility for the technology they produce. Scientists are <u>not</u> instruments of society, they are MEN with conscience. Technology is <u>not</u> neutral, it is directed. The men who build the bombs must bear the ultimate responsibility of their use.[28]**

The pamphlet was explicit in connecting research at MIT directly to war, avoiding the assumption that academic work must be safely academic, completely sequestered in the ivory tower. "The man who designs the gun is just as guilty as the man who pulls the trigger, for without him, there would be no gun."[29] The most relevant passage is the pamphlet's concluding statement, which emphasizes in no uncertain terms how collaboration—the particular academic model most celebrated at MIT, and the basis for the Institute's many interdisciplinary centers, Kepes's included—had become synonymous with complicity. The language is significant, for the actual word "collaboration" takes on a special charge: "JOIN US IN DEMANDING THAT MIT ACCEPT RESPONSIBILITY FOR ITS PRESENT COLLABORATION WITH AN IMMORAL AND IMPERIALIST WAR, AND THAT FUTURE COLLABORATION CEASE."[30]

Kepes was committed to the aesthetics of collaboration, and the progressive possibilities it might offer, but what about the ethics? To what ends were these methods actually directed? As students diagrammed so clearly and spelled out with full capitalization, it was collaboration that tied together the individuals and institutions sustaining the war in Vietnam. Was Kepes, in making such innovative models possible in the arts, complicit with these arrangements—or, alternatively, was he able to change the meaning of collaboration, to convert the purposes it served?

The confusion over these questions indicates the highly elliptical patterns of systems and networks embedded in Kepes's communicative model. Reinhold Martin argues that the "logic of military-industrial control and communication in Kepes's curatorial strategy at São Paulo" ultimately rendered Kepes an "agent of the organizational regime."[31] By contrast, I understand these systems and networks not as means of preventing change but as means of creating it; I explore the ways Kepes navigated MIT with the hope of altering its organizational regime. He purposely suspended himself between contradictions: between artists and scientists, of course, but also between faculty and fellows, administrators and students, and generations old and new. He gained a unique form of agency through ambiguity, through a precise lack of clearly articulated position. He attempted to figure a mutual interdependence and interconnectedness that would link these groups, thus changing them in the process. Kepes's ambiguous position, however, also left him entrapped in controversy.

This entanglement consumed Kepes in the late 1960s. Turmoil even engulfed aspects of his artistic life that he had long imagined to be fully protected. Even his painting was not safe, as a historical anecdote demonstrates. Kepes considered the act of painting to be wholly private, an activity separate from his work at MIT; he created richly textured canvases during summer retreats to Cape Cod or at a separate studio he kept in Boston, away from campus. (He comments: "I sometimes dream about being just a painter, painting and forgetting everything else").[32]

But the studio is not separate. In April of 1969, as the plans for São Paulo began to unravel, Kepes received a letter from Stefan P. Munsing, Director of the Art in Embassies program established by the Museum of Modern Art and later administered by the US Department of State. The program placed works of American art in government offices around the globe. "We have been enjoying the warmth of your painting 'Melted Glow' which you so generously donated to the Art in Embassies Program," the letter states. "We recently had it 'on duty' at the Department of State in the interim offices of the new Secretary of State."[33] The letter's timing places *Melted Glow*, a 1964 oil painting with an ominous title that evokes the glare of a nuclear explosion (fig. 5.8), in the offices of William P. Rogers, President Nixon's Secretary of State from 1969 to 1972. Rogers had limited influence on foreign policy decisions; Nixon instead crafted his disastrous plans for Vietnam with Henry Kissinger, then his National Security Advisor and later Rogers' successor. Nonetheless, the proximity of the work of art to world affairs is more than suggestive. The letter's timing also indicates that the Nixon administration had just begun its covert bombing of Cambodia and Laos, an attempt to obliterate bases used

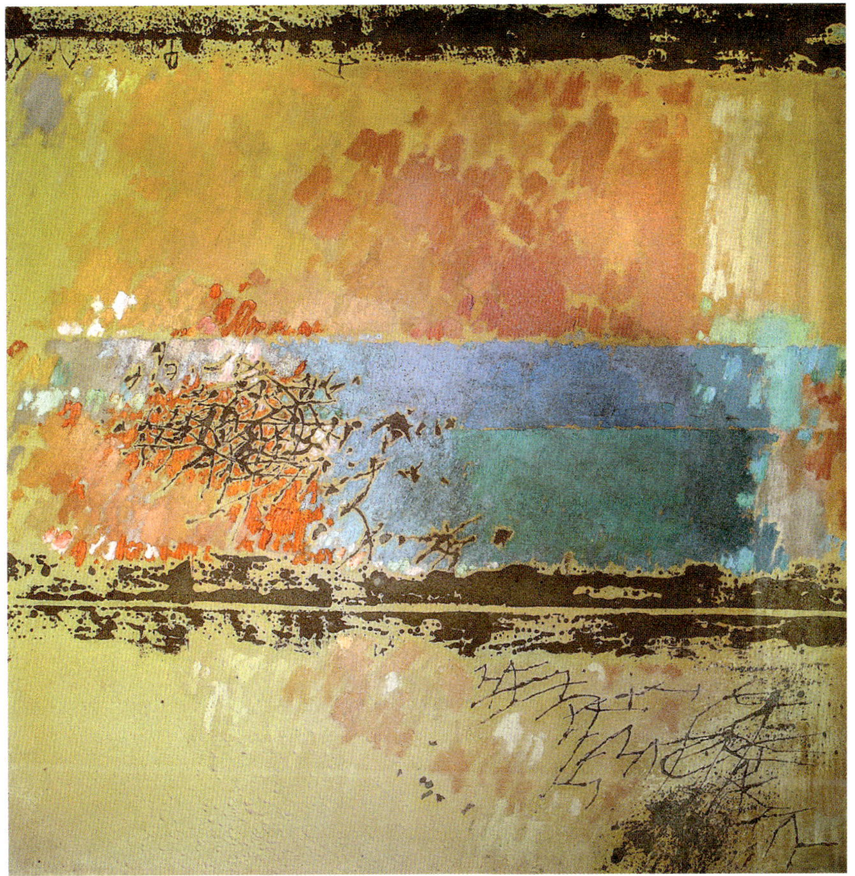

5.8

Gyorgy Kepes, *Melted Glow*, 1964. Oil on canvas. © The Estate of Gyorgy Kepes.

by the North Vietnamese Army and the National Liberation Front; Kissinger adamantly supported the campaign and Rogers adamantly opposed it (the two men hated each other). Nixon sided with Kissinger, and the bombs fell. Kepes's painting, hanging in the offices of the Secretary of State, was not just watching from the wall, not just a passive witness to history. It was also actively "on duty," fulfilling its charge to illuminate the spaces where these very dark war plans were debated.

This episode indicates the disconnect between Kepes's ideals and the real forms they took. Kepes participated in the Art in Embassies program out of an honest belief in art's power to cultivate peace and understanding—the same mission he articulated through *Explorations*. His sincerity is incontrovertible; I do not mean to reject it. He would have never imagined that *Melted Glow* could be connected to the US military's brutal bombing campaign in Southeast Asia. This dissonance suggests the tension between theorizing ideals and actually realizing them, between creating systems of exchange and becoming unintentionally trapped within them.

Aestheticizing Politics, Politicizing Aesthetics

First, however, it is crucial that we make Kepes's known political positions clear, especially given the tendency to assume that his positions mirrored his institutional affiliation. I have previously described Kepes's politics after his arrival in the United States in 1937 as variously "camouflaged" and "contained"—suppressed by necessity in the consensus culture that pervaded academic life in the early Cold War. MIT was widely caricatured in the 1950s and 1960s for this culture. In a 1958 *New Yorker* cartoon, reprinted in MIT's *Technology Review* as proof of how the world sees the Institute, a procession of scientists in lab coats and thick glasses march to a chant in the tune of the US Marines' official hymn (fig. 5.9). "From the cyclotron of Berkeley to the labs of M.I.T., We're the lads that you can trust to keep our country strong and free."[34] Not only does the cartoon mock militarism, it also parodies the bland conformity of the Institute. This image was hardly unique to popular news commentary. It also appeared in art magazines. Jonathan Benthall, writing in *Art International*, similarly described the Institute: "M.I.T. seems the acme of all that is wrong with American society. Its endless corridors and the ponderous columns of its main façade confirm the impression that it is a kind of academic Pentagon."[35]

Kepes certainly looked at home in this setting, as a portrait of him with a curatorial model of *Explorations* suggests (fig. 5.10). Kepes always had a conservative demeanor and sartorial formality—suit and tie were his daily uniform—at odds with the popular conception of the radical artist. Confusion over his appearance was common in the media; one report explained that Kepes's "dark pin-striped suits make him look more like a banker than an avante-garde [*sic*] artist."[36]

And yet, by the 1960s, this camouflage and containment actually gave way to commitment. Kepes became a vocal critic of military power. Extensive evidence documents his activities. He supported the pacifist candidate H. Stuart Hughes, a professor of history at Harvard, over Ted Kennedy for the Massachusetts US senate seat vacated by President-elect John F. Kennedy. Hughes ran against Ted Kennedy from the left; the core of his platform was immediate nuclear disarmament and an end to the war in Vietnam. Kepes's name is emblazoned across "Hughes for Senate" letterhead; he donated his paintings to a "Hughes for Senate" fundraiser.[37] Kepes was also a sponsor of the 1965 March on Washington for Peace in Vietnam, one of the earliest demonstrations on the National Mall against US intervention in Southeast Asia.[38] Organized by the Committee for a Sane Nuclear Policy (then co-chaired by Hughes), the march helped galvanize mainstream antiwar sentiment.[39] That same year Kepes signed an open letter addressed to US Secretary of State Dean Rusk (predecessor to Secretary Rogers) and published in the *New York Times* that condemned Rusk's support for increased US intervention.[40] In 1967, he joined Scientists and Engineers for McCarthy, a group that worked for Eugene McCarthy's 1968 presidential campaign, based almost entirely on immediate US withdrawal.[41]

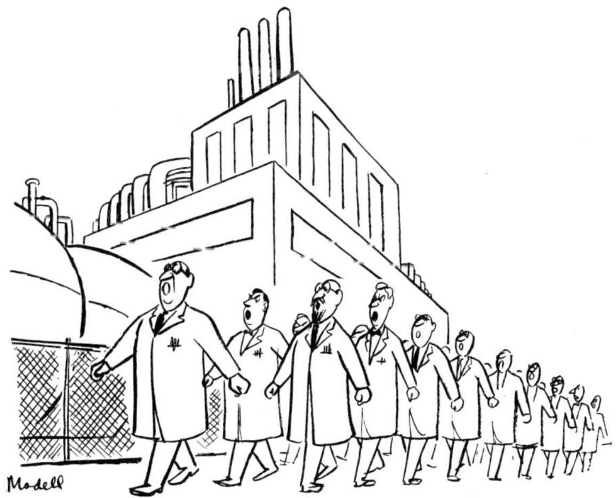

"From the cyclotron of Berkeley to the labs of M.I.T., We're the lads that you can trust to keep our country strong and free."

5.9

Frank Modell, cartoon published in the *New Yorker*, 18 January 1958. Reprinted in *Technology Review*, April 1958. Frank Modell/The New Yorker Collection/ The Cartoon Bank.

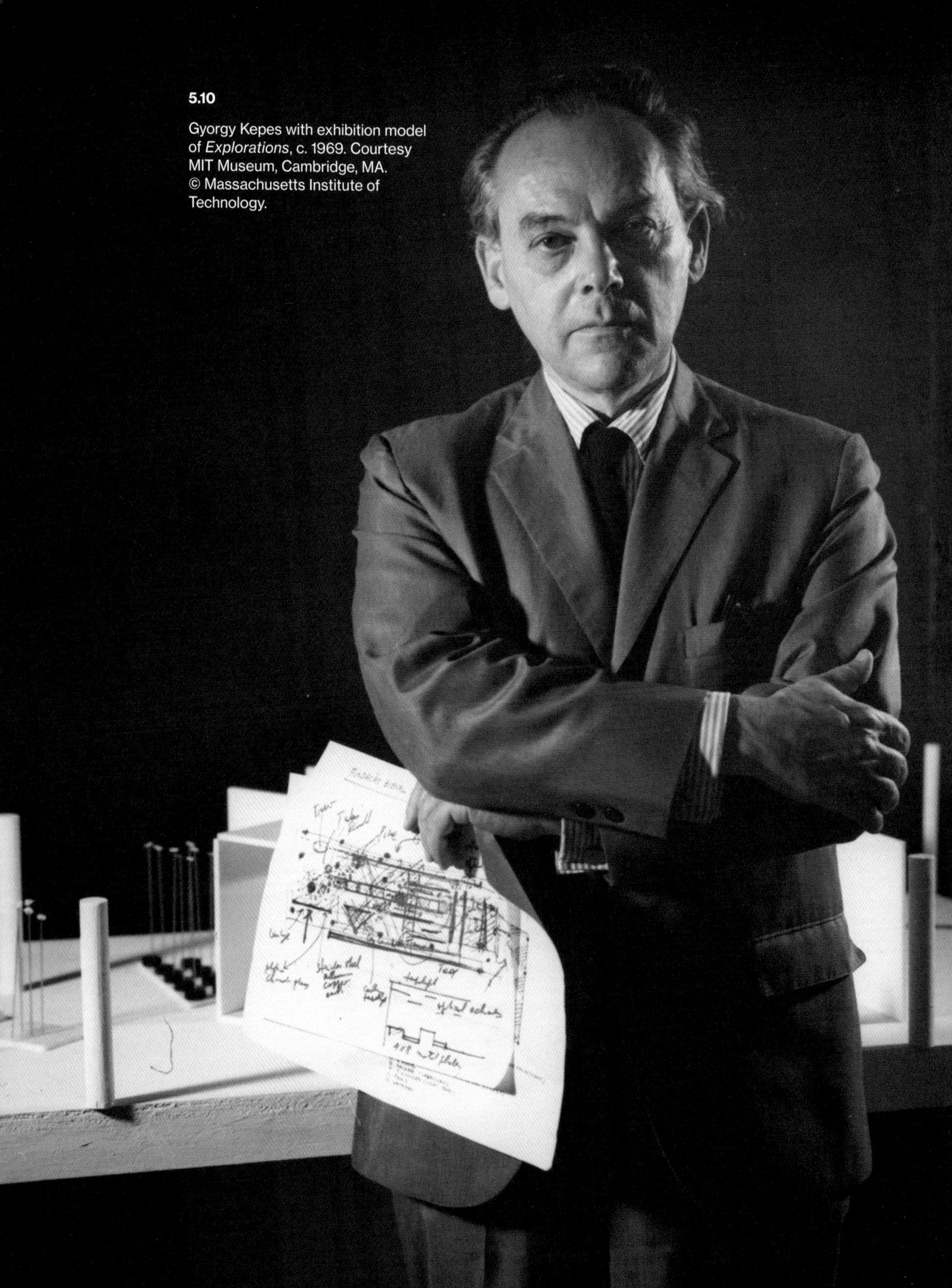

5.10

Gyorgy Kepes with exhibition model of *Explorations*, c. 1969. Courtesy MIT Museum, Cambridge, MA. © Massachusetts Institute of Technology.

Kepes put his name on a 1971 statement in MIT's campus paper declaring, in no uncertain terms: "The War in Indochina must be stopped!"[42] "Dear Colleague," began one letter circulated by the MIT Peace Coalition in 1973:

> Will it take one year, four years, or more before the US government ceases to bomb civilians, hospitals and other targets in Indochina?
>
> IT IS OUR MONEY THAT FINANCES THESE DEEDS, OUR TECHNOLOGY THAT MAKES THEM POSSIBLE, OUR CONGRESS THAT AUTHORIZES THEM, OUR ARMED FORCES THAT CARRY THEM OUT, OUR PRESIDENT WHO ORDERS THEM, AND OUR WILL THAT IS INVOKED.
>
> We must object – we must join with others to protest and use what power we have to bring this war to an end.[43]

Kepes, along with Noam Chomsky and MIT's other famous tenured radicals, signed the letter. So too did his colleagues in the sciences, like Philip Morrison, Bruno Rossi, and Jerrold R. Zacharias—figures who were not always opposed to MIT's military involvements, having themselves previously participated in them.

Kepes was also invited to contribute to a portfolio of prints organized by Artists Against Racism and the War, a group of Boston-based artists and activists. The release of the prints was timed to coincide with a series of protest actions the group organized in May 1968. In the print Kepes created for the portfolio, he employs his decalcomania process—a thick smear of pigment left on the lithograph plate—to create an abstract impression (fig. 5.11). The result evokes a splatter of blood, a simple metaphor for the times.[44]

This purpose of this recitation of Kepes's declarations is to make clear, without a doubt—and in opposition to suggestions to the contrary—that Kepes was committed. Decades later, he recalled with regret that he was not sufficiently "active" in his opposition: "I was in spirit very much against US involvement in Vietnam, but was not active in my resentment to the war."[45] Kepes misremembers, perhaps because of his own sense of the futility of protest. His regret indicates not lack of action but lack of impact. Given Kepes's personal history—his status as an immigrant and the investigations led by the House Un-American Activities Committee into his background—his clear statements are even more significant.

It is because of these many pointed and persuasive declarations that Kepes's Center becomes so difficult to interpret. Was it more radical or reactionary in its aesthetics and politics? How do we make sense of the establishment image projected by Kepes, given his political beliefs—and does it matter that he held these beliefs from a position fully entangled within the establishment? Can one work against an institution's agenda while also working within the institution? In what follows, I examine the Center's entanglement and the ideological confusion it generated.

5.11

Gyorgy Kepes, *The Fifteen Days of May:
Untitled*, 1968. Lithograph. Harvard
Art Museums/Fogg Museum, Gift of
Impressions Workshop, M15038. Photo:
Imaging Department © President and
Fellows of Harvard College. © The Estate
of Gyorgy Kepes.

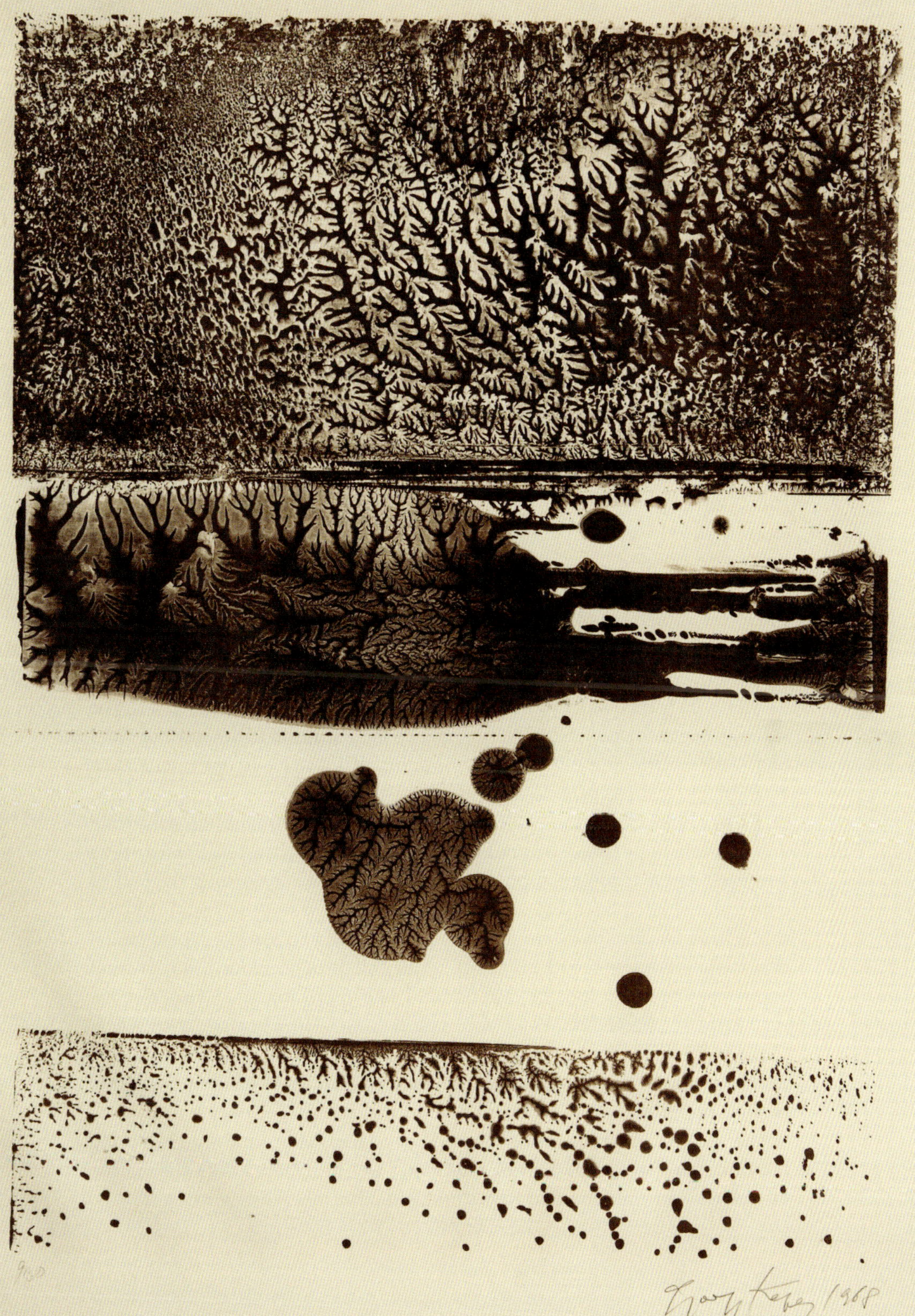
Georgy Kepes 1968

Idioms of Collaboration

Kepes first publicly announced his ambitions for the Center in a 1965 statement titled "The Visual Arts and Sciences: A Proposal for Collaboration," published in an issue of the interdisciplinary journal *Daedalus* devoted to "Science and Culture." "I propose the formation of a closely knit work community of eight to ten promising young artists and designers," he declares. Such a work community would be "located in an academic institution with a strong scientific tradition." It evokes the "research lab" and "university of vision" he had elaborated in his research notes years earlier. A footnote further indicates that Kepes had submitted such a proposal to MIT's administration.[46]

Kepes's conception of the Center as a work community follows iconic precedents from the historical avant-garde, from the Institute of Artistic Culture or INKhUK to the Bauhaus. It draws from two models Moholy proposed in his posthumous 1946 book *Vision in Motion*: an "institute of light," where artists would use science and technology as tools for aesthetic exploration; and a "parliament of social design," where workers would "embody all specialized knowledge into an integrated system through cooperative action."[47] Kepes also claimed broader precedents, like the Institute for Advanced Study at Princeton—the Center's name ("Advanced Visual Studies") was intended to evoke the famed art-and-science institute, which had hosted such luminaries as Albert Einstein and Erwin Panofsky. But perhaps the most significant model—one that Kepes did not acknowledge—was local; Kepes also looked to MIT's own unique institutional arrangements. As early as the 1940s, the Institute began forming research centers like the Radiation Laboratory or "Rad Lab," later known as the Research Laboratory of Electronics; each center was established to solve particular problems that defied single disciplinary concerns. In a report from the 1950s, MIT's administration visualized the Institute as a "university polarized around the sciences," with a constellation of these advanced research centers orbiting core departments but resisting the pull of conventional disciplines (fig. 5.12).[48] Kepes aimed to insert the visual arts into this picture by developing an institution that would transcend departmental designations but also broader fields of inquiry, all through group projects using approaches he had pioneered over decades, like interthinking and interseeing.

In theory, the Center would accomplish this goal by allowing artists access to the technologies and techniques developed at MIT, which they could use to explore projects that would exceed the conventional limits of the arts—issues of civic significance that extended beyond the gallery and museum. By "recognizing common problems of adjoining or related fields," and then exploring their solution through aesthetics, the Center would achieve a "dovetailing" of disciplines—an ambition that Kepes understood as inherently virtuous.[49]

In his published proposal, Kepes phrased these interdisciplinary relationships in positive terms. But just as *The New Landscape* contained evidence of the polemical nature of exchange, it is possible to find evidence of antagonism in the relationships behind CAVS. An unpublished proposal for the Center submitted to MIT's administration from the School of Architecture and Planning argues that the arts are not enrichment or entertainment, but a legitimate field of study. "Art is not merely a humanistic ornament to the education of scientists and engineers. It has its own frontier of discovery."[50] Another proposal submitted to MIT's administration also argues that "men of science cannot afford to disregard men of art, whose imprint on other times and places proved to be the most lasting cultural manifestation of all," presumably more lasting than scientific and technological achievement. The arts, explains the proposal, teach "value judgments," providing students with a "sense of the qualities of life." The lack of these values, according to the proposal, was the essential problem not only at MIT but in society at large: "this is the crisis our culture is experiencing."[51]

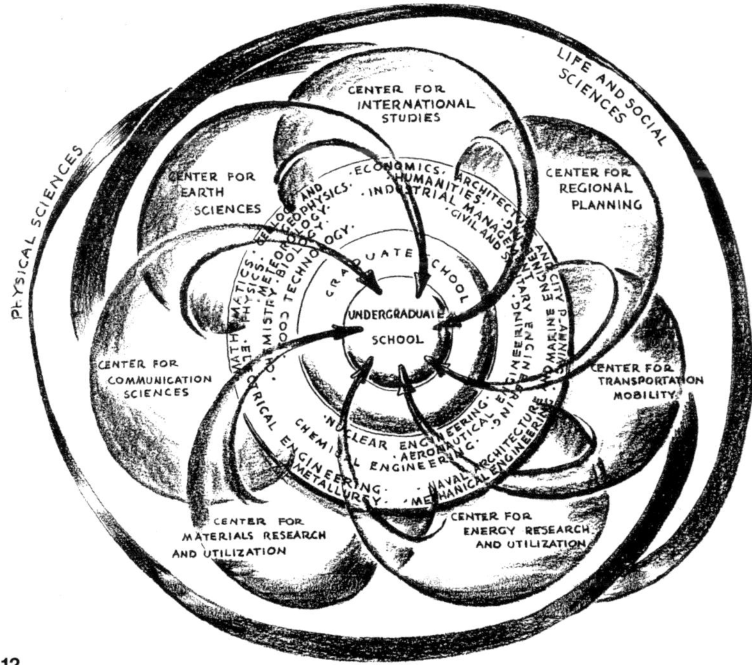

5.12

Diagram from an MIT report, 1958.
Reprinted in *Technology Review,*
December 1959. Courtesy MIT Museum,
Cambridge, MA.

All of these bureaucratic documents indicate the theoretical ambitions of the Center; its purpose was directed not to the creation of works of art but to the creation of "whole new techniques of collaboration," as Kepes puts it. Above all, he hoped to innovate methods and methodologies. Kepes was not concerned with technology as a material apparatus but, rather, a mental apparatus, a set of approaches or way of understanding the world. He elaborates in an interview:

> Technology today does not simply imply a physical implement, "a machine" mechanical or electronic, but a systematic, disciplined, collaborative approach to a chosen objective. There is a new technology that Daniel Bell has called "intellectual technology" – that is what artists must accept and understand.[52]

These formulations—interthinking, interseeing, and now "intellectual technology"—anticipate interdisciplinary approaches that are now commonplace.

To promote these ideas, Kepes contacted MIT's most respected scientific faculty, seeking their support. In a letter, he explains a desire to "explore ways in which the project can correspond to M.I.T.'s needs"—or how his Center might benefit scientists. He held conversations with metallurgist Cyril Stanley Smith, psychologist Hans-Lukas Teuber, mathematician Manuel Cerillo, astronomer George Clark, biophysicist Walter Rosenblith, engineers Harold Edgerton and Jerome Wiesner, and the physicists Francis Bitter, Philip Morrison, Bruno Rossi, Charles H. Townes, and Jerrold Zacharias.[53] Kepes notes that "the response was positive," with these experts all offering "valuable suggestions and promises of collaboration." He proposed a series of seminars "on art-science relationships" which would serve "as the initial clearing house of ideas" and would "adumbrate further work patterns and collaboration-catalyzing idioms." Kepes's language here is a crucial aspect of the project: interdisciplinarity is rhetorical. Working together was often a matter of finding ways to talk about working together. The seminars would be "a first step toward discovering common boundary phenomena and establishing complementary activities," he writes.[54]

This agenda is made most clear in a 1966 proposal titled "A Collaborative Approach at the Center for Advanced Visual Studies," a document Kepes prepared for the Old Dominion Foundation (now known as the Andrew W. Mellon Foundation) in order to secure funding.[55] The proposal contains an elegant explanation of the Center's core mission:

> The Center's goal, initially, is to develop "idioms of collaboration" among artists through experimental work on common tasks challenging enough to focus their creative energies and interests. In order to stimulate collaboration between artists and scientists, these tasks must also challenge the technical abilities, resources, and inventiveness of the M.I.T. community, attracting the cooperation and involvement of people from the widest range of disciplines and broadest spectrum of knowledge.

The Center's participants were intended to be highly diverse, rather than representing a single artistic skill or scientific approach—they would be "sometimes complementary, sometimes contradictory, but always supporting each other."[56]

Kepes offered a poetics of interdisciplinarity. On the occasion of the joint dedication that formally established the Center for Advanced Visual Studies alongside the Center for Theoretical Physics—an event that demonstrates the Center's aspirations—Kepes authored a promotional booklet that presented his elegant rhetoric. He conjures the language of systems: "The world as a set of structural systems does not divide into the two distinct territories." He rejected the notion of the ivory tower, calling for MIT to become "a community with a continuous horizon of humane and scientific interest," a "cultural *communitas*." This new community would employ the approaches of *"complementation*, that is to say, the collaboration of friendly opposites."[57]

The booklet, as well as a poster for the symposium, used a Gestalt diagram to demonstrate the figure-ground reversal; interlocking black-and-white forms fluctuate between positive and negative relationships, each one dependent on the other (fig. 5.13). It was a simple visual metaphor for the Center's mission, with figure and ground representing art and science, and their interdependence (it also appears on the book jacket of Kepes's *Arts of the Environment*).

Similar expressions of interdisciplinary exchange as stylistic were pervasive in public commentary. In a report on the joint dedication published in *Artforum*, Grace Marmor Spruch—not a typical art critic, as she was then a research scientist in physics at New York University—described the settings of each Center as strangely inverted. The Center for Advanced Visual Studies assumed the look and feel of a scientific laboratory:

> Flashing lights, filters, projectors, brass and steel constructions, wires, magnets, motors, an exhibition on light and motion. A polyethylene tube coiling like some immense transparent python, ingesting its meal of air from a floor pump. A stairway leading down to a lower level to accommodate huge balloons suspended from the ceiling, reminiscent of accelerator rooms in physics laboratories.[58]

Photographic documentation also shows the Center to be cold and severe, filled with technical models and enlargements of Kepes's photographic experiments (figs. 5.14–5.15). In contrast, the Center for Theoretical Physics was opulent, filled with the luxurious décor appropriate for a museum:

> The entrance was like that of a plush art gallery. A wire sculpture hung from the ceiling. Carpeting flowed out of the offices into the corridor from wall to distant wall. Paintings.... "This place looks like a visual arts center," someone was heard to say.[59]

These contradictory descriptions make clear how interdisciplinarity was as much an aesthetic in its own right.

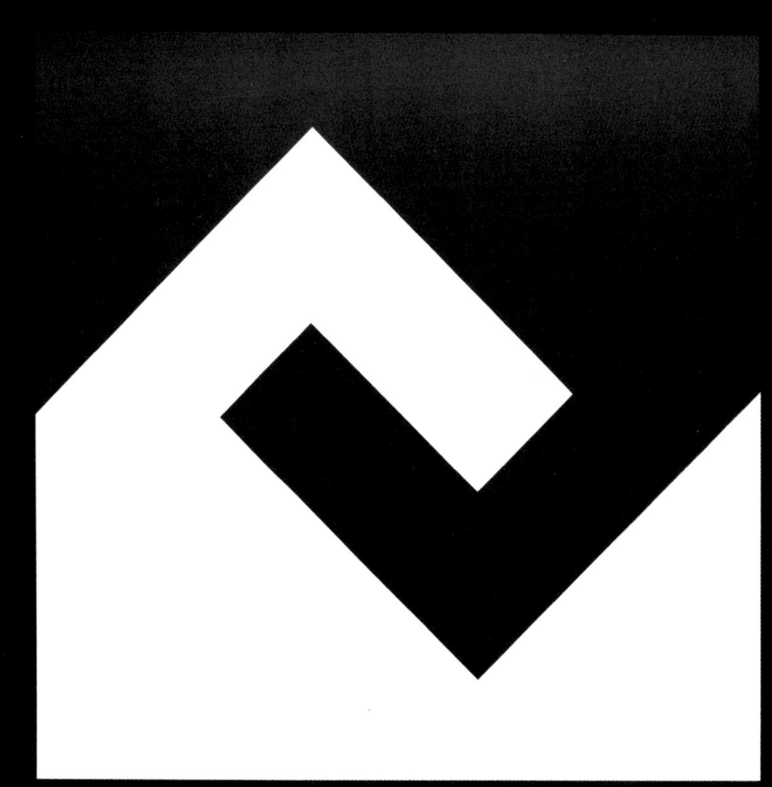

Art
Science
Technology

A symposium on the occasion of
the joint dedications of the
Center for Advanced Visual Studies
and the
Center for Theoretical Physics
Massachusetts Institute of Technology
The public is invited

Thursday, March 21
10:00 am
Kresge Little Theatre
Welcome

Jerome B. Wiesner
Provost, Institute Professor
Professor of Electrical Engineering, MIT

10:15 am
Session I
Art, Technology, and Form Making

Moderator
Wayne V. Andersen
Associate Professor of the History of Art
MIT

Panel:
Cyril S. Smith
Institute Professor
Professor of Metallurgy, MIT

Robert Wilson
Director
National Accelerator Laboratories

Harold Tovish
Sculptor
Fellow of the Center for Advanced Visual
Studies, MIT

Otto Piene
Sculptor
Fellow of the Center for Advanced Visual
Studies, MIT

2:00 pm
Kresge Little Theatre
Session II
Art, Technology, and Communication

Moderator
Henry Millon
Associate Professor of the History of
Architecture, MIT

Panel:
Jerome Y. Lettvin
Professor of Communications Physiology
MIT

Stan VanDerBeek
Film maker

Ivan Sutherland
Associate Professor of Electrical
Engineering
Harvard University

Billy Kluver
President
Experiments in Art and Technology

Friday, March 22
10:00 am
Kresge Little Theatre
Session III
Art and Science

Moderator
Elting E. Morison
Professor of History
Yale University

Panel:
George Wald
Professor of Biology
Harvard University

Charles Eames
Designer

Philip Morrison
Professor of Physics, MIT

Robert Rauschenberg
Painter

James S. Ackerman
Professor of Fine Arts
Chairman of the Department
Harvard University

2:00 p.m.
Kresge Little Theatre
Session IV

Chairman
Julius A. Stratton
President Emeritus, MIT
Chairman, Ford Foundation

Speakers:
Victor F. Weisskopf
Institute Professor, Professor of Physics
Head of the Department, MIT

Gyorgy Kepes
Professor of Visual Design
Director of the Center for Advanced Visual
Studies, MIT

Conrad H. Waddington
Professor of Animal Genetics
University of Edinburgh, Scotland

R. Buckminster Fuller
Architect
Designer

5.13

Poster for the joint dedication of the MIT
Center for Advanced Visual Studies
and the Center for Theoretical Physics, 1968.
Design: Jackie Casey. Courtesy Center for
Advanced Visual Studies Special Collection,
MIT Program in Art, Culture and Technology.
© Massachusetts Institute of Technology.

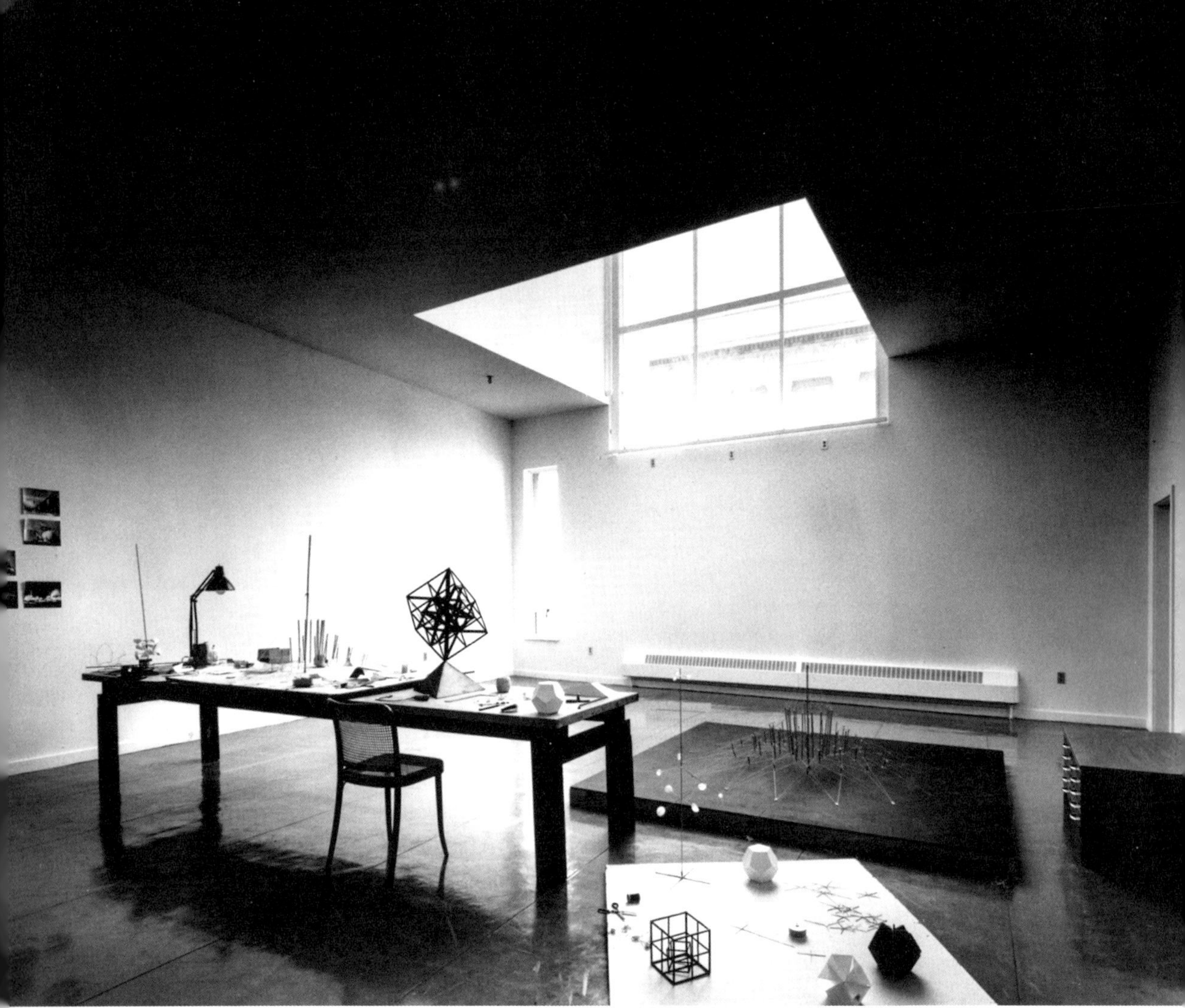

5.14 and 5.15

Interior photographs, Center for Advanced
Visual Studies, c. 1969. Photos: Nishan
Bichajian. Courtesy Center for Advanced
Visual Studies Special Collection,
MIT Program in Art, Culture and Technology.
© Massachusetts Institute of Technology.

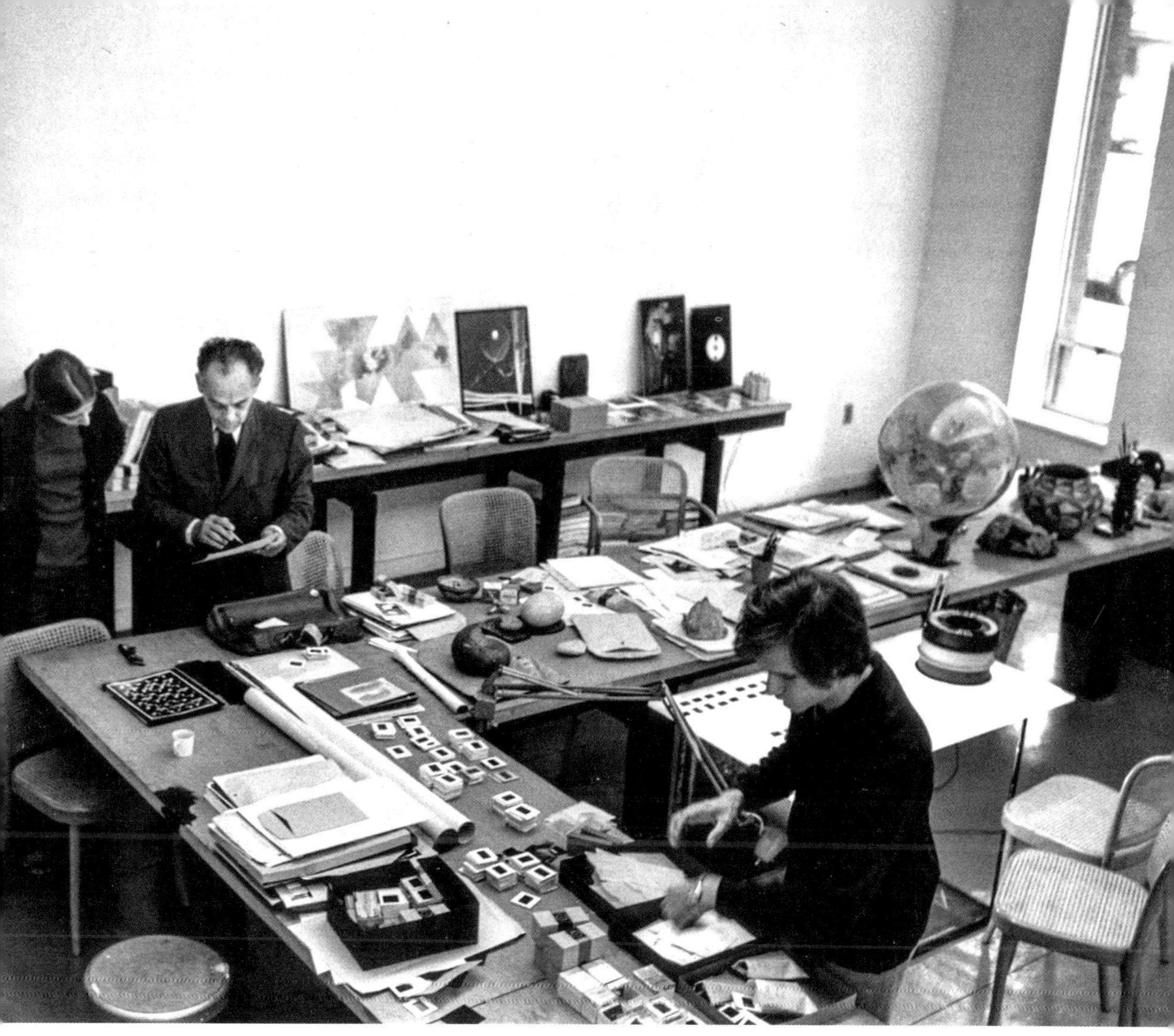

Problems of Interaction

The rhetoric is elegant, the theorizations are appealing, but how did this "institute of light," this "social parliament," this *new* New Bauhaus, actually function in the mid-1960s? How did Kepes's model of communicative politics—a system of approaches he variously describes as new idioms of collaboration and communication, as interdependence and intercommunication—work in practice? In his *Daedalus* proposal, Kepes indicates that communication is never easy. "In such a cooperative effort the value will come not only from an exchange of complementary ideas, but also from the friction of the conflicts that inevitably arise when such a group of individuals, each with his own angle of approach, works toward a common goal."[60] But his proposal belies the actual amount of friction that consumed the Center in its early years.

Memoranda written by Kepes and issued to the fellows indicate many banal problems. Kepes complained frequently that the fellows left the Center's doors open, invited unauthorized guests, and placed too many long-distance telephone calls.[61] Tools from the CAVS workshop repeatedly went missing.[62] Limited studio space and office storage were acute issues; so too were the fellows' conflicting and uncoordinated schedules.[63] Kepes was often annoyed that his artists failed to use their proper titles as members of the Center in public statements.[64]

But some of the problems were more existential, and none more so than the persistent lack of money. Kepes repeatedly complained about "the awful chore of fundraising."[65] He sent what he called "an SOS letter" to James R. Killian, Jr.: "Though I tell everyone that the Center is happy and healthy, we are actually quite poor."[66] He complained to the fellows about the Center's finances, and made a plea for proposals which might be used to secure funding from "civic and government agencies, foundations, and business organizations." He upbraided the fellows: their lack of participation in grant applications was a cause of financial difficulties. "I know, and you have to forgive me if I rub it in," he writes, "that we lost some chances because I didn't have any concrete material from you that I could use to convince people that we have ideas and deserve support."[67] Kepes even devised an elaborate moneymaking scheme—what he described as a "system used in the Bauhaus in Germany"—that would generate income from the fellows. The Center would keep a percentage of any commission paid to a fellow for outside work. Nothing came of the plan—no doubt it struck the fellows as exploitative.

Other problems were more fundamental to the Center's conceptual basis. For example, CAVS fellow Charles Frazier complained privately to Kepes that the group experience envisioned for *Explorations*—the fact that there would be no wall labels—meant the work of other fellows would overpower his contribution. Frazier called these interpersonal tensions "problems of interaction." He even hinted that Piene's inflated sculptures plagiarized his work; in sharing his proposals to the work community in the spirit of collaboration, he had been copied. "I can't afford to give my ideas and have them followed up by Otto.... To paraphrase your memorable remark, 'MIT is not a gas station,' neither am I a gas station."[68] Frazier refers here to Kepes's explanation that the Center was not just a site for filling up on technical expertise or scientific knowledge—"a gas station"—but an arena for meaningful exchange.[69]

Kepes was annoyed. "I was troubled by your letter. I wish we could avoid any conflicts in this community for it is hard enough to survive without inner tensions." But the interpersonal drama did not suggest to Kepes that communication was inherently flawed. Instead it suggested that more communication was necessary. "I assume the only way to create a mutual understanding in our common goals is to have maximum communication with each other." He reiterated his vision:

> I very much hope that this exhibition will be more than, as I expressed earlier, "an anthology of individual work," and that is why I asked all of you to collaborate and work out some interlocking relationship between your individual contributions.... I have to beg you all to recognize and accept that the Center's only meaning is in a convergence of goals and a cooperation of effort, and that sometimes implies some individual sacrifices.[70]

War Is Interdisciplinary, Again

Kepes could manage the day-to-day issues at the Center with pointed memoranda and curt letters, but the Center suffered from a more fundamental problem: the "intellectual technologies" that supported its mission, like interthinking and interseeing, were, by their very definition, methodologies that related the arts to other fields—fields that were already compromised by MIT's military involvements. Indeed, the term "intellectual technology" that Kepes borrows from Daniel Bell was actually a precise reference to approaches employed in military research. In *The Reforming of General Education*, Bell defines "intellectual technology" as shorthand for "the development of game theory, decision theory, simulation, linear programming, cybernetics, and operations research, many of which are tooled, as it were, by the computer."[71] These methods were all implicated in military research and development, and as Peter Galison has argued, in reference specifically to cybernetics, such militaristic associations "do not so simply melt away."[72]

We might also compare how the Institute visualized its interdisciplinary centers as a "university polarized around the sciences" to the picture of WAR

INCORPORATED, and its sinister pentagon of power (figs. 5.6 and 5.12). Or consider a spread from the *Old Mole*, a radical underground newspaper that emerged in Cambridge during the late 1960s (fig. 5.16). The paper's name referred to Karl Marx's metaphor of revolution as an "old mole," as the paper's masthead explains more fully: "We recognize our old friend, our old mole, who knows so well how to work underground, suddenly to appear: the revolution."[73] The spread, titled "Spider Webs in the Ivory Tower," shows the dizzying relationships between area universities like MIT and Harvard and other institutions, as organized under categories like "Finance," "Government," and "Major Corporations." How was Kepes's Center integrated into this tangled web? Where does interdisciplinarity actually lead, if we trace its many exchanges and dialogues?

In fact, the MIT faculty whom Kepes had contacted while planning his Center make an impressive roster of cold warriors. To take just one example: physicist Jerrold Zacharias was a veteran of MIT's Radiation Laboratory or "Rad Lab," where he helped to invent radar during the Second World War. He was one of four directors of the Manhattan Project at the Los Alamos National Laboratory, where he helped to invent the atomic bomb. He participated in numerous military projects in the 1950s, including MIT's "summer studies"—special research sessions on advanced military topics that took place during breaks in the academic calendar. The most relevant of these might be Project Lamp Light, which investigated continental defense and early warning systems—and also bore a name of special interest to Kepes. In an intriguing semantic slip, it was "light" that Zacharias offered as a topic for reconciling disciplines at MIT. In his letter to Stratton summarizing preparations for the Center and its founding symposium, Kepes notes: "Professor Zacharias offered an appealing suggestion for uncovering the kind of scientific-esthetic interrelations with which this seminar would be concerned." Zacharias proposed an exhibition on "the parallel and often overlapping relations of art and science to light phenomena."[74]

The presence of military power at MIT became an institutional problem on a very specific date: not May 1968, but 4 March 1969. On that day, MIT students, under the auspices of a group they called the Science Action Coordinating Committee, or SACC, joined with MIT faculty, under the auspices of a group they called the Union for Concerned Scientists, or UCS, for a campuswide "research stoppage." The movement, named March 4th in reference to public demonstration (as in, to march forth), replaced the day's normal research activities with discussions about the extent of MIT's participation in Vietnam, specifically through the Institute's two "special labs" devoted to

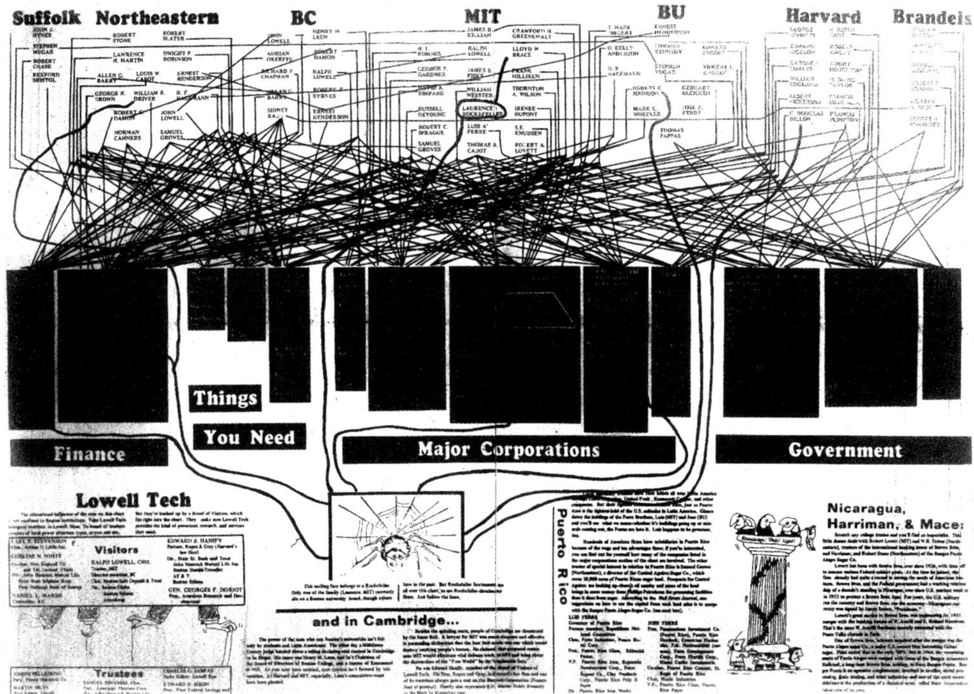

5.16

Centerfold spread titled "Spider Webs in the Ivory Tower," from the *Old Mole*, 14 October to 23 October 1969.

military research and development: the Instrumentation Laboratory and the Lincoln Laboratory.[75] One resonant line from the remarks made that day by historian Howard Zinn of Boston University succinctly distills the theme of the protest in just three words: "War is interdisciplinary."[76]

Kepes's closest scientific colleagues—the ones he had surveyed in preparing his Center—represented a full range of responses to the March 4th movement. Bruno Rossi and Victor F. Weisskopf both signed a faculty statement in support of the research stoppage. Zacharias signed a rival statement in opposition. Such a split suggests how Kepes, regardless of his own sympathies, had to carefully navigate a divided academic landscape. His Center required the backing of the scientific establishment, but that establishment was already deeply fractured over the ethical questions of its research.

In an attempt to contain these debates before they consumed the Institute, MIT's administration responded swiftly. Howard Johnson, MIT President, organized a Review Panel on Special Laboratories to examine the problem of military research on campus. Known as the "Pounds Panel" for its chair, Dean William F. Pounds of the Sloan School of Management, the group met for many fractious public hearings before issuing a final report in the fall that documented the startling extent of defense contracts at MIT. In 1968, $120 million, more than half of MIT's annual operating budget of $214 million, poured into the two special labs (adjusted for inflation, that figure represents more than $810 million of an annual operating budget of more than $1.45 billion).[77] The report also itemized what all this money supported. The Instrumentation Lab used government funds to create navigation and guidance systems for NASA's Apollo program and for the Poseidon, Polaris, and Sabre ballistic missile projects. The Lincoln Lab used government funds to develop Moving Target Indication or MTI radar, a technology that could detect people in motion through foliage and was put to use in the jungles of Vietnam (it is significant, given Kepes's concerns, that MTI radar is a visual technology).

It would be wrong to assume that this research, distant as it may appear from the humanistic goings-on of a university campus, has nothing to do with art. Making connections across disciplines was implicitly the purpose of Kepes's Center. Making such connections explicit—revealing what they actually represented—was precisely the point of the March 4th movement.

In fact, Kepes collaborated with both of the special labs. A daily logbook recording work toward the São Paulo Biennial is our smoking gun. One entry lists Kepes's contact with the founder and director of the Instrumentation Lab, Stark Draper, whose forceful defense of the Lab's agenda polarized opinion on campus (he was known as MIT's very own Dr. Strangelove).[78] A notation in

Kepes's hand records a "phone call with director Instrumentation Lab"; Draper "offered help, including underwriting special project expenses." Kepes's call occurred just weeks after 4 March, at a moment when Kepes would have been fully aware of the Lab's military activities.[79]

The logbook also indicates that Kepes met with Louis L. Sutro, a scientist at the Instrumentation Lab specializing in optical technologies such as stereoscopic TV (another visual technology that surely intrigued Kepes). Sutro also "offered collaboration." A later entry further indicates that Kepes, accompanied by CAVS fellows Harold Tovish and Ted Kraynik, visited the Lab in order to discuss possible engineering options for a series of "cybernetic forms"— mechanical devices using feedback—for inclusion in *Explorations* (the exhibition catalog acknowledges Sutro for his contribution). On Kepes's behalf, Sutro even planned to speak with Draper after the meeting "about sponsorship" that might benefit Kepes's Center. Kepes's visit occurred two weeks after his earlier call with Draper—again, after 4 March.[80]

These collaborations were reciprocal; not only were artists engaging MIT's most controversial scientific contexts, but MIT's most controversial scientists were also engaging artistic contexts. Just that month Draper published an article for *Leonardo*, the "International Journal of the Contemporary Artist," on the role of the sciences in education.[81] An editor's note appended to the article explains its relevance to the artistic community: "Many contemporary artists are turning to science and technology for inspiration and for new materials and techniques. It is believed that they will be interested in the problems of modern higher education for [the] training of professional engineers and technologists in the United States."[82] In the text, Draper describes the "atmosphere of creativity" at the Instrumentation Laboratory, and embraces the Lab's "active projects with industry and government" as a successful model for advanced research in science and technology. He implies that such a model might also be used for advanced research in the arts, an idea likely borrowed from Kepes.[83]

In addition to his call with and visit to the Instrumentation Lab, Kepes also visited the Lincoln Lab, where he met with Jack Nolan, Group Leader of the Computer Systems division. Nolan had joined the Lab in the 1950s to work on SAGE, the system of radars and anti-aircraft weapons spanning North America and powered by digital computers that was first initiated by Zacharias's summer study, Project Charles. Nolan's work in the 1960s involved the development of a communications system called SIMPLEX that would connect all levels of the military in a computerized network. In a letter Nolan sent to Kepes on Lincoln Laboratory letterhead following their meeting, he enthusiastically endorses Kepes's interdisciplinary agenda, gushing:

> I greatly enjoyed your visit to the Laboratory.… The ideas you expressed … were both cogent and stimulating.… I am strongly interested in the potential role of the visual arts in modern society; your new center's explorations of the relations between art and science strike me as an important avenue toward this goal. I would appreciate any opportunity to participate in the activities of the center – e.g., in seminars, study groups, etc. – where my professional background in computer sciences and graphics may be pertinent.… Again, thank you for a stimulating afternoon of discussion on a difficult but important inter-disciplinary problem.[84]

Kepes replied with similar enthusiasm:

> I greatly appreciate your kind comments on my visit to your laboratory. The chance to have some exchange of ideas with you and your colleagues and the opportunity to see some of the work at Lincoln Laboratory was a major experience for me. Naturally, I would be most happy if you could become involved in our work, and I will let you know as soon as an opportunity arises.[85]

Kepes did not acknowledge Nolan in the *Explorations* catalog, and he appears infrequently in the records of the Center's activities; it is not clear how meaningful their exchange really was. On the other hand, Nolan had already left Lincoln Lab by 1969 in order to head the Massachusetts College of Art, a surprising career change indicative of the rapidly closing gap between the two cultures (Nolan's interests spanned the arts and sciences: he was not only a computer scientist, but also a painter of abstract watercolors). Perhaps Nolan's departure for a job in the arts was also a way of registering protest against the uses and abuses of his expertise at MIT.[86]

As such correspondence indicates, interdisciplinary relationships were not just symbolic—not merely part of the sophisticated visual and verbal rhetoric Kepes developed—but also substantive. They provided opportunities for the exchange of legitimacy, of prestige and power, but also actual knowledge. And, of course, there was money changing hands. The "umbilical cord of gold" famously identified by Clement Greenberg in 1939 as the source sustaining the modernist avant-garde was now tying together a new formation, a Cold War avant-garde.[87] Let me emphasize again that this avant-garde was literally militaristic.

We can trace the events of March 4th and this exchange of prestige and power, knowledge and money, more directly to *Explorations*. Kepes actually wrote a memorandum to the fellows on the day under consideration: 4 March 1969. But the memo was not about the protest events. Actually, the very existence of the document indicates that Kepes *did not* participate in the research stoppage—he was too busy writing memos about São Paulo. His message asks the fellows to make a good impression on MIT's administration, specifically former MIT President and then Chairman of the MIT Corporation, James R. Killian, Jr., who had "reservations" about whether the Center's curatorial contribution would be "impressive enough" to justify the cost. "As he [Killian] is the only man from the Administration who follows the art scene and its literature," explains Kepes, "he also has a personal scale of values concerning the issues involved. His scale of values is not necessarily ours, but we have to recognize his attitude as an essential factor."[88] The clash between the agenda of March 4th and Kepes's memo of 4 March is shocking: in order to play the politics necessary to secure MIT funding, Kepes asked the fellows to help him flatter a figure in the administration whose "scale of values" was not necessarily the same as theirs; indeed, a figure whose scale of values was actually determined not by the arts at all but by science and technology and their military applications—Killian was invested in such pursuits throughout his career (he served Eisenhower as the first Presidential Science Advisor, a position he used to endorse the doctrine that came to be known as Mutually Assured Destruction, or MAD). Students and faculty were protesting precisely these types of relationships and what they saw as their damaging effect on MIT's educational mission.

Although the US contribution to São Paulo failed to materialize, these relationships still persisted in the form the exhibition eventually took at MIT and at the National Collection of Fine Arts. The *Explorations* catalog acknowledges the show's many industrial sponsors, noting that their offer of funds constituted "tangible evidence of a commitment to collaborate with artists."[89] A few of these sponsors were simultaneously supporting war research at MIT. The catalog lists both Raytheon and General Electric as sponsors of the exhibition; the Pounds Panel report lists both as sponsors of the Poseidon ballistic missile project.[90] Many other companies supporting *Explorations* were also beneficiaries of defense contracts and had previously or were currently engaged in military research and development, among them Fairchild Camera and Instrument Company, Corning Glass Works, Sylvania Electric Products, Polaroid Corporation, Xerox Corporation, and Honeywell.[91] Kepes had no qualms about begging for support from such companies. In a letter to EG&G, the defense contractor established by MIT engineer Harold Edgerton, he justifies the cause: "To create an exhibition that is presented with sophisticated technology, as you know, requires a rather extensive budget, which we do not have. We are desperately in need of sympathetic support from industries."[92] Such support included money: Standard Oil of New Jersey, the petroleum giant, and Rohm and Haas, makers of Plexiglas—the acrylic material that was first mass-produced for use in aircraft and submarines during World War II—contributed funds.[93] And it included materials: photosensitive glass from Corning, photocells from Raytheon, tape lights from Sylvania, and polarized plastic from Polaroid—the same material Kepes used to make his *Photoelastic Walk*.[94]

Serving the Human Purpose?

This tangled web of relationships may seem to be an abstract network far removed from the actual activities of Kepes's Center. But let me reiterate: the Center was not separate. Later that year it even became the explicit target of student protests. Furious at the slow pace of change after 4 March, the SACC organized a demonstration on Alumni Day, a homecoming reunion comprised of talks and panels following commencement that June. According to press coverage in *The Tech*, "The actual trigger for the student protests that Monday was the panel that included … Gyorgy Kepes."[95]

Titled "The Human Purpose," the panel (fig. 5.17) explored how MIT and its alumni serve humanity through science and technology. In addition to Kepes, panelists included Governor Francis Sargent of Massachusetts and Governor Luis Ferré of Puerto Rico, both MIT alumni, and MIT Professors Jay W. Forrester and John F. Collins. Forrester, a computer engineer, developed the Whirlwind computer that ran SAGE in the 1950s; in the 1960s, he pioneered systems dynamics, a field that made use of new computer modeling techniques. Collins was a former mayor of Boston; Forrester had collaborated with Collins on applications of systems dynamics to the maintenance of actual cities like Boston, a field he called urban dynamics.

The tentative outline of the day's events distributed to participants indicates the panel's technocratic agenda. Governor Sargent described how "massive state problems" in the Commonwealth of Massachusetts might be solved with new methods; Forrester specifically explained the use of computer models in such methods; and Collins gave an overview of MIT alumni involved in such a program.[96] Kepes inserted the arts into this agenda, arguing for the "need to help the human condition by recognizing the role of art in the transformation of the urban environment." Closing remarks by Killian summarized the proceedings and the value of the Institute's mission: "M.I.T., by attacking problems that bridge broad fields, succeeds also in producing a new kind of educated man to play a key role in our society."[97]

Students thought the panel was a sham—a glorification of interdisciplinary methods at the expense of solving anything at all, a deceptive distraction from very real human problems. A group of local activists called the "Ad Hoc Coalition to Convert MIT into Serving the Boston Communities" tried to force the panel to discuss real issues like social and economic inequality. But, as photographic documentation of Alumni Day indicates, their protest was quickly subsumed by that persistent question over MIT's war research. "How Does MIT Serve 'The Human Purpose?'" asks one protest sign, a sign that directly questions the name and basis of Kepes's panel (fig. 5.18). Another

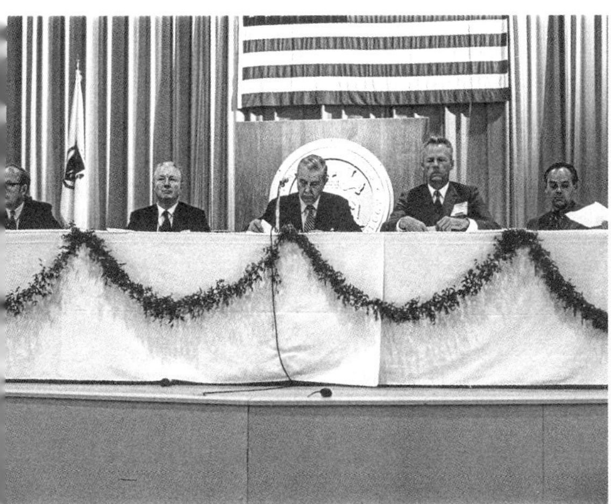

5.17

Alumni Day panel at MIT, 16 June 1969. Kepes is seated at far right. Courtesy MIT Museum, Cambridge, MA. © Massachusetts Institute of Technology.

5.18

Alumni Day protest at MIT, 16 June 1969. Courtesy MIT Museum, Cambridge, MA. © Massachusetts Institute of Technology.

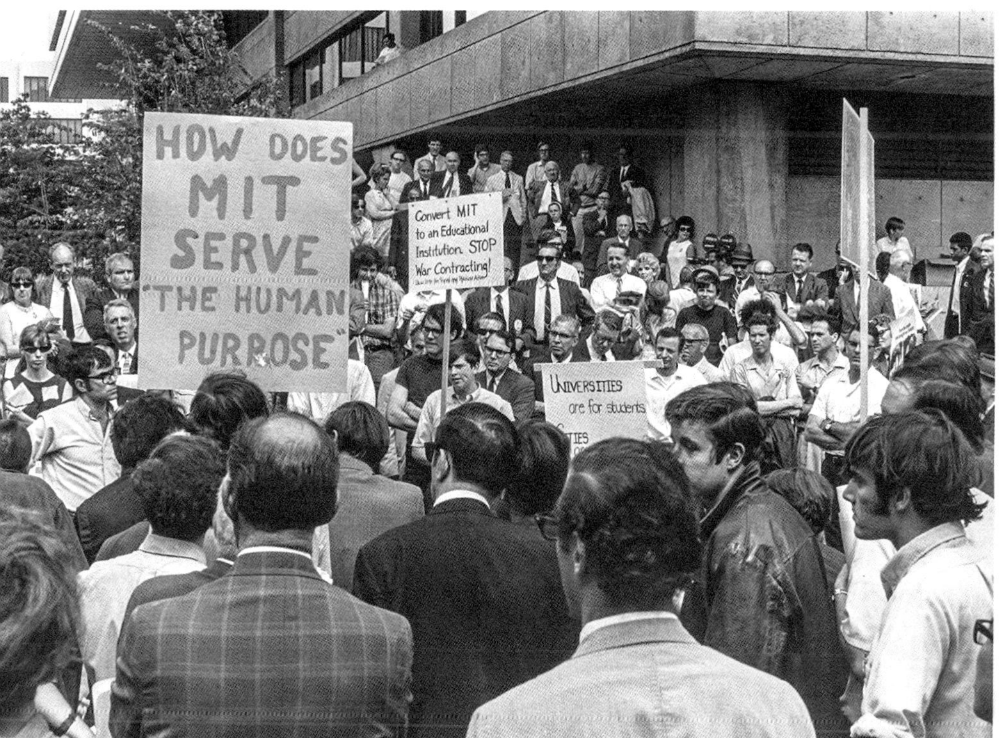

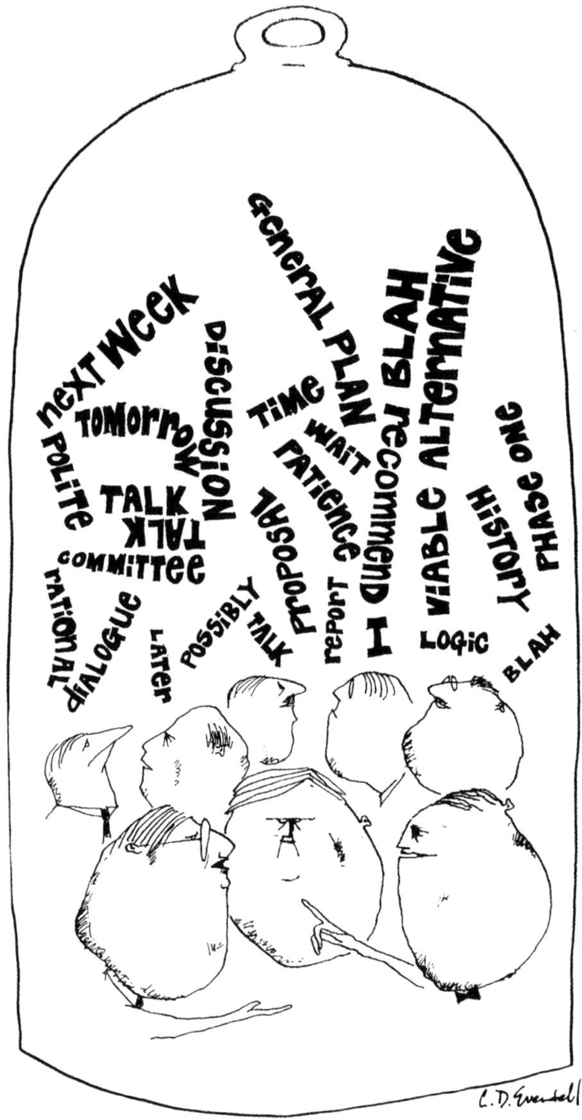

5.19

C. D. Everdell, cartoon from
the *Old Mole*, 26 September
to 9 October 1969.

placard declares: "Convert MIT to an Educational Institution. STOP War Contracting!" In tying the panel to war contracting, the protestors were onto something; "The Human Purpose" was organized by an MIT alumnus named Claude W. Brenner, an astronautical engineer who was then working on nuclear weapons testing for EG&G.[98] To be blunt: a discussion on the "human purpose" was organized by a scientist working on military research that was closer to destroying humanity than serving it.

The Alumni Day protest achieved nothing. At the evening banquet following the event—a reception Kepes likely attended, as he had received a formal invitation—Institute President Howard Johnson affirmed MIT's agenda: "This institution will live up to its responsibilities in this nation's defense."[99] The only tangible effect was a scathing response from alumni. Letters sent to the editor of MIT's student newspaper after the events reveal their disgust. "MIT is infested with a lot of left-wing radicals," wrote one.

> [T]he beautiful buildings and facilities were contaminated by so-called students who had every appearance of being animals rather than civilized educated young men. The so-called students were a conglomeration of long-haired hippies, kooks, freaks, and fairys [*sic*] of all sorts. The famed "Bowery Bums" command more respect than the assortment of animals we saw littering the beautiful MIT campus.[100]

I want to emphasize that this criticism is expressed in aesthetic terms, as a defilement and desecration of MIT's high modern purity, a contamination of its "beautiful buildings" and "beautiful campus."

These divisions between students and faculty were given visual form in the weekly satirical cartoons that ran in the *Old Mole*. One shows talking heads ensconced in a bell jar, like the type used to contain a rare object for display (fig. 5.19). The chattering class jabbers and babbles: "rational," "dialogue," "proposal," "report," "patience"—the words just float aimlessly. All talk, no action. The image lampoons the futility of communication and collaboration—the model that Kepes so believed in, and the particular means of engagement on display at the Alumni Day panel—as nothing more than hot air.

Conversion

And yet, despite the complex ways in which the "interdependence" and "interconnectedness" that Kepes advocated actually associated the Center with the darker activities of MIT's faculty, it would be wrong to assume that these collaborations were completely compromising. I really do not mean to reject Kepes's project—and it should not be dismissed so quickly. The majority statement in the Pounds Panel report introduced the idea of "conversion": the transformation of military research into nonmilitary research. Conversion would maintain MIT's commitment to scientific and technological expertise, but would use this expertise to solve problems free from associations with the war in Vietnam; addressing such issues would subsequently attract new, more neutral, forms of investment. The majority statement argued for incremental changes toward this goal. The Institute's funding was tied so tightly to defense projects that it seemed impossible to immediately abandon all existing research and development initiatives. The statement reassured MIT's administration: "It is important to note that the objective of this recommendation is not the ultimate elimination of all defense work in the special laboratories."[101]

By contrast, the more radical position, one articulated in a minority statement appended to the Pounds Panel report and authored by MIT graduate student activist Jonathan P. Kabat with signatures from fellow graduate student Jerome Lerman and Professor Noam Chomsky, introduced the idea of "total conversion," which Kabat defined as "the conversion of the laboratories and the society from wasteful to socially useful production."[102] To achieve this goal, the statement paradoxically argued for "closer ties with the special laboratories." Among the socially productive uses imagined in total conversion was the investigation of scientific and technological solutions for problems like "widespread poverty and malnutrition, grossly deficient school systems, poor and inequitable medical care, urban decay, inadequate public transportation, [and] air and water pollution."[103]

Total conversion was also a model implicitly embraced by Kepes and actualized at his Center. It was completely consistent with the transformative role of aesthetics he developed in his projects at MIT, from *The New Landscape* to *Vision + Value*. Kepes's "collaborative idioms" were designed to shift and shape not only the arts, but also the sciences—including, we can assume, their research agendas and funding models. In a memo to the fellows dated October 1969, the same month the Pounds Panel released its final report, Kepes is clear: "As you may know, there is a major shift at MIT from the emphasis on science of the last 10 to 15 years to areas which promise social contributions in a direct sense."[104] He asked fellows to propose projects that could use the arts as conversion, a means of transforming science and technology from wasteful to socially useful production, or, as Kepes here calls it, "social contributions in a direct sense." Perhaps this was the motivation for Kepes's collaborations in MIT's most compromised laboratories.

Still, conversion is a highly ambiguous proposition, its direction of exchange hard to decipher: who converts what, and why? William Irwin Thompson, a writer then teaching in MIT's humanities program, describes conversion as a red herring. It would not make the sciences more humane by relating them to the arts—the point of conversion as Kepes envisioned it—but instead make the arts more inhumane by relating them to the sciences. Commenting on the founding symposium that established Kepes's Center for Advanced Visual Studies concurrently with the Center for Theoretical Physics, Thompson writes:

> In the expensive symposium in dedication to the new Center, M.I.T. was not trying to listen to the artists but was trying to beguile them with all the temptations at its disposal. The technocrats were seeking to convert the artists into new celebrants of the system; in making technological art possible, they were insuring the stability of their technological society.… Men who are specialists in the management of complex organizations learn quite early that the best way to preserve the stability of powerful routines is to give on the surface the illusion of reform and constant innovation. The symposium which celebrated the new M.I.T. with its Centers for Advanced Visual Studies and Theoretical Physics was just such brilliant camouflage for the "military-industrial complex."[105]

As Thompson describes it, MIT's administration colluded with military power as part of a morally and ethically questionable grand bargain: by generously funding the arts and humanities, the Institute could justify its even more generous funding of activities with dubious educational value, like those projects then proceeding at the two special labs. The "guilty technologists," as Thompson calls them, used their "servile apologists, the technocratic artists," to rationalize the support of defense research.[106] One culture—art—would camouflage the corrupt activities of the other: science.

This cynical interpretation of interdisciplinary exchange is relevant to the crisis in the arts and humanities today, and the possible compensatory function of both on college campuses, especially as academic culture leans further toward what we now call STEM subjects (science, technology, engineering, and mathematics). Thompson anticipates our present condition:

> The humanist at M.I.T. thus finds himself in a situation that is no doubt prophetic of the condition of the citizen in the technological society of the *magnus annus*, A.D. 2000. To the degree that the humanist succeeds in technologizing the humanities (by turning them into the social sciences), he destroys the humanities; to the degree that he ignores the technological world and teaches as one might at Cardinal Newman's Oxford, he ensures the conviction in his students' minds that the humanities are simply irrelevant to the mastery of our new complex society; to the degree that he succeeds in communicating the relevance of the traditional humanities to our society, he finds himself welcomed by the administration as valuable camouflage, and resented by his students, who correctly point out that while he makes a great noise, he is still powerless to affect the inhumane training of the whole Institute. The naïve humanist thinks that in teaching the humanities to M.I.T. students he is helping a major American institution deal with the problems of our civilization, but it does not take long for the students to educate the teacher to see that the Institute is, as Eldridge Cleaver would say, not part of the solution, but part of the problem.[107]

Kepes's Center was entangled in these debates. The art critic Jonathan Benthall explains that CAVS had "the opportunity to tackle and modify technological rationality on its own ground," that is, to change the context that perfected technological rationality in the first place. But Benthall also explains the inherent danger of this opportunity: "As a benign foreign body planted near the core of American technocracy, the Center runs the double risk of either being rejected altogether as irrelevant, or—more subtly—of being assimilated too readily, as a public sign by technocrats that they can find space for art within their system."[108]

These arguments were even present in the very same issue of the journal *Daedalus* in which Kepes put forth his proposal for the Center (fig. 5.20). *Daedalus* was a sympathetic venue for Kepes's statement. The institution that published it, the American Academy of Arts and Sciences, encompassed the agenda of interdisciplinarity in its very name. This mission was also evident in the journal's logo, which Kepes actually designed in the 1950s: the pattern represented an aerial perspective of the labyrinth built by the inventor Daedalus, himself a master of science but also art, on the isle of Crete in Greek mythology.[109] The vantage was intended to visualize not the isolation created by a maze of unrelated specializations but the connection between them, the relationships that allowed one to trace through all fields of knowledge, and thus out from the labyrinth.[110]

In keeping with this total picture, the "Science and Culture" issue of *Daedalus* included not only essays in support of interdisciplinary exchange, compiled in a section labeled "On Coherences and Transformations" (where Kepes's proposal appeared), but also essays against interdisciplinary exchange, compiled in a separate section labeled "On Disjunction and Alienation." Here, Herbert Marcuse offered a critique that reads as if it is directed against Kepes. Titled "Remarks on a Redefinition of Culture," Marcuse's essay explains the "two cultures" conceit as a false construct, describing how their unification results not in an equivalence between them but instead in the complete assimilation of the first culture into the second. Interdisciplinary exchange forces the arts to abandon their antagonistic function and instead take on an affirmative one, serving "to fortify the hold of the Establishment over the mind."[111] While there had once been a time when "the mind could find an Archimedean point outside the Establishment from which to view it in a different light," a place of critical distance that the arts provided, such a position was now lost. "This Archimedean point seems to have disappeared."[112] Marcuse's argument was also about money: "The overwhelmingly generous financial support which the physical sciences enjoy today is support not only for research and development in the interest of humanity but also in the opposite interest"—the destruction of humanity.[113]

And yet, despite this critique of technological rationality and its colonization of the arts, Marcuse also suggests some theoretical possibilities—however impossible to realize—for critical intervention, or what I have called conversion. He argues for the "liberation" of science from the scientists; he hopes that "science would have a new function."[114] He elaborates:

> One can envisage the establishment of an academic reservation where scientific research is undertaken entirely free of any military connections, where the inauguration, continuation, and publication of research is left entirely to an independent group of scientists committed to a humanist pursuit.... Today, even these modest-sensible ideas are scorned as naïve and romantic, are heaped with ridicule. The fact that they are condemned before the omnipotent technical and political apparatus of our society does not necessarily demolish the value which they may possibly have.[115]

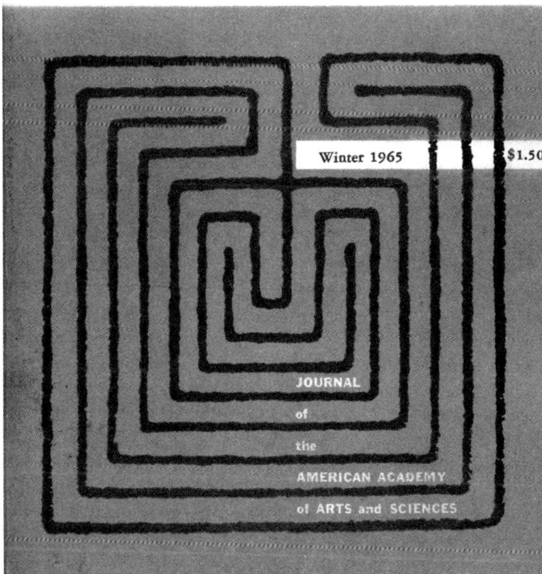

5.20

Cover of the journal *Daedalus* 94:1 (Winter 1965). Cover logo designed by Gyorgy Kepes. © 1965 by the American Academy of Arts and Sciences.

In his 1969 *Essay on Liberation*, Marcuse further argues that science and technology are not inherently destructive at all; it is a corrupt culture that makes them destructive. He writes, with purposeful exasperation: "Is it still necessary to repeat that science and technology are the great vehicles of liberation, and that it is only their use and restriction in the repressive society which makes them into vehicles of domination?"[116]

Perhaps in response to Marcuse, "conversion" would also enter into Kepes's own rhetoric.[117] In 1972, the Museum of Modern Art in New York held an international symposium organized by design curator Emilio Ambasz and titled "The Universitas Project." The purpose of the event was to explore the idea of the *Universitas*—the same concept Kepes had fantasized about in his *Vision + Value* volumes.[118] To debate these ambitions, hundreds of intellectuals descended on MoMA, including close colleagues of Kepes like Jay Forrester, now former CAVS fellow Jack Burnham, and contributors to Kepes's many books, such as René Dubos and Karl Deutsch. Other participants in the event, like Louis Althusser, Jean Baudrillard, Manuel Castells, Michel Foucault, and Alain Touraine, reflected a new generation of critics aligned against the technocratic establishment. Kepes attempted to follow them. In comments made during a discussion session, he argued for the role of the arts as a force of conversion capable of transforming the university into "a university of subversion, an institution that faces reality and can be angry, passionately angry about issues that we have to be angry about." He suggested that this anger "has to find its focus," that it should not be distracted by "technological fetishism" but instead directed toward "social revolution." The politics in Kepes's proposition are vague, even while he embraces a model of conversion:

> I do not speak on any political platform or anything like that, but there is a system, which evidently didn't live up to the new demands of the scale. And the system has to be changed. And this change has to find its tools…. We have different conditions now, and we have to find a different technology of revolution.[119]

Even at this moment, Kepes avoids any specific "political platform." Perhaps he felt that creating a "university of subversion" and a "technology of revolution" required him to maintain a neutral position. This position, however, only became increasingly difficult to preserve.

"Political Issues Sao Paulo"

We are now prepared to return to São Paulo and to *Explorations*, and to contemplate again the core questions raised by the exhibition: Was collaboration and communication a form of conversion, or a form of cooptation? Was science and technology a means of liberation, or only a means of control? The impossibility of answering these questions in any definitive way forced Kepes and the artists he had invited to participate in his exhibition into profound struggles. They grappled with the idealism of Kepes's communicative model, and what it actually represented at MIT in 1969. They both embraced Kepes's project but also rejected its particular actualization; like Kepes, they were tied in a dialectical knot.

In the Center's archives, in a folder labeled "Political Issues Sao Paulo," a pile of letters addressed to Kepes reveals these tensions. They suggest the difficulty artists faced in determining an appropriate mode of action—and how the concepts of communication and collaboration became contested in their own right. The first letter to Kepes is from Hans Haacke, the first to withdraw from the exhibition. "I have finally come to the conclusion that I cannot participate in an exhibition that represents the United States abroad," Haacke declares.

> **The U.S. Government is fighting an immoral war in Vietnam and is vigorously supporting fascist regimes in Brazil and in other parts of the world. At present any exhibition under U.S. Government auspices is designed and/or liable to promote the image and the policies of this government. It becomes a tool of international public relations, irrespective of the honorable intentions of the organisers and participants. With repressive tolerance the energies of artists are being diverted into serving the needs of a policy that these very artists rightfully despise. If they do not want to become involuntary accomplices they have no choice but to refuse the showing of their work in a national representation abroad.[120]**

In his reply, Kepes respected Haacke's decision even though he disagreed, arguing that communication was always preferable to no communication: "what one says" is always more important than "where one says it."[121]

Haacke's letter set off a barrage of protest statements, especially in France, where an international community of artists—a real human community—argued against participation. The American critic Dore Ashton, then in Paris, sent a letter to Kepes urging that withdrawal would be "a great gesture of solidarity with artists all over the world," specifically with artists who "have hoped that the power of America" could be used "for once in concert with the forces of good on an international scale."[122] The French critic Jean Clay explained that Brazilian expatriates then in Paris were unanimously opposed to participation (he cites Lygia Clark and Hélio Oiticica). He elaborated "how art is used" in Brazil, arguing that the Biennial would legitimate Costa e Silva's regime; there would be "[p]ictures with the winners, commentaires [*sic*] on the 'new' Brazil [as the] mother of art," and press claiming the country is now "backed by the whole world."[123] Clay also sent Kepes an anonymous dossier of supporting documents, prepared for a meeting of "about 300 artists" in Paris, that carried a startling image on its cover depicting the hand of Uncle Sam—or, more accurately, given what we now know about the operation planned for the 1964 coup, "Brother Sam"—grasping greedily for Latin America (fig. 5.21).[124]

In response to the dossier, Kepes drafted a statement he hoped the remaining artists would sign as an affirmation of communication and collaboration, a statement he could then release to the press and publish in the planned exhibition catalog as proof of his politics.[125] The lengthy text reveals Kepes's confused position, pulled as he was in opposite directions, and the contorted justifications that maintaining such a conflicted position required. He agreed with protest, but also with participation. He wrote against military dictatorship abroad, despite MIT's proximity to military power at home. He was against war, but also entangled in it:

> We artists, who are against all that prohibits human ways or stifles creative liberty, learned with increasing alarm about the present circumstances in Brazil – the plight of our fellow artists and intellectuals caused by the curtailment of democratic liberties. We artists from the United States have some understanding that the misplaced weight of our superwealth [*sic*] and power is not unrelated to the tragic difficulties our fellow artists face in Brazil. For the invited artists who have learned about conditions in Brazil and have solidarity with the Brazilian artists, there are two alternative stands. They can either boycott the exhibition, or participate with an unmistakable position toward Brazil's present situation....
>
> We recognize as they do that the present political rulers in Brazil could utilize the exhibition to claim that cultural conditions in Brazil are not as oppressive as they, in reality, are. Brazil's rulers could no doubt use an exhibition with international participation as a window dressing to hide the absence of cultural liberty in their country....
>
> We believe that in the long run our chance to communicate deeply-held ideas without compromise can have a far greater possible impact on Brazilian life than can be accomplished by a boycott and also could best contribute to the international, cultural commonwealth.... Our exhibition is intended as an "idea" exhibition; this idea is a democratic community expressed through a symbiosis of creative statements. Such concert is patently opposed to dictatorship oppressive to intellectual and civil liberties....
>
> We recognize the justification of confrontation with all inhuman political power systems, but we deeply feel that this must complemented by constructive, life-building efforts. There is an old Chinese saying that it is better to light a single candle than to curse the darkness. We believe that both are needed and we, rightly or wrongly, chose the first one.[126]

Kepes uses the metaphor of light to affirm collaboration as the only means of politics; for him, to not participate was to extinguish all light, to allow the darkness to descend. He was unable to imagine that collaboration might be co-opted in any way.

The statement prompted a very polarized response. On the one hand, it was far too political for MIT's administration, which was annoyed about Kepes's explicit discussion of power, even though Kepes did not cite MIT.

While the statement never appeared in the press in full, critic Grace Glueck quoted the final paragraph of Kepes's statement in a report for the *New York Times*.[127] Presumably in reference to Kepes's description of "the justification of confrontation with all inhuman political power systems"—a statement which seems somewhat harmless but which may have read as a veiled critique of MIT as "inhuman"—Institute President Howard Johnson circulated Glueck's article to MIT's Jerome Wiesner with an attached memorandum: "I see Kepes kept the one ¶ we had doubts about + of course it was quoted."[128]

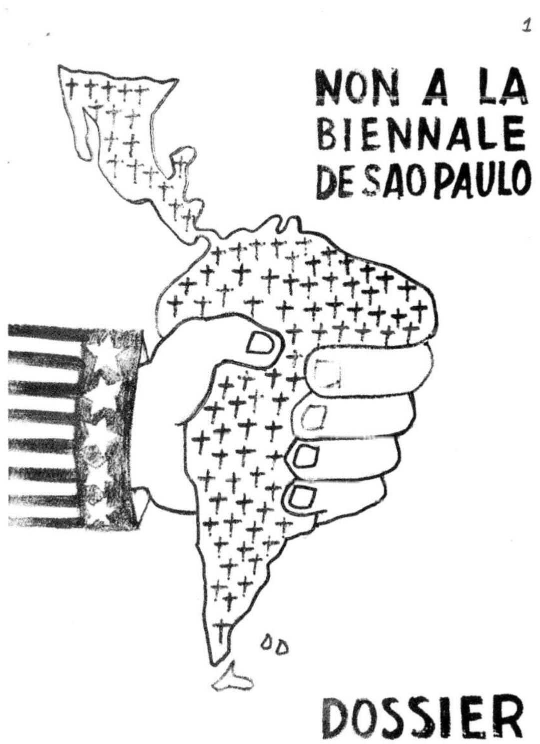

5.21

"Non a la Biennale de Sao Paulo," dossier photocopy, 1969. Courtesy Center for Advanced Visual Studies Special Collection, MIT Program in Art, Culture and Technology.

On the other hand, the statement was not political enough for Kepes's critics, especially Clay, who disagreed less with its content than its very existence—the model of communication and collaboration that the statement represented. "You will have it [the statement] in the north american [*sic*] catalogue. Very good! But who will read that? A few hundred, rather rich, people—no more." Clay realized a fundamental limitation to any intervention based on communication: there must be someone listening. "I am absolutely convinced that you will not have any ability to publish it in Portuguese in any paper.... Your position will be totally hidden."[129] Communication is not transparent; social, cultural, and political forces control and contain it. Collaboration is not neutral; it is directed.

About a week before he composed his statement on the Biennial, Kepes also wrote a private message to CAVS fellow Vassilakis Takis which reveals the ethical and moral confusion he experienced.[130] Takis and Kepes frequently exchanged candid letters. Kepes's confessional to Takis reveals an abiding faith in the "free communication of ideas" as "the only guarantee that our troubled and sometimes ugly life could be made a little happier and more human." He continues:

> I am certain that if the aspiration of the Center is expressed convincingly through the work exhibited, and made uncompromisingly clear in my statement in the catalog ... we could contribute more constructively to the Brazilian and the international artistic community than by withdrawing our participation.

And yet, Kepes also acknowledged some threat of co-optation:

> But I have no illusions.... For some people, no doubt, participation in an international exhibition in a country which is under brutal military dictatorship could be read as an approval of the regime. As our material would represent the US section, and as this country's relationship with Brazil's present rulers is justly questioned, our participation could also be read, by those who want to do so, as an acceptance of this country's [the US] role in relation to the Brazilian regime.[131]

Kepes's letter makes clear the complexity of these decisions, and how carefully he was considering the Center's actions. Despite this deeply felt letter, he still refused to really see the connections that had become obvious to all. Kepes continues with an assertion of the "facts" of the Biennial, but these facts are more aspirational than actual. He declares that that the US section "is organized by our Center without any strings attached. It is not a government exhibition but is cosponsored by the Smithsonian and MIT." The Smithsonian is, of course, a government institution; he bends the truth. And as the March 4th movement had already made abundantly clear, there were many strings—an entire spider web—attached to MIT.

Kepes would not acquiesce; his loyalty to MIT and to a country that had provided refuge was steadfast. He built his case for communication through more communication. He reached out to Brazilians whose "political stands ranged from the far left through the various shades of liberals," all of whom "without exception … felt very strongly that the most important thing for the Brazilian intellectuals was to have communication with the outside."[132] He even obtained a letter from none other than Brazil's former president Juscelino Kubitschek (João Goulart, the Brazilian leader ousted in the 1964 coup, had served as Kubitschek's vice president in the 1950s).[133] Kubitschek endorsed US participation without reservation:

> It is true that we live at present under a regime in which all liberties have complexly disappeared. But the Brazilian artistic and intellectual community, and men of culture in general, are in the forefront of those who guard against and draw attention to this state of affairs.…
>
> I feel, that, since the matter in question is an art movement, the presence of the Center for Advanced Visual Studies would not be interpreted otherwise than as a tribute paid by American artists to their Brazilian colleagues.
>
> The appearance of the Mit [*sic*] Center will, I am sure, excite great interest and curiosity in Brazilian intellectual circles, because of the tradition it represents, and especially because of the prestige of its director, Professor Kepes.

The fact that Kepes was able to obtain a letter from Kubitschek is remarkable; Kepes had hoped for "civic art" and here he was communicating with a former world leader, practicing the art of global geopolitics. The letter is also remarkable in revealing that a Brazilian opposition figure—Kubitschek was blacklisted after the 1964 coup, and briefly exiled—endorsed US participation in the Biennial even while artists worldwide protested against it. The letter suggests a break between the perspective of artists and that of politicians, but also a break between viewpoints abroad and within Brazil. "In short," writes Kubitschek, "I consider the representation of the Center of Advanced Visual Studies at the Sao Paulo Bienal [*sic*] to be of the greatest importance."[134]

Kubitschek's letter contains semantic slips, however. He argues that artists "guard against" and "draw attention to" political crises. But the artists still participating in Kepes's exhibition did not create work that was political in any overt way. It would be challenging to interpret the phenomenological technics of the works in the show—like those of Kepes's *Photoelastic Walk*—as guarding against or drawing attention to the crisis in Brazil.

Perhaps in an attempt to bolster his political credentials, perhaps because he was aware that the art in his exhibition—his art included—was not actually engaged, Kepes participated in a communication sent directly to President Costa e Silva, the Brazilian dictator. Kepes had his name appended to a telegram denouncing Costa e Silva's repressive politics. Prompted by the forced retirement of 68 professors accused of leftist sympathies, the cable decried the "massive expulsion of Brazilian professors and scientists from their universities" and defended "academic freedom, democratic procedures and the integrity of the sciences and humanities."[135] The telegram included some three hundred names from dozens of universities across the country, including CAVS artists Jack Burnham and Otto Piene and even Kepes's secretary at the time, Joan D. White. The cable was significant for including the former Ambassador to Brazil, Lincoln Gordon, one of the original architects of "Operation Brother Sam" and formerly a supporter of the 1964 coup.

But Kepes's letter from Kubitschek, and even his statement to Costa e Silva, made little difference. Soon, letters from other artists announcing their decision to withdraw from the exhibition arrived at the Center. The most damning was a powerful, now-famous letter from Robert Smithson:

> To celebrate the power of technology through art strikes me as a sad parody of NASA. I do not share the confidence of the astronauts. The rationalism and logic of the engineer is too self-assured. Art aping science turns into a cultural malaise. The righteousness of "team spirit" is no solution for art.... I am sick of "lighting candles," I want to know what the "darkness" <u>is.</u>

Smithson's letter is a forceful indictment of Kepes's mission; with one line, he extinguishes Kepes's metaphor of light. He continues:

> As rockets go to the moon the darkness around the Earth grows deeper and darker. The "team spirit" of the exhibition could be seen as an endorsement of NASA's Mission Operations Control Room with all its crew-cut teamwork. Some Brazilians see the "advances" of technology as a military byproduct. I am withdrawing from the show because it promises nothing but a distraction amid the general nausea. If technology is to have any chance at all, it must become more self-critical. If one wants teamwork he should join the army. A panel called "What's Wrong with Technological Art" might help.[136]

Other letters followed, many reflecting on the nature of communication itself. John Goodyear writes: "the hoped for communication with Brazilian intellectuals will not materialize. A meaningful communication with either party to the struggle is clearly hopeless through participation." Goodyear indicates that once communication enters the world, it can be used and abused, changed and manipulated. "Your clear and courageous statement condemning their [Brazil's] misuse of power will not improve channels of communication.... I am convinced that your good intentions run too great a risk of being distorted."[137]

But the letters were hardly unanimous, and numerous artists also wrote to Kepes expressing support. Newton Harrison argued that withdrawing was hypocritical. "A work of art is an exercise in freedom. To remove this exercise from an environment already significantly lacking freedom serves only to give power by default to that which I/we disagree with."[138] Charles Frazier also complained about censorship: "the 'Fascists' have won the day. Not only have they succeeded in silencing their own artists, … they have strangled the cry of outrage that might have rung out defiantly from within their own frontiers."[139] Other artists condemned not censorship but what they perceived as the orthodoxy of the New York art world. "I find the politicing [*sic*] of certain artists connected with the show no more desirable than the politics of the Brazilian Government," argues Les Levine.[140] Vera Simons recalled meeting the Brazilian artist Rubens Gerchman, then living in New York, whose work the Brazilian government had censored. "He urged me to withdraw from the Bienal [*sic*] and warned of the possible destruction of my work in Brazil if I did not," she reported to Kepes. "That is intimidation from the one side."[141] Kepes followed up with a letter to Gerchman accusing him of trying "to coerce them [Kepes's artists] into not participating." He even claims that such intimidation is "not unlike" the tactics of the Brazilian regime.[142]

The point is that the Center's utopianism had actually resulted in an endless stream of confusing, not clarifying, exchanges, an infinite circle that seemed to lead nowhere at all. The tiresome letter-writing was finally for naught; the chain of positions and rebuttals, statements and counterstatements, added up to nothing. There simply were not enough participants to justify an exhibition. The show was not viable. The cause had to be abandoned. Kepes had the final word:

> We feel as deeply concerned at the lack of creative and democratic liberties as those who have withdrawn, and the intention of the fourteen of us had been to participate in the exhibition and place on record our deep resentment against cultural suppression. To our great disappointment we have to conclude that our basic concept to create a community exhibition has been crippled by the withdrawal of nine participants of the community.[143]

After withdrawing from São Paulo, Kepes worked to guarantee that the exhibition could still be realized at MIT and at the National Collection of Fine Arts. He would be more careful. Concerned "about the artists' political moodiness," he would screen participants.[144] He ruled out four of the original artists because "their response would have been obviously negative."[145] From everyone else, he would obtain "an absolute assurance that they know the circumstances and accept the rules of the game."[146] But Kepes was also devastated by these events.[147] Writing to Jerome Wiesner later that summer, he disclosed a profound disappointment:

> Now in the aftermath of all the struggle and tensions, I am tying up the loose ends and "burying the dead." I am sure you know that it was not an easy time for me. I had a great deal of hopes and expectations for the Sao Paulo Exhibition and confidence that we could make a real contribution, and am heartbroken about the way things developed.[148]

The episode seemed to refute his essential ambitions, for communication and collaboration did not create the type of change Kepes once imagined possible.

Coda: The Return of Revolution

The events of the late 1960s put new pressure on the principles guiding Kepes's program. By the end of the decade, the world was spinning too fast. Each image that flickered across the TV screen seared the eyes: bombs blasting over jungles, helicopters landing with a deafening roar, human flesh burnt by napalm. Each violent convulsion reverberated again at home: students amassed in protest, police armed with riot gear, a silent majority seething with anger. In this moment, Kepes's hopeful ideals seemed hopelessly out of touch; even honest principles like collaboration and communication rang hollow, just empty noise: platitudes and clichés.

A culminating episode reveals this painful gap between imagining ideals and making them come true—between dreaming and undreaming. The episode indicates the distance Kepes had traveled since first arriving on American shores as an erstwhile revolutionary, and his own surprise at where he ended up.

That fall, Kepes resolved to give new life to the Center; he wanted "to offer something to MIT which would be meaningful."[149] With help from fellows Ted Kraynik, Otto Piene, Wen-Ying Tsai, and Stan VanDerBeek, he planned to open the Center's doors to the community at large for a series of "informal seminars," the goal of which would be "to stimulate communication and exchange of ideas."[150] But the atmosphere that fall was charged with tension. In November, police attacked student protestors in front of the Instrumentation Lab (fig. 5.22). Later that month, the *Old Mole* ran a cover (fig. 5.23) showing the flag of the National Liberation Front unfurled over a balcony in the grand lobby of MIT's historic Rogers Building on Massachusetts Avenue; the cover was accompanied by a headline echoing chants that could be heard across campus: "6 5 4 3 LET'S SMASH MIT!" The issue was intended to galvanize the November Action, an attack coordinated by MIT's most radicalized students against what they saw as MIT's continued complicity with the war. A strike poster later reproduced in the paper shows the faculty as empty suits, powerless to make meaningful change: tweed puppets with no eyes, easily manipulated by Uncle Sam (fig. 5.24).

5.22

Protest at the Instrumentation Laboratory,
Massachusetts Institute of Technology,
5 November 1969. Courtesy MIT Museum,
Cambridge, MA.

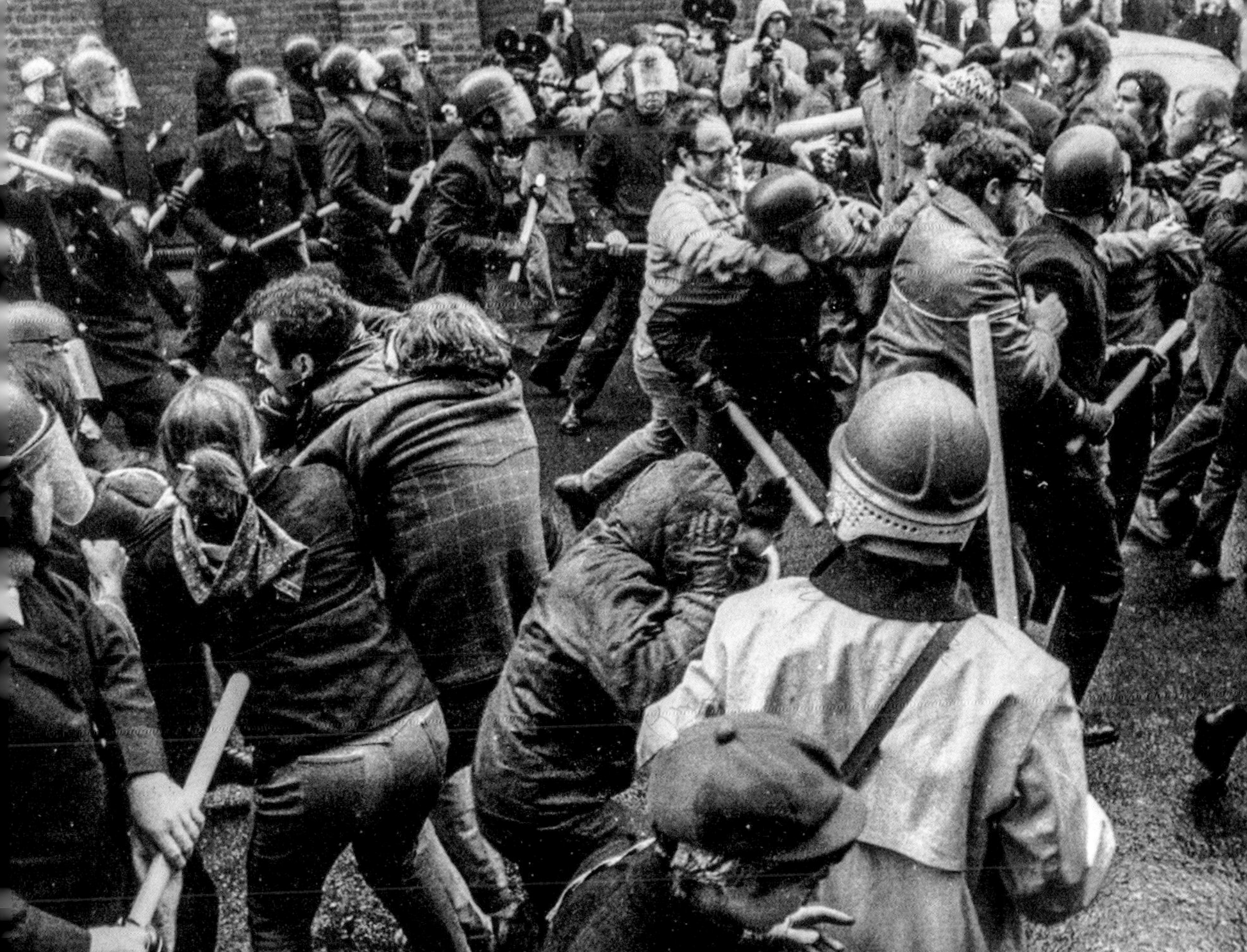

BHP110503-11/5/69-CAMBRIDGE,MASS.-Helmeted police with nightsticks,plow
into the militant anti-war demonstrators in from of the Instrumentation
Lab at M.I.T., 11/5.Some of the demonstrators,rear,right,are attacking
two policemen.The police dispersed the cursing demonstrators. UPI TELE.

This was not a great time for another peaceful "exchange of ideas." After leading the informal seminar at the Center that November, Otto Piene discovered a pamphlet left behind by a visitor. It was signed with the name of a mysterious group: the "Council for Conscious Existence," or CCE.[151] The leaflet caricatured Kraynik, Piene, and a brainy Kepes, in lab coat and thick glasses, as they inflate balloons with helium (fig. 5.25). The drawing mocks Piene's ongoing development of what he called Sky Art. Since his arrival at the Center in 1967, Piene had worked with MIT astrophysicist Walter H. G. Lewin on outdoor environmental sculpture made from polyethylene forms.[152] Sky Art had become a recurring feature of the MIT landscape. Just weeks prior, on 15 October, Piene deployed it at a massive antiwar rally called the Moratorium to End the War in Vietnam. Long, rainbow-like balloon arches floated hundreds of feet into the air, guiding the MIT community from the steps of the Rogers Building, down Massachusetts Avenue, across the Harvard Bridge, and onto Boston Common, where more than 100,000 students and faculty gathered in peaceful protest (fig. 5.26). Piene even inscribed the words "Peace" and "Paz" on the big balloons.

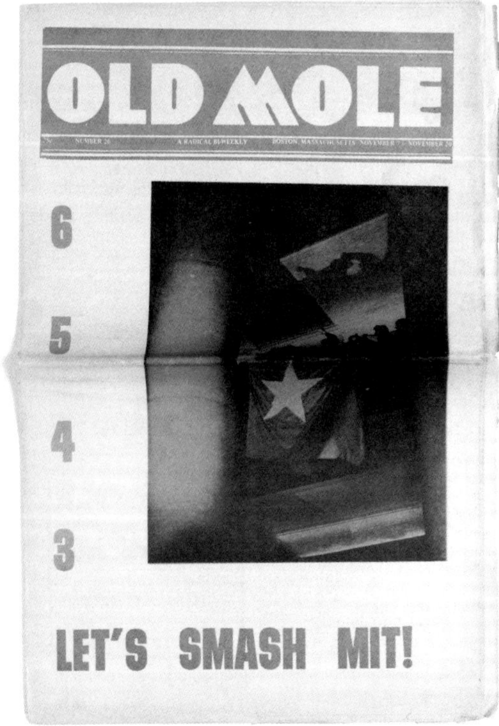

5.23

Cover of the *Old Mole*,
7 November to
20 November 1969.

5.24

Strike poster made for a
Massachusetts Institute
of Technology student
protest, published in the
Old Mole, 15 May to
28 May 1970.

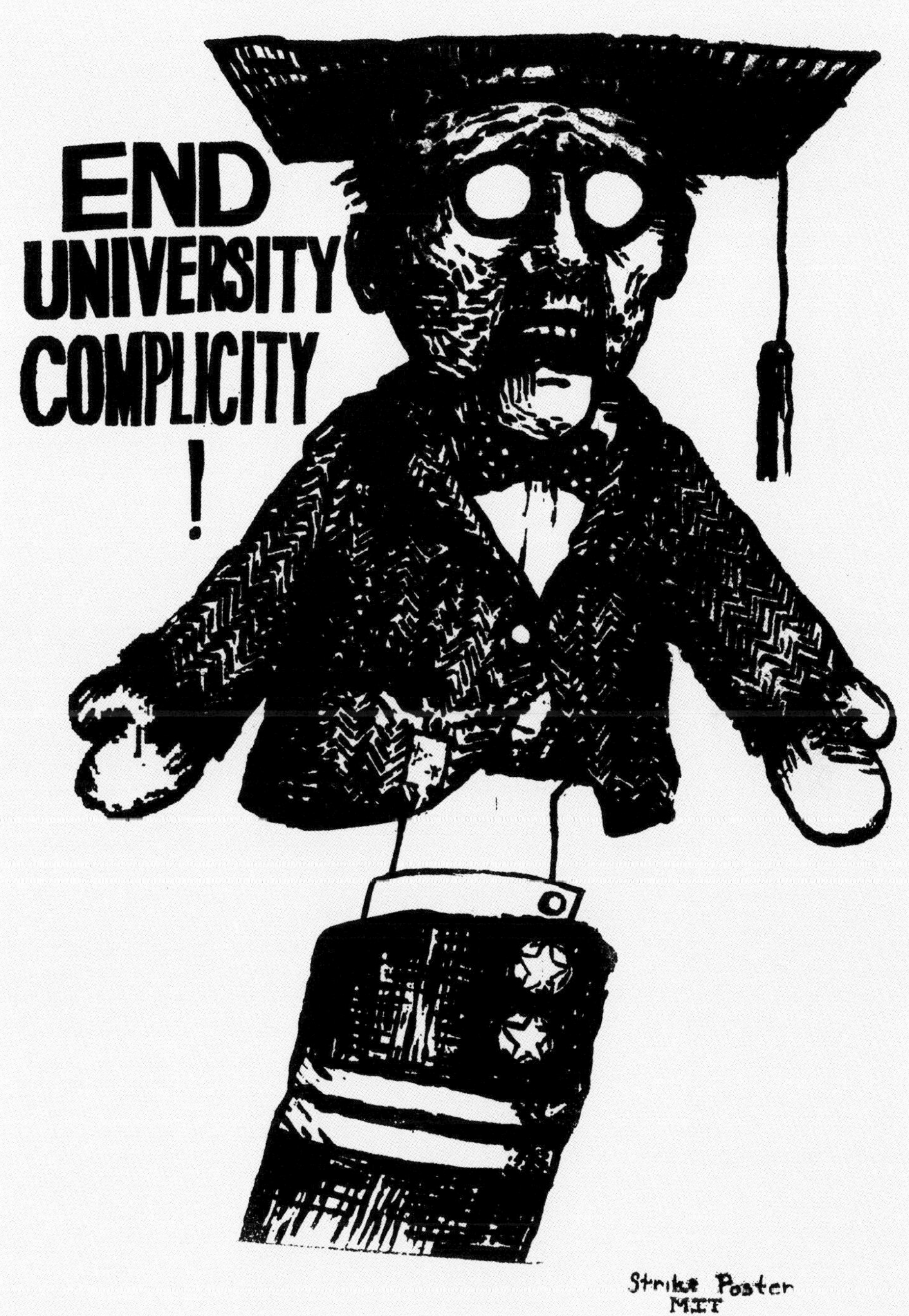

END
UNIVERSITY
COMPLICITY
!
Strike Poster
MIT

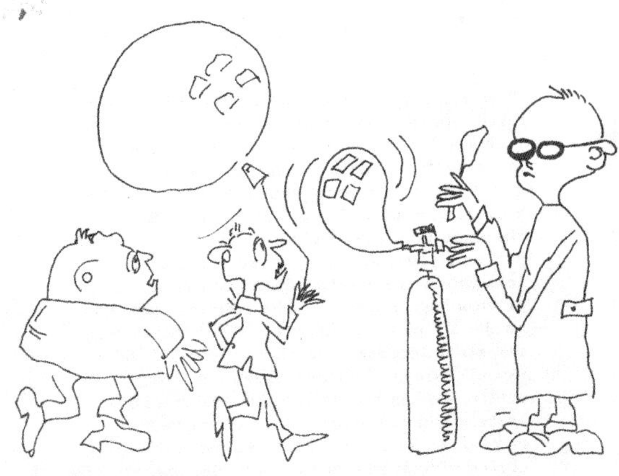

Ted Kraynick, Otto Piene, and Gyorgy Kepes
pretend to organize the participation of spectators. This is nothing but the first step in
transforming the spectators of empty culture
into its organizers, to refill the inanity of the
spectacle through the obligatory participation
of the spectator, the passive agent par excellence.

22   November 20, 1969

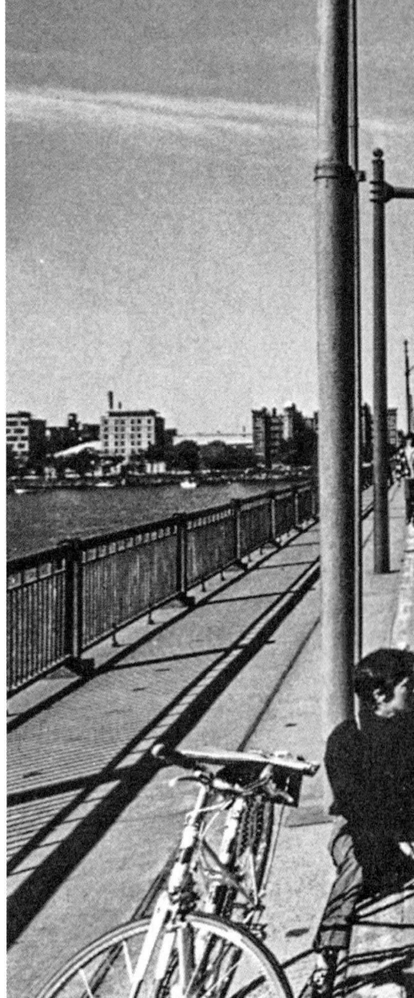

5.25

Pamphlet created by the Council
for Conscious Existence (CCE), 1969.
Gyorgy Kepes papers, 1909–2003,
bulk 1935–1985. Archives of American Art,
Smithsonian Institution.

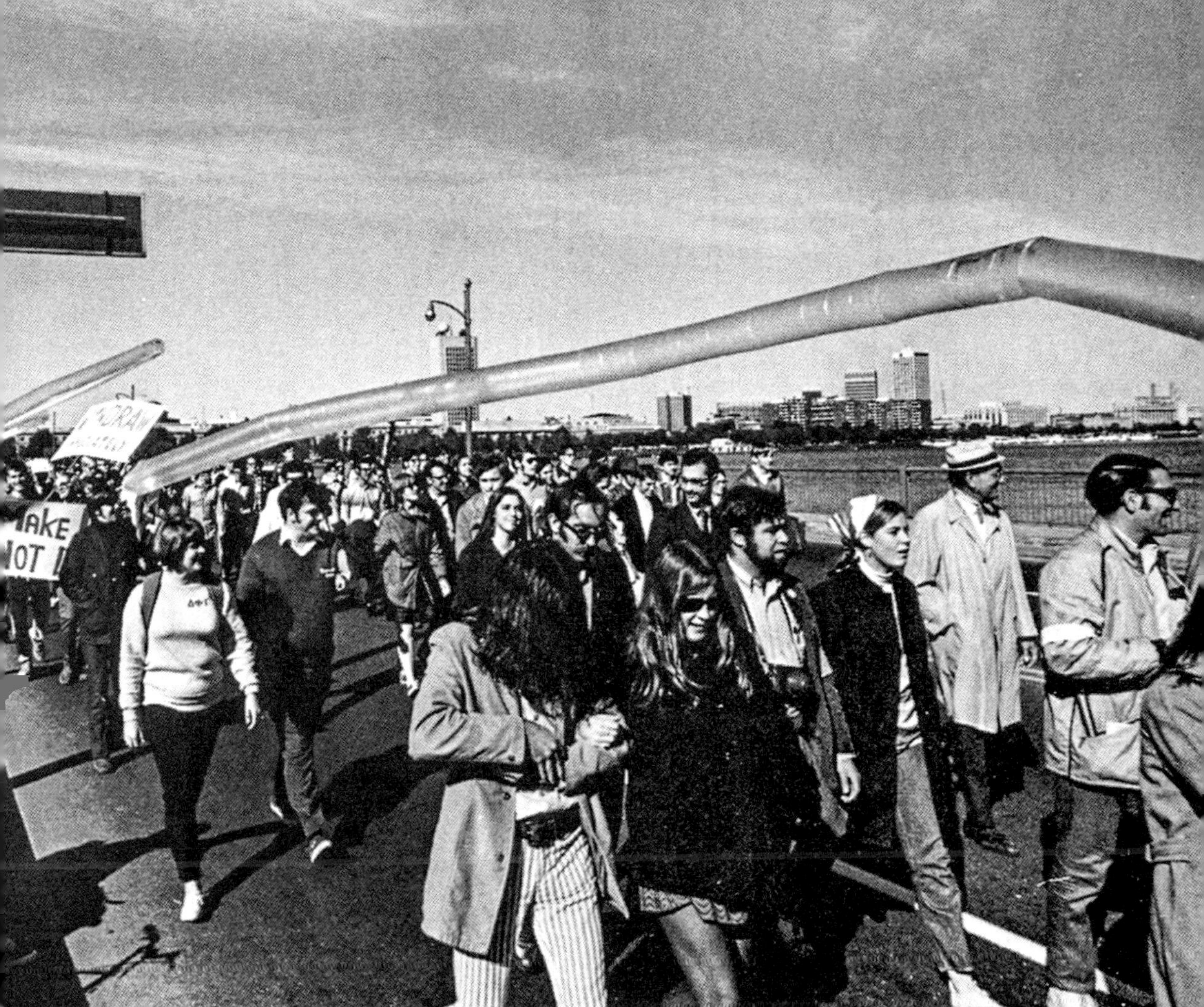

5.26

Protestors cross the Charles River under
Otto Piene's *Sky Art* during the Moratorium
to End the War in Vietnam, 15 October 1969.
From *Technique* 86, the Massachusetts
Insitute of Technology class of 1970 yearbook.
Cambridge, MA: Massachusetts Institute of
Technology, 1970. Courtesy MIT Museum,
Cambridge, MA.

The image in the CCE pamphlet rejects Piene's Sky Art—like all the "talk" and "dialogue" floating around Cambridge that year—as so much hot air. The text within the pamphlet is soaked with contempt. It uses the obtuse constructions of neo-Marxian prose to condemn the Center's balloon activities as a spectacle of science and technology—and as Spectacle, in the Debordian sense. The pamphlet reads:

> **As artists throw up ever more monstrous representations of the society of emptiness, one point is unmistakable: the next revolution in art must be the art of revolution, the transformation of everyday life. In the laboratories of individual creativity, a revolutionary alchemy transmutes into gold the most vile metals of everyday-lifeness.**[153]

The text is adapted from one of the founding treatises of the Situationist International, authored in French in 1967: Raoul Vaneigem's *The Revolution of Everyday Life*. Full passages are lifted almost verbatim from Vaneigem and translated (albeit rather poorly) into English, likely here for the first time.[154]

The CCE had its origins in the student protests that erupted at Columbia University in the spring of 1968. A former Columbia PhD candidate named King Collins, together with other dropouts, formed a collective called the "Radical Action Cooperative," or RAC, that disrupted classrooms and distributed propaganda pamphlets. The group was organized against SDS, which Collins considered too centrist (at the time, Collins had a loose affiliation with the American Section of the Situationists International; the Situationists later disavowed the RAC, and subsequently the CCE, because of disagreements with Collins). Months later, Collins and his coterie of the disaffected, now styled as the CCE, traveled from New York to Cambridge, Massachusetts, in order to penetrate the heart of American higher education with their peculiar brand of offensive political theater.[155] In spring of 1969 the group gained notoriety after infiltrating a sociology course at Harvard; they scrawled vulgarities on the blackboard while forcing an esteemed professor, trapped in his lecture hall, to eat a banana in front of his class.[156]

Collins considered these stunts intellectually sophisticated. The CCE was devoted to working through Situationist theory and then putting theory into practice through critical interventions. "The writing of Debord and Vaneigem excited us," he explains. "We translated laboriously from the French and the text came out slowly in pieces, but it revealed the thought of Hegel, Marx, and the Frankfurt School and all the creativity and sophistication of European radicals and artists."[157] Two members of the CCE, Wilma and Sid Lewis, decided to stage one of the group's infamous interventions against Kepes and his Center. They drafted the pamphlet left in Piene's seminar.

The document is striking in part because the group's stated mission mimics that of the Center. CAVS and CCE—the parallel bureaucratic acronyms are alone suggestive. Kepes's rejection of traditional artistic contexts could be understood, perhaps, as an interest in the "transformation of everyday life." His work with scientists was a retreat, of sorts, to the "laboratories of individual creativity." Conversion was a form of "revolutionary alchemy." The pamphlet even evoked the metaphor of light as enlightenment that was so essential in Kepes's thought.

But any sense of common cause was lost on the CCE.[158] The pamphlet continues with vituperative rants directed at each artist depicted in its cartoon image. The first, addressed to Kraynik, quotes a description of his *Video-Luminar Light Mural* (fig. 5.27), a work presented at *Explorations* that created viewer interaction through closed-circuit video cameras. The CCE understood the work's participatory dynamic as ideological interpellation, a means of suturing viewers into Spectacle. They write to Kraynik:

> You are syphilis, a plague eating into consciousness; and having accepted your own half imitation of life, you ask us to join you in the anti-chamber [*sic*] of the cybernetic welfare state. You are even so arrogant, so infected with your own urine, that you present us with images of our own destruction.[159]

The text oozes disgust. The second diatribe was addressed to Piene. It quotes Grace Marmor Spruch's account of the founding symposium for the joint dedication of the Center for Advanced Visual Studies and the Center for Theoretical Physics, specifically referring to Piene's remarks at the event that "the greatest service technology could do for art would be to enable the artist to reach a proliferating audience" through new tools, namely television. The CCE did not accept this proposition:

> You are the advanced guard of the cybernetic welfare state – the reconsecration of order, no longer with God as ruler, but with technology raised to myth in the perfect order of zombies. You are not an artist; you are a W H O R E for power, decorating the society of consumption – not for just the jaded ruling class but for everyone. You are so benighted that you publicly suggest the use of TV so that all can "participate" in the unilateral reception of images of their own alienation.[160]

Lastly, the pamphlet delivered a single line of invective to the Center's founder. "And to you, Gyorgy Kepes[,] whose dream it was to gather this scum, fuck you."[161]

The pamphlet's hyperbolic absurdity, its crude and puerile tone, might be amusing if it were not so chilling. In response, both Kepes and Piene offered carefully worded rebuttals. "Irrespective of the language you used, there are some issues you touched upon that I think should be clarified," wrote Kepes.

5.27

Ted Kraynik, *Video-Luminar Light Mural*, 1969 (detail). Photo: Nishan Bichajian. Courtesy Center for Advanced Visual Studies Special Collection, MIT Program in Art, Culture and Technology. © Massachusetts Institute of Technology.

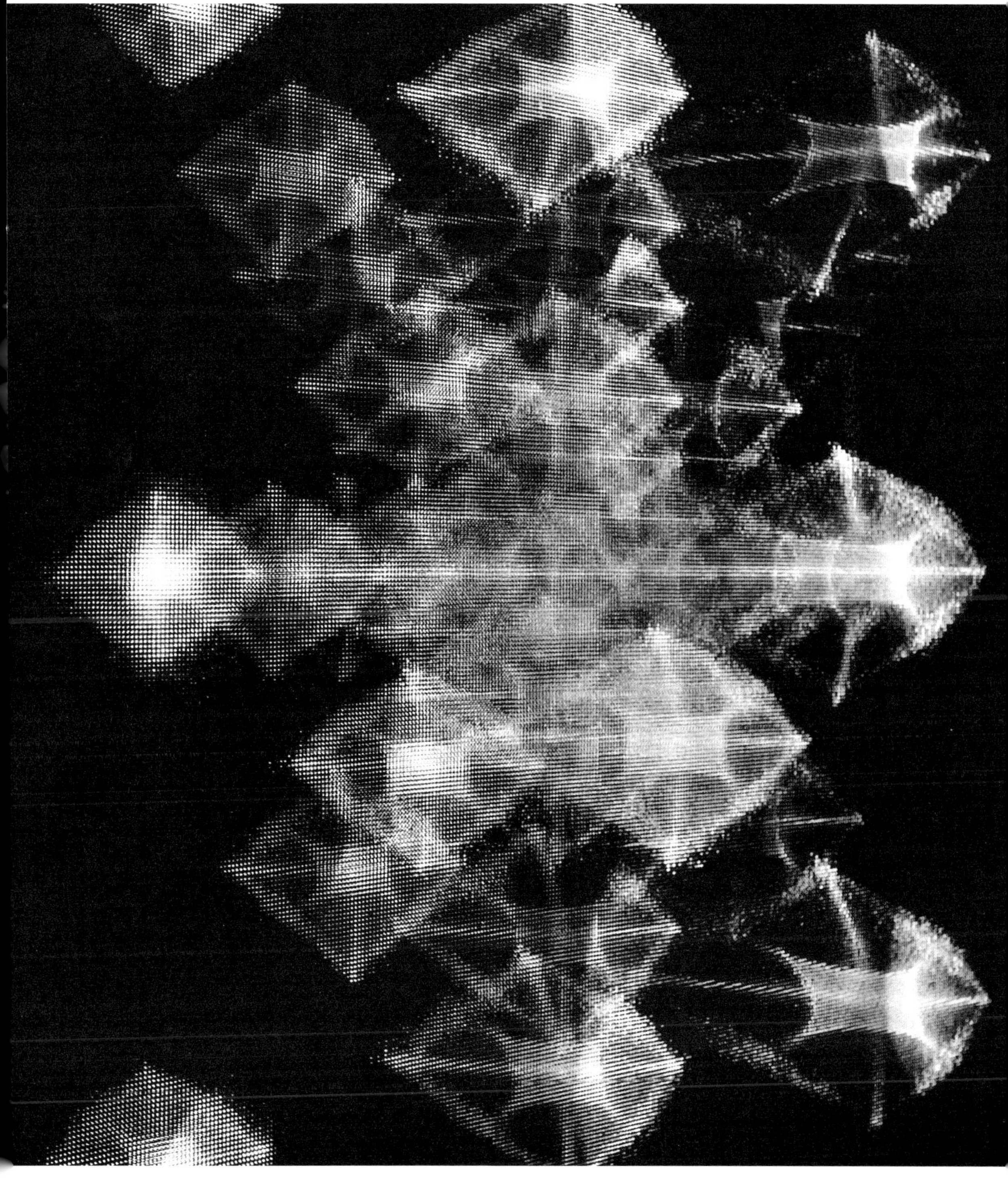

Even under this extreme attack, he maintained the faith. He repeated his standard refrain, one offered in the face of all conflicts, one he had already written so many times to so many others. He calmly invited his critics to talk, to communicate, to collaborate: "If you are sincerely interested in the problems of our time, and the artist's role in the modern world, it may be profitable for both of us to have an exchange of ideas concerning these issues. Please call me at a time convenient for you so that we can sit down and give each other a chance to exchange points of view."[162]

Piene similarly offered to open up a conversation, but unlike Kepes—who could not face confrontation—Piene pushed back, rejecting the CCE's caricature of the Center's aesthetics and politics. "After reading your pamphlet I wonder why you did not sit down and discuss it with us. I could have easily corrected a number of obvious misunderstandings. Besides that I am interested in 'the art of revolution.' Are you a group of reactionary conservatives in fashionable disguise?"[163] His puzzlement illustrates the ideological confusion of the period.

In the end, these tiresome appeals for dialogue were really beside the point. Institutional affiliation and its imprimatur—the organizational acronyms, the letterhead emblazoned with "Office of the Director" in clean sans-serif font, and the secretary to type one's correspondence (Kepes dictated his letters, including his reply to the CCE, to his secretary Joan D. White)—inherently constitute power. The CCE suggested that all power inherently corrupts. Kepes had hoped to change MIT from within—by navigating its systems and networks, converting its agendas, transforming its protocols—but in the process, he found himself trapped in a new military-industrial-aesthetic complex. Wilma Lewis wrote a final reply on behalf of the CCE to Kepes (she ignored Piene's letter entirely) summarizing, in a vulgar way, these points: "Your head is in your ass. We have not misunderstood. When there is no secretary to type your letter, perhaps then, scum, you will understand."[164]

Nothing more came of the exchange. The pamphlet was forgotten, the letters were filed away; it was all just par for the course, one of the many conflicts that Kepes encountered and endured. There was no time for distractions. There was work to be done. Kepes was resolute. Life at the Center had to go on.

This book has explored how Kepes attempted to figure a new paradigm for the artist at a moment when there seemed to be "no more revolution." Now, at the end of the 1960s, the revolution, in all its rage and fury, had returned. Kepes was not part of it.

Artificial Natures

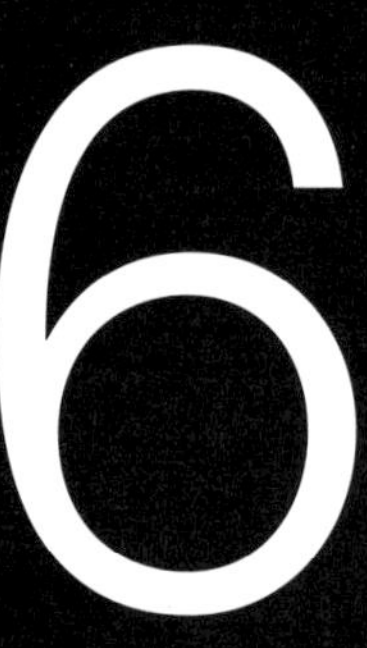

6

Estrangement from nature (the first nature), the modern sentimental attitude to nature, is only a projection of man's experience of his self-made environment as a prison instead of as a parental home.

When the structures made by man for man are really adequate to man, they are his necessary and native home; and he does not know the nostalgia that posits and experiences nature as the object of its own seeking and finding. The first nature, nature as a set of laws for pure cognition, nature as the bringer of comfort to pure feeling, is nothing other than the historico-philosophical objectivation of man's alienation from his own constructs.

Georg Lukács, *The Theory of the Novel*, 1920[1]

A Second Nature

Arts of the Environment, published in 1972 as the final installment of the *Vision + Value* series, opens with an unusual picture essay: Kepes uses the format of the book to compare and contrast similar but different photographic images—the very same method he used over decades (fig. 6.1). Nuclear bombs detonate in disturbing detail on one page—Kepes uses the photographs depicting bomb blasts at the very moment of ignition that he had collected for his Light Book from MIT engineer Harold Edgerton—while the arc of a solar prominence erupts from the surface of the sun, consuming the Earth, rendered to scale as a tiny dot at far right, on the next. Turn the page (fig. 6.2): smoke billows above a forest fire; a tornado tears across the landscape; streaks of anti-aircraft gun-fire illuminate the night sky; a conflagration consumes Tokyo after an aerial attack during the Second World War.

Kepes's picture essay is terrifying. The images demonstrate how culture, depicted on one side of the book's two-page spreads, rivals nature, depicted on the other. They suggest that scientific and technological interventions into the environment are as powerful as natural events; that human-made disasters are now equal in measure to natural ones. Titled "Toward a New Environment," the essay frames the texts that follow, and the book as a whole, as a collective manifesto for a new environmental art, one that would match the scale of environmental tragedy with creative imagination, one that would convert the destructive force of science and technology into a constructive force through aesthetic spectacle.

This chapter examines those spectacles. Could environmental art address actual environmental catastrophe, real ecological crises, like air and water pollution? Could the Center for Advanced Visual Studies (CAVS) counteract the destruction that science and technology had already caused by instead using science and technology aesthetically?

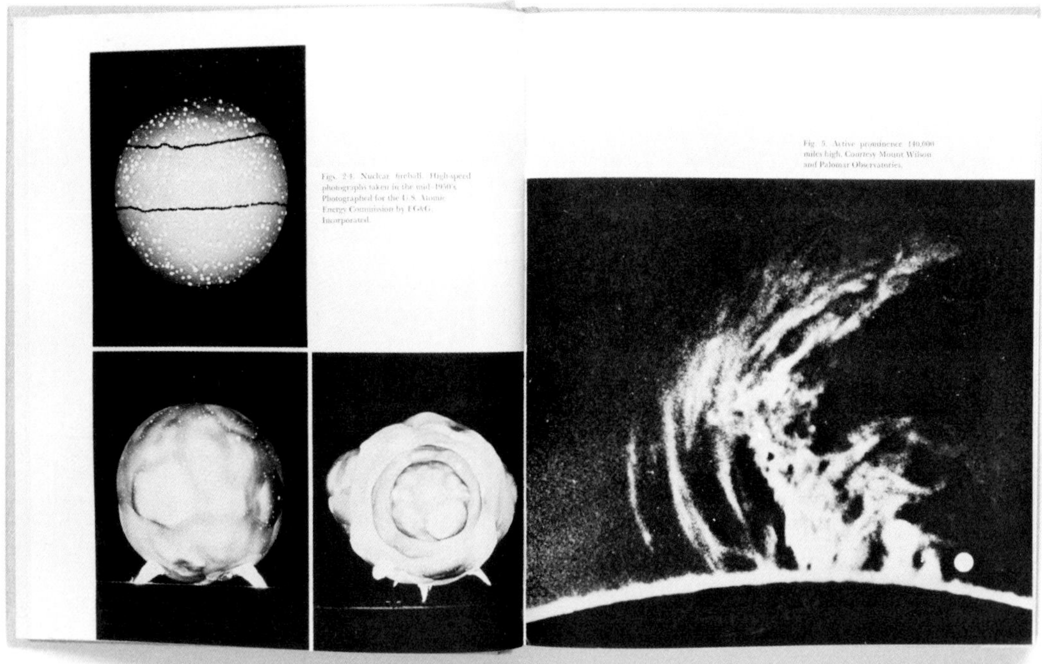

6.1

Two-page spread from *Arts of the
Environment*, edited by Gyorgy Kepes.
New York: George Braziller, 1972.
Courtesy George Braziller.

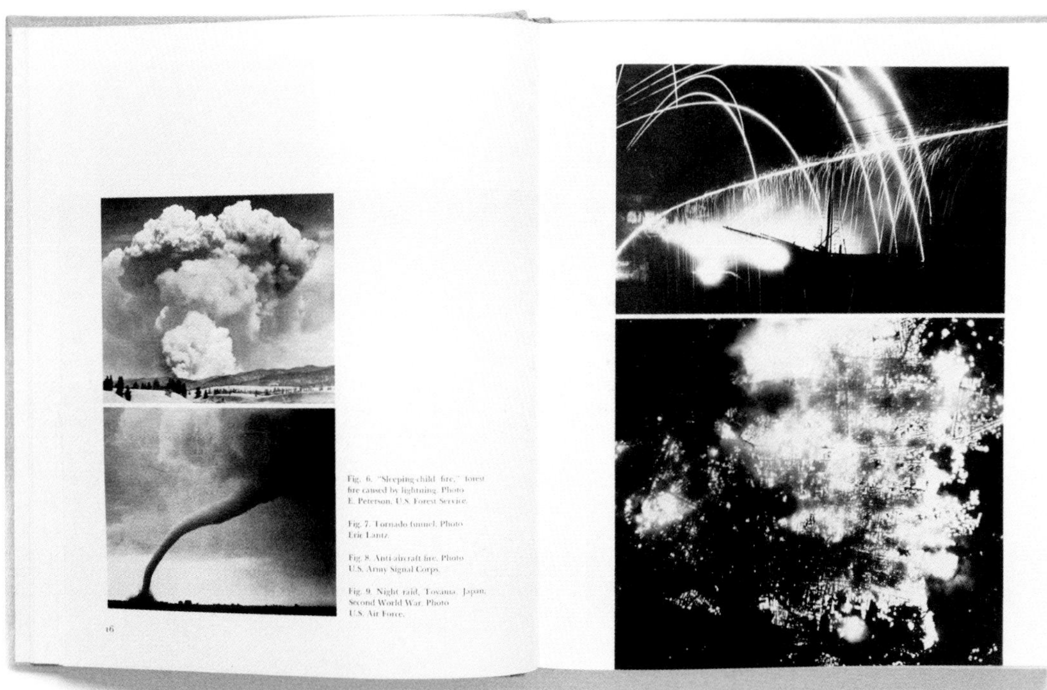

6.2

Two-page spread from *Arts of the
Environment*, edited by Gyorgy Kepes.
New York: George Braziller, 1972.
Courtesy George Braziller.

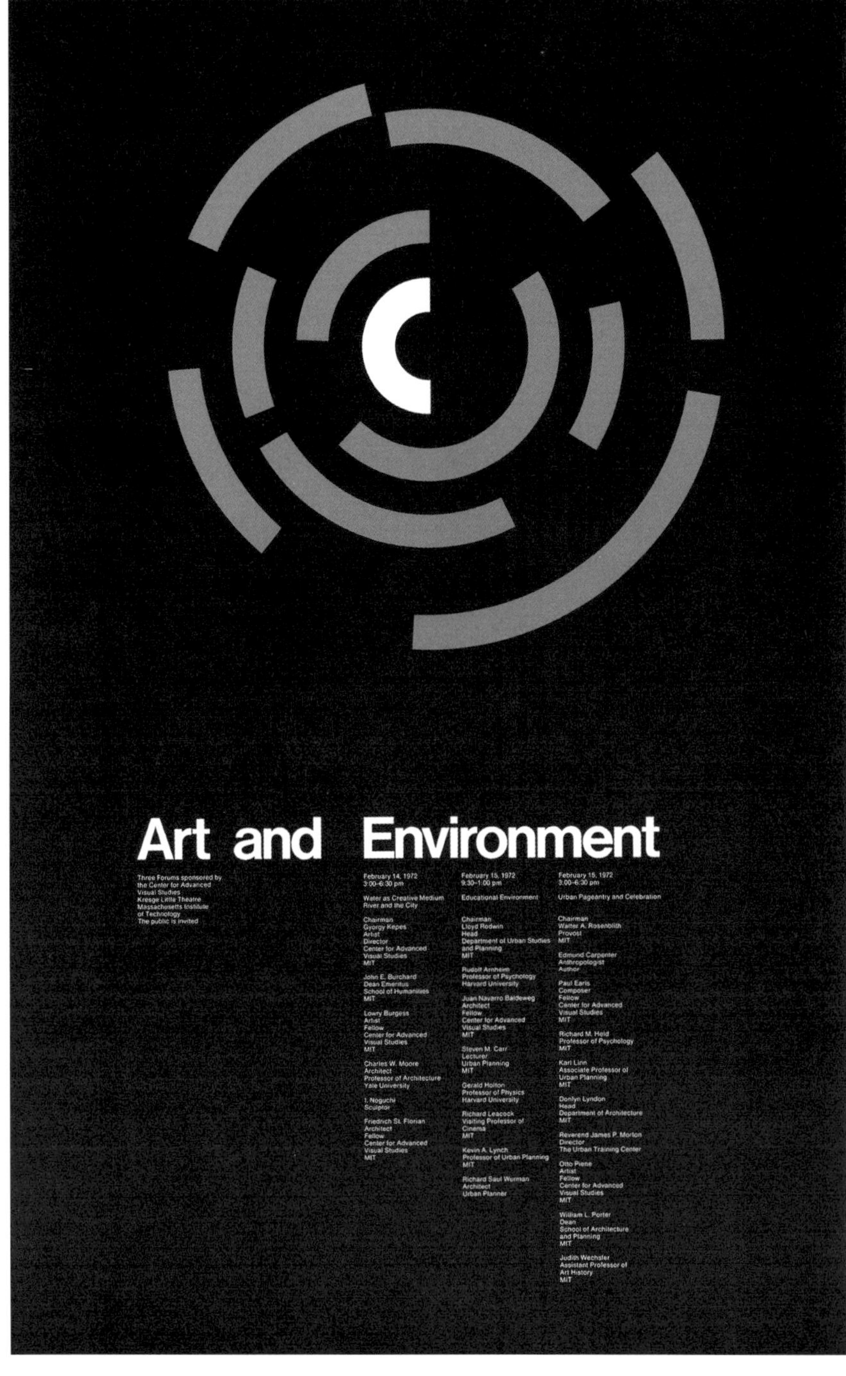

Art and Environment

Three Forums sponsored by
the Center for Advanced
Visual Studies
Kresge Little Theatre
Massachusetts Institute
of Technology
The public is invited

February 14, 1972
3:00–6:30 pm

Water as Creative Medium
River and the City

Chairman
Gyorgy Kepes
Artist
Director
Center for Advanced
Visual Studies
MIT

John E. Burchard
Dean Emeritus
School of Humanities
MIT

Lowry Burgess
Artist
Fellow
Center for Advanced
Visual Studies
MIT

Charles W. Moore
Architect
Professor of Architecture
Yale University

I. Noguchi
Sculptor

Friedrich St. Florian
Architect
Fellow
Center for Advanced
Visual Studies
MIT

February 15, 1972
9:30–1:00 pm

Educational Environment

Chairman
Lloyd Rodwin
Head
Department of Urban Studies
and Planning
MIT

Rudolf Arnheim
Professor of Psychology
Harvard University

Juan Navarro Baldeweg
Architect
Fellow
Center for Advanced
Visual Studies
MIT

Steven M. Carr
Lecturer
Urban Planning
MIT

Gerald Holton
Professor of Physics
Harvard University

Richard Leacock
Visiting Professor of
Cinema
MIT

Kevin A. Lynch
Professor of Urban Planning
MIT

Richard Saul Wurman
Architect
Urban Planner

February 15, 1972
3:00–6:30 pm

Urban Pageantry and Celebration

Chairman
Walter A. Rosenblith
Provost
MIT

Edmund Carpenter
Anthropologist
Author

Paul Earls
Composer
Fellow
Center for Advanced
Visual Studies
MIT

Richard M. Held
Professor of Psychology
MIT

Karl Linn
Associate Professor of
Urban Planning
MIT

Donlyn Lyndon
Head
Department of Architecture
MIT

Reverend James P. Morton
Director
The Urban Training Center

Otto Piene
Artist
Fellow
Center for Advanced
Visual Studies
MIT

William L. Porter
Dean
School of Architecture
and Planning
MIT

Judith Wechsler
Assistant Professor of
Art History
MIT

This chapter considers such a premise through the unrealized environmental designs created at Kepes's Center between its founding in 1967 and Kepes's retirement in 1974. Kepes established the Center not only to develop new "collaborative idioms" between artists and scientists but also to apply these idioms to the creation of "civic-scale" public art, an art situated in the actual landscape and addressing issues of civic significance. In the founding statement for his Center published in the journal *Daedalus*, Kepes names the exploration of these "new aspects of environmental art" as one of the Center's key areas of research.[2]

These projects were theorized through systems-based principles like homeostasis, as set forth in emerging environmental discourses, like ecology, but also as developed from ideas that originated in cybernetics, systems theory, and information theory and which had long interested Kepes; in this sense, they developed directly from Kepes's earlier work. The ideas at the core of these projects were also widespread in the arts in this period; the Center's activities relate to land art, for example, which also drew from ecological thought.[3] In practice, however, the Center's projects were more idiosyncratic, planned as massive architectural and sculptural interventions into specific landscapes around MIT, often as landmarks like beacons or towers.

These proposals were codified as four distinct projects, each of which resulted in particular designs: the Boston Harbor Project, focused on the construction of a light tower in the Harbor's waters; the Charles River Project, devoted to reimagining the river as an aesthetic experience; the Boston Harbor Islands Project, directed at plans for developing the islands dotting the waterways around the city; and the Long Wharf Project, concerned with the renovation of a historic pier on the city's waterfront. These efforts generated ideas for specific structures that were put to paper in written plans but also envisioned in photographs, many of which were subsequently presented in a two-part exhibition at Boston's Institute of Contemporary Art in 1975 and 1976, and at MIT in shows like *Art in Civic Scale*. These efforts also sustained academic events, such as a 1972 symposium on *Art and Environment* (fig. 6.3).

6.3

Poster for the MIT symposium *Art and Environment*, 1972. Design: Jackie Casey. Courtesy Center for Advanced Visual Studies Special Collection, MIT Program in Art, Culture and Technology. © Massachusetts Institute of Technology.

As we explored in chapter 5, the Center had become entangled in controversy almost immediately after Kepes established it. Its first endeavor, the US representation for the 1969 São Paulo Biennial, was a disaster, generating a backlash consistent with a wider rejection of art-and-technology and art-and-science spectaculars in the Vietnam Era. In response, the Center shifted priorities, focusing on what Kepes imagined as a more benevolent program. This shift was consistent with a larger adjustment at MIT, a homeostatic adjustment, where "conversion"—meaning the conversion of research devoted to weapons and war into research with socially enriching applications—became a priority, a correction for a university that had veered off course. Studying solutions to environmental problems was a crucial part of this newly converted research agenda.

While the Center followed this agenda by attempting to use the arts for aesthetic conversion, it also attempted to make aesthetic projects tangibly beneficial by moving them outside, into the actual environment, where they might become functional and practical. Lowry Burgess, a CAVS fellow involved in these projects, recalls Kepes's belief in "the great power of art to purify," and this now meant purifying contaminated soil, water, and air. "We could rival the defense department, we could have the budgets of anybody, we could have a higher position," Burgess claims.[4] Addressing environmental issues was consistent with modeling the Center as a humanistic alternative to the research institutes comprising MIT; it would use MIT's financial and intellectual resources for alternative ends, ends that specifically opposed military power.

Kepes's concern for the environment was, however, longstanding. He designed covers for *The Atlantic* through the 1960s and one, from a 1961 issue devoted to "Our National Parks in Jeopardy" (fig. 6.4), explicitly addressed the topic—the magazine's cover story examined environmental damage at national parks across the country. The magazine predates Rachel Carson's *Silent Spring*, which set the mainstream environmental movement into motion (but would not be published for another year). Kepes's cover used a photograph of a lake seen through a stand of trees; it may be an anonymous source image, but it looks to me like a view of the woods around Long Pond, in Wellfleet, Massachusetts, on Cape Cod, where Kepes spent his summers enjoying nature and painting luminous canvases in a small cabin designed by Marcel Breuer in the 1940s. He contrasted the photograph, printed in intense green, with a red "X" marking a denial, a refusal of nature.

But Kepes's interest in the environment was even more longstanding, more fundamental to his philosophy of seeing. Nature is a constant feature in *The New Landscape in Art and Science*, for example (the word "landscape"

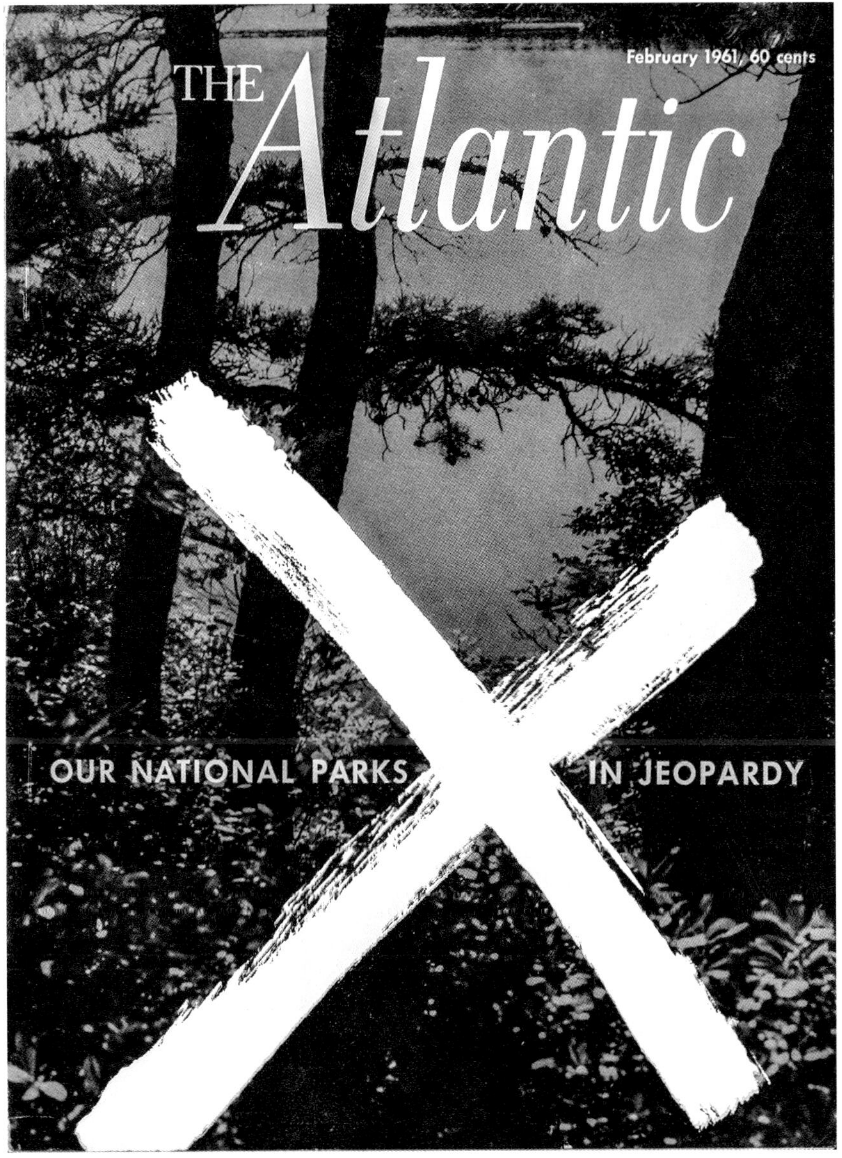

6.4

The Atlantic, February 1961. Cover designed
by Gyorgy Kepes. Photo credit: Gyorgy Kepes/
Atlantic Monthly/TNS.

was intended to evoke a view of nature, nature seen through very unnatural means). The book is filled with quotations from Ralph Waldo Emerson's *Nature*, from Walt Whitman's *Leaves of Grass*, and from Henry David Thoreau on Walden Pond (Kepes's home in Cambridge was only ten or so miles from the famed lake, and mere footsteps from a similar body of water, Fresh Pond). Actually, Kepes quotes Whitman's "primal sanities of nature" in both *The New Landscape* and a 1972 report to the National Endowment for the Arts (NEA) on the Center's environmental art, indicating how consistent Kepes's philosophy was, spanning some twenty years.[5] Kepes understood these influences as more significant than art-historical genealogies that tied him so inextricably to Moholy and the Bauhaus. He declares: "my response to the American spirit came through Whitman Emerson Thoreau … + my blood boil[s] where their deeper roots are … misread," when "interpretation is limited to stereotype[d] art world ideas—[on the] one hand abstract expressionism—on the other Bauhaus Constructivists." Kepes was more naturalist than technologist; he looked backward as well as forward; he looked toward science and technology, but also away from it.

Environmental Art

The first of these proposals might be Kepes's very own "Tower of Light," an unrealized commission designed in the early 1960s for a large-scale urban renewal project in downtown Baltimore called the Charles Center, which included a landmark skyscraper designed by Mies van der Rohe. The earliest descriptions of the tower call it a "tree of light" and a "garden of light."[6] Kepes described it more thoroughly in his founding statement for the Center, published in *Daedalus*, which quotes directly from a report on the project. The hundred-foot tower would be topped with a massive gold screen, some twenty-five feet wide, that would reflect light from powerful lamps. Shafts of light would continue past the reflective screen, illuminating the sky above. During the day, the tower's screen would reflect sunlight. It would clarify the "pattern of light and shadow" in the urban landscape.[7] It would glow like an artificial sun, replicating nature through very unnatural means (fig. 6.5).

Such wonders transfixed Kepes. His notebooks from the period are filled with actual pieces of nature: dried leaves (from the oak and tulip trees), pressed between heavily inscribed pages of notes and drawings—simple mementos from the outdoors (fig. 6.6). But he also understood nature as something that could be artificially simulated. "Natural environment presents us with some luminous mobile kinetic spectacles," he writes in one research note. Continuing:

> aurora borealis
> sunset – sunrise
> rainbow
> pageantry of season
> the spring carpet of fields
> the fall leaves in wind
> bursting into glow
> our city night + day scape is a dense optical texture stretching immense distance
> covering immense territories without beginning + end, shape or focus …
> Piccadilly, Time[s] Square, Broadway etc.
> but they are accidental – without guided quality[8]

The text shifts from an idealized vision of the natural landscape to an idealized vision of the urban landscape. The Center's designs aimed to simulate these spectacles by guiding, shaping, orchestrating the existing optical pattern of this built environment. Kepes would even explore the possibility of transforming the illumination at public spaces like New York's Times Square.[9]

While his Tower of Light was never constructed, Kepes used it to focus the activities at the Center. He pitched the concept to prospective fellows as a project for collaboration. (Otto Piene responds to an invitation from Kepes: "I thought about your plan to suggest my being invited as a fellow to work on the light tower and found out that I like your project very much. . . . The work on the great light project could be a challenge to everyone involved.")[10] The general idea of a tower anchoring Boston Harbor also became the centerpiece of the Center's earliest environmental efforts—what became the Boston Harbor Project.

A note composed by Piene in fall of 1968 records the Center's first attempts at conceptualizing this project. Peine suggests it might result in "an object," meaning a monumental structure; "a phenomenon," meaning an optical experience (something mimicking the "Northern Lights," offers Piene); or maybe "an object/phenomenon"—a hybrid structure and experience.[11] Many of these vague notions took form in photographs created with the help of Nishan Bichajian, a former student of Kepes's who became his longtime assistant (he had collaborated with Kepes at MIT since the 1950s); Kepes specifically indicates one of Bichajian's primary tasks at the Center as overseeing "experimental work with light," including "the light tower."[12] With Bichajian's help, fellows combined photographs of physical models with images of Boston, as in a tower designed by Michio Ihara and Mauricio Bueno, and photographed by Bichajian (fig. 6.7).

Kepes's own such proposals imagined a mile-long programmed luminous wall made of light beams cast from lamps or lasers (fig. 6.8). He called it "light architecture." He created multiple versions (fig. 6.9); some show lights

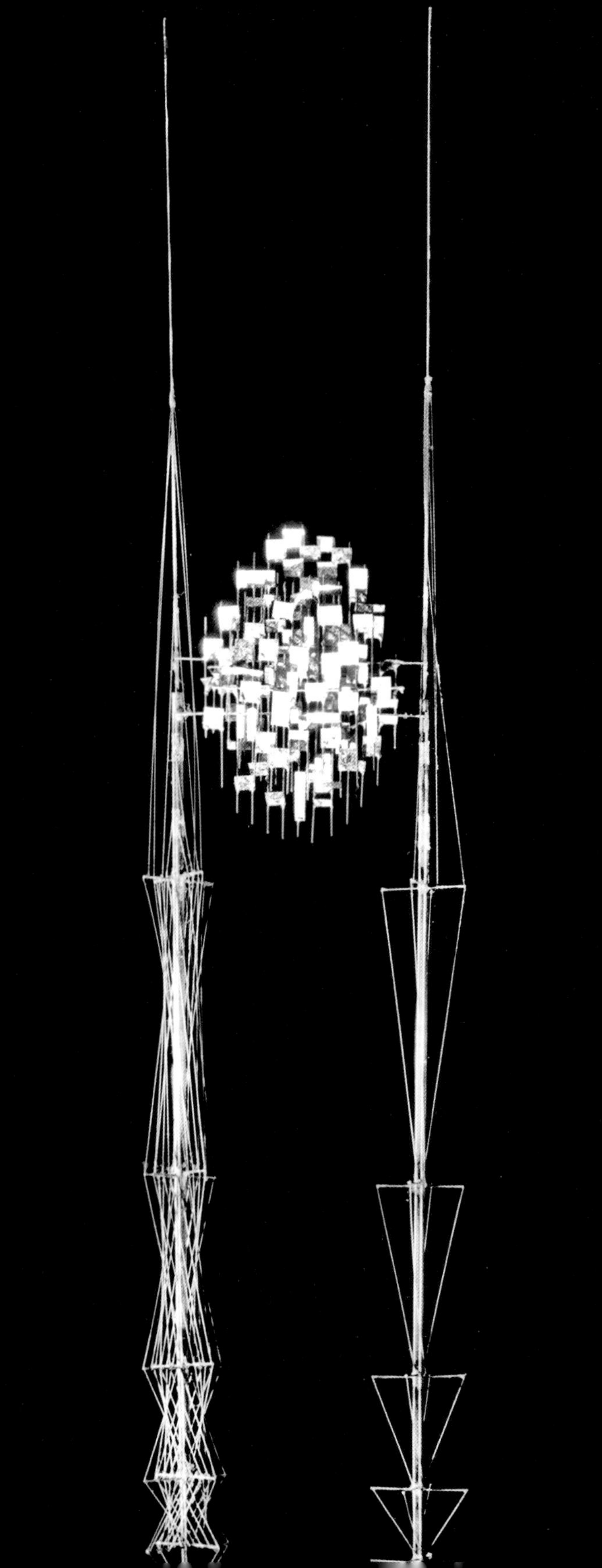

organized in gridlike arrays, with variations in illumination creating different optical forms—"transformations of space." Lights could be programmed to flash, creating an optical system that would change in response to environmental conditions.[13]

Another project created by Kepes and photographically imagined with Bichajian's help involved giant spherical buoys capped with reflective concave surfaces (fig. 6.10). Powerful lights aimed at the buoys would scatter light projections into the air; sprays of water pumped upward—artificial clouds— would catch these "animated surfaces," creating fluctuating "light patterns," like simulated rainbows.[14] Kepes's photographic models attempt to evoke these patterns, mimicking lighting effects through the formation of caustic curves, areas of intensified brilliance that form from overlapping rays of curved light bounced off hemispherical surfaces.

These proposals became more elaborate with the Charles River Project. In a 1972 report submitted to the NEA, Kepes explains the project as the exploration of "water as an artistic media" and the river itself "as an environmental-artistic problem." The goal was to recover aesthetic experiences that environmental damage had destroyed—experiences like "the joys of light and color, the freedom of open space, the refreshing purity of water," all of which were now negated by the "menace of the smog-draped sun, the degrading impact of crowding, or the blight of polluted rivers and lakes."[15] Recovering these lost qualities would then compel the public toward environmental awareness—aside from the sky, the river was the "only remaining major link the urban population has to natural phenomena," Kepes claims.[16] He understood these efforts as a way to escape the "maze of ecological blind alleys." Kepes updates his old metaphor of blindness, one he originally developed in response to MIT's infinite corridor, to the environment.[17]

6.5

Gyorgy Kepes with Michio Ihara, *Light Tower for the Charles Center*. Gyorgy Kepes papers (M1796). Dept. of Special Collections and University Archives, Stanford Libraries, Stanford, Calif.

6.6 (following page)

Gyorgy Kepes, notebook with dried leaves. Gyorgy Kepes papers (M1796). Dept. of Special Collections and University Archives, Stanford Libraries, Stanford, Calif. © The Estate of Gyorgy Kepes.

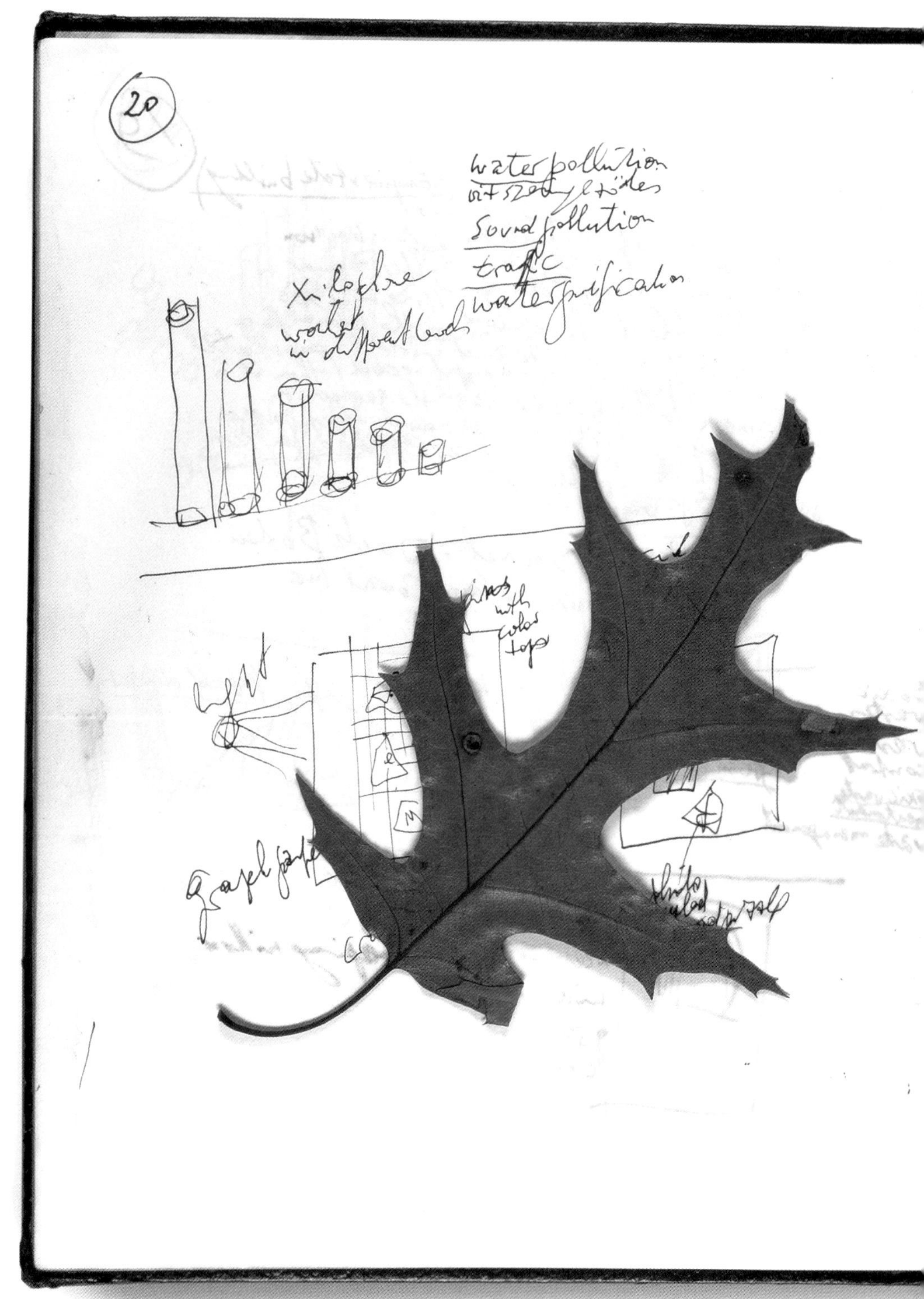
20
water pollution
Sound pollution
traffic
Xilophne
water purification

(21)

studies & paintings
Polaroid 20×24
reverse photos left → right

collage materials
(1) crumpled crossword puzzles
(2) cracked mirrors
    face treat
    black + white
    with early leaves

water on table
mirror

(15)
(16) Bra
(17) thermometer
(18) pages from the New landscape
(19) stencil letters A B C   X . Y .
(20) stenciled X + Y
(21) * Clouds (photos projected slides)
     L. K. Mexico
(22) out of biographical pages photos

orange pink

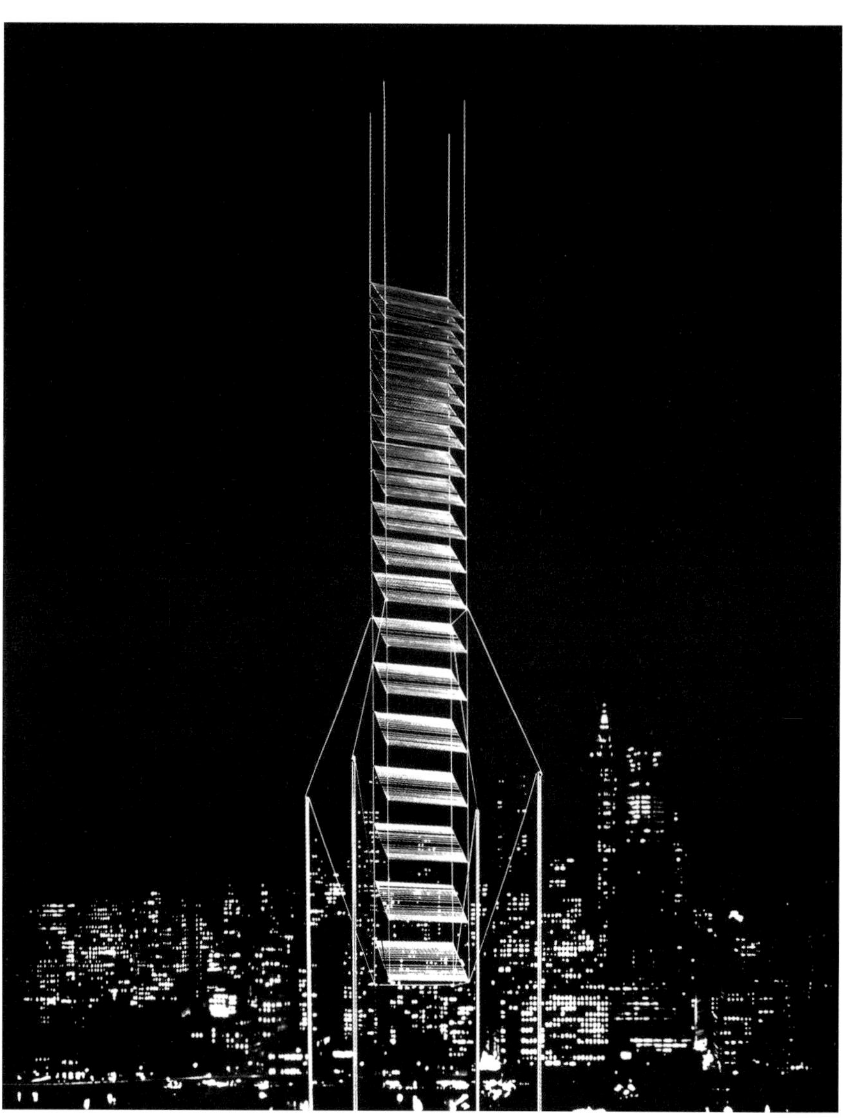

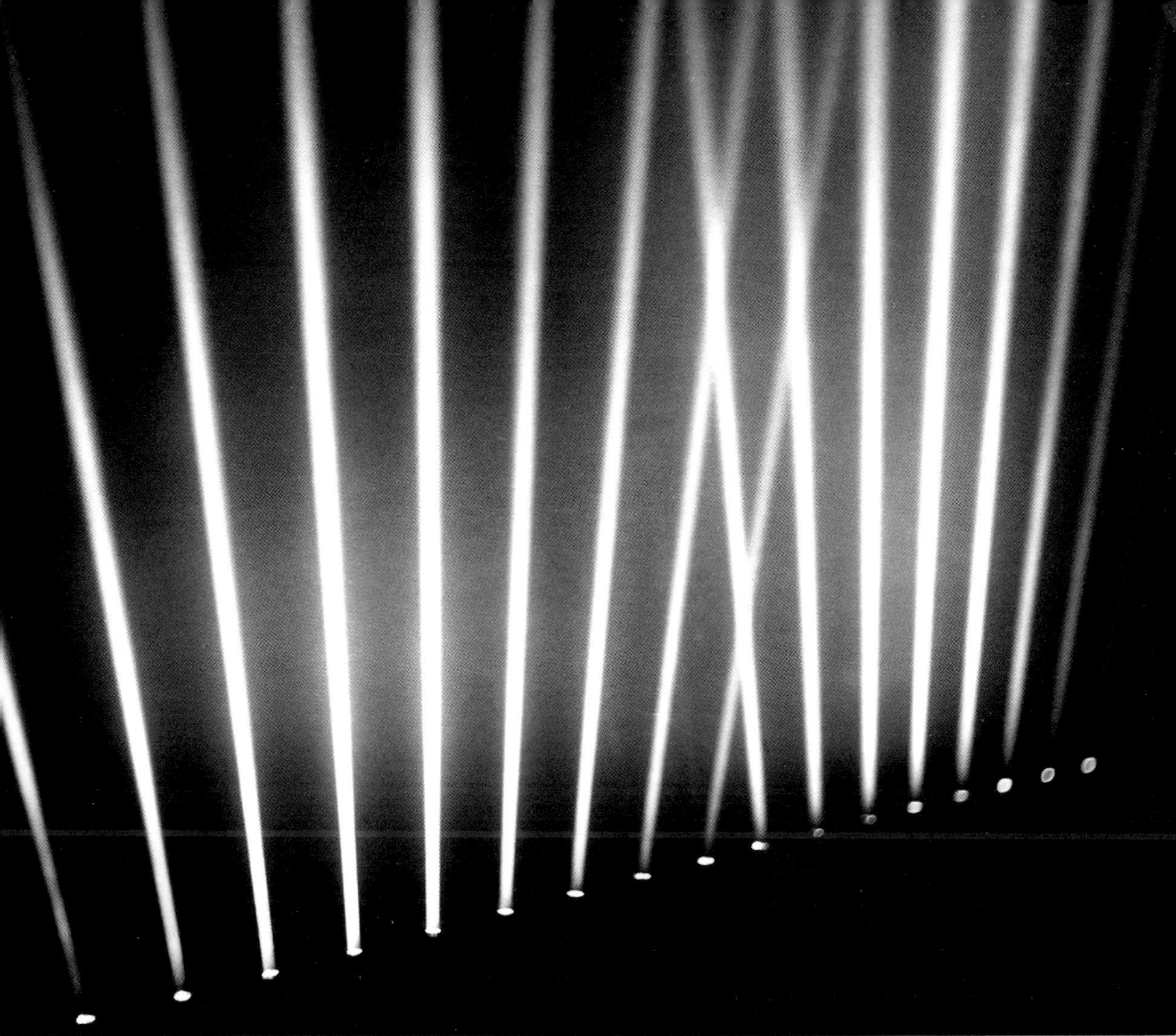

6.7

Michio Ihara and Mauricio Bueno, *Light Tower*, 1970. Photo: Nishan Bichajian. Courtesy Center for Advanced Visual Studies Special Collection, MIT Program in Art, Culture and Technology. © Massachusetts Institute of Technology.

6.8

Gyorgy Kepes, simulated light architecture for Boston Harbor, 1966. Photo: Nishan Bichajian. Courtesy Center for Advanced Visual Studies Special Collection, MIT Program in Art, Culture and Technology. © Massachusetts Institute of Technology.

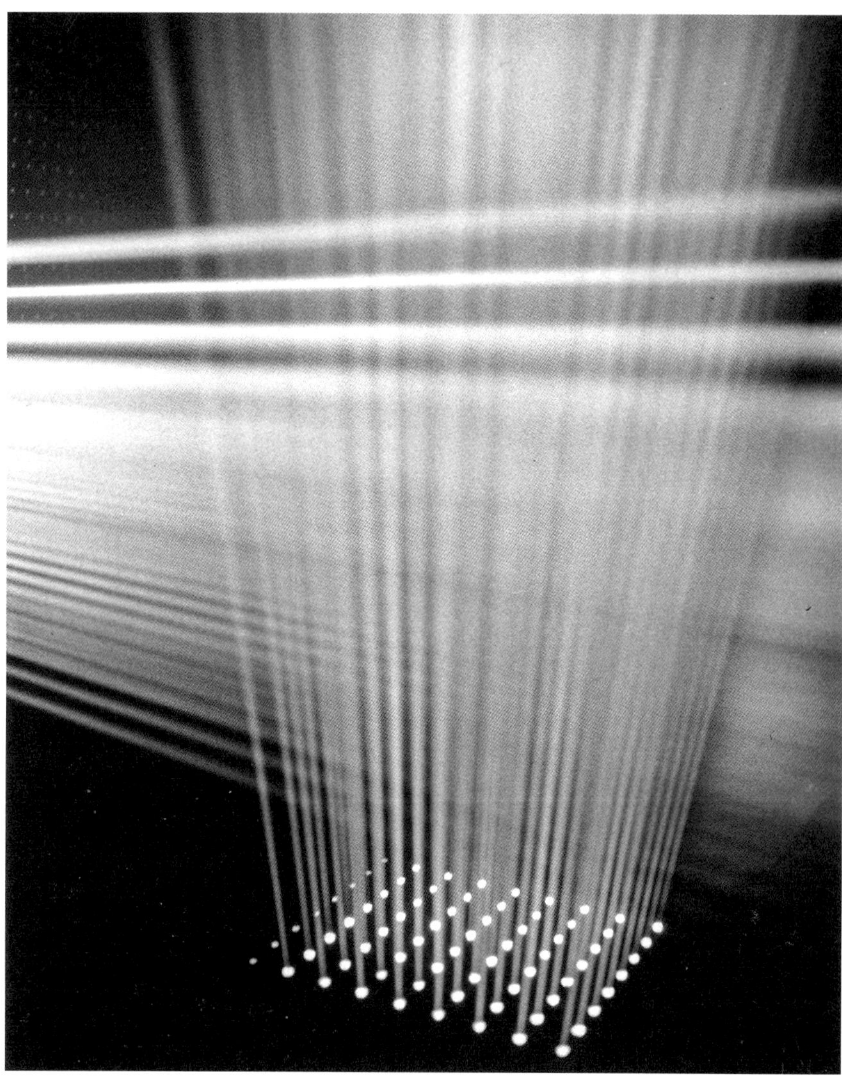

6.9

Gyorgy Kepes, simulated light architecture for
Boston Harbor, 1966. Photo: Nishan Bichajian,
1966. Courtesy Center for Advanced Visual
Studies Special Collection, MIT Program in
Art, Culture and Technology. © Massachusetts
Institute of Technology.

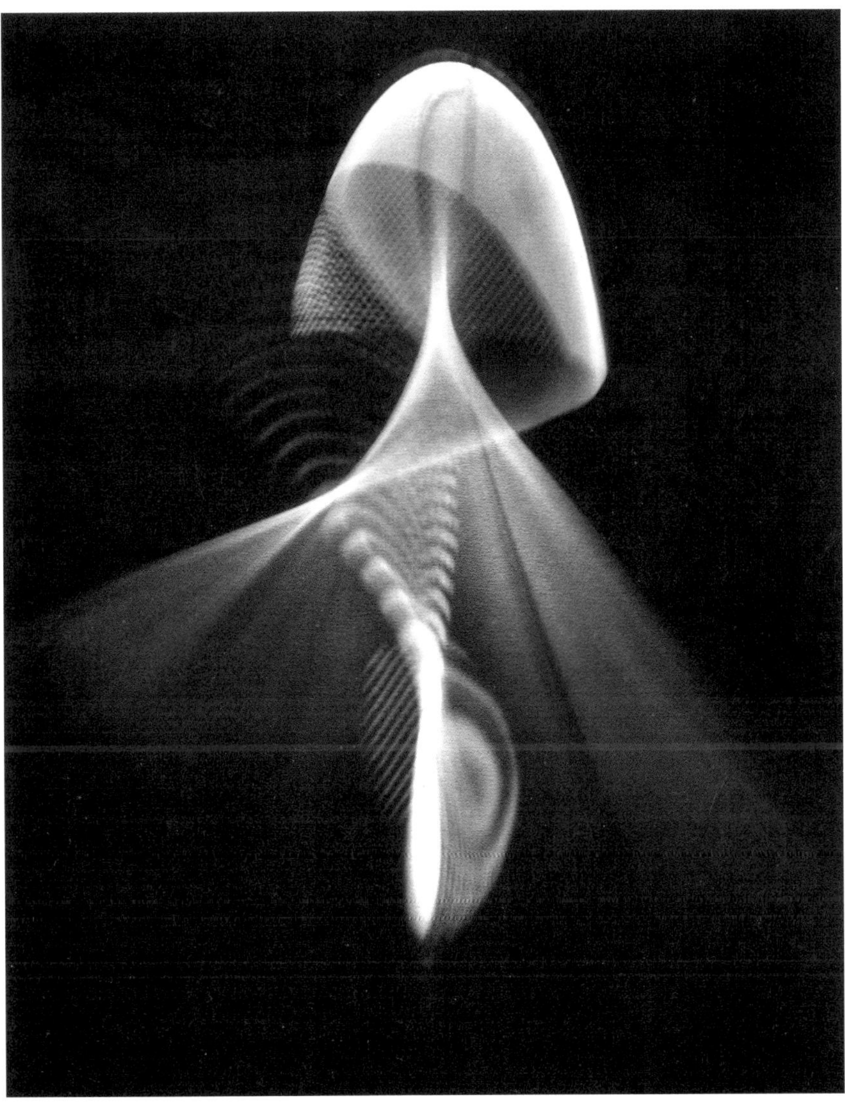

6.10

Gyorgy Kepes, *Caustic Curves*. Photo:
Nishan Bichajian. Courtesy Center for
Advanced Visual Studies Special Collection,
MIT Program in Art, Culture and Technology.
© Massachusetts Institute of Technology
and the Estate of Gyorgy Kepes.

The Charles River Project, like the Boston Harbor Project, was based on elaborate studies, or what Kepes called "[s]tudio scale experiments."[18] He imagined a "floating buoy garden," made from buoys "acquired from the Army engineers" or "the Navy" and then put to aesthetic use—an example of conversion.[19] Keiko Prince created what she called a fountain maze (fig. 6.11). Water pumped through hundreds of underground holes would erupt into a liquid architecture, "a maze of tunnels" through which children could run and play.[20] Shooting some twelve to fifteen feet in to the air, the fountain maze would "produce rainbows with its spray."[21]

Most of the Center's fellows used light as a medium for activating the environment. Ted Kraynik proposed a "Collimated Light Divider" or "Collimated Light Buoy" that would project illumination from a circular structure through floating vertical slots; an interior ring would slowly rotate, causing the slots to open and close in succession, thereby scattering light on the water's surface in a shifting pattern (fig. 6.12). As Kraynik describes it, the work could also report on environmental conditions; "the lights inside the structure can be made to vary from red to green indicating degree of pollution as read by various stations."[22]

Interactivity was a defining feature of these proposals. They rendered visible invisible phenomena, like currents of wind and water, through interactive feedback with ecological systems—the pull of the river or the flow of air. Another proposal, designed by one of Kepes's graduate students, used a chain of searchlights on floating cables to create optical displays that would fluctuate in response to the river's flow, creating a "man-made aurora borealis," as Kepes describes it, as the cables rose and fell with waves (fig. 6.13). A kinetic *Wind-Sound-Light Sculpture* designed by another graduate student would come to life in response to air. A set of photographs show the work both at rest and in motion (figs. 6.14–6.15). A series of curved and flat propellers connected by spokes to a vertical axis would catch the wind, turning at varying velocities. Each rotation would generate electronic noise; colored lights affixed to the sculpture and equipped with audio sensors would respond, flashing in sync with the sounds the propellers created. In effect, the sculpture would generate a system not only between the work and the environment but also within itself; it would self-reflexively respond to its own environmental response through a series of intersecting systems, a system of systems, all based on feedback as a means of linking relationships.

What do we make of all these unusual projects? First, let me be cynical. They are little more than models and manipulated photographs, simple darkroom tricks—just thought experiments (the Center would realize none of

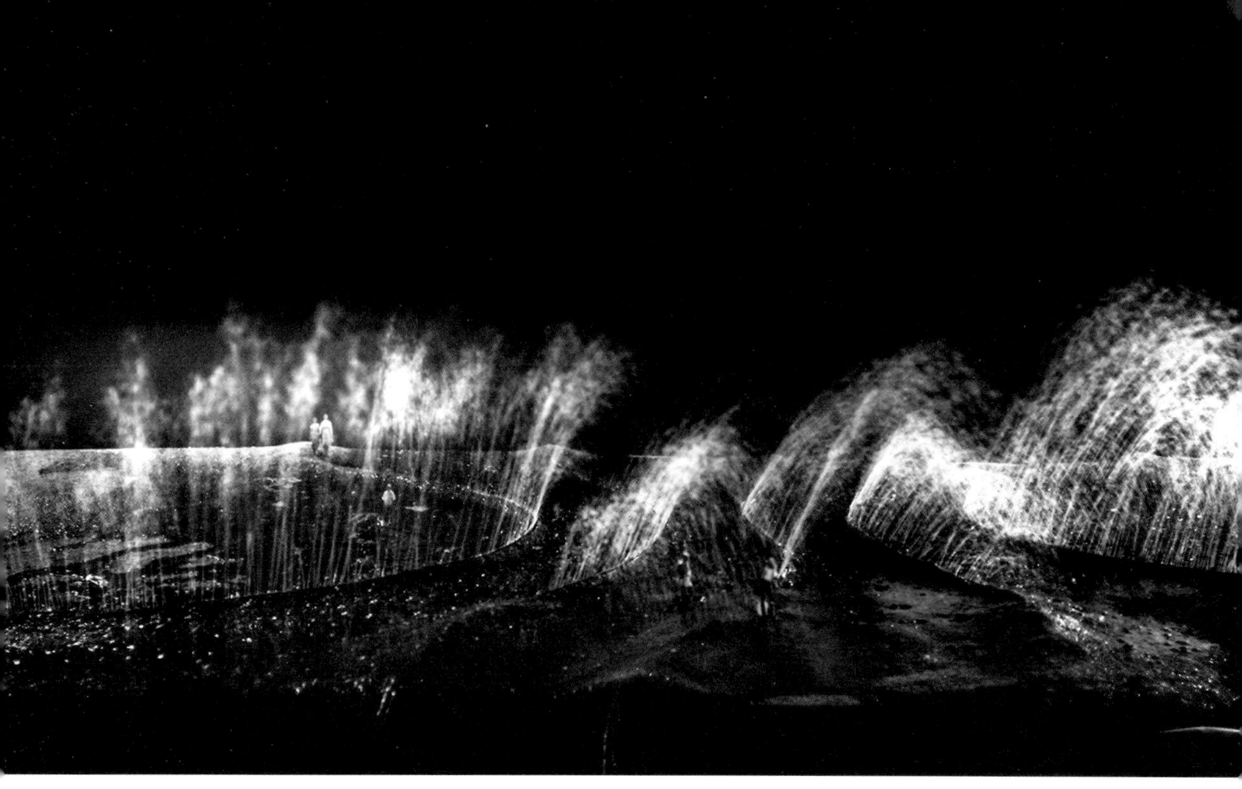

6.11

Keiko Prince, *Fountain Maze*, 1972. Courtesy
Center for Advanced Visual Studies
Special Collection, MIT Program in Art,
Culture and Technology. © Massachusetts
Institute of Technology.

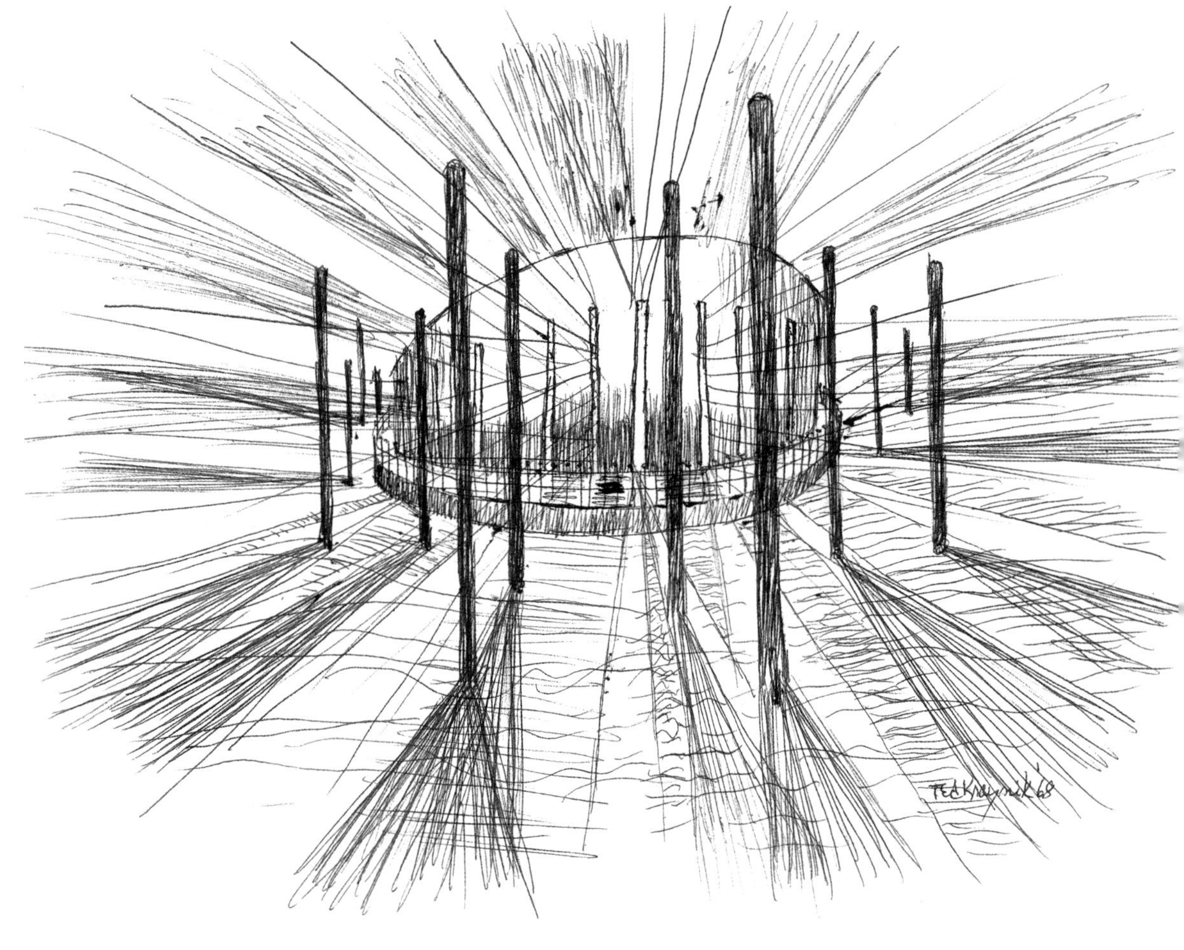

6.12

Ted Kraynik, *Rotating Collimated Light Divider*,
1968. Courtesy Center for Advanced Visual
Studies Special Collection, MIT Program in
Art, Culture and Technology. © Massachusetts
Institute of Technology.

them, though some fellows would later create variations, sometimes at far reduced scale). Kepes's own images representing light scattered by "floating buoys" are nearly identical to the photograms, or what he called photogenics, that he created with mirrors and lenses at the New Bauhaus in Chicago in the 1940s (fig. 6.16). Back then, decades ago, Kepes framed these laboratory demonstrations as investigations into optical phenomena, part of his light pedagogy; now he presented them as monumental interventions into the urban landscape. This drastic increase in scale, without a corresponding increase in the means with which to actually realize such interventions, suggests the discrepancy between imagining environmental art and actually creating it. Art on a civic scale lacked a similarly scaled mechanism for construction, for obtaining civic approval, civic support, and civic financing.

Consider a series of studies documenting the possible "light effects" for Kepes's Boston Harbor projects (figs. 6.17–6.18). They are visually striking—dazzling light reflected and refracted, molded and formed. Kepes, with Bichajian's help, created such effects by scattering light through mesh materials, bouncing it off metallic surfaces, chopping it and bending it. The images make captivating photographs, but what do they signify as environmental art? They are too abstract to be legible as works in the real environment—too ethereal to read as towers, walls, or murals. They exist only on paper, on both photographic paper and the kind used for reports, memoranda, and funding applications. They are abstractions, mere sensations. They are imaginary.

Kepes was aware of such problems. He suggests that the Center's projects were "limited only by the artist's imaginative power," but then admits that they were also limited by the realities of "verbal projection"—the fact that they can exist only in words—which could be "misleading."[23] These projects all have a "utopian ring," they "seem to be fantasies."[24]

But the Center's proposals are also susceptible to a more damning ideological critique: they all affirm the mythical and redemptive purity of nature by simulating it, while also, simultaneously, affirming the mythical and redemptive power of science and technology. They posit nature as virtuous but use scientific and technical means to create its false imitation, an empty simulacrum. Moreover, by accommodating science and technology—the very cause of ecological damage in the first place—they function as fancy packaging for a very reactionary vision. Military power had been the Center's first troubling ideological entanglement, and it seems the environment was its second.

These ideological entanglements evidently failed to register at the Center. The graduate student project for a "man-made aurora borealis" (fig. 6.13) specifically recalls the architect Albert Speer's design for a Cathedral of Light

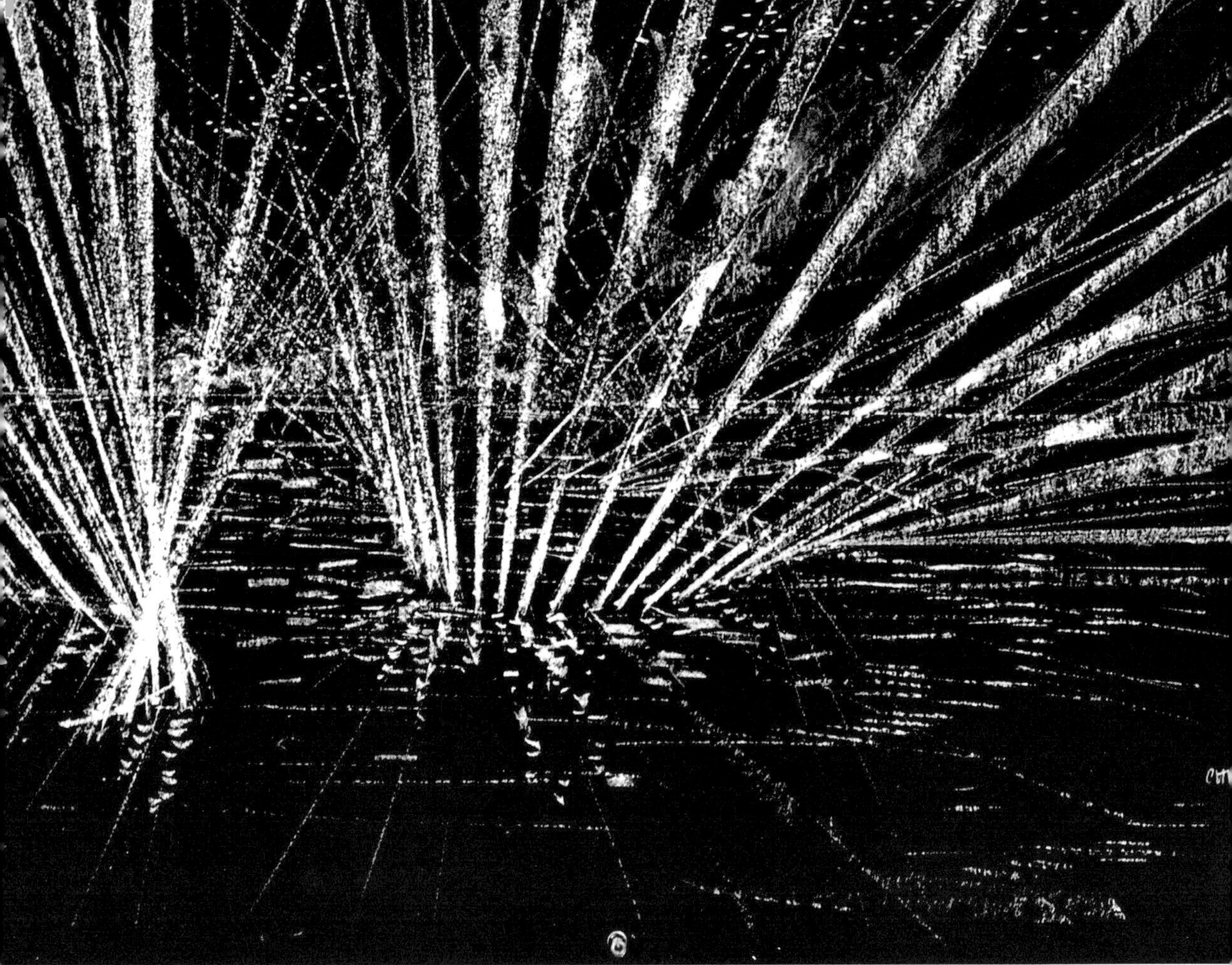

6.13

Chumpon Surintraboon, MIT graduate
student project designed under Gyorgy Kepes,
c. 1967. Gyorgy Kepes papers (M1796).
Dept. of Special Collections and University
Archives, Stanford Libraries, Stanford, Calif.

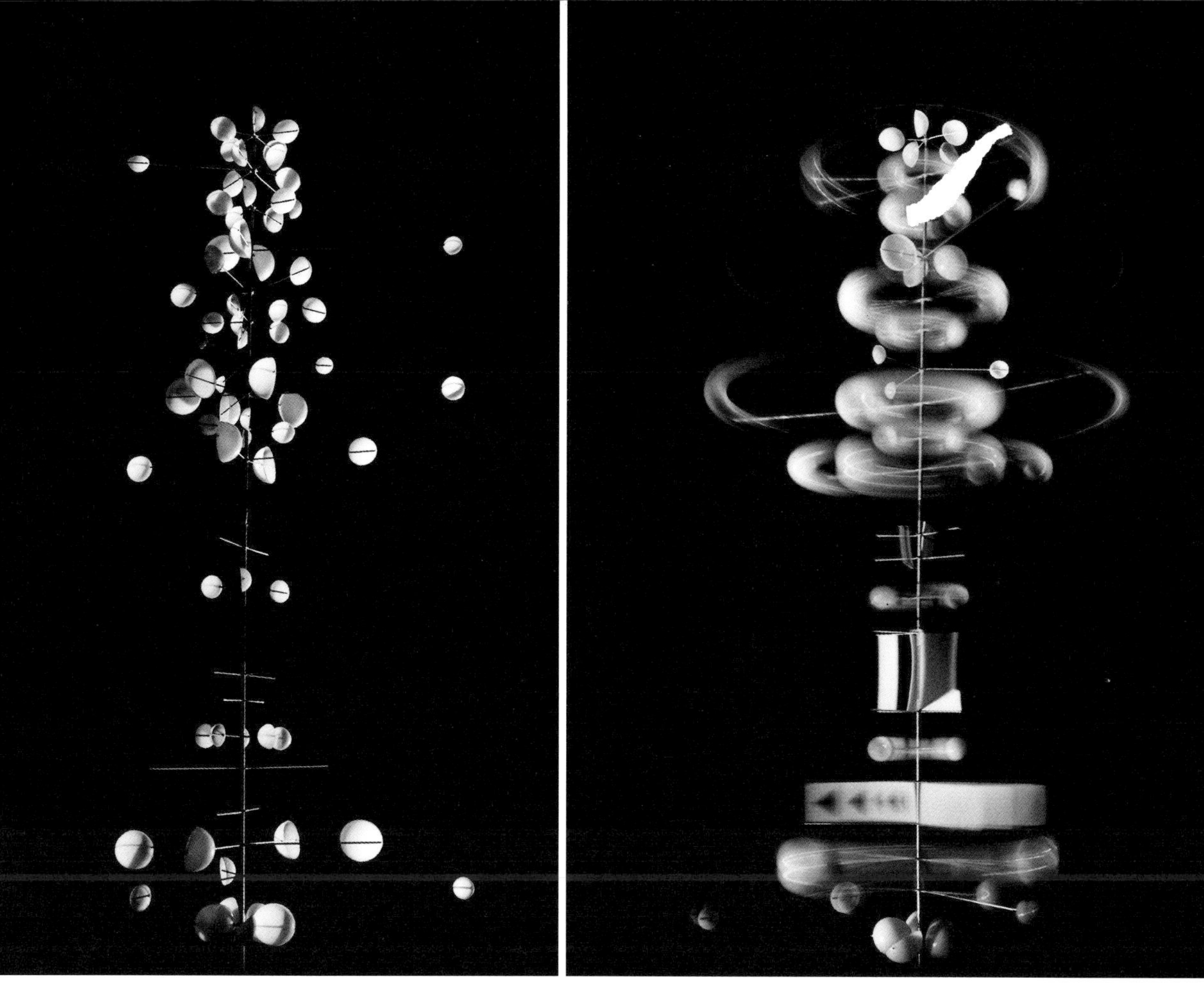

6.14, 6.15

Wind-Sound-Light Sculpture, MIT graduate
student project designed under Gyorgy Kepes,
1967. Photos: Nishan Bichajian. Courtesy
Center for Advanced Visual Studies Special
Collection, MIT Program in Art, Culture
and Technology. © Massachusetts Institute
of Technology.

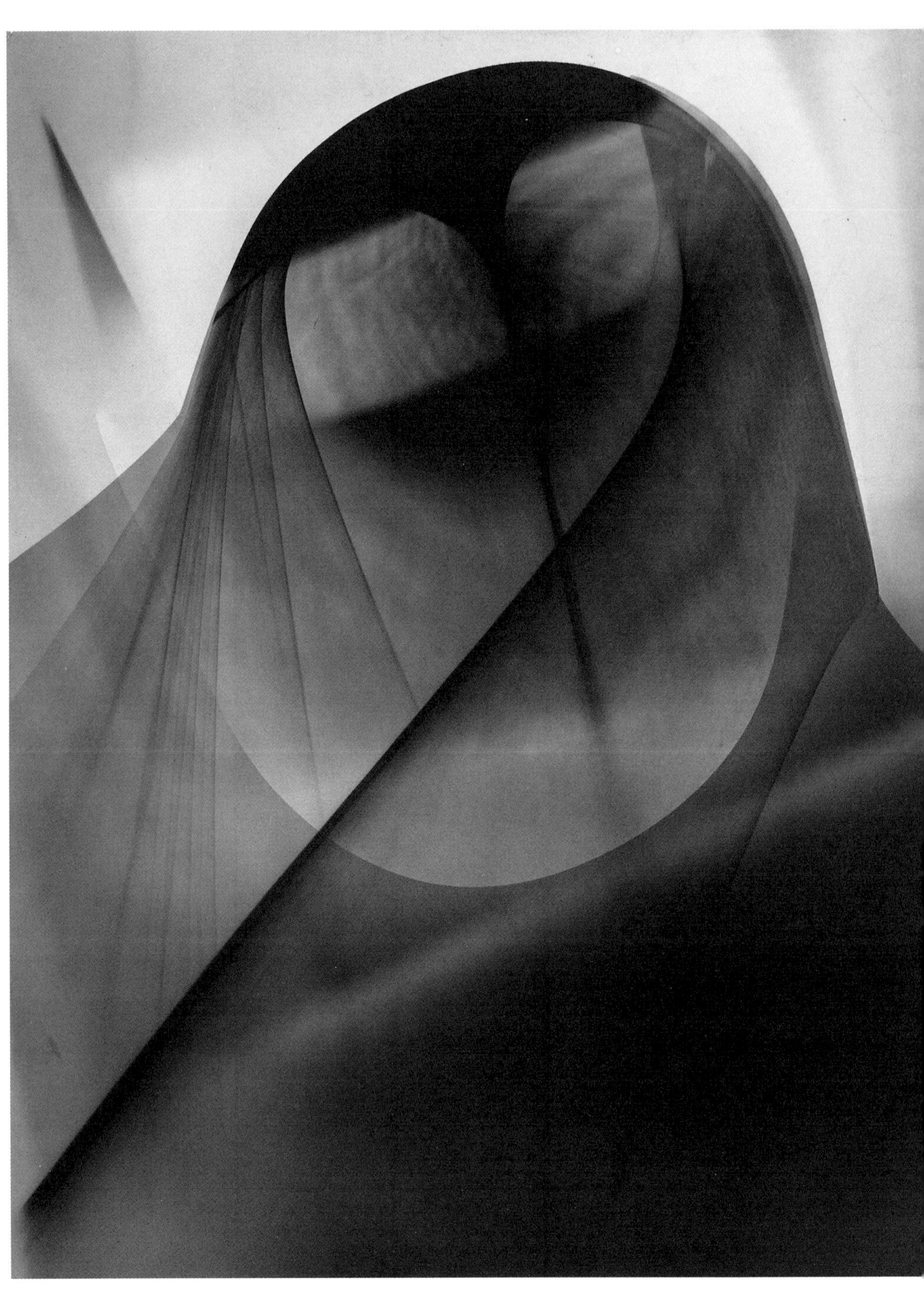

in Nuremberg, Germany; Speer aimed powerful searchlights—the kind used to track enemy aircraft so flak guns could fire at them—up toward the sky, creating a monument imposed on the landscape. Beginning in 1934, Speer's Cathedral of Light framed the annual rallies of the Nazi Party. Kepes's own proposals for "light architecture" (figs. 6.8–6.9) are intrinsically linked to this precedent—they, too, use searchlights aimed skyward; they, too, create light monuments imposed on the landscape. I have no doubt that his proposals were intended to recover a natural world, to animate invisible phenomena like wind and waves, but to what degree are they also, inadvertently, accidentally, unknowingly (could Kepes have not known about Speer's light architecture?), but still essentially connected to these darker precedents? What are the ideological implications of using science and technology to create such spectacles, given these origins—origins that Kepes and the fellows surely renounced, but which nonetheless structure the Center's projects?[25]

And yet, despite asking these questions—questions that lack satisfactory answers—I also want to move beyond them, beyond critique, beyond the cynicism, in an attempt to consider the optimism of these projects. I want to take their utopianism seriously. Doing so requires a certain suspension of disbelief, but it also allows us to explore how Kepes and the fellows imagined aesthetic conversion might work, both in theory and in practice.

In fact, these plans were not just theoretical; Kepes and the fellows really attempted, as best they could, to realize them. Kepes contacted the Boston Redevelopment Authority to discuss the "ambitious, utopian project" and the possibility of constructing a work in Boston Harbor for the 1976 Bicentennial celebrations.[26] After Kepes presented his ideas, the Authority responded positively; it all looked very "creative and exciting."[27] Nothing more came of the exchange. The Boston Harbor Islands Project did not result in built projects either, but CAVS fellow Friedrich St. Florian would still boast of "the fact that we are working closely with the Metropolitan Area Planning Council, the MDC [Metropolitan District Commission] and the US Army Corps of Engineers." He announces that the Center has "already received the support of the Governor's office and of Boston 200 [an organization supporting Bicentennial celebrations]."[28] These relationships signify attempts to make the utopian manifest, but nothing more came of them.

6.16

Gyorgy Kepes, *Abstraction*, 1940.
© The Estate of Gyorgy Kepes.

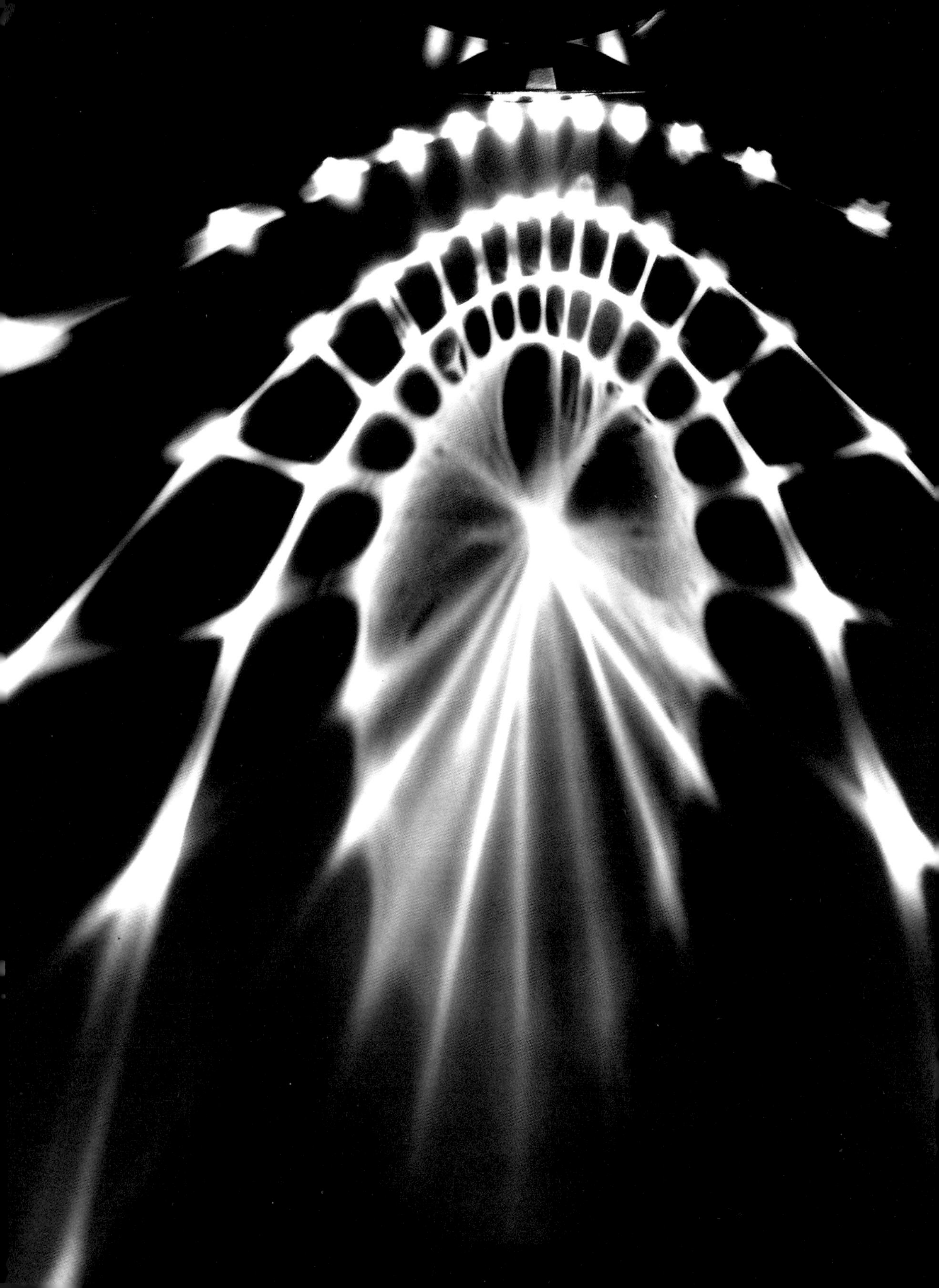

6.17

Gyorgy Kepes, *Light Effects*, 1970. Photo: Nishan
Bichajian. Courtesy Center for Advanced
Visual Studies Special Collection, MIT Program
in Art, Culture and Technology. © Massachusetts
Institute of Technology and the Estate of
Gyorgy Kepes.

6.18

Gyorgy Kepes, *Light Effects*, 1970. Photo:
Nishan Bichajian. Courtesy Center for
Advanced Visual Studies Special Collection,
MIT Program in Art, Culture and Technology.
© Massachusetts Institute of Technology
and the Estate of Gyorgy Kepes.

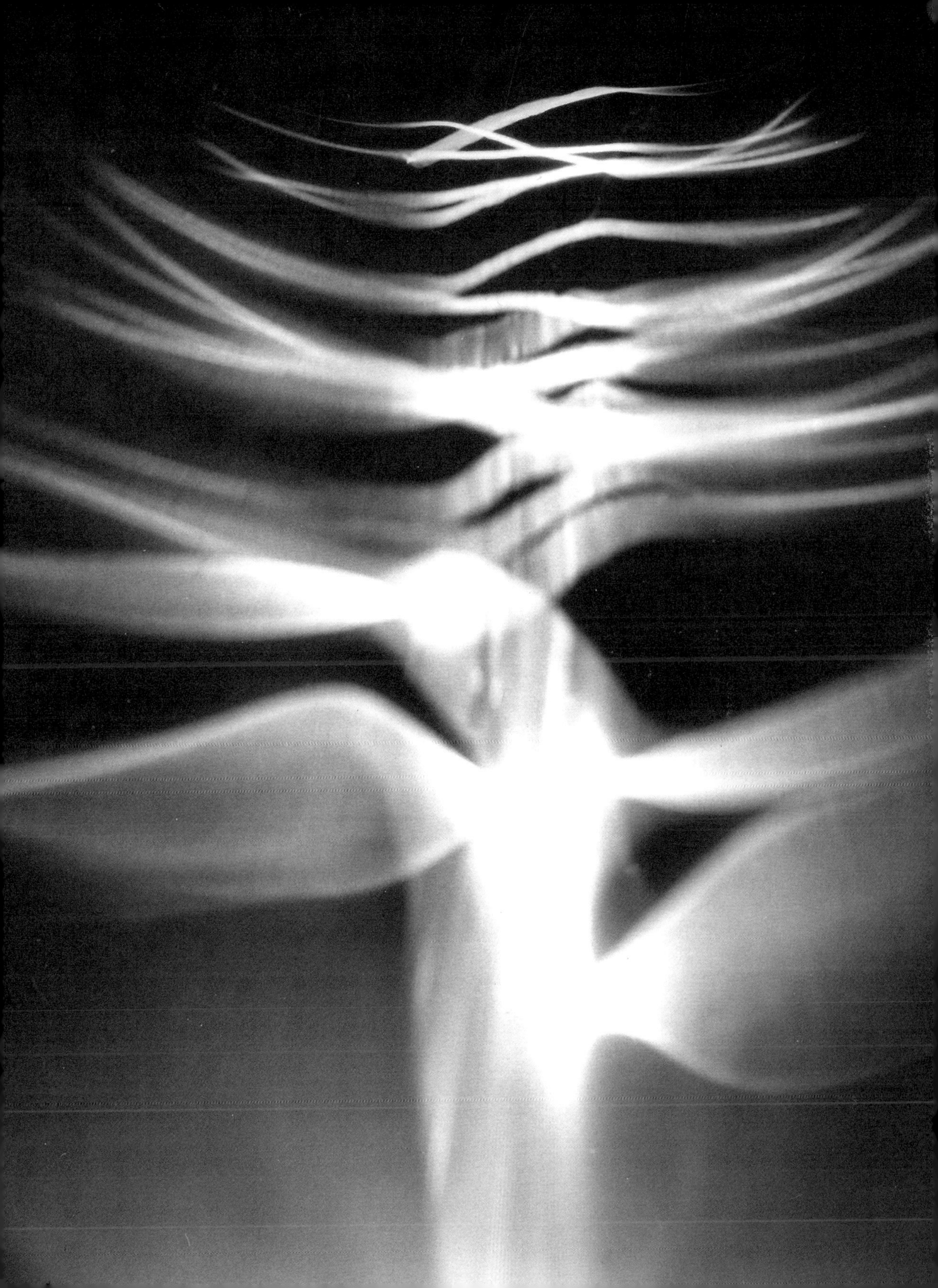

The Center also incorporated actual scientific and technological devices like ecological sensors, pollution abatement equipment, and even such pedestrian facilities as sewage treatment plants into their works. The core of the Charles River Project, for example, was a proposal by Thomas McNulty and the CAVS fellows for "the construction of marginal conduits" which would capture contaminated runoff and channel it to a chlorination plant for treatment.[29] The plan involved discussion with the MDC about the feasibility of actually building such a conduit. From Kepes's research notes:

> solid waste
> cities are suffocating + poisoned in their own refuse
> air, water pollution, noise, smog, waste of nuclear
> entangled in transportation
> toxic fallout …
> means of removing the atmospheric pollutant sulfur dioxide from the gas and converting it … has been worked by the Monsanto Chemical Co …
> This necessary + possible self-cleaning, self-correcting process – could in some cases be displayed in gigantic scale collaboration of artists, engineers, planners … could make such operations spectacles.
> Waterworks, nature filtering water from particles, could be made into demonstrations of dynamic kinetic events[30]

Kepes's note, presented in his characteristic mode of free-form association, suggests the use of technology as a self-cleaning and self-correcting instrument. His comments are ironic: Monsanto was notorious for its production of toxic chemicals like DDT and Agent Orange, and for contaminated sites across the country—but maybe this very fact made Monsanto an ideal target for the Center's model of environmental conversion.

In an essay titled "The Artist's Role in Environmental Self-Regulation," published in *Arts of the Environment*, Kepes elaborates on these "combinations of pollution-abatement technology and vital aesthetic experiences." He explains how water treatment plants are separated from the cities they serve, thereby limiting public awareness of the water cycle. He elaborates how "the water-purification process is a potential display feature that can be an exciting artistic experience in itself," especially through the construction of "immense transparent structures that give visibility to hydraulic processes." He describes the "ballet of water racing through obstacles of filters," through an elaborate array of pipes "in a variety of thicknesses," all creating "impressive sculptural forms that have never existed before."[31] Kepes relates these ideas to the Roman aqueducts and the fountains of Tivoli. Elsewhere he had in mind more ancient waterworks; a note describes "early reservoir construction" and forms of "irrigation," including "reservoirs fed by canals or natural drainage" dated "4000 B.C. in Euphrates valley" and "2000 B.C. India."[32] Another note:

celebrate water
water purification work – surrounded by water park
water sprays in different light
arch of lighted water
fountain
water in all transformation steam
in winter ice
temperature – boiling water
cascade – from boiling to freezing
water garden
invoking participation involvement
geyser
colonnades
transparent containers
fluid obstacles
splashes – illuminated with strobe lights
strobelights [*sic*] modulated with sound[33]

These ideas suggest how Kepes imagined water as medium, as a creative substance, even while the Center's waterworks actually addressed real contamination.

Kepes also sent letters to potential industrial partners about these plans. He contacted Cummins Diesel about a concern over the "degeneration of our urban environment" and an interest in "redirecting the environmental transforming processes to a constructive life-enhancing direction"—the company, a manufacturer of engines, must have been perplexed.[34] They did not reply. Kepes was willing to talk to anyone, even if they would not listen; he also contacted the president of Eastern Gas & Fuel Associates, which manufactured fuel distribution and water treatment equipment, about "the large scale light form for Boston Harbor" and the Center's "rather Utopian notion"; he sent another letter to Electro Optical Systems, Inc., about the use of equipment "for a large scale fountain."[35] He heard nothing in response.[36]

Self-Regulation

In their sensorial effects and interactivity, and the way they addressed actual pollution, the projects imagined by Kepes and the fellows at the Center all relied on variations of systems discourse—a theory of what Kepes called "environmental self-regulation." Kepes understood pollution abatement, for example, as a means of achieving homeostasis—and homeostasis not as the maintenance of order and control but as a means of changing the system, of adjusting its course and repairing its damage: as a self-cleaning, self-correcting process. His report on the Charles River Project describes "self-regulating devices" that are "constantly eliminating useless toxic matter from the body" or "converting" it into something useful—the same process he proposes for "our extended body, our social and man-transformed environment," a mode of self-regulation that would embrace "all scales and levels, ranging from the immediate urban surroundings to [the] national and global scale."[37] In his notes, he distinguishes between this desired homeostasis and "pseudo-homeostasis," a false correction or deceptive fix. Against legitimate forms of regulation, Kepes imagined "artificial balancing" and "ersatz" or "surrogate" homeostatic mechanisms. One example was the abuse of chemical substances; rather than treating the cause of psychological states like anxiety and depression, we use "drugs to numb," "drugs to intoxicate," "drugs to energize," "drugs to eliminate consciousness." The page reveals homeostasis as bodily and individual, but also as a social, national, even global means of change, as a course correction.[38]

Elsewhere in his notes he elaborates the concept of pollution as literal and metaphorical, as toxic emissions but also as a more profound social and cultural phenomenon. He explains how science and technology, and the empirical worldview they support, have generated unanticipated byproducts: "mechanical scientific exact values, socially synchronized activities, absolute common standards paid for: fall out [*sic*]." The note describes this fallout: "pollution of air, water, mind; epidemic of emotional explosion; chronic neurosis; emotionally starved, morally starved."[39] The implication is that scientific and technological change have detrimental effects not only in a strict ecological sense, the fallout experienced as particulate matter in air or chemicals poisoning water, but also as mental and emotional disturbances, as cultural fallout. Elsewhere, he describes the acid—the "poisons poured out by our cars"—that attacks "the physical surface, our urban environment ... but [is also] bitten deeper into our sensibilities, dulled, alienated sensibilities, result of fallout of a mistreated environment."[40] Environmental art would aim to clean actual toxic emissions but also this mental and social pollution, the dulled aesthetic sensibility Kepes identified as equivalent to ecological damage.

One chaotic page of scribbled notes expresses the urgency of these issues: "ecological excrement," "defecation of wells," "four letter words." In a frantic scrawl: "life pollution! Auto toxication due to scale conflict"—a reference to self-destruction in a society out-of-scale with its own changes, a society unable to contain the fallout its scientific and technological achievements had created. Such life pollution was present not only in the levels of toxic substances that might be measured in rivers and streams, but also in our "homogenized life" and the resulting "soul searching" that it prompts, the "search for new rituals," for "collective purification," for "creative ritual."[41]

These ideas were most clearly compiled in an essay Kepes authored for *Arts in the Environment* titled "Art and Ecological Consciousness," one of his most forceful statements on environmental aesthetics.[42] As was his way, Kepes marshaled an eclectic range of sources, including the "warnings" of poets and thinkers centuries ago. He quotes John Ruskin: light, once clean and pure, is now "umbered and faint"; air is now "defiled with languid coils of smoke"; and water is "dimmed and foul." Ruskin linked such damage to industrialization, to science and technology. "Ah, masters of modern science[,]... you have divided the elements, and united them; enslaved them upon the earth, and discerned them in the stars."[43] Kepes also quotes William Morris, who laments the factories that "blacken rivers, hide the sun and poison the air with smoke and worse," and asks why Manchester, perhaps the most industrial of Britain's cities, cannot learn "how to consume its own smoke" while producing so many "useless guns" (perhaps the remark resonated as an oblique description of the Institute's military projects).[44] But Kepes also refers to unexpected sources. Improbably, he quotes Jimi Hendrix, specifically citing Hendrix's "Purple Haze": "Excuse me while I kiss the sky." He quotes (actually, misquotes) Mick Jagger of The Rolling Stones: "They stole my imagination"—a line Kepes interprets as a vision of nature denied by social forces. He contrasts recent scientific and technological achievements—"bioengineering, genetic engineering, the pill, distant sensors, cyborgs, and an ever-increasing communications network"—with the inability to use these achievements to solve environmental problems. In his wildly associative style, he refers to descriptions of the Earth as seen from outer space; to Emerson's meditations on lighthouses; to Thoreau and Turner and Constable. He concludes the essay with an impassioned embrace of the individual's ability to effect change: "The unchartered space is within ourselves, in our still unfathomed ethical potentials, in our still untapped imaginative power."[45]

Kepes's understanding of environment suggests the allegorical associations he attached to the Center's activities—but it also takes us back to the military, back to war, back to Vietnam. Kepes is explicit in "Art and Ecological Consciousness" about relating science and technology to military power; he describes advances in methodologies like "computer game theory, theories of servo-mechanism, [and] systems approaches," but argues that "most sophisticated systems applications of technical know-how yet devised are those that have been used to invent means of tearing and burning the flesh from our brothers"—an allusion to Vietnam—and not used to benefit humankind.[46] Among Kepes's research materials is a 1968 paper by Arthur W. Galson, a biologist and botanist who argued against the use of Agent Orange, the herbicide employed to defoliate jungles in an attempt to expose enemy guerrillas. The paper, titled "Unanticipated Environmental Hazard Resulting from

Technological Intrusions," explains the lasting impact of these chemicals in ecological systems (Agent Orange in the environment continues to cause cancer and birth defects). In bold orange highlighter, Kepes marked a passage in Galson's article about the chemical picloram, another of the so-called "rainbow herbicides" used in Vietnam (the reference to rainbows was intended to indicate how the military distinguished herbicidal agents by color-coding barrels; it also resonates with Kepes's fixation on rainbows in the natural world). The use of picloram to "destroy conifer forests in Vietnam may lead to prolonged infertility of the soil in the treated regions," causing "significant ecological damage."[47] The passage is important not because of the particularities of picloram use but because it demonstrates how Kepes reads environmental crisis as intertwined with the Vietnam War, as one and the same.

Kepes's research materials contain other sources that similarly extend ecological issues into issues of military power. A pamphlet titled "On Chemical and Biological Warfare" contains a section on herbicidal war:

> **The only really effective instrument against chemical and biological warfare is a well-informed and socially responsible public opinion.... Only when people realize that they are to some extent responsible for the action which their governments take in their name will they be able to affect their government in any way at all.[48]**

Another pamphlet documents the applications of science and technology in weapons, specifically through new chemicals and biological pathogens; Kepes marked a passage in red that described the use of "non-lethal riot-control gas in the war in South Vietnam" as a means of what the pamphlet termed "gas warfare."[49]

Kepes also created more balance sheets—recall the idiosyncratic lists of opposites he generated in the 1940s and 1950s—but applied his dialectics of seeing to environmental concepts, to terms from nature. Under categories defined by the four classical elements (earth, water, air, fire), presented here as positive, Kepes listed negative associations:

elements + −
<u>water</u>
polluted
water hose riot control
<u>fire</u>
flamethrower
fireballs
Hiroshima
Dresden
London
immolation
H. bomb

<u>air</u>
pollution
smoke
air raids
sky
gas mask
<u>earth</u>
defloration
concentration camps
mass grave
nuclear fallout
blacktop concrete[50]

Kepes's environmental cosmology references actual environmental pollutants like smoke, but also a series of more abstract associations. He reads water through the photographs of firehoses aimed against Civil Rights demonstrators in Birmingham, Alabama. He reads fire through images of nuclear fireballs and hydrogen bomb tests, the burning of Dresden, and the London Blitz. He reads air as air war, as the gas masks used to breathe during chemical attacks. And he reads earth in the images of mass graves and concentration camps seared into memory. Another note on such "elements" refers again to the infamous self-immolation of Buddhist monks as profound antiwar statements: "fire—ritual as Vietnam war protest."[51]

What do these wide-ranging ideas mean for the activities at the Center? How does Kepes's homeostatic self-regulation actually work? Kepes describes the Center's most ambitious projects as combining "actual pollution-abating devices with ecological consciousness-forming artistic characteristics."[52] Even what might appear to be purely aesthetic spectacles were imagined as promoting literal and figurative homeostatic adjustment. Some works used pedagogy, teaching participants about ecology through simulated environments. Juan Navarro Baldeweg proposed the construction of pneumatic bubbles that would sustain artificial climates—"grassland, tropical forest, or desert"—in unlikely settings, like a rainforest preserved in an "arctic landscape" (fig. 6.19). Such projects were intended to "increase social and individual awareness and experience of the major terrestrial ecosystems."[53]

Other projects used participation to encourage such adjustment. Michio Ihara proposed a series of floating barges (fig. 6.20) as a site for music concerts, theatrical productions, firework displays, and celebratory gatherings. Still others attempted to integrate these pedagogical and participatory aspects with actual ecological monitoring. "Distant sensors coupled to transmitters" would convey the "conditions of air, noise, water, heat pollution, and traffic congestion" on a "central display device."[54] Ted Kraynik's "Synergic Light Buoys"

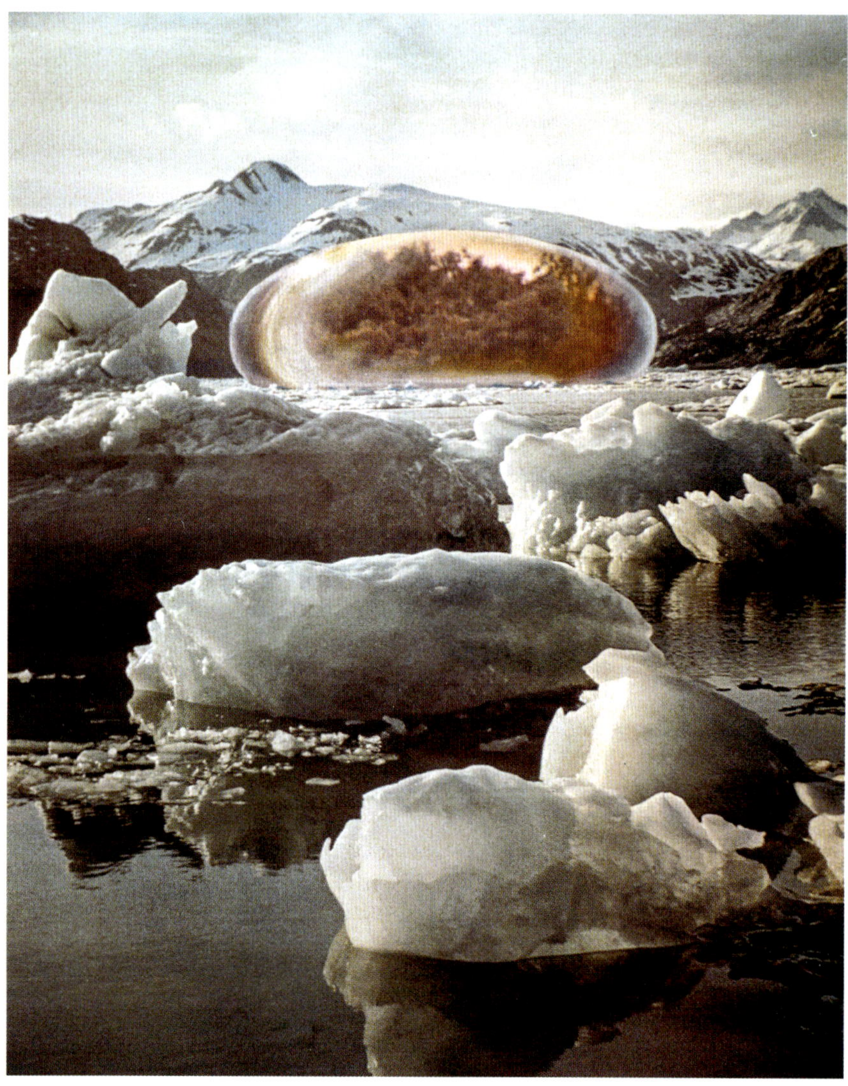

6.19

Juan Navarro Baldeweg, *Proposal for Increasing
Ecological Experiences*. Courtesy Center for
Advanced Visual Studies Special Collection,
MIT Program in Art, Culture and Technology.
© Massachusetts Institute of Technology.

(fig. 6.21), for example, would measure and visually present environmental indicators through illuminated vertical masts, like a giant bar graph—or, as he describes it, like a "symbolic throbbing heart describing the life processes of the urban landscape."[55] Each buoy would measure a particular urban activity and project it through light. The buoys would, in turn, move in response to the Harbor's water, responding not only to the urban environment but also to their more immediate one. They would, Kraynik says, echo "the masts of ships" and create "a very scintillating effect."[56] Kraynik explains the concept of synergy, popularized in the 1960s by R. Buckminster Fuller, as a description of the "inter-relationships" between people but also "their relationship to the structure in which they operate," to a larger system, a system of systems.[57] He was preoccupied by such ideas, even drafting an academic paper he titled "Synergic Art from Cave to Computer."[58]

Mauricio Bueno designed towers that would similarly capture environmental information and represent it on what Kepes calls a "display landmark" or "data fountain" set in a public space (fig. 6.22). Such information could include the "spatial sources of pollution, physical and social causes, balance sheets of gain and loss in the use of DDT, pollution of the air by automobiles, and so forth." The purpose of tracking such data was, again, to create "civic and ecological consciousness," not only through the information communicated but also through the immersive experience of its visualization and spatialization. Bueno's tower was clunky and mechanical—imagined as an imposing monument, like "radar antennas scanning space"—but its form was really secondary to function.[59] Data constituted an aesthetic object. The art was information. Moreover, these environmental facts would inspire actions, performances, and participatory events. Kepes explains: like the "obelisks of Renaissance Rome," or the "Roman fora" and "Greek agora" that fostered democracy by creating areas for "discussion of the common condition," these works would mark sites for "pageantries," an "Olympics of creative ideas, adventures, spectacles, sky festivals, collective murals, and collective multimedia light plays."[60]

In Kepes's thinking during this period, these events all represented what he described as "missing Sundays," a lost day of rest and relaxation, of recovery (he also imagined Sundays as days of public spectacle and performance; consider the tradition of promenading). Dismantling the name for the day of the week also renders a day of sun, a day explicitly directed toward nature. In a research note, he laments the lack of Sundays: "environmental dirt," "fever," "virus germ," "out of place element"—all these forms of pollution are "missing Sunday."[61] Another note frames natural phenomena like the "sunset sunrise," the "rainbow," as "the Sunday of nature, the dew on the glen blades."[62] Kepes's

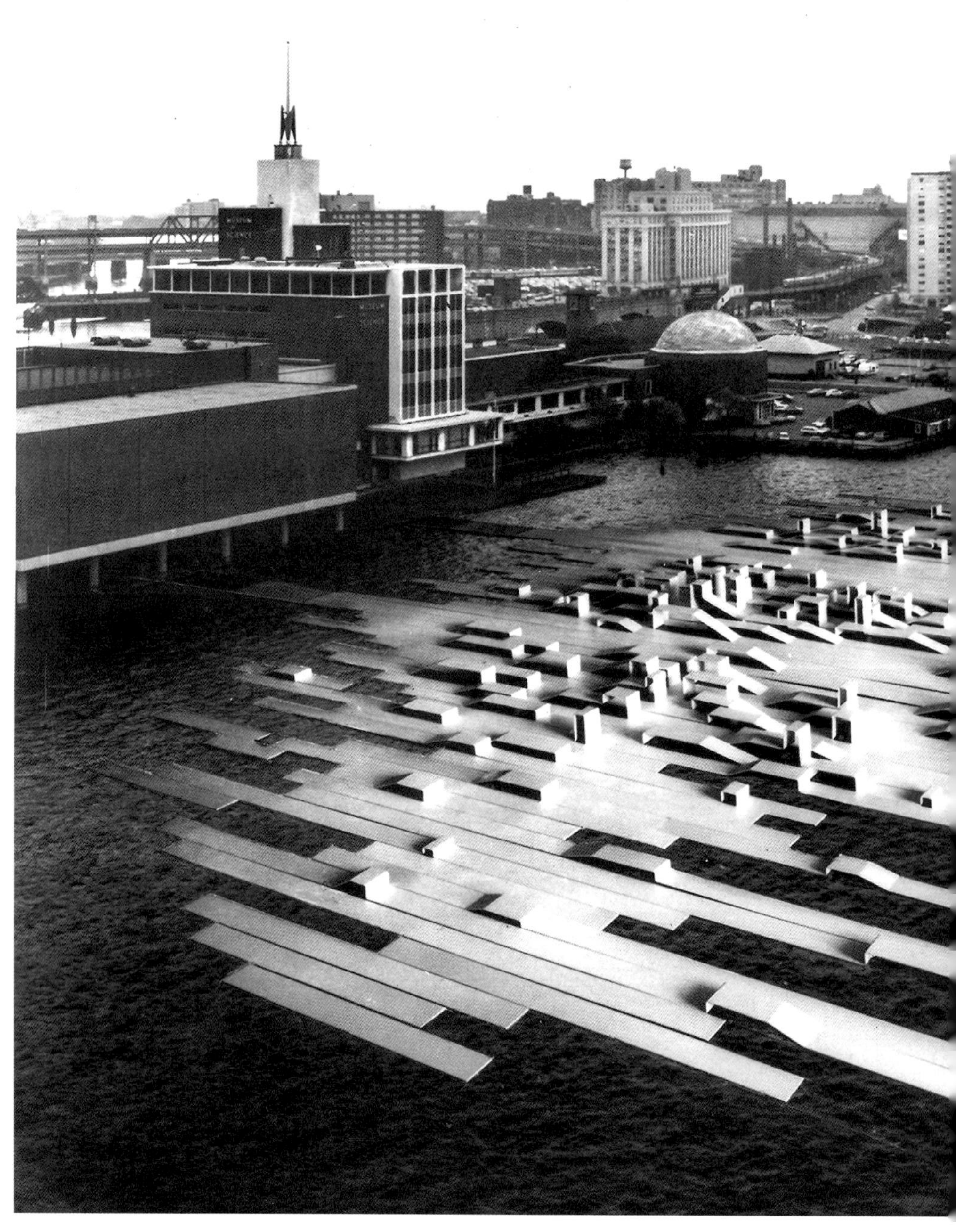

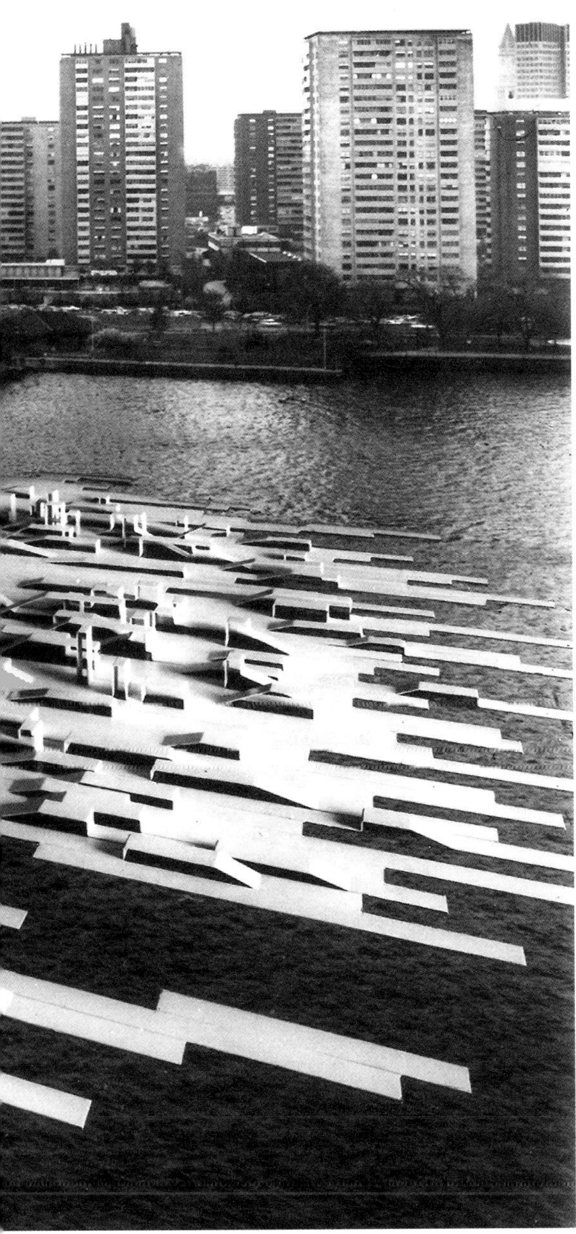

6.20

Michio Ihara, *Charles River Floating Walkway Proposal,* 1973. Photo: Nishan Bichajian. Courtesy Center for Advanced Visual Studies Special Collection, MIT Program in Art, Culture and Technology. © Massachusetts Institute of Technology.

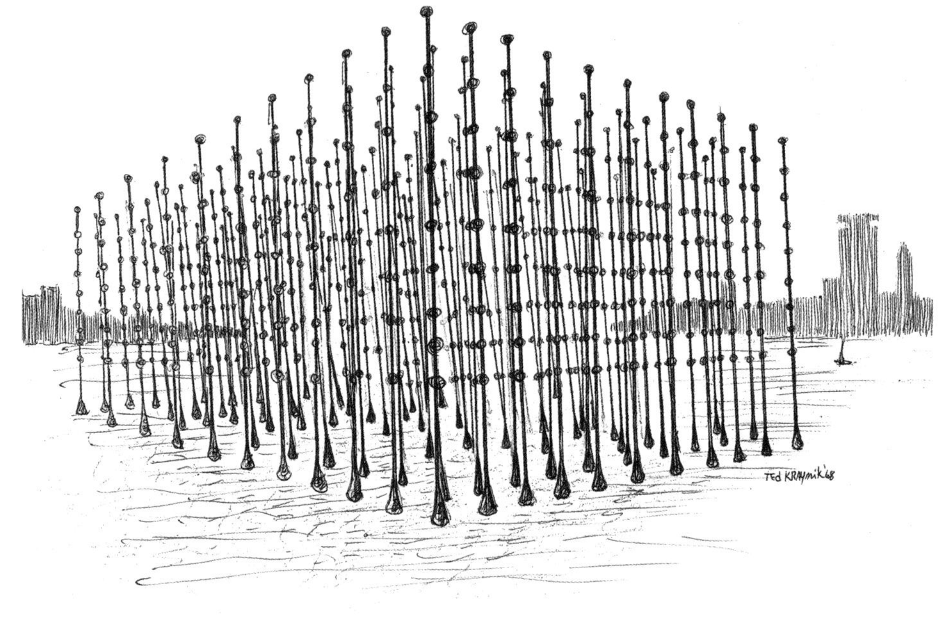

6.21

Ted Kraynik, *Synergetic Light Buoys*, 1968. Courtesy Center for Advanced Visual Studies Special Collection, MIT Program in Art, Culture and Technology. © Massachusetts Institute of Technology.

6.22

Mauricio Bueno, *Pollution Monitoring Tower*, 1969. Photo: Nishan Bichajian. Courtesy Center for Advanced Visual Studies Special Collection, MIT Program in Art, Culture and Technology. © Massachusetts Institute of Technology.

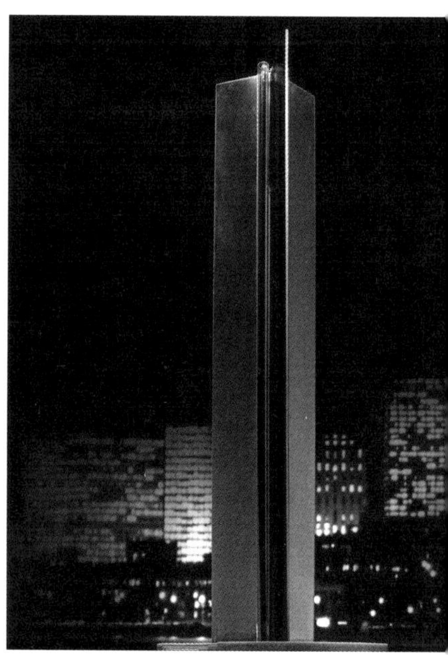

own sketches imagined an "educational raft" or a series of "floating fountains" suspended in the river, with water pouring over "transparent shelters" that could serve as sites for these performances, for missing Sundays; one such plan involved a "floating sail mural" that might function like a backdrop for such programs (fig. 6.23).[63]

In this way, the Center's projects were an eclectic mix of practices, evoking the cybernetic towers of Nicholas Schöffer and the megastructures of Superstudio, but also the happenings of Allan Kaprow (Kepes had planned to follow *Arts of the Environment* with a volume titled *Arts of Participation*. It would have included texts by the likes of Kaprow).[64] Kepes also linked these environmental works to the exploded scale of postwar painting. A note labeled "environmental art" explains: "increasing overpowering dimension almost room size of canvases—envelopes the spectator not only by size but by activities." He cites "R. Rauschenbergh [*sic*] large collage" as an example.[65]

The Center's projects also had earlier precedents, from the historical avant-garde. They recall Vladimir Tatlin's Tower, though they lack Tatlin's political charge. They recall the architecture of the Bauhaus, or even Moholy's *Light-Space Modulator*, now rendered at a "civic-scale," as a computerized *Gesamtkunstwerk*. Kepes cites Erwin Piscator and Bertolt Brecht for the performances these environmental works might facilitate.[66]

But Kepes traced a lineage for environmental art back still further. The Center's futuristic displays would re-create very antiquated experiences: In *Arts of the Environment*, he compares Bueno's towers and Kraynik's buoys to the Campanile in Venice's Piazza San Marco, and the parades and pageants the Campanile anchored as a symbolic landmark, as in a nighttime procession through the Piazza depicted by Francesco Guardi in a painting of 1748. The painting is included in *Arts of the Environment* next to photographs of Kraynik's proposal. In a note labeled "environmental art happening," Kepes again looks back to remote examples: "one can hardly imagine more total visual involvement than a venetian 17 century ... meeting of the council in the great hall," a reference to the Great Council meeting in Venice's Palazzo Ducale—a surprisingly distant reference.[67] The allusions were not just to architectural spaces but also to more elusive experiences, like "that rich sense of community which we see in some of Brueghel's paintings."[68] The fellows also saw their work through these precedents; Friedrich St. Florian describes a proposed bridge crossing the Charles River, part of the Charles River Project, as "not unlike Ponte Vecchio," the famous medieval walkway filled with storefronts that spans the Arno River in Florence.[69] Keiko Prince references Tsukimi, the Japanese celebration of the autumnal moon, as the basis for a shelter, also

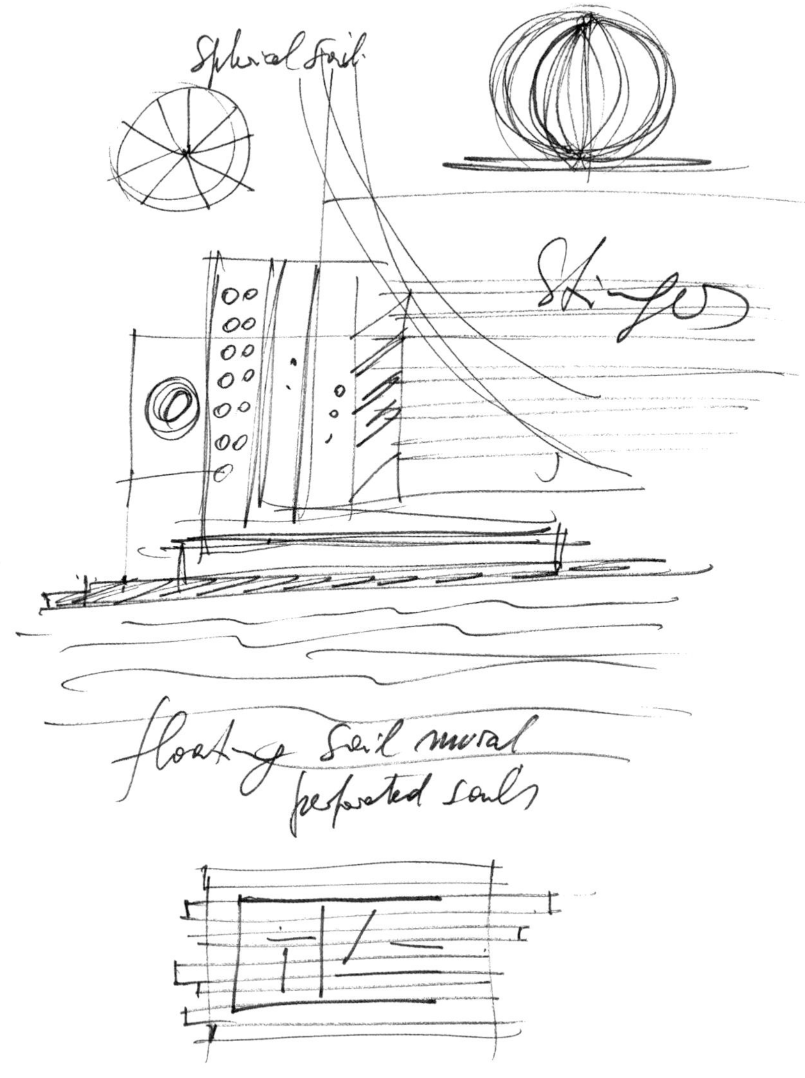

6.23

Gyorgy Kepes, drawing, c. 1970. Courtesy
Center for Advanced Visual Studies Special
Collection, MIT Program in Art, Culture and
Technology. © Massachusetts Institute of
Technology and the Estate of Gyorgy Kepes.

designed for the Charles River Project, that would capture light from both the sun and the water's surface on its roof, functioning like a sundial.[70] Or, as Ted Kraynik puts it: the goal is to "bring art to the people in a manner similar to the effect created by a Gothic cathedral."[71]

Kepes's book also included images of the fountains at Tivoli and the cascading pools at the Alhambra. His photo essay at the start of *Arts of the Environment* includes paintings by John Constable and J. M. W. Turner—hardly standard sources for environmental art of the postwar period. He similarly describes the Center's projects through the words of philosopher of science Sir Francis Bacon, who defined his experiments as both *frucifera* (meaning practical, or bearing fruit) but also *lucifera* (meaning illuminating, or generating light)—in other words, as both science and art.[72]

Kepes's notes and writings are filled with many such observations, providing a deep cultural history of environmental aesthetics. These references reveal his conception of environmental art, like light, as allegorical. Kepes also extended this allegory of light back toward the environment, toward the Center's projects. It is the "dream of all painter[s]" to capture light, he notes—he is referring specifically to Constable and Turner—but now, "we need a new attitude," we need a "light tower."[73] As I have argued, Kepes believed that the romantics—not just Constable and Turner and American Transcendentalists like Emerson and Thoreau, but also William Wordsworth and William Blake—envisioned a natural world that was disappearing; they captured a fading light, a light eclipsed by industrialization, by the haze of pollution that darkened sky and sun. One spread from *Arts of the Environment* even compares and contrasts Gustave Doré's print of polluted London—first used as part of Kepes's Light Book material—against the light of the Temple of Apollo in Delphi and Machu Picchu in the Andes (fig. 6.24), sites Kepes imagines as examples of pristine air and luminous clarity; they represent a pure vision of environmental experience. Elsewhere he describes St. Francis of Assisi delivering sermons "to birds and flowers," and argues that artists must speak now to contemporary iterations of nature, not to birds and flowers but to "today's extended world," to "far-off galaxies, ultra-microscopic forms," to "turbulence patterns, and invisible radiation." These images—a litany of the visual contents contained in *The New Landscape*—were "as stirring as sunsets, flowers, and the white mantle of winter snow."[74] They indicate what Kepes had earlier described as "replenishing in an imaginary plane" what is destroyed in the real one. This idea is the essence of the Center's mission, the purpose of its plans and proposals, all of which exist only on this "imaginary plane."[75] Perhaps St. Francis is a perfect model for Kepes as Kepes roamed MIT's infinite corridor, sermonizing.

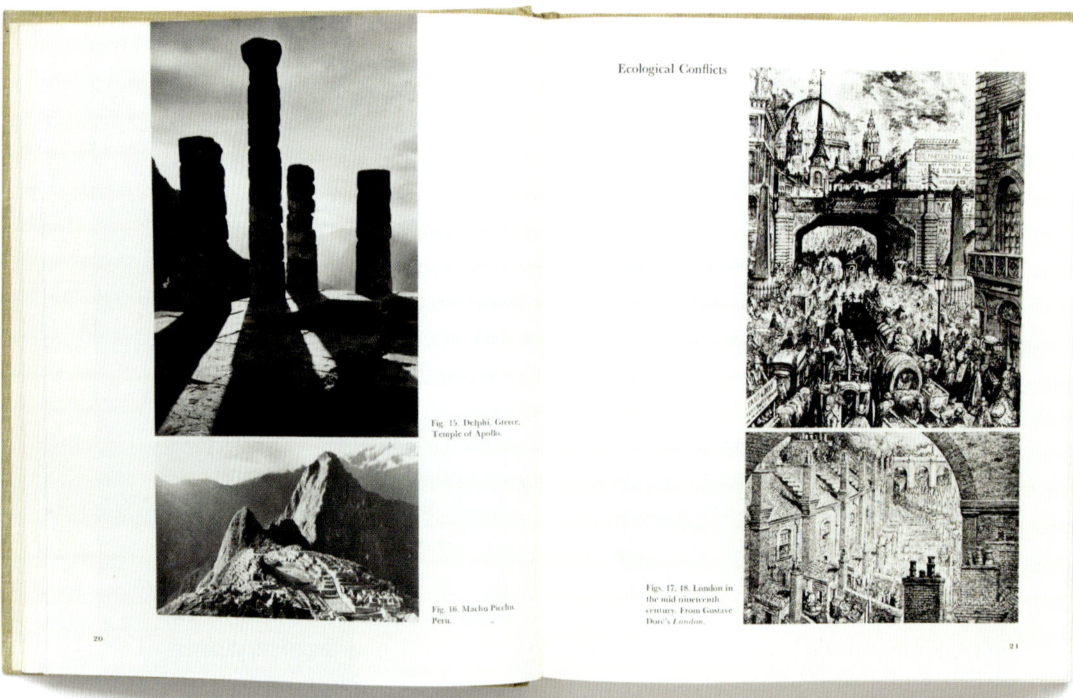

6.24

Two-page spread from *Arts of the Environment*, edited by Gyorgy Kepes. New York: George Braziller, 1972. Courtesy George Braziller.

I am fascinated by these unexpected images and imaginings, the conceptual leaps Kepes makes to a distant past but also a very real present, to meanings that range from rainbow herbicides to rainbows in stormy skies, as captured by romantic poetry and painting. This metaphorical and analogical concatenation of associations is a method he employs across all his projects. But the utopianism in Kepes's and the Center's proposals is also romantic in a bad way—terribly naïve, hopelessly escapist. Let me be cynical about these projects, and about their actual potential as environmental interventions, once again.

First of all, they were questioned even at MIT, among Institute administrators who were supposed to believe in the Center's mission. Part of the problem was the paradoxical nature of these proposals, based as they were on correcting the deleterious effects of science and technology by using science and technology. The illogic of such a proposition was perplexing. After the *Art and Environment* symposium organized by the Center, Kepes sent a letter to MIT Provost Walter Rosenblith—a neurophysiologist who had chaired a session at the event on "Urban Pageantry and Celebration"—apologizing for "the clumsiness of our words" and what he worried came across to Rosenblith as a "Luddite attitude" among the Center's artists. Kepes feared that Rosenblith would mistake the Center's nuanced position. He writes: "I can assure you that the very existence of the conference is a testimony to … my acceptance of the potentials of scientific technology as a major and hopeful challenge." The exchange indicates the delicate balance Kepes had to maintain: he hoped to counter environmental destruction caused by science and technology while also appealing to scientists and technologists. The result was his very awkward argument, his attempt to have it both ways: as he writes to Rosenblith, "we live in an epic age with epic tasks which demand the new scale tools of scientific technology for their resolution"—even though the epic tasks to be resolved by these new scale tools, namely ecological damage, were actually caused by those very same tools in the first place.[76] In an untitled text on the Charles River Project, Kepes describes the illogical logic of self-regulation as a search for "man's defense against himself," an attempt at "safeguarding ourselves from our own evils."[77]

These contorted justifications were not convincing, and they became increasingly strained. A series of hastily drafted memos documenting a meeting between Rosenblith and William Porter, Dean of MIT's School of Architecture and Planning, more than a year later records Rosenblith's continued confusion; during the discussion, Porter attempted to explain the "role of education" at the Center and the meaning of "civic scale." Rosenblith indicated remarks made by "WP," William Porter, that cast doubt on the Center's leadership: "K's [Kepes's] hopelessly naïve view of Society + University." The next page records a simple inscription in pencil: "Naiveté."[78]

Even the Center's own fellows became disenchanted. Even Jack Burnham, who was initially an eager participant in the life of the Center and had been a major proponent of interdisciplinary ventures in the visual arts—he famously organized the 1970 exhibition *Software* at the Solomon R. Guggenheim Museum, published on art and technology in books like *Beyond Modern Sculpture* (for which he shared a publisher with Kepes in George Braziller), and codified his approach to systems aesthetics in a canonical article of the same name—was now disillusioned.[79] In an early letter to Kepes, Burnham writes: "I am very sympathetic with your plans."[80] This feeling was once mutual. Kepes, too, privately lauded Burnham as "one of the significant spokesman for the new generation of artists."[81]

But the feeling did not last. Burnham remembers: "Slowly it began to dawn on me that the Center's underlying purpose was not primarily to do visual research or to make art, but to produce lavishly illustrated catalogues and anthologies that would impress foundations."[82] The Center produced many such anthologies, including a pair of glossy catalogs published in conjunction with exhibitions showcasing the Center at Boston's Institute of Contemporary Art in 1975 and 1976, both timed for the US Bicentennial. The second such catalogue was subtitled "Environmental Art." While these materials documented the Center's otherwise ephemeral—because unrealized—projects, they were also promotional: something the Center could use to "impress foundations," as Burnham puts it. Kepes was savvy about circulating them with the specific aim of impressing funding agencies ("I have marked the areas that have some relevance to our common concerns," he would write).[83] The implication of Burnham's critique is that the Center was merely creating aesthetic value for itself, as if the work the Center produced—its photographic proposals, its reports and plans—was just a resource to use for securing more money with which to produce still more proposals, still more reports and plans. The Center's "self-regulation" was entirely narcissistic, a model of self-legitimation through self-aggrandizement.

Seen through the lens of these critiques, the very basis of the Center's environmental art comes into focus as a cynical form of strategic and tactical positioning, an opportunistic way for the Center to be relevant, to make its ill-defined interdisciplinary activities seem timely. After all, the environment, as public issue, as problem to be solved, was suddenly mainstream. The Environmental Protection Agency was established in 1970. Nixon signed landmark environmental legislation like the Clean Water Act during his presidency. Ecological concerns were real—it is wrong to discount efforts at addressing them—but environmentalism was also a compelling framing device that made the Center's civic-scale projects seem worthy of support. It also positioned them in a manner consistent with MIT's concept of conversion.

A memorandum from Kraynik dated 1971 regarding ideas for the Charles River Project, sent in carbon copy to Kepes, offers specific ways for the fellows to frame their work in this way. "Do not emphasize the art and aesthetic aspects of the venture," he urges, suggesting that art was less likely to receive financing than science. "Emphasize the practical aspect," specifically how projects would "educate and promote an awareness of environmental + pollution problems." The Center's work must "be functional," it should only "<u>incidentally</u> do something to add beauty to the environment." His argument was simple: only projects with instrumental applications would obtain funding. "If we agree to this then we can apply <u>to the Environmental Education Office in an area</u> to which they give a number one priority" (Kraynik refers to an organization then run through the US Office of Education, a predecessor to the Department of Education). He attached a proposal for his *Collimated Light Buoy* to the memo, positioning it in this way. "The point is that if we come up with ideas which serve the function of displaying information we are in a better position to recruit support."[84]

Kepes similarly framed the contributions from fellows for his reports to funding institutions, urging that descriptions of work completed had to sound legitimate—even if the Center's projects were never legitimately completed. When compiling his report on the Charles River Project for the NEA, he castigated none other than Kraynik for an overly conceptual description: "I am afraid that I cannot accept it." Kepes reminded Kraynik that the NEA expected "a report on what we did and not what we thought about." He wanted "sketches" and visual material to make the report more convincing, more "tangible, visible, sensible." As he explains, "a good deal is expressed in beautiful words, but our existence is only important when these good intentions are transformed into living realities."[85] Or as he states in a second stern letter to Kraynik, sent when he still was not satisfied with Kraynik's report, the Center's activities must be "defendable."[86] The irony, of course, is that these reports and proposals were all little more than "beautiful words" and "good intentions," even if accompanied by a few sketches and Bichajian's photographic studies; Kepes's letter reads as if he is justifying the Center's efforts not only to Kraynik, or to the NEA, but also to himself.

I am struck by an image documenting all the Center's photographic plans displayed in the Center's own hallways for an exhibition titled *Art in Civic Scale* (fig. 6.25). There are panels depicting Kepes's designs for Boston Harbor, the caustic curves created from floating buoys, and an enlarged photograph of the *Wind-Sound-Light Sculpture*—plus projects from other fellows, like the *Cybernetic Sculpture System* designed by Wen-Ying Tsai. These images all serve as a solipsistic affirmation of the Center's mission; the "civic scale," with its very grand public ambitions, is reduced to a very private gesture, one meaningful only for those who created it. They are the only ones who would even see it.

For Burnham, then, the Center's technological environments amounted to little more than abstractions, speculations, mere fantasies:

> Certainly the Center never really had any concrete program, outside of fulfilling the director's vague dreams of creating urban spectaculars. During my first month and a half we met twice weekly to discuss Kepes' ambitions for erecting a colossal light tower in the middle of Boston Harbor. Somehow the conversations and exchange of ideas remained maddeningly vague. I began to ask specific questions:
> Did the Center have funds for such a project or any idea of costs? *No.*
> Given that the Boston Harbor was directly in the flight patterns of Logan Airport, had the Center checked on the feasibility of the project with the local Civil Aeronautics Board, or with the Boston Harbor Authority? *No.*
> Did they understand the problems of laying underwater electric conduit or the costs? *No.*
> What was the civic purpose of the light monument? *No one really knew.*[87]

Burnham's commentary is incisive, but also ironic. The Center was not always as impractical as he remembers. Friedrich St. Florian, for example, circulated a memorandum to the fellows regarding "Constraints on Boston Harbor Development" that included an attached technical report. As the memo clearly states, "The proximity of Logan International Airport remains however the single most significant limitation to all kinds of artistic propositions, particularly those dealing with employment of strong light sources."[88] Of all the proposals imagined for the Boston Harbor, Burnham's own was also the most uninspired. He suggested a work titled "Plug-In" which would use illuminated buoys and Sylvania-brand Panelescent lights, both powered by battery (no underwater electric conduit required), to illuminate the flight approach to Logan Airport. The Airport's runways would be visually extended into the Harbor through these floating elements. "Plug-In" was practical, I suppose, something useful and functional, maybe more likely to receive approval from the Civil Aeronautics Board (though there is no evidence that Burnham even checked, despite his attack on others for not checking), but it was also boring— nothing more than a glorified approach lighting system, like the kind used to land aircraft at modern airports everywhere.[89]

6.25

Installation view of *Art in Civic Scale,*
CAVS, MIT, 1971. Photo: Nishan Bichajian.
Courtesy Center for Advanced Visual Studies
Special Collection, MIT Program in Art,
Culture and Technology. © Massachusetts
Institute of Technology.

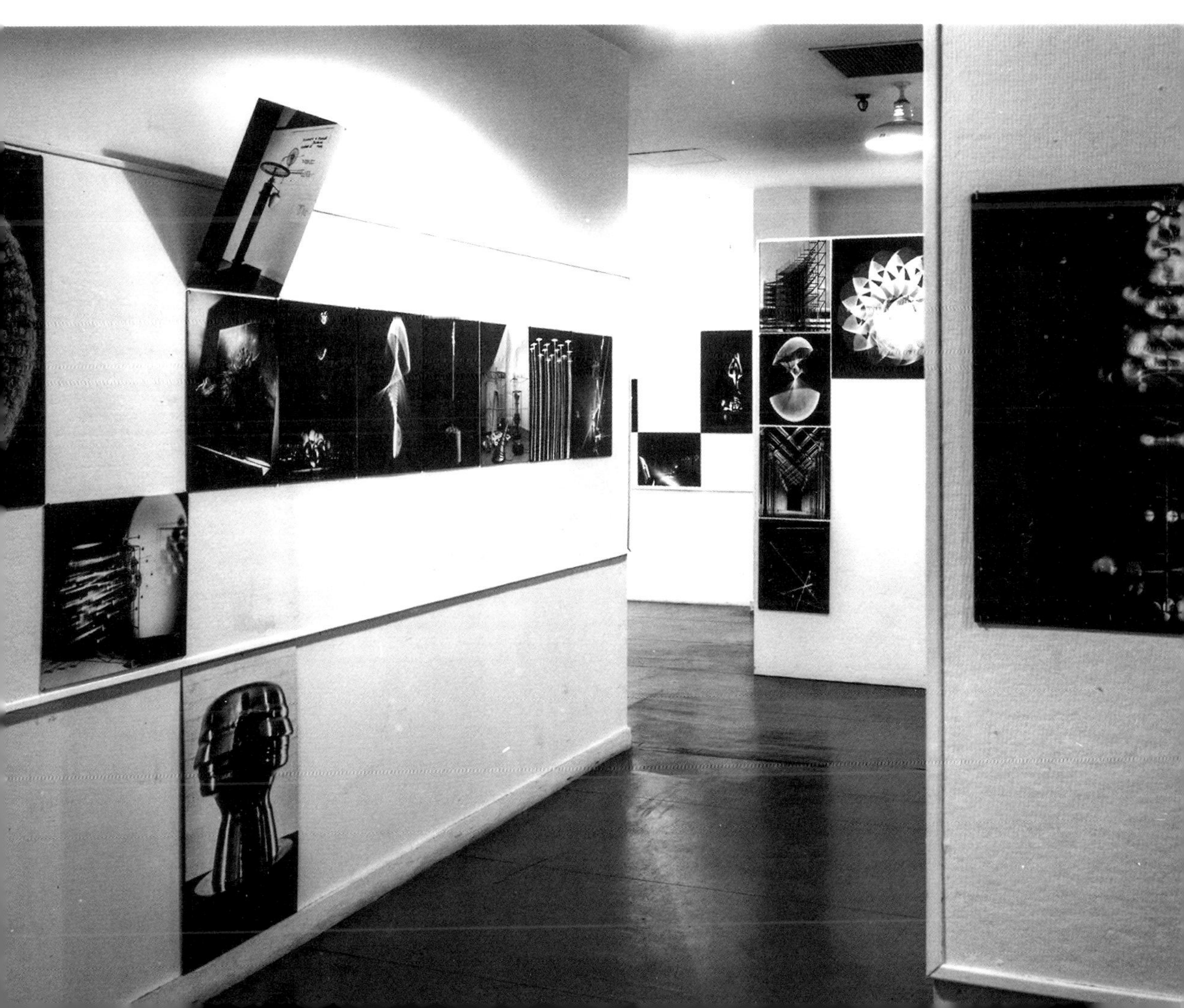

Nonetheless, Burnham's remarks help to clarify the Center's shortcomings. In a 2004 interview, Burnham is even more sarcastic, with a sardonic tone that suggests a remarkable change of heart from his earliest enthusiasm. He attacks Kepes's "naïve fetishism," his "Bauhaus romanticization of technology," his "worshipful eulogizing of scientific photographs," his "panegyrics," all of which seemed "vapidly visionary" and "sloppily sentimental and misleading." Kepes imagined the Center as an "aesthetic think-tank," filled with "tweedy academics in horned-rimmed glasses and white lab coats," but the result was really just "artists playing."[90] (Harold Tovish similarly remembers Kepes's attempts to engage MIT's scientists in the Center's activities—it "must have seemed like child's play" to them.)[91] For Burnham, the Center's grand environmental projects were little more than "computer images, luminous blinkies, and helium balloons."[92]

There was also a more compromising issue: attempts to simulate nature through science and technology risked only further destroying nature. After an exhibit of the Center's Long Wharf proposals, an artist named Roger Kizik wrote to CAVS with a long list of critiques. Some echoed frequent criticisms—Kizik calls the Center's work an "arrogant display of elitist egotism," for example, and attacks specific projects as modernist kitsch, each civic-scale light tower as nothing more than an "International Style tinker toy," something "appropriate for any number of faceless suburban shopping malls." But he also argued that the Center's environmental mandate was hypocritical, that the construction of projects purportedly preserving environments, even in an abstract sense, would actually damage them. In reference to a plan conceived by Lowry Burgess that included a large carved "landing stone" mimicking the effects of liquid water through undulating forms, Kizik asks: "Can an 'ocean' of stone, no matter how resolutely carved, offer more to delight and stir the senses than the sea itself?" The Center's projects would also insert wildly futuristic statements into sites most valued for their connections to the past. "Why should the last historically precious and vulnerable remnant of what Long Wharf was be asked to share company with such a crass, moronic blight"?[93] Such comments serve as a reminder that the Center's activities really were abstractions, speculations, mere fantasies, offering a utopian vision that often overlooked the specificities of their most local environments, the natural and historical ecologies they would actually engage.

• • •

Our evaluation of the Center's technological environments ultimately reflects how we interpret the romantic gesture, the utopianism. On a page in his research notes, Kepes copied a quotation from Konrad Lorenz, an ornithologist and pioneer in the field of ethology, the study of animal behavior: "A bird which has no material to build [a] nest may perform the movement of nest building in the air."[94] The remark resonates as a description of the Center's civic-scale art, an environmental art that would not be built in the actual environment. With no materials available, with inadequate funding and improbable ideas, Kepes and the fellows nonetheless performed the movement of building, creating light towers that would remain "in the air." Do unrealized, and unrealizable, projects represent failure—and is this failure caused by not creating these works, or by not understanding the ideological implications of designing them in the first place? Was the Center theorizing and practicing homeostasis or a self-deceptive, self-deluded pseudo-homeostasis?

In the end, at the final reckoning, I want to believe in the utopianism. I want to believe in the romantic gesture, even while recognizing its limitations. In a personal notebook from the 1970s, Kepes records a memory. He retrospectively recalls when he once appeared on a radio show devoted to discussions about design. Kepes spoke about color perception. Not long after the program aired, he received a letter dictated by a listener who happened to be blind. The letter read: "<u>please give us rainbow</u>."[95] To Kepes, the anecdote represented what he believed he was offering through his environmental aesthetics: he was making the environment visible to those who, perhaps literally but certainly figuratively, could not see it. A visible environment is one that would inspire ecological consciousness, an awareness of destruction and degradation, and would thus also prompt individual, social, and cultural transformation, a conversion, a self-regulation. In the same notebook, Kepes justifies this aesthetic approach: "nature's respect is not romantic/sentimental self indulgence, but a needed reestablishment of lost loyalties," "a return."[96]

Imagination is powerful. Appreciating the Center's technological environments requires relinquishing complete cynicism and the impulse of critique, and instead embracing the idealistic belief that a transformation might happen, that conversion is possible. Maybe the best works of art are the ones that exist only on this plane of the imagination.

Afterimages

epilogue

Crisis of Confidence

The central thesis in this study is that Gyorgy Kepes embodied a new model for aesthetic practice in a scientific context. He left behind the avant-garde, his faith in revolution, and instead took refuge in the Cold War avant-garde. He entered the control tower and assumed new functions, many of which departed dramatically from the familiar roles associated with the artist in the postwar period. He became court painter to MIT's power elite, a medieval mystic for the Cold War, a sagelike seer, an "oracle"; he offered a mystical and mythological engagement with vision as an antidote to the strict rationality and logic that permeated the Institute. This context and these roles provided unusual opportunities but also great perils; militarism was a totalizing force at MIT, omnipresent and all-powerful. Kepes could not escape it.

In this way, the artistic persona Kepes embodied was torn: he attempted to resist a technocratic culture while still becoming part of it. Kepes was not ambivalent; he was conflicted, torn between seemingly incompatible, irresolvable demands. The architectural historian Sigfried Giedion described this distinctly midcentury character as the "man in equipoise," a figure who "can control his own existence by the process of balancing forces often regarded as irreconcilable," thereby maintaining a precarious stability.[1] Others have claimed that Kepes personifies a reactionary glorification of science and technology precisely because of these ties to technocracy, and technocracy's ties to warfare. Instead, this study has explored Kepes's attempt to change this culture from his unique position within it, even if such a mission appeared to others—and still appears to many—as hopelessly naïve.

Kepes's history is past but also present. Science and technology dominate the academy of today. The rhetoric around interdisciplinary research is ubiquitous. The crisis of confidence that Kepes faced is one we now all face; declining enrollment and depressed investment in arts and humanities subjects—or the institutional support of merely spectacular forms of art, of the humanities as entertainment—indicate a contemporary predicament indebted to this earlier moment.

Kepes's history tells us that the relationship between the two cultures is highly charged. If this relationship was an antagonistic battle of opposites in Kepes's time, it is only more so today. To take one example: according to data compiled by the American Academy of Arts and Sciences, and published in a 2013 report authored by Harvard University, sustained academic interest in the arts and humanities has slid precipitously. Completed bachelor's degrees in humanities subjects dropped by half nationwide between 1966 and 2010, falling from 14 percent to 7 percent of all degrees awarded.[2] Data compiled by the Modern Language Association shows a similar drop in bachelor's degrees in humanities subjects as a percentage of all degrees awarded (fig. 7.1).

At Stanford University, where I began this study and where the proximity to Silicon Valley continues to structure an emphasis on science and technology, this downward trend felt acute. Stanford has made a substantial institutional investment in new arts programs over the last ten years, including new arts buildings and new campus collections; but even so, I often felt that these commitments were not straightforward in their aims and purposes. Institutional support for the arts is not equivalent to true engagement. The arts are made available not necessarily as intellectual calling but as useful spectacle, a diversion and distraction from the more serious activities taking place on campus. In an age when all knowledge must be a means toward an end, when education is vocational, it is hard not to see the arts on offer as empty; they provide ornamentation and adornment. They make our campuses appear well-rounded and balanced. These pressures and these functions motivated my interest in Kepes's role, decades ago, at MIT.

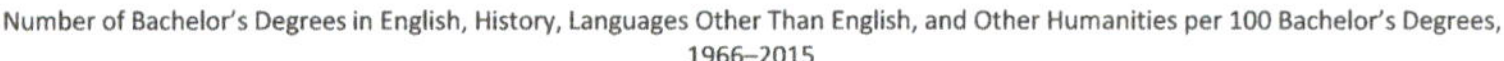
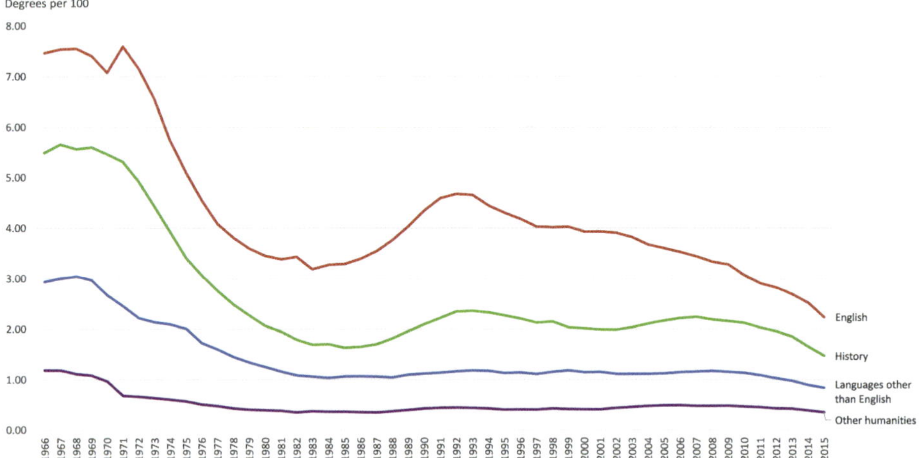

7.1

Number of Bachelor's Degrees in English, History, Languages Other Than English, and Other Humanities per 100 Bachelor's Degrees, 1966–2015. Data source: National Science Foundation. Courtesy of the Modern Language Association, https://mlaresearch. mla.hcommons.org/2017/06/26/the-decline-in-humanities-majors/.

The notion that the arts and humanities are in crisis while science and technology have become all-powerful continues to inspire widespread academic anxiety. Issues of the journals *Daedalus*, *Representations*, and *Critical Inquiry* have addressed the topic.[3] Many critics place blame on the corporatized model that has become the default in higher education. Even defenses of the arts and humanities are now subjected to instrumental reason; a humanities education is not valuable on its own merits—not because we are human—but because it teaches marketable skills that students can leverage. These debates are not just academic. They appear not only in headlines for the *Chronicle of Higher Education*, but in the mainstream media. A *New York Times* article describes efforts to slash liberal arts funding at public universities: "Several Republicans have portrayed a liberal arts education as an expendable, sometimes frivolous luxury that taxpayers should not be expected to pay for."[4] Such comments echo almost verbatim the remarks quoted by John E. Burchard in his unpublished preface to *The New Landscape in Art and Science*, from congressional hearings in the 1950s concerning the apparent irrelevance of the arts and humanities.

More central for this project, MIT also recently authored its own report on the place of the arts in an academic institution devoted to science and technology. The Institute's 2011 "Arts and MIT" white paper embraced Kepes as a model not only of the Institute's past but also of its future:

> MIT pioneered collaborative relationships that are now considered normal.... This wisdom of encouraging this kind of flow among disciplines is increasingly obvious in our current intellectual and culturally diverse environment. Today the Institute is building on its strengths in collaborative research to tackle problems greater than any single discipline can resolve and find solutions unimaginable without transdisciplinary morphing. Artists, like scientists, accept paradox, error, and uncertainty as necessary elements in the process of discovery. Scientists and engineers conceptualize through visual metaphors and evaluate their work in aesthetic terms. Increasingly, they recognize the common ground that underlies their endeavors and the artistic enterprise.[5]

A *New York Times* article similarly describes Kepes's legacy institution at MIT, the Center for Art, Science & Technology (CAST)—an academic unit directly linked in name and function to Kepes's earlier Center for Advanced Visual Studies (CAVS). "CAST, as it is known, has revitalized an M.I.T. model begun in the late 1960s of bringing in artists to humanize technology and create more expansive-thinking scientists. M.I.T. is at the forefront of this cross-disciplinary movement."[6]

Both MIT's white paper and press coverage of the Institute's arts initiatives show how easy it is to embrace Kepes as hero, as a solution to the crisis, without fully interrogating the roles he actually played—what interdisciplinarity meant not just in theory but also in practice. The "Arts and MIT" report relies on the same seductive rhetoric that Kepes first formulated in reports for the Institute decades ago. It argues for the intrinsic value of collaborative pursuits—but it overlooks the backlash they generated, the protests they prompted, and the ethical and moral dilemmas they created. It ignores how polemical the relationship between art and science actually was—and how entangled with military power these relationships became. Kepes could never escape the accusation that his work was used by the Institute for its own ends, as a strategic and tactical resource; perhaps his history now serves the same function: as an archive to mine, another source of legitimation.[7]

Does the survival of the arts require engagement with science and technology? Or should the arts resist such pressures? Is it important to embrace disciplinary rivals—even if doing so requires compromising on core principles—or better to maintain autonomy? My goal has been to raise these questions, to make clear the importance of thinking interdisciplinarity historically. Doing so allows us to understand the uses and abuses of the arts, the complexity of commonplace notions around exchange, and the agendas that may also be at play in promoting such dialogues.

Undreaming

Quoting the words of Otto Piene, Kepes once described his ambition at MIT as "undreaming" the Bauhaus. He writes:

> To utilize the vision of the last generation, the present generation has to learn to see purposes; the past is over. Its dreams are real only insofar as they are remembered on the awakened immense scale of the present. A thoughtful and sensitive artist, Otto Piene, once wrote me a letter that touched the heart of the matter. "The difference between Moholy's time and our time to me," Piene said, "is the difference between dreaming dreams and undreaming them, i.e., letting them come true. I think the courage to face reality is there – and the awareness of the difficulties as well."[8]

Dreaming is an escape. It is separate. We dream when action is impossible. In undreaming the Bauhaus, Kepes attempted to realize what was once only imagined. He hoped to make manifest Bauhaus ideas that had originated decades earlier. In undreaming utopia into existence—or at least attempting to do so— Kepes encountered a resistance more intense than he anticipated. Undreaming often became a nightmare.

Kepes experienced this backlash personally, in the vitriolic responses he received from such forceful critics as Sibyl Moholy-Nagy, Jack Burnham, and the Council for Conscious Existence. He reflects on the repeated threat to his philosophy of seeing: "[M]any things that I was completely confident about in 1939 … were fading a little," he notes. He once was certain that his projects would create change "in real terms and in the real full sense of those terms." But after retiring from MIT, he had doubts: "I am fading a little in my confidence, but still haven't given up my original convictions.... I did not give up my beliefs, but I had to edit in a certain minor key."[9]

One last quotation from Kepes's research papers: in a note on "visual fundamentals," Kepes refers to the writings of French art historian Émile Mâle, specifically quoting from Mâle's 1899 study of Gothic art, later translated as *Religious Art in France*; Mâle explains that the illiterate masses in the medieval world "learned through their eyes almost all they know of their faith."[10] What is most compelling about Kepes is this faith, a sometimes blind faith, in vision. Despite the repeated doubts, the backlash, the protests, the ethical and moral dilemmas, Kepes maintained faith in the power of the eye.

Notes

Acknowledgments

1   This letter is located in the Gyorgy Kepes papers (M1796), Dept. of Special Collections and University Archives, Stanford Libraries, Stanford, Calif. [hereafter Kepes papers, Stanford] box 37, folder 1.

Preface

1   Previous descriptions of this photograph suggest that a camera is aimed upward at Kepes; given the framing of the image, a mirror placed on the floor below is more likely. The Hungarian photographer Eva Besnyö, a friend of Kepes's in Berlin, produced a similar portrait above a mirror that reveals the camera in her hand, affirming this interpretation. See Anne H. Hoy, "Chronology," in *Gyorgy Kepes: Light Graphics* (New York: International Center of Photography, 1984), no pagination; and Elizabeth Finch, "Languages of Vision: Gyorgy Kepes and the 'New Landscape' of Art and Science" (PhD dissertation, City University of New York, 2005), 104–108.

2   See Leah Dickerman, "El Lissitzky's Camera Corpus," in Nancy Perloff and Brian Reed, eds., *Situating El Lissitzky: Vitebsk, Berlin, Moscow* (Los Angeles: Getty Research Institute, 2003): 153–176.

Introduction

1   From the Kepes papers, Stanford. I encountered this item before manuscript materials were assigned box and folder numbers.

2   In *The Artist as Producer: Russian Constructivism in Revolution* (Berkeley: University of California Press, 2005), Maria Gough asks (8): "What is the role and efficacy of the vanguard artist in revolution?" I ask a similar but different question: What is the role and efficacy of the vanguard artist *after* revolution?

3   I pull biographical details from Judith Wechsler, "Gyorgy Kepes," in *Gyorgy Kepes: The MIT Years: 1945–1977* (Cambridge, MA: MIT Press, 1978); and the chronologies in *Gyorgy Kepes: Works in Review* (Boston: Museum of Science, 1973) and *Gyorgy Kepes: Light Graphics* (New York: International Center of Photography, 1984).

4   Kepes in Rudolph de Harak, "Gyorgy Kepes Revisits the Visual Landscape," in Steven Heller and Marie Finamore, eds., *Design Culture: An Anthology of Writing from the AIGA Journal of Graphic Design* (New York: Allworth Press, 1997), 143; originally published in the *AIGA Journal of Graphic Design* 5, no. 2 (1987).

5   Kassák's circle first formed after the First World War around an earlier journal called *Ma* or "Today."

6   For the history of Kassák and the *Munka* and *Ma* circles, see Lee Congdon, *Exile and Social Thought: Hungarian Intellectuals in Germany and Austria, 1919–1933* (Princeton: Princeton University Press, 1991), 139–176.

7   For a brief description of Kepes's Berlin milieu, see Richard Whelan, *Robert Capa: A Biography* (Lincoln: University of Nebraska Press, 1985). Kepes introduced Capa to photography and gave him his first camera.

8   Gyorgy Kepes, "The Artist's Response to the Scientific World," in *Technology and Culture in Perspective: An Occasional Paper Published by the Church Society for College Work* (Cambridge, MA: Church Society for College Work, 1966), 36.

9   Kepes in Rudolph de Harak, "Gyorgy Kepes Revisits the Visual Landscape," 144.

10   International Red Aid was founded in Russia as Международная организация помощи революционерам, meaning "International Organization for Aid to Revolutionaries." In Germany, it was subsequently transliterated as Mezhdunarodnaia Organizatsiia Pomoshchi Revoliutsioneram or MOPR.

11   Oral history interview with Gyorgy Kepes, 18 August 1968, Archives of American Art, Smithsonian Institution, transcript available at http://www.aaa.si.edu/collections/interviews/oral-history-interview-gyorgy-kepes-11952.

12   Kepes actively eliminated these commissions from his *oeuvre*. They are rarely mentioned in interviews and catalogs, and their covers are infrequently reproduced. Kepes also appears to have edited them out of the documentation of his career. His papers at Stanford include a collection of transparencies and negatives, including covers for *Das Neue Russland*. But one cover, for which Kepes used an image of Joseph Stalin, is missing, presumably because Kepes destroyed it. He likely did not want to be remembered for such a work. See the Kepes papers, Stanford, box 48, folder 8.

13   This drawing is now lost but is reproduced in *Gyorgy Kepes: Light Graphics*, where it is titled *Against War* and dated 1936; it is also reversed, presumably because it was reproduced from a negative. In *The New Masses* (19 July 1938), it is titled *Bread* and dated 1938.

14   Oral history interview with Gyorgy Kepes, 7 March 1972–11 January 1973, Archives of American Art, Smithsonian Institution, page 11 of transcript.

15   "Ke" is listed on the issue's masthead as well as below the drawing, suggesting it was an intentional attribution, not a typographic error. Kepes used this pseudonym only in this single instance. He was presumably motivated to disguise his identity because of his status as an immigrant and the journal's affiliation.

16   Juliet Kepes Stone, interview with the author, November 2013. Kepes never identified as Jewish after his arrival in the United States, instead joining a Unitarian Universalist church in Cambridge, Massachusetts. The Unitarian movement, however, has roots not only in New England but also in Hungary and the region of what is now Romania known as Transylvania. It was also popular among MIT's faculty, which included many immigrants.

17 Oral history interview with Gyorgy Kepes, 7 March 1972–11 January 1973, Archives of American Art, Smithsonian Inhstitution, page 11 of transcript.

18 Kepes, "The Artist's Response to the Scientific World," 36.

19 C. P. Snow terms the "two cultures," referring to the sciences and the humanities, specifically literary culture, in his well-known Rede Lecture of 1959. See C. P. Snow, *The Two Cultures and the Scientific Revolution* (Cambridge: Cambridge University Press, 1959). Kepes's interest in relating art and science predates Snow's commentary.

Snow's argument is not neutral; he advocates a reconciliation of the "two cultures" at a moment when science was already ascendant over the arts and humanities, and would soon dominate in terms of both institutional funding and intellectual prestige. His argument justified privileging the sciences to the detriment of the arts and humanities or, even worse, making the arts and humanities more like the sciences, all in the name of establishing supposed parity between fields.

20 From the book jackets to Kepes's *Vision + Value* volumes (New York: George Braziller, 1965–1966, 1972). Such language repeats in Kepes's writings of the 1960s and 1970s.

21 Kepes uses these terms widely in the 1960s. See, for example, Gyorgy Kepes, "The Visual Arts and Sciences: A Proposal for Collaboration," *Daedalus* 94, no. 1 (Winter 1965): 122–123; and Gyorgy Kepes, ed., *The Nature and Art of Motion* (New York: George Braziller, 1965), xi.

22 Kepes uses "visual culture" in his "Introduction to the Issue 'The Visual Arts Today,'" *Daedalus* 89, no. 1 (Winter 1960): 3 ("In this issue of *Daedalus* various disciplines are brought to bear on our visual culture"). This issue of *Daedalus* was later published in book form as Gyorgy Kepes, ed., *The Visual Arts Today* (Middletown, CT: Wesleyan University Press, 1960). This is one of the earliest examples of the term in its current usage. Kepes uses "new media" in "Kinetic Light as a Creative Medium," *Technology Review* 70, no. 2 (December 1967): 29. This is not to suggest that visual culture and new media are programmatically derived from Kepes, however.

23 In *PM: An Intimate Journal for Art Directors, Production Managers, and Their Associates* 6, no. 3 (February–March 1940), no pagination. This source should not be confused with the New York newspaper also titled *PM*.

24 As my study focuses on Kepes's career in the United States, I spell his name without diacritics (Gyorgy Kepes), in the anglicized form Kepes used from the 1940s through his professional life at MIT. After retirement, he often returned to the original Magyar spelling, especially as he strengthened ties with his country of origin.

25 Jacques Ellul, *The Technological Society*, trans. John Wilkinson (New York: Alfred A. Knopf, 1964); originally published in French in 1954.

26   Alain Touraine, *The Post-Industrial Society: Tomorrow's Social History: Classes, Conflicts and Culture in the Programmed Society*, trans. Leonard F. X. Mayhew (New York: Random House, 1971); originally published in French in 1969.

27   Daniel Bell, *The Coming of Post-Industrial Society: A Venture in Social Forecasting* (New York: Basic Books, 1973).

28   See Manuel Castells, *The Rise of the Network Society*, 2nd ed., vol. 1 of *The Information Age: Economy, Society and Culture* (London: Blackwell, 2000).

29   For a study of the computer as a design object, see John Harwood, *The Interface: IBM and the Transformation of Corporate Design 1945–1976* (Minneapolis: University of Minnesota Press, 2011).

30   The other archetypal Cold War university was Stanford. Rebecca S. Lowen considers Stanford and the notion of a "Cold War university" in *Creating the Cold War University: The Transformation of Stanford* (Berkeley: University of California Press, 1997). David Kaiser presents a parallel history of MIT in "Elephant on the Charles: Postwar Growing Pains," in David Kaiser, ed., *Becoming MIT: Moments of Decision* (Cambridge, MA: MIT Press, 2010), 103–122. More recently, Matthew Levin uses the term in reference to the University of Wisconsin, Madison, in *Cold War University: Madison and the New Left in the Sixties* (Madison: University of Wisconsin Press, 2013).

31   Heinrich Wölfflin, *Principles of Art History: The Problem of the Development of Style in Later Art*, trans. M. D. Hottinger (London: G. Bell and Sons, 1932), 11; originally published in German in 1915. Kepes saved a typed copy of this quotation in his papers, under which he wrote in hand: "Not everything is needed at all times—we may add to it. Not borrowing system of symbols from past, but if the artist could find the visual possibilities of his times." Kepes papers, Stanford, box 82, folder 5.

32   Marshall McLuhan, *Understanding Media: The Extensions of Man* (Cambridge, MA: MIT Press, 1964), 65. McLuhan later mentions Kepes by name (129), suggesting that it may be Kepes whom he had in mind for his conception of the artist "in the control tower."

33   From one of Kepes's notebooks in the Kepes papers, Stanford, box 20.

34   "Abstract, but Romantic," *Time* (7 March 1960).

35   See Michel Foucault, *Discipline and Punish: The Birth of the Prison*, trans. Alan Sheridan (New York: Pantheon Books, 1977); originally published in French in 1975.

36   Gyorgy Kepes, *Language of Vision* (Chicago: Paul Theobald, 1944), 12.

37   Ghamari-Tabrizi introduces the term in *The Worlds of Herman Kahn: The Intuitive Science of Thermonuclear War* (Cambridge, MA: Harvard University Press, 2005). Pamela M. Lee cites Ghamari-Tabrizi's terminology in "Aesthetic Strategist: Albert Wohlstetter, the Cold War, and a Theory of Mid-Century Modernism," *October*

138 (Fall 2011): 15–36. See Lee's forthcoming *Think Tank Aesthetics: Midcentury Modernism, the Cold War and the Neoliberal Present*.

38   Donna Haraway explores these analogies in her *Crystals, Fabrics, and Fields: Metaphors of Organicism in Twentieth-Century Developmental Biology* (New Haven: Yale University Press, 1976). Haraway considers the Theoretical Biology Club; Kepes had loose associations with some participants in the group. Kepes also had loose associations with the "Unity of Science" movement, led by Rudolf Carnap and Otto Neurath, which attempted to integrate knowledge across disciplines through the philosophy of the Vienna Circle of logical positivists. It migrated to the United States through semiotician Charles W. Morris, who occasionally taught at the School of Design alongside other University of Chicago faculty aligned with the movement: Carl Eckart, a physicist; S. I. Hayakawa, an advocate of "General Semantics"; and Ralph W. Gerard, a physiologist and behavioral scientist who was an early proponent of cybernetics. The movement later surfaced at Harvard. On the connections between the Unity of Science movement and the Bauhaus and New Bauhaus, see Peter Galison, "Aufbau/Bauhaus: Logical Positivism and Architectural Modernism," *Critical Inquiry* 16, no. 4 (Summer 1990): 709–752; and Peter Galison, "The Americanization of Unity," *Daedalus* 127, no. 1 (Winter 1998): 45–71. For the reception of the cybernetics group, see Steve J. Heims, *Constructing a Social Science for Postwar America: The Cybernetics Group, 1946–1953* (Cambridge, MA: MIT Press, 1991); and Fred Turner, *From Counterculture to Cyberculture: Stewart Brand, the Whole Earth Network, and the Rise of Digital Utopianism* (Chicago: University of Chicago Press, 2006), among others.

39   Kepes uses the phrase "communication crisis" on the dustjackets to his seven *Vision + Value* volumes.

40   Kepes's engagement departs from existing accounts of the sciences in modern and contemporary art history. For example, in her *Fourth Dimension and Non-Euclidean Geometry in Modern Art* (Princeton: Princeton University Press, 1983), Linda Dalrymple Henderson explores the social history of science as a force of influence on the avant garde. In her *Machine in the Studio: Constructing the Postwar American Artist* (Chicago: University of Chicago Press, 1996), Caroline A. Jones considers the impact of science on the identity and self-presentation of the artist in the postwar period. Both studies focus largely on the cultural representation of science rather than collaborations with actual scientists.

41   All three participated in Kepes's *Vision + Value* series. Smith contributed to Gyorgy Kepes, ed., *Structure in Art and in Science* (New York: George Braziller, 1965). He also founded the "Philomorph Group" with Cambridge-area scientists in order to study patterns in nature, and he published widely on the arts; see Smith, *A Search for Structure: Selected Essays on Science, Art, and History* (Cambridge, MA: MIT Press, 1981). Bronowski contributed to Kepes's *Structure in Art and in Science* and published on the role of the arts and humanities in a scientific culture; see his *Science and Human Values* (New York: Julian Messner, 1956), first presented at MIT as a series of lectures in 1953. Forrester contributed to Gyorgy Kepes, *Arts of the Environment* (New York: George Braziller, 1972). Rossi contributed to Gyorgy Kepes, *The New Landscape in Art and Science* (Chicago: Paul Theobald,

1956). Smith, Forrester, and Rossi were on the faculty at MIT; Bronowski was a visiting professor at the Institute in 1953.

42 Bruno Rossi, *High-Energy Particles* (New York: Prentice-Hall, 1952), vii.

43 It is notable that Kepes never collaborated with Teller, despite their shared Hungarian roots. Kepes's daughter explains that her father "hated Teller" because of Teller's promotion of the hydrogen bomb and testimony against J. Robert Oppenheimer during the hearings of the House Un-American Activities Committee. Juliet Kepes Stone, interview with the author, November 2013.

44 Kepes became close to Wiesner through Wiesner's role as president of the Institute. Philip Morrison, an MIT faculty member, contributed to Gyorgy Kepes, ed., *Module, Proportion, Symmetry, Rhythm* (New York: George Braziller, 1966). He was interested in design; he wrote and narrated the script for the 1977 version of Charles and Ray Eames's film *Powers of Ten*.

45 For a classic history of these issues, see Stuart W. Leslie, *The Cold War and American Science: The Military-Industrial-Academic Complex at MIT and Stanford* (New York: Columbia University Press, 1993). See also Kelly Moore, *Disrupting Science: Social Movements, American Scientists, and the Politics of the Military, 1945–1975* (Princeton: Princeton University Press, 2008); Sarah Bridger, *Scientists at War: The Ethics of Cold War Weapons Research* (Cambridge, MA: Harvard University Press, 2015); and Matthew Wisnioski, *Engineers for Change: Competing Visions of Technology in 1960s America* (Cambridge, MA: MIT Press, 2012).

46 Kepes's collaborators index the full range of scientific inquiry. Other contributors to Kepes's books included engineer Lancelot Law Whyte, geneticist C. H. Waddington, inventor William J. J. Gordon, psychologist James J. Gibson, psychologist Abraham Maslow, engineer Dennis Gabor, and cyberneticist Norbert Wiener. These scientists had various relationships to military research: Whyte provided financing for the development of the jet engine, Waddington conducted systems analysis for the Royal Air Force, Gordon developed a problem-solving strategy called "synectics" for use by military and civilian research-and-development teams, Gibson derived psychological theories of vision from studies of sight in combat aircraft, Maslow conducted research on the authoritarian character structure, Gabor proposed models of technological forecasting for systems think tanks, and Wiener developed cybernetics based on his study of antiaircraft guns.

47 This communicative model of politics resonates with Jürgen Habermas's model of the public sphere, as put forth in *The Structural Transformation of the Public Sphere: An Inquiry into a Category of Bourgeois Society*, trans. Thomas Burger (Cambridge, MA: MIT Press, 1989); originally published in German in 1962.

48 Jeffrey Herf coins the term "reactionary modernism" in *Reactionary Modernism: Technology, Culture, and Politics in Weimar and the Third Reich* (Cambridge: Cambridge University Press, 1986). Herf argues that modernism, specifically Enlightenment rationality, is not itself reactionary; only the uses to which it is put are

reactionary. Herf's study of a different historical moment resonates with the context of the early Cold War.

49   The concept of "paranoid reading" is elaborated in Eve Kosofsky Sedgwick's "Paranoid Reading and Reparative Reading, or You're So Paranoid, You Probably Think This Essay Is about You," in her *Touching Feeling: Affect, Performativity, Pedagogy* (Durham: Duke University Press, 2003). For another art-historical study that rejects paranoid reading, see Darby English, *1971: A Year in the Life of Color* (Chicago: University of Chicago Press, 2016).

50   Paul N. Edwards, *The Closed World: The Computer and the Politics of Discourse in Cold War America* (Cambridge, MA: MIT Press, 1996).

51   See Fred Turner, *The Democratic Surround: Multimedia and American Liberalism from World War II to the Psychedelic Sixties* (Chicago: University of Chicago Press, 2013); and Jamie Cohen-Cole, *The Open Mind: Cold War Politics and the Sciences of Human Nature* (Chicago: University of Chicago Press, 2014). These studies create new possibilities for histories of the Cold War emphasizing openness.

52   Theodore Roszak, *The Making of a Counter Culture: Reflections on a Technocratic Society and Its Youthful Opposition* (Garden City, NY: Doubleday, 1968), 9.

53   Bell, *The Coming of Post-Industrial Society*, 354n14.

54   Ibid., 381; 356. Bell even ties nuclear physics to advances in painting (381n14), specifically the tendency for artists to form schools like scientists: "In this respect, science is like many an intellectual or artistic community where painters or writers seek each other out and, when they are united by a common interest, will reinforce each other's work. One can take an analogous movement such as abstract painting in the 1950s when such artists as Hoffman, Pollock, de Kooning, Still, and Motherwell extended the qualities of 'painterliness'—i.e., the effects of the *texture* of the paint as a dimension of the painting itself   to the formal limits of painting, and thus, in a sense, exhausted a paradigm." Emphasis in original. Bell is drawing from Clement Greenberg. See note 61 in this chapter. For other readings of Cold War aesthetics, see Joshua Shannon's discussion of photography in relation to the technocratic culture of the Cold War in *The Recording Machine: Art and Fact during the Cold War* (New Haven: Yale University Press, 2017); and John Jay Curly's discussion of a broader "Cold War visuality" in *A Conspiracy of Images: Andy Warhol, Gerhard Richter, and the Art of the Cold War* (New Haven: Yale University Press, 2013).

55   Lancelot Law Whyte, "The Unity of Visual Experience," *Bulletin of Atomic Scientists* 15, no. 2 (February 1959): 72. Whyte cites Kepes in reference to Kepes's *The New Landscape in Art and Science*. Whyte later participated in Kepes's edited volume *Structure in Art and in Science*. The scientist Cyril Stanley Smith, another close Kepes contact, edited the issue with Martyl Langsdorf, the artist who created the journal's Doomsday Clock.

56   This caption is printed in Kepes, *The Nature and Art of Motion*, 7.

57 Ulam invented, together with John von Neumann, the Monte Carlo method of systems analysis, which was employed in war simulations. Kepes's caption also acknowledges P. R. Stein, a Los Alamos lab technician, with whom Ulam published his mathematical findings in a technical article: Stanislaw Ulam, with P. R. Stein, "Non-linear Transformation Studies on Electronic Computers," *Rozprawy Matematyczne* 39 (1963): 1–66. This article followed a classified technical report published by Los Alamos (LADC-5688 1962) and later declassified and published in Stanislaw M. Ulam, *Analogies between Analogies: The Mathematical Reports of S. M. Ulam and His Los Alamos Collaborators*, ed. A. R. Bednarek and Françoise Ulam (Berkeley: University of California Press, 1990). The image is actually a photograph produced with a Polaroid camera mounted to the computer's oscilloscope screen.

58 I borrow the phrase from Eva Cockcroft, "Abstract Expressionism: Weapon of the Cold War," *Artforum* 12, no. 10 (June 1974): 39–41.

59 Ibid.

60 I borrow the phrase from Harold Rosenberg, "The American Action Painters," *Art News* 51, no. 8 (December 1952): 22–23, 48.

61 See Clement Greenberg, "Modernist Painting," in *Forum Lectures* (Washington, DC: Voice of America, 1960). In fact, Greenberg is here asserting the autonomy of artistic media as similar to the autonomy of scientific disciplines, a belief Kepes, invested as he was in interdisciplinarity, did not share. See also note 54 in this chapter.

62 I borrow the phrase from Clement Greenberg, "Avant-Garde and Kitsch," *Partisan Review* 6, no. 5 (Fall 1939): 34–49.

63 In chronological order, the key biographical and contextual studies of Kepes include Marga Bijvoet, *Art as Inquiry: Toward New Collaborations Between Art, Science, and Technology* (New York: Peter Lang, 1997); Anne Collins Goodyear, "The Relationship of Art to Science and Technology in the United States, 1957–1971: Five Case Studies" (PhD dissertation, University of Texas at Austin, 2002); Elizabeth Finch, "Languages of Vision: Gyorgy Kepes and the 'New Landscape' of Art and Science" (PhD dissertation, City University of New York, 2005); Bill Arning, "Gyorgy Kepes' *Vision + Value* 1965–1972: A Practical Training Program to Create a New Super-Perceiving Problem-Solving Artist" (PhD dissertation, Tufts University, 2008); Elizabeth Anne Roach, "A Positive, Popular Art: Sources, Structure, and Impact of Gyorgy Kepes's *Language of Vision*" (PhD dissertation, Florida State University, 2010); and, more recently, Márton Orosz's Hungarian study, "A látás revíziója. Művészet mint humaniesta tudomány Kepes György korai életművében" (PhD dissertation, Eötvös Loránd University, 2013). The crucial references from Kepes's lifetime include the essay by Judith Wechsler in the exhibition catalog *Gyorgy Kepes: The MIT Years: 1945–1977*; and the chronologies in the exhibition catalogs *Gyorgy Kepes: Works in Review* and *Gyorgy Kepes: Light Graphics*. For one of many studies of the wider design culture of the period, see Eva Díaz, *The Experimenters: Chance and Design at Black Mountain College* (Chicago: University of Chicago Press, 2015).

64   Reinhold Martin, *The Organizational Complex: Architecture, Media, and Corporate Space* (Cambridge, MA: MIT Press, 2003).

65   Ibid., 67. Emphasis in original. Martin suggests that Kepes's aims are "always already discolored" by their recourse to science and technology (70). He also presents Kepes as foil to a more progressive figure, namely Robert Smithson. See Reinhold Martin, "Organicism's Other," *Grey Room* 4 (Summer 2001): 46. A number of scholars have followed this construction. Anne Collins Goodyear sees Kepes as the bureaucratic alternative to Billy Klüver and his Experiments in Art and Technology (E.A.T.), and Yates McKee sees Kepes as the bureaucratic alternative to the environmental group Pulsa. See Anne Collins Goodyear, "Gyorgy Kepes, Billy Klüver, and American Art of the 1960s: Defining Attitudes toward Science and Technology," *Science in Context* 17, no. 4 (2004): 611–635; and Yates McKee, "The Public Sensoriums of Pulsa: Cybernetic Abstraction and the Biopolitics of Urban Survival," *Art Journal* 67, no. 3 (Fall 2008): 47–67.

66   Anna Vallye, "The Strategic Universality of *trans/formation*, 1950–1952," *Grey Room* 35 (Spring 2009): 41. See also Vallye's "Design and the Politics of Knowledge in America, 1937–1967: Walter Gropius, Gyorgy Kepes" (PhD dissertation, Columbia University, 2011).

67   Arindam Dutta, ed., *A Second Modernism: MIT, Architecture, and the "Techno-Social" Moment* (Cambridge, MA: MIT Press, 2013); see especially Dutta's "Linguistics, Not Grammatology: Architecture's *A Prioris* and Architecture's Priorities," 2–69.

68   Orit Halpern provides a positive gloss on Kepes's engagements with science and technology, emphasizing how Kepes anticipates contemporary digital culture. See her *Beautiful Data: A History of Vision and Reason since 1945* (Durham: Duke University Press, 2014).

69   Juliet Kepes Stone, interview with the author, November 2013.

70   Katherine Kuh, "Gyorgy Kepes," *Arts and Architecture* 69, no. 6 (June 1952): 19. Kepes never learned to drive. Juliet Kepes Stone, interview with the author, November 2013.

71   Otto Piene, untitled essay in *Gyorgy Kepes: Works in Review*, 20.

72   Gyorgy Kepes, undated note titled "light," Kepes papers, Stanford, box 16, folder 1. The top of the notation reads: "Was asked to speak on photography, Will not speak about on [*sic*] photography."

73   My approach is inspired in part by models from the sciences, like Bruno Latour's "actor-network theory." See his *Reassembling the Social: An Introduction to Actor-Network Theory* (Oxford: Oxford University Press, 2005).

74   Gyorgy Kepes, undated and untitled note, Kepes papers, Stanford, box 82, folder 6. For a canonical acount of academics in exile, see Lewis A. Coser, *Refugee Scholars in America: Their Impact and Their Experiences* (New Haven: Yale University Press, 1984).

75   From one of Kepes's notebooks in the Kepes papers, Stanford, box 20.

76 Richard Kostelanetz, *Moholy-Nagy: An Anthology* (New York: Da Capo Press, 1970), 214.

77 A photocopy of this passage is included in the Kepes papers, Stanford, box 28, folder 25.

Chapter 1

1 Max Horkheimer and Theodor W. Adorno, *Dialectic of Enlightenment: Philosophical Fragments*, ed. Gunzelin Schmid Noerr, trans. Edmund Jephcott (Stanford: Stanford University Press, 2002), 148; originally published in 1947.

2 Kepes's design charts refer to Walter Gropius and László Moholy-Nagy's series of *Bauhausbücher* or Bauhaus books. The top image is Kepes's version of Wassily Kandinsky's *9 Points in Ascendance*, published in Kandinsky's *Punkt und Linie zu Fläche* (*Point and Line to Plane*, 1926), the ninth in the Bauhaus series. The second image is Kepes's version of Piet Mondrian's *Composition* of 1915, published in Mondrian's *Neue Gestaltung: Neoplastizismus Niew Beelding* (*New Design: Neoplasticism New Movement*, 1925), the fifth in the series. The third image is Kepes's version of a Suprematist composition by Kazimir Malevich, as published in Malevich's *Die Gegenstandslose Welt* (*The Nonobjective World*, 1927), the eleventh in the series. The fourth image may draw from Paul Klee's *Pädagogisches Skizzenbuch* (*Pedagogical Sketchbook*, 1925), the second in the Bauhaus series. Finally, the last image is from a pamphlet Kepes produced for the Container Corporation of America titled "Folding Cartons +." It is reproduced in Katherine Kuh, *Advance Guard of Advertising Artists* (Chicago: Katherine Kuh Gallery, 1941). Kepes reproduced the Kandinsky, Mondrian, and Malevich in his *Language of Vision* (Chicago: Paul Theobald, 1944), 22, 47, thereby framing his volume as an American installment of the series. In fact, the German translation of *Language of Vision*, not published until 1971, appeared in a series edited by Hans Wingler called the *Neue Bauhausbücher*.

3 Gyorgy Kepes, from the pamphlet *Illustration* (Chicago: Collins, Miller & Hutchings, Inc., February 1941), no pagination. Find this magazine in the Institute of Design Collection, University of Illinois at Chicago [hereafter ID Collection, UIC], box 6, folder 172.

4 Collins, Miller & Hutchings, Inc., published its *Illustration* pamphlet with portfolios from artists based in Chicago through the 1950s.

5 Kepes describes the "language of the eye" in the first text he published after arriving in the United States: "Educating the Eye," *More Business* 3, no. 11 (November 1938). Kepes published *Language of Vision* in 1944 but first used the phrase "language of vision" in "The Task of Visual Advertising," from the Gyorgy Kepes insert in *PM* 6, no. 3 (February–March 1940), 10. For more on the book, see Michael J. Golec's helpful article, "A Natural History of the Disembodied Eye: The Structure of Gyorgy Kepes's 'Language of Vision,'" *Design Issues* 18, no. 2 (Spring 2002): 3–16.

6   Mattei Calinescu, "The Idea of the Avant-Garde," in *Five Faces of Modernity* (Durham: Duke University Press, 1987), 93–148; first version published in 1977. For an earlier genealogy of "avant-garde," see Renato Poggioli's *The Theory of the Avant-Garde*, trans. Gerald Fitzgerald (Cambridge, MA: Harvard University Press, 1968).

7   Such a transformation echoes Peter Bürger's canonical division between the avant-garde of the prewar period and the neo-avant-garde of the postwar period in his *Theory of the Avant-Garde*, trans. Michael Shaw (Minneapolis: University of Minnesota Press, 1984), 58; originally published in German in 1974. This thesis inspired a generation of art historians to recuperate the criticality of the neo-avant-gardes; emblematic is Branden W. Joseph, *Random Order: Robert Rauschenberg and the Neo-Avant-Garde* (Cambridge, MA: MIT Press, 2003). My study shifts attention from criticality as such, exploring why one might abandon radical aspirations in the first place.

8   This condition resonates with Kenneth E. Silver's reading of the return to order following the First World War. See his *Esprit de Corps: The Art of the Parisian Avant-Garde and the First World War, 1914–1925* (Princeton: Princeton University Press, 1989).

9   Judith Wechsler, "Gyorgy Kepes," in *Gyorgy Kepes: The MIT Years, 1945–1977* (Cambridge, MA: MIT Press, 1978), 10.

10   Kepes interviewed by Katherine Kuh, "On Public Art," *Saturday Review* 3, no. 2 (18 October 1975): 63. Kepes flew over Chicago as part of the city's official camouflage committee, which he joined at the request of Mayor Edward Kelly.

11   Gyorgy Kepes, undated and untitled note in the Gyorgy Kepes papers, Archives of American Art, Smithsonian Institution, Washington, DC [hereafter Kepes papers, AAA], reel 5303, frame 480. Kepes later published variations of this text.

12   See László Moholy-Nagy, *The New Vision: From Material to Architecture*, trans. Daphne M. Hoffmann (New York: Brewer, Warren, and Putnam, 1932).

13   For a discussion of bombing strategy, see Peter Galison, "War against the Center," *Grey Room* 4 (Summer 2001): 6–33. Barton J. Bernstein highlights the moral dimension of strategic aerial bombardment, and the use of the atomic bomb, in "The Atomic Bombings Reconsidered," *Foreign Affairs* 74, no. 1 (January–February 1995): 135–152.

14   Paul Virilio, *War and Cinema: The Logistics of Perception*, trans. Patrick Camiller (London: Verso, 1989), 26.

15   I thank John Harwood for bringing this aspect of the bombsight's function to my attention.

16   See, for example, the "U.S. Strategic Bombing Survey, Overall Report (European War)" (Washington, DC: US General Publishing Office, 1945).

17   W. G. Sebald, *On the Natural History of Destruction*, trans. Anthea Bell (London: Penguin, 2003), 19.

18   See Galison, "War against the Center." Galison suggests that campaigns abroad made Americans consider their own vulnerability.

19   See Shell's intriguing study *Hide and Seek: Camouflage, Photography, and the Media of Reconnaissance* (New York: Zone Books, 2012). For more on aerial perspective generally, see K. Saint-Amour, "Applied Modernism: Military and Civilian Uses of the Aerial Photomosaic," *Theory, Culture, and Society* 28, no. 7–8 (2011): 241–269; and his "Modernist Reconnaissance," *Modernism/Modernity* 10, no. 2 (April 2003): 349–380. See also these two authoritative accounts: Jean-Louis Cohen, *Architecture in Uniform: Designing and Building for the Second World War* (Paris: Éditions Hazan, 2011); and Jason Weems, *Barnstorming the Prairies: How Aerial Vision Shaped the Midwest* (Minneapolis: University of Minnesota Press, 2015). For an architectural investigation that considers not camouflage but, rather, blackout, see Sandy Isenstadt, "Groping in the Dark: The Scotopic Space of Blackouts," *Senses and Society* 7, no. 3 (2012): 309–208. Roy R. Behrens first explored many of these topics. See, for example, *Camoupedia: A Compendium of Research on Art, Architecture and Camouflage* (Dysart, IA: Bobolink Books, 2009). For the origins of camouflage in natural history during the nineteenth century, see Alexander Nemerov, "Vanishing Americans: Abbott Thayer, Theodore Roosevelt, and the Attraction of Camouflage," *American Art* 11, no. 2 (Summer 1997): 50–81.

20   "How Chicago May Hide from Bombers!," *Chicago Herald American*, 13 January 1942. See also the account in "Kelly and Army Plan Hiding of City from Foe," *Chicago Daily News*, 8 May 1942.

21   See C. H. R. Chesney, *The Art of Camouflage* (London: Robert Hale, 1941); and Robert P. Breckenridge, *Modern Camouflage: The New Science of Protective Concealment* (New York: Farrar and Rinehart, 1942).

22   See especially the discussion of Boeing Plant II in Jean-Louis Cohen, "Camouflage, or the Temptation of the Invisible," in *Architecture in Uniform*, 187–215.

23   "The Camoufleurs," *Time* (6 August 1945): 27.

24   Martin Heidegger, "The Age of the World Picture," in *The Question Concerning Technology and Other Essays*, trans. William Lovitt (New York: Harper and Row, 1977), 134.

25   "Art by Accident," *Architectural Review* (September 1944): 68.

26   For other wartime courses, see "National Defense Courses" pamphlet, ID Collection, UIC, box 3, folder 64.

27   Moholy was annoyed that Kepes waited until after Pearl Harbor to begin camouflage instruction. László Moholy-Nagy to Gyorgy Kepes, 19 November 1942, Bauhaus-Archiv, Berlin. Correspondence from the Bauhaus-Archiv was provided without identifying box or folder numbers. The School of Design's camouflage and other war courses are also described in two excellent accounts: Maggie Taft, "Better Than Before: László Moholy-Nagy and the New Bauhaus in Chicago," in Mary Jane Jacob and Jacquelynn Baas, eds., *Chicago Makes Modern: How Creative Minds Changed Society* (Chicago: University of Chicago Press, 2012), 31–43; and

Robin Schuldenfrei, "Assimilating Unease: Moholy-Nagy and the Wartime/Post-war Bauhaus in Chicago," in Schuldenfrei, ed., *Atomic Dwelling: Anxiety, Domesticity, and Postwar Architecture* (London: Routledge, 2012), 87–126.

28   "Outline of the Camouflage Course at the School of Design in Chicago 1941–1942," ID Collection, UIC, box 3, folder 64.

29   Ralph Waldo Emerson, *Nature* (Boston: James Munroe and Company, 1836), 13.

30   "Outline of the Camouflage Course at the School of Design in Chicago 1941–1942," ID Collection, UIC, box 3, folder 64.

31   Ibid.

32   László Moholy-Nagy, "Make a Light Modulator," *Popular Photography* 2, no. 7 (March 1940). See also László Moholy-Nagy, *Vision in Motion* (Chicago: Paul Theobald, 1947), 198–203.

33   Gyorgy Kepes and Nathan Lerner, "Creative Use of Light," in Dagobert D. Runes and Harry G. Schrickel, eds., *Encyclopedia of the Arts* (New York: Philosophical Library, 1946), 559. The book was likely assembled around 1942, even though it was not published until the war concluded.

34   Lerner quoted in Moholy-Nagy, *Vision in Motion*, 200.

35   "Kelly and Army Plan Hiding of City From Foe," *Chicago Daily News*, 8 May 1942.

36   This notebook is housed at the Illinois Institute of Technology. Other students likely produced similar notebooks.

37   "Outline of the Camouflage Course at the School of Design in Chicago 1941–1942," ID Collection, UIC, box 3, folder 64.

38   George Kepes [Gyorgy Kepes], "Summary of the Introductory Lecture for the Camouflage Course," dated 16 September 1942, ID Collection, UIC, box 6, folder 173.

39   "Outline of the Camouflage Course at the School of Design in Chicago 1941–1942," ID Collection, UIC, box 3, folder 64.

40   George Kepes [Gyorgy Kepes], "Summary of the Introductory Lecture for the Camouflage Course," dated 16 September 1942, ID Collection, UIC, box 6, folder 173.

41   See Charles W. Morris's typescript titled "The Intellectual Program of the New Bauhaus" and dated 1937, ID Collection, UIC, box 7, folder 195. For a comprehensive account of the early New Bauhaus, see, among others, Lloyd C. Engelbrecht, "The Association of Art and Industries: Background and Origins of the Bauhaus Movement in Chicago" (PhD dissertation, University of Chicago, 1973). See also Alain Findeli, "Moholy-Nagy's Design Pedagogy in Chicago (1937–1946)" *Design Issues* 7, no. 1 (Autumn 1990): 4–19.

42   As listed in the "Outline of the Camouflage Course at the School of Design in Chicago 1941–1942," ID Collection, UIC, box 3, folder 64. A brochure for

"National Defense Courses" taking place in winter 1942 lists only a selection of these guest speakers.

43 These methods were also militaristic; information theory, for example, first emerged through studies of cryptography—in other words, forms of camouflage.

44 See the report titled "Tests on Pattern Lighting Chicago Metropolitan Area," National Archives and Records Administration [hereafter NARA], record group 171, entry 114, box 2. There is no indication of particular participants in the program.

45 Kepes interviewed in Kuh, "On Public Art." Kepes is presumably referring to the Chicago Metropolitan Area's Department of Techniques.

46 Judith Wechsler explains in "Gyorgy Kepes," 10. Wechsler indicates that these projects anticipate those later undertaken at the CAVS. See chapter 6.

47 Oral history interview with Robert O. Preusser, January–October 1991, Archives of American Art, Smithsonian Institution, transcript available at http://www.aaa.si.edu/collections/interviews/oral-history-interview-robert-o-preusser-13337.

48 "The Camoufleurs," *Time* (6 August 1945): 27.

49 László Moholy-Nagy to F. J. Kelly, US Office of Education, 13 March 1942, ID Collection, UIC, box 6, folder 183.

50 László Moholy-Nagy to Milton C. Mapes, 21 April 1942, NARA, record group 171, entry 10, box 119, folder "Gen. Correspondence 1940–1942, 515–516."

51 "Research Problems in Camouflage at the School of Design in Chicago," NARA, record group 171, entry 10, box 119, folder "Gen. Correspondence 1940–1942, 515–516."

52 L. D. Gasser to László Moholy-Nagy, 4 May 1942, NARA, record group 171, entry 10, box 119, folder "Gen. Correspondence 1940–1942, 515–516."

53 See the OCD's official report, "History of the Engineer Section," dated 8 March 1944, in NARA, record group 171, entry 116, box 1, folder "History of Engineer Section and of Camouflage Unit."

54 L. D. Gasser's generic response is included in NARA, record group 171, entry 10, box 119, folder "Gen. Correspondence 1940–1942, 515–516."

55 Greville Rickard, "Report on Camouflage Educational Program," NARA, record group 171, entry 116, box 1.

56 See such letters in NARA, record group 171, entry 116, box 1, folder labeled "Correspondence Connected with 4 courses at Ft. Belvoir."

57 Memorandum on "Eligibility for OCD sponsored certificates," included in Rickard, "Report on Camouflage Educational Program," NARA, record group 171, entry 116, box 1.

58 Many sought active duty positions in camouflage, though such positions were rare. Letters between Kepes, his student Robert Preusser, and Moholy regarding Preusser's problems securing a camouflage position indicate the issue. See correspondence in the ID Collection, UIC, box 7, folder 197.

59 See Rickard, "Report on Camouflage Educational Program," in NARA, record group 171, entry 116, box 1.

60 Undated letter from the War Department to OCD, NARA, record group 171, entry 116, box 1, folder "Letters—War Dept. Camouflage School at Ft. Belvoir."

61 These included Columbia, Cornell, Harvard, New York, Princeton, Yale, Rutgers, and Stanford Universities; The Universities of California, Colorado, Illinois, Kansas, Maryland, Michigan, Minnesota, New Hampshire, Oklahoma, and Pennsylvania; Dallas College, Ohio State University, and Oregon State College; Carnegie Institute of Technology, Georgia Institute of Technology, and Massachusetts Institute of Technology; Alabama Polytechnic Institute and New York State Institute of Agriculture; Cleveland School of Art, Cleveland Museum of Art, Cranbrook Academy of Art, Kansas City Art Institute, Pratt Institute, Rhode Island School of Design, and, of course, Chicago's School of Design. See Rickard, "Report on Camouflage Educational Program," in NARA, record group 171, entry 116, box 1.

62 Ibid., 2.

63 See the course schedule in the Edward M. Farmer Papers, Hoover Institution Archives, Stanford University, box 1.

64 In a copy of the course listing, Edward M. Farmer notes: "Lecture was exactly based on outlines of book Breckenridge: <u>Modern Camouflage</u>." Farmer Papers, Hoover Institution Archives, Stanford University, box 1.

65 Rickard, "Report on Camouflage Educational Program," 2, NARA, record group 171, entry 116, box 1.

66 From the collated responses in NARA, record group 171, entry 116, box 1, folder "Questionnaires to Universities [*sic*] Directors on Ft. Belvoir courses."

67 Ibid.

68 Ibid.

69 Gyorgy Kepes to Robert Preusser, 23 June 1942, ID Collection, UIC, box 7, folder 197.

70 László Moholy-Nagy to Gyorgy Kepes, 19 November 1942, Bauhaus-Archiv Berlin.

71 Gyorgy Kepes to Robert Preusser, undated [early fall 1942], ID Collection, UIC, box 7, folder 197.

72 See NARA, record group 171, entry 116, box 1, folder "Questionnaires to Universities [*sic*] Directors on Ft. Belvoir Courses."

73 László Moholy-Nagy to Walter Paepke, Container Corporation of America, 1 October 1942, ID Collection, UIC, box 6, folder 183.

74   See László Moholy-Nagy to Walter A. Jessup, Carnegie Corporation of New York, 13 October 1942, ID Collection, UIC, box 6, folder 183.

75   László Moholy-Nagy to F. P. Keppel, Carnegie Corporation of New York, 7 January 1943, ID Collection, UIC, box 6, folder 183.

76   Julian Huxley to László Moholy-Nagy, 12 February 1942, ID Collection, UIC, box 6, folder 184.

77   Julian Huxley to Dean James M. Landis, Office of Civilian Defense, 12 February 1942, ID Collection, UIC, box 6, folder 184. There is a copy of this letter enclosed with the letter from Huxley to Moholy-Nagy quoted above.

78   Ibid.

79   Fred Turner, *The Democratic Surround: Multimedia and American Liberalism from World War II to the Psychedelic Sixties* (Chicago: University of Chicago Press, 2013), 40–57. See also Ellen Herman, *The Romance of American Psychology: Political Culture in the Age of Experts* (Berkeley: University of California Press, 1995).

80   Edward J. Kelly to "George Kepes" [Gyorgy Kepes], 13 March 1942, Kepes papers, AAA, reel 5303, frame 0155.

81   Gyorgy Kepes, undated and untitled note, Kepes papers, AAA, reel 5312, frame 0582. This note and the one following are also cited in Anna Vallye, "Design and the Politics of Knowledge in America, 1927–1967: Walter Gropius, Gyorgy Kepes" (PhD Dissertation, Columbia University, 2011), 210–211.

82   Gyorgy Kepes, undated and untitled note, Kepes papers, AAA, reel 5312, frame 0609.

83   "Report of NDRC ad hoc Committee on Camouflage," dated 25 February 1942, page 1, NARA, record group 227, entry 133, box 13.

84   Vannevar Bush, foreword to Irvin Stewart, *Organizing Scientific Research for War: The Administrative History of the Office of Scientific Research and Development* (Boston: Little, Brown, 1948), ix.

85   "Report of NDRC ad hoc Committee on Camouflage," dated 25 February 1942, page 6, NARA, record group 227, entry 133, box 13.

86   Ibid., page 33.

87   Greville Rickard, "Scrap Book of Greville Rickard," NARA, record group 171, entry 118, box 14.

88   Compton quoted in C. Guy Suits and George R. Harrison, eds., *Applied Physics: Electronics; Optics; Metallurgy* (Boston: Little, Brown, 1948), 201.

89   See the official summary in "Camouflage Projects," ibid., 254–263. A complete description can be found in the declassified "Summary Technical Report of Division 16," NDRC, vol. 2, "Visibility Studies and Some Applications in the Field of Camouflage."

90 See Elizabeth Hutchinson, ed., *Louis Comfort Tiffany and Laurelton Hall: An Artist's Country Estate* (New York: Metropolitan Museum of Art, 2006), 217.

91 H. Richard Blackwell, "Contrast Thresholds of the Human Eye," *Journal of the Optical Society of America* 36, no. 11 (November 1946): 624.

92 "Summary Technical Report of Division 16," NDRC, vol. 2, "Visibility Studies and Some Applications in the Field of Camouflage," 12.

93 From a 1942 report of the board of trustees of the Tiffany Foundation, Tiffany Foundation Records, Archives of American Art, Smithsonian Institution, Washington, DC, reel N69-25.

94 See the list of OSRD contractors in Larry Owens, "The Counterproductive Management of Science in the Second World War: Vannevar Bush and the Office of Scientific Research and Development," *Business History Review* 68, no. 4 (Winter 1994): 515–576.

95 As quoted on the Foundation's website: http://louiscomforttiffanyfoundation .org/info_about.asp

96 Siebert Q. Duntley, preface to the "Summary Technical Report of Division 16," NDRC, vol. 2, "Visibility Studies and Some Applications in the Field of Camouflage," illegible pagination.

97 Suits and Harrison, *Applied Physics*, 255.

98 Blackwell, "Contrast Thresholds of the Human Eye," 632.

99 Duntley, preface to "Summary Technical Report of Division 16," NDRC, vol. 2, "Visibility Studies and Some Applications in the Field of Camouflage," illegible pagination.

100 National Research Council, *Psychology for the Fighting Man* (Washington, DC: Infantry Journal, 1943), 24. See also the reference to this book in Turner, *The Democratic Surround*, 154.

101 National Research Council, *Psychology for the Fighting Man*, 30.

102 Max Horkheimer, *The Eclipse of Reason* (New York: Continuum, 1974), 141–142; originally published in 1947.

103 Kepes, *Language of Vision*, 68.

104 Field Manual 5-20A, "Camouflage of Individuals and Infantry Weapons" (War Department, February 1944), 3. This replaced Field Manual 5-20, which was in use during Kepes's tenure at Fort Belvoir.

105 Kepes, *Language of Vision*, 66.

106 This desire to destroy oneself through camouflage is the central idea in Roger Caillois's canonical 1935 essay "Mimicry and Legendary Psychasthenia," trans. John Shepley, in Annette Michelson et al., eds., *October: The First Decade, 1976–1986* (Cambridge, MA: MIT Press, 1987), 58–74.

107 Gyorgy Kepes to Fred Keck, 23 October 1942, Bauhaus-Archiv Berlin.

108 László Moholy-Nagy to Gyorgy Kepes, 19 November 1942, Bauhaus-Archiv Berlin.

109 Ibid.

110 Ibid.

111 Oral history interview with Gyorgy Kepes, 7 March and 30 August 1972 and 11 January 1973, Archives of American Art, Smithsonian Institution, page 17 of typescript. It is interesting that Devin Fore evokes Kepes to argue that abstract photography in the interwar period reflected an ideological demotivation; perhaps we can read this change more specifically as one caused by the militarization of contemporary life before and during the Second World War. See his *Realism after Modernism: The Rehumanization of Art and Literature* (Cambridge, MA: MIT Press, 2012), 39.

112 Investigation of Un-American Propaganda Activities in the United States, Special Committee on Un-American Activities, House of Representatives, Seventy-Eight Congress, Second Session on H. Res. 282, Appendix Part IX, Communist Front Organizations, 292.

113 Ibid., 302.

114 Gyorgy Kepes to Krisztina Passuth, undated [circa 1980s], Kepes papers, Stanford, box 50, folder 9. Passuth was researching *Moholy-Nagy* (New York: Thames and Hudson, 1985).

115 Gyorgy Kepes interviewed by Robert Brown, 7 March and 30 August 1972 and 11 January 1973, AAA, pages 19 and 20 of typescript.

Chapter 2

1 Erich Fromm, *Escape from Freedom* (New York: Holt, Rinehart and Winston, 1941), 19–20.

2 The Hayden Gallery opened in 1950 in MIT's Charles Hayden Memorial Library; it is now a gallery within MIT's List Visual Arts Center. The lattice structure may have been in part the idea of Thomas McNulty, Kepes's collaborator on the installation.

3 Despite the similarity between Kepes's *New Landscape* and the Independent Group's *Growth and Form*, there is little evidence that Kepes was aware of the Independent Group's activities while assembling his exhibition. A note from Richard Hamilton to Kepes ("Thank you for your letter. The catalogue should follow this in about a week. R. Hamilton"), accompanied by a brochure for the *Growth and Form* exhibition, suggests that Kepes had asked about Hamilton's show after it opened—and thus after his *New Landscape* show also opened. Kepes certainly knew about the Independent Group's activities while assembling his book. See the Kepes papers, Stanford, box 83, folder 3. For readings that relate Kepes's exhibition to other intellectual traditions, specifically biocentrism and organicism, see the work of Oliver A. I. Botar, especially "György Kepes' 'New

Landscape' and the Aestheticization of Scientific Photography," in *A Fényjátékosok: Kepes György és Frank J. Malina a tudomány és a művészet metszéspontján* [*The Pleasure of Light: Gyorgy Kepes and Frank J. Malina at the Intersection of Science and Art*], exh. cat. (Budapest: Ludwig Museum Budapest, 2010), 124-143; and Charissa N. Terranova, *Art as Organism: Biology and the Evolution of the Digital Image* (London: I. B. Tauris, 2015).

4   These are Lichtenberg's words from 1777, as quoted in translation by Yuzo Takahashi in "Two Hundred Years of Lichtenberg Figures," *Journal of Electrostatics* 6 (1979): 3. Lichtenberg's technique uses the same principle as xerographic photocopying and laser printing. Lichtenberg's images are briefly discussed in Siegfried Zielinski, *Deep Time of the Media: Toward an Archaeology of Hearing and Seeing by Technical Means*, trans. Gloria Custance (Cambridge, MA: MIT Press, 2006), 165.

5   Similar photographic experiments predate von Hippel; see Ben Burbridge, ed., *Revelations: Experiments in Photography*, exh. cat. (London: MACK, 2015). See also Martin Kemp's discussion of the Lichtenberg figure in, for example, "Trees of Knowledge," *Nature* 435, no. 7044 (16 June 2005): 888.

6   See Gyorgy Kepes, "The New Landscape," *Arts and Architecture* (May 1951): 21–23. Kepes's publisher suggested using a Lichtenberg figure on the cover of his book, though Kepes preferred the art-and-science comparison described below.

7   Kepes hoped to publish the book with the Technology Press, predecessor to the MIT Press, in closer conjunction with the exhibition; contractual obligations to Paul Theobald prevented him from doing so and delayed publication. Kepes eventually threatened to sue Theobald over these difficulties. Kepes's relationship with Theobald is well documented in the Paul Theobald Papers at the Art Institute of Chicago.

8   Gyorgy Kepes, *The New Landscape in Art and Science* (Chicago: Paul Theobald, 1956), 367.

9   Gyorgy Kepes, "Function in Modern Design," in Thomas J. Wilson et al., *Graphic Forms: The Arts as Related to the Book* (Cambridge, MA: Harvard University Press, 1949), 7.

10   Commenting on similar projects like the *Mnemosyne Atlas*, Spyros Papapetros suggests that "links appear to be broken via the mechanical reproduction of the photograph as black-and-white illustrations." As I see it, the photograph actually creates links more than breaks them, for without the leveling effect of the black-and-white illustration there would be nothing at all to connect the phenomena Kepes compares and contrasts. See Papapetros, "Between the Academy and the Avant-Garde: Carl Einstein and Fritz Saxl Correspond," *October* 139 (Winter 2012): 85.

11   Cybernetics appears with specificity rarely in Kepes's extensive research material, and it is my contention that it has overdetermined prior interpretations; one of my goals is to introduce the wider array of sources and discourses Kepes marshaled.

12   See this brief characterization in Kevin Lotery, *"an Exhibit*/an Aesthetic: Richard Hamilton and Postwar Exhibition Design," *October* 150 (Fall 2014): 96; and in T'ai Smith's *Bauhaus Weaving Theory: From Feminine Craft to Mode of Design* (Minneapolis: University of Minnesota Press, 2014), 162.

13   Reinhold Martin, *The Organizational Complex: Architecture, Media, and Corporate Space* (Cambridge, MA: MIT Press, 2003), 79. See also Orit Halpern's survey of Kepes's work, which similarly emphasizes its relation to cybernetics and technology as a mediated form of vision, in her *Beautiful Data: A History of Vision and Reason since 1945* (Durham: Duke University Press, 2014), 79–99.

14   Kepes quoted in Robert C. Morgan, "Sermon for Tranquility: An Interview with Gyorgy Kepes," *Afterimage* 10, no. 6 (January 1983): 7. This important line is also cited in Elizabeth Finch, "Languages of Vision: Gyorgy Kepes and the 'New Landscape' of Art and Science" (PhD dissertation, City University of New York, 2005); and Anne Collins Goodyear, "The Relationship of Art to Science and Technology in the United States, 1957–1971: Five Case Studies" (PhD dissertation, University of Texas at Austin, 2002).

15   Gyorgy Kepes, undated and untitled note, Kepes papers, Stanford, box 82, folder 5.

16   Gyorgy Kepes, "The New Landscape," *Graphis* 7, no. 36 (1 March 1951): 200.

17   Spyros Papapetros, *On the Animation of the Inorganic: Art, Architecture, and the Extension of Life* (Chicago: University of Chicago Press, 2012), 150.

18   For "paradigm repetition," see Benjamin H. D. Buchloh, "The Primary Colors for a Second Time: A Paradigm Repetition of the Neo-Avant-Garde," *October* 37 (Summer 1986): 41–52. Kepes was fully aware he was repeating earlier modernist projects; as he writes in a notebook: "I have [a] certain déjà vu feeling with many of the avant garde [*sic*] explorations." From one of Kepes's notebooks in the Kepes papers, Stanford, box 20.

19   For rival interpretations of the "closed world" and "open mind," see Paul N. Edwards, *The Closed World: The Computer and the Politics of Discourse in Cold War America* (Cambridge, MA: MIT Press, 1996); and Jamie Cohen-Cole, *The Open Mind: Cold War Politics and the Science of Human Nature* (Chicago: University of Chicago Press, 2014). Fred Turner's *From Counterculture to Cyberculture: Stewart Brand, the Whole Earth Network, and the Rise of Digital Utopianism* (Chicago: University of Chicago Press, 2006) and especially his *The Democratic Surround: Multimedia and American Liberalism from World War II to the Psychedelic Sixties* (Chicago: University of Chicago Press, 2013) consider the openness of the "closed world." Others have interpreted specific discourses like cybernetics, which are thought to be synonymous with such a culture, as instead open; see, for example, Andrew Pickering, *The Cybernetic Brain: Sketches of Another Future* (Chicago: University of Chicago Press, 2010).

20   Erwin Panofsky, *Tomb Sculpture: Its Changing Aspects from Ancient Egypt to Bernini* (London: Thames and Hudson, 1964), 26–27. I thank Bryan J. Wolf for first bringing this concept to my attention.

21   Yve-Alain Bois, "On the Uses and Abuses of Look-alikes," *October* 154 (Fall 2015): 127–149. See also Pamela M. Lee's consideration of pseudomorphosis in her

forthcoming study *Think Tank Aesthetics: Midcentury Modernism, the Cold War and the Neoliberal Present*.

22   The same image is captioned "Deer Tongue" on page 114 of *The New Landscape in Art and Science*. The presence of typed labels affixed to the original panels indicates that textual modifiers were present in the exhibition.

23   Martin, *The Organizational Complex*, 73.

24   Edward Weston to Gyorgy Kepes, 26 January 1952, Kepes papers, AAA, reel 5303, frame 0306. Kepes also asked Weston whether he would write a text for the volume about "what the concept pattern means to you." Gyorgy Kepes to Edward Weston, 1 February 1952, Kepes papers, AAA, reel 5303, frame 0307.

25   The same image appears rotated and inverted elsewhere in Kepes, *The New Landscape in Art and Science*, 274, and in *Arts and Architecture* (May 1951): 22; it is rotated, but not inverted, in Kepes's *The Visual Arts Today* (Middletown, CT: Wesleyan University Press, 1960), 180. Such variability indicates that Kepes manipulated images based on the demands of layouts. There is, of course, no single orientation for a scientific specimen.

26   Bruno Rossi, "The Esthetic Motivation of Science," in Kepes, *The New Landscape in Art and Science*, 67. For another reading of the mystical and mythological force of scientific photographs, see the chapter "PhotoPhysics" in Terri Weissman's *The Realisms of Bernice Abbott: Documentary Photography and Political Action* (Berkeley: University of California Press, 2011). Weissman considers the images made by Bernice Abbott at MIT starting in the late 1950s.

27   F. H. Merrill and A. von Hippel, "The Atomphysical Interpretation of Lichtenberg Figures and Their Application to the Study of Gas Discharge Phenomena," *Journal of Applied Physics* 10 (December 1939): 873. Von Hippel's Lichtenberg figures were also popularized for a general audience. An article in *Life* explained their utility. "The scientists at M.I.T.—however entranced they may be by the beauty of the pictures—put them to practical uses." See "Speaking of Pictures ... These Show Science as True and Beautiful," *Life* 11, no. 4 (28 July 1941): 9.

28   No doubt many a future art historian has received this message from parents. Von Hippel encountered art history at the University of Munich during the summer term of 1921. "I ... became enchanted by the lectures and seminars of Professor Wölflin [*sic*] on the 'Renaissance.' The famous Art Galleries of Munich, the 'Alte' and 'Neue' Pinakothek and the 'Glyptothek,' provided an inexhaustible supply of original works.... Therefore I signed up as an art student." Arthur R. von Hippel, *Life in Times of Turbulent Transitions*, ed. Frank von Hippel (Anchorage: Stone Age Press, 1988), 43. After von Hippel completed his term at Munich, his father sent him advice: "complete studies in science or starve" (44). Von Hippel subsequently completed his doctorate in physics. Before pursuing a career as a physicist, however, he traveled to Italy to view the works he had seen in Wölfflin's lectures. He writes: "I knew that I could enjoy art deeply but not produce it. I therefore returned [to Germany] to become a scientist" (48).

29   Gyorgy Kepes, undated and untitled notes, Kepes papers, Stanford, box 83, folder 5 and box 83, folder 4.

30   Kepes, *The New Landscape in Art and Science*, 20.

31   Gyorgy Kepes, outline for "Revision of Vision," undated [1947?], Kepes papers, AAA, reel 5312, frame 0421.

32   Kepes, *The New Landscape in Art and Science*, 18; emphasis in original.

33   Gyorgy Kepes, from an audio recording of a lecture delivered at Yale University on 9 March 1966. Yale and New Haven Audio and Visual Recordings (RU 803 Series Accession 2012-A-009). Manuscripts and Archives, Yale University Library.

34   Gyorgy Kepes, interview, 2 December 1988, VHS recording, available at https://www.youtube.com/watch?v=3NSGJLtXWr8.

35   John E. Burchard, foreword to Kepes, *The New Landscape in Art and Science*, 14.

36   John E. Burchard to Gyorgy Kepes, 16 October 1954, MC 76, MIT, box 6, folder 28.

37   Quoted in John Ely Burchard, "Foreword, The New Landscape, 10/19/54," MC 76, MIT, box 6, folder 28. The typescript bears the heading: "to precede the other six pages if Professor Kepes thinks best."

38   Ibid.

39   Walter Gropius, "Reorientation," in Kepes, *The New Landscape in Art and Science*, 95. Gropius's contribution is a version of a text that originally appeared elsewhere.

40   Or, "art without knowledge is nothing." The phrase derives from the writings of fourteenth-century architect Jean Mignot, in reference to the Milan cathedral, which was considered a work of art and science, specifically geometry. For the art-historical treatment of the phrase, see James Ackerman, "'Ars Sine Scientia Nihil Est': Gothic Theory of Architecture," *Art Bulletin* 31, no. 2 (June 1949): 84–111. For Ackerman's use of the term "scientia," see his "On *Scientia*," *Daedalus* 94, no. 1 (Winter 1965): 14–23; this essay appears next to Kepes's proposal for the CAVS.

41   William W. Wurster to Gyorgy Kepes, 30 August 1945, Kepes papers, AAA, reel 5303, frame 0175. Kepes's hire is also discussed in Finch, "Languages of Vision," 192.

42   *Massachusetts Institute of Technology Bulletin* 81, no. 1 (October 1945), 139.

43   *Massachusetts Institute of Technology Bulletin* 83, no. 1 (October 1947), 142.

44   For Selmer-Larsen's course, see *Massachusetts Institute of Technology Bulletin* 80, no. 4 (June 1945), 91.

45   For Kepes's role in architectural training at MIT, see especially Anna Vallye's research, including her essay "Gyorgy Kepes's 'Universities of Vision'" in *Émigré Design Cultures: Histories of the Social in Design* (London: Bloomsbury, 2017); her

essay "The Middleman: Kepes's Instruments," in Arindam Dutta, ed., *A Second Modernism: MIT, Architecture, and the "Techno-Social" Movement* (Cambridge, MA: MIT Press, 2013); and her "Design and the Politics of Knowledge in America, 1937–1967: Walter Gropius, Gyorgy Kepes" (PhD dissertation, Columbia University, 2011). Vallye also cites a few of Kepes's notes I discuss below.

46   Gyorgy Kepes, undated note titled "Education of Vision, General Introduction," Kepes papers, AAA, reel 5312, frame 0505.

47   Gyorgy Kepes, undated note titled "University of Vision," Kepes papers, AAA, reel 5312, frame 0902.

48   Gyorgy Kepes, undated and untitled note, Kepes papers, AAA, reel 5312, frame 0898.

49   The bibliographies are located in the Kepes papers, AAA, reel 5312, frame 0636 and following. There are also a few additional pages from these bibliographies in the Kepes papers at Stanford.

50   See Heinrich Wölfflin, *Renaissance und Barock* (Munich: Ackermann, 1888); Gottfried Semper, *Der Stil in den Technischen und Tektonischen Künsten oder Praktische Ästhetik* (Frankfurt am Main: Verlag für Kunst und Wissenschaft, 1860–1863); Erwin Panofsky, *Die Perspektive als "Symbolische Form"* (Leipzig: Teubner, 1927); John Dewey, *Art as Experience* (New York: Minton, Balch, and Co., 1934); Mary Hamilton Swindler, *Ancient Painting: From the Earliest Times to the Period of Christian Art* (New Haven: Yale University Press, 1929); and Miriam Schild Bunim, *Space in Medieval Painting and the Forerunners of Perspective* (New York: Columbia University Press, 1940).

51   John Burnet, "An Essay on the Education of the Eye with Reference to Painting," in *A Treatise on Painting* (London: James Carpenter, 1837), 33, 35. Perhaps Burnet's opening sentence compelled Kepes; it lamented the position of arts education during the Industrial Revolution in ways resonant with the postwar period: "In a country so largely connected with manufactures as this is, we cannot but wonder why the education of the eye has not been more generally cultivated" (1).

52   Gyorgy Kepes, undated note titled "Visual Fundamentals, Introduction," Kepes papers, AAA, reel 5312, frame 0524.

53   Kepes, *The New Landscape in Art and Science*, 20. Kepes frequently used this expression. For more on these figures, see Amy F. Ogata, *Designing the Creative Child: Playthings and Places in Midcentury America* (Minneapolis: University of Minnesota Press, 2013).

54   In their breadth and depth, Kepes's bibliographies anticipate Robert Smithson's library; indeed, Smithson studied many of the same titles. Unfortunately, no complete record of Kepes's library remains. For Smithson's library, see Ann Reynolds, *Robert Smithson: Learning from New Jersey and Elsewhere* (Cambridge, MA: MIT Press, 2003), 297–345.

55   Karl von Frisch, *Bees: Their Vision, Chemical Senses, and Language* (Ithaca: Cornell University Press, 1950).

56   Gyorgy Kepes, undated note titled "The New Landscape," Kepes papers, Stanford, box 81, folder 4.

57   Ibid.

58   Kepes saw the Ames demonstrations at a 1947 symposium at Princeton University; see Thomas H. Creighton, *Building for Modern Man: A Symposium* (Princeton: Princeton University Press, 1949). For more on Ames, see W. C. Bamberger, *Adelbert Ames, Jr.: A Life of Vision and Becomingness* (Whitmore Lake, MI: Bamberger Books, 2006). For more on Kepes's use of the Ames demonstration and the context for such studies in a field known as transactional psychology, see Anna Vallye, "The Strategic Universality of *trans/formation*, 1950–1952," *Grey Room 35* (Spring 2009): 28–57.

59   Kathleen Lonsdale, *Crystals and X-Rays* (London: Bell, 1948), 52–53.

60   Gyorgy Kepes, undated note titled "Education of Vision," Kepes papers, AAA, reel 5312, frame 0548.

61   Gyorgy Kepes, *Language of Vision* (Chicago: Paul Theobald, 1944), 202.

62   Gyorgy Kepes, undated note titled "Balance Sheet," Kepes papers, AAA, reel 5312, frame 0463.

63   Gyorgy Kepes, undated note titled "Education of Vision," Kepes papers, AAA, reel 5312, frame 0499.

64   Gyorgy Kepes, undated note titled "Visual Fundamentals," Kepes papers, AAA, reel 5312, frame 0729.

65   Gyorgy Kepes, undated note titled "Education of Vision," Kepes papers, AAA, reel 5312, frame 0785.

66   See Rudolf Koch, *The Book of Signs, Which Contains All Manner of Symbols Used from the Earliest Times to the Middle Ages by Primitive Peoples and Early Christians* (New York: Dover, 1955); Hermann Weyl, *Symmetry* (Princeton: Princeton University Press, 1952); and Matila Ghyka, *The Geometry of Art and Life* (New York: Sheed and Ward, 1946).

67   See D'Arcy Wentworth Thompson, *On Growth and Form* (Cambridge: Cambridge University Press, 1942). Kepes consulted this revised edition.

68   James Bell Pettigrew, *Design in Nature* (London: Longmans, Green, 1908), xxi.

69   Ibid., xxiii.

70   Theodore Andrea Cook, *The Curves of Life, Being an Account of Spiral Formations and Their Application to Growth in Nature, to Science and to Art, with Special Reference to the Manuscripts of Leonardo da Vinci* (New York: Henry Holt, 1914), 7.

71   Ibid., 407.

72   Ibid., 1.

73   Ibid., 2.

74   Gyorgy Kepes, undated note titled "Education of Vision," Kepes papers, AAA, reel 5312, frame 0786.

75   Hermann Rorschach, *Psychodiagnostics: A Diagnostic Test Based on Perception*, trans. Paul Lemkau and Bernard Kronenberg, ed. W. Morgenthaler (Berne, Switzerland: Hans Huber, 1942); originally published in 1921, edition edited by W. Morgenthaler in 1932. Kepes also compared actual inkblots and the biomorphic forms of a Joan Miró painting in *Language of Vision*.

76   Ibid.

77   Such are the responses Rorschach lists in his descriptions of sample tests.

78   See the table in ibid., 50–51.

79   Ibid., 130.

80   Leonardo da Vinci, quoted in Ernst Gombrich, *Art and Illusion: A Study in the Psychology of Pictorial Representation* (Princeton: Princeton University Press, 1960), 188, 189. Gombrich's study of optical illusions and related phenomena would have interested Kepes, even if the core of Gombrich's argument *vis-à-vis* naturalism in painting was not relevant. Gombrich later contributed an essay to Kepes's 1960 edited volume *The Visual Arts Today*.

81   Gyorgy Kepes, undated and untitled note, Kepes papers, AAA, reel 5312, frame 0518. See also Vallye, "Design and the Politics of Knowledge in America," 251.

82   From one of Kepes's notebooks in the Kepes papers, Stanford, box 20.

83   From one of Kepes's notebooks in the Kepes papers, Stanford, box 20.

84   See Frederic C. Bartlett, *Remembering: A Study in Experimental and Social Psychology* (New York: Macmillan, 1932).

85   Herbert Read, *Education through Art* (London: Faber and Faber, 1943), 185.

86   For more on Jan Evangelista Purkyně's 1819 study of phosphenes, see Zielinski, *Deep Time of the Media*, 192–203. Read refers specifically to a mandala from *The Secret of the Golden Flower: A Chinese Book of Life*, trans. Richard Wilhelm, with a commentary by C. G. Jung (London: Kegan Paul, Trench, Trubner, and Company, 1935). In fact, this particular mandala image, presented anonymously as an image created by one of Jung's male patients, was created by Jung himself; the same image is included in C. G. Jung's illustrated manuscript *The Red Book Liber Novus*, ed. Sonu Shamdasani, trans. Mark Kyburz, John Peck, and Sonu Shamdasani (New York: W. W. Norton, 2009).

87   Charles Morris, "Man-Cosmos Symbols," in Kepes, *The New Landscape in Art and Science*, 99.

88 Ibid. Morris's comments also relate to his book *The Open Self* (New York: Prentice-Hall, 1948). See also Turner, *Democratic Surround*, 157–158.

89 See Heinz Werner, *Comparative Psychology of Mental Development* (New York: International Universities Press, 1948); originally published in 1940. Werner widely influenced the arts; Kaira M. Cabañas cites his impact on the Brazilian art critic Mário Pedrosa in "Learning from Madness: Mário Pedrosa and the Physiognomic Gestalt," *October* 153 (Summer 2015): 55–57. Werner's research reveals a troubling fetishization of "the primitive"; similarly, Kepes's projects include problematic examples of cultural appropriation. For instance, the first edition of *The New Landscape in Art and Science* included a photograph of Japanese calligraphy that was printed upside down, revealing how little Kepes actually knew about the cultures from which he took his images. See the correspondence between Kepes and Lolita Theobald in the Paul Theobald Papers at the Art Institute of Chicago.

90 Werner, *Comparative Psychology of Mental Development*, 263.

91 See Viktor Lowenfeld, *Creative and Mental Growth: A Textbook on Art Education* (New York: Macmillan, 1947), 224–225. Lowenfeld is cited in Kepes's notes.

92 See Werner, *Comparative Psychology of Mental Development*, 71. Werner quotes Kandinsky directly.

93 Gordon Allport, foreword to Werner, *Comparative Psychology of Mental Development*, xi.

94 Heinz Werner, "On Physiognomic Perception," in Kepes, *The New Landscape in Art and Science*, 282.

95 Gyorgy Kepes, undated note titled "Visual Fundamentals," Kepes papers, AAA, reel 5312, frame 0526.

96 Theodor W. Adorno, *Minima Moralia: Reflections from Damaged Life* (London: Verso, 2005), 50; originally published in 1951.

97 Arthur R. von Hippel, *Dielectrics and Waves* (New York: John Wiley and Sons, 1954), 239.

98 Arthur R. von Hippel, ed., *Dielectric Materials and Applications* (Cambridge, MA: Technology Press, 1954), 283–290.

99 As von Hippel explains in ibid., x.

100 Gyorgy Kepes, undated and untitled notes, Kepes papers, Stanford, box 9, folder 6.

101 As described in the table of contents to *Technology Review* 55, no. 2 (December 1952). The image was created by none other than MIT electrical engineer John G. Trump, Donald J. Trump's uncle.

102 Gyorgy Kepes, undated note titled "Education of Vision," Kepes papers, AAA, reel 5312, frame 0899.

103 Gyorgy Kepes, undated and untitled note, Kepes papers, AAA, reel 5312, frame 0903.

104 Kepes insisted on the inclusion of this color spread, which his publisher considered costly and unnecessary. See the correspondence between Kepes and Paul and Lolita Theobald in the Paul Theobald Papers at the Art Institute of Chicago.

105 Gyorgy Kepes, undated and untitled note, Kepes papers, AAA, reel 5312, frame 0784.

106 Gyorgy Kepes, undated note titled "Visual Fundamentals," Kepes papers, AAA, reel 5312, frame 0544.

107 "Architecture and the Allied Arts," typescript of lecture presented at the Cooper Union, Kepes papers, AAA, reel 5313, frame 0023 and following.

108 Gyorgy Kepes, undated note titled "C. U. [Cooper Union] lecture," Kepes papers, AAA, reel 5313, frame 0816.

109 Daniel Bell, *The End of Ideology: On the Exhaustion of Political Ideas in the Fifties* (Glencoe, IL: Free Press, 1960).

110 On the relationship between scientists and McCarthyism, see Jessica Wang, *American Science in an Age of Anxiety: Scientists, Anticommunism, and the Cold War* (Chapel Hill: University of North Carolina Press, 1999).

111 Julius Stratton to Gyorgy Kepes, 7 August 1966, MC 431, MIT, box 18, folder "JAS—Kepes, Gyorgy." Their correspondence was initiated on account of the establishment of Kepes's CAVS; Stratton worked with Kepes to realize the Center.

112 See the hearings before the Committee on Un-American Activities, House of Representatives, 1951.

113 For more on the red scare at MIT, see Flo Conway and Jim Siegelman, *Dark Hero of the Information Age: In Search of Norbert Wiener, the Father of Cybernetics* (New York: Basic Books, 2005), 255–271; and Sylvia Nasar, *A Beautiful Mind: The Life of Mathematical Genius and Nobel Laureate John Nash* (New York: Simon and Schuster, 1998), 152–154. For a larger such history, see Ellen Schrecker, *No Ivory Tower: McCarthyism and the Universities* (New York: Oxford University Press, 1986). See also Howard Zinn's reflections about FBI investigations in Cambridge in "The Politics of History in the Era of the Cold War: Repression and Resistance," in Noam Chomsky et al., *The Cold War and the University: Toward an Intellectual History of the Postwar Years* (New York: New Press, 1997).

114 Hearing before the Committee on Un-American Activities, House of Representatives. Communist Methods of Infiltration (Education), 25, 26, and 27 February 1953 (Washington, DC: US Government Printing Office, 1953), 1023.

115 Juliet Kepes Stone, interview with the author, November 2013. A FOIA request for an FBI file on Kepes was not responsive; it is possible that the FBI did not target Kepes specifically but instead interrogated him as part of investigations into other MIT faculty members.

116 Juliet Kepes Stone, interview with the author, November 2017.

117 Gyorgy Kepes to Julius Stratton, 30 November 1966, MC 431, MIT, box 18, folder "JAS—Kepes, Gyorgy."

118 Hilla Rebay to Gyorgy Kepes, 5 December 1951, Kepes papers, Stanford, box 50, folder 12.

119 Rebay dropped this paragraph before sending the letter to Kepes. Hilla Rebay to Gyorgy Kepes, 2 December 1951 (draft), Guggenheim Museum. The Guggenheim Museum provided this correspondence without box or folder numbers.

120 *Art Education for Scientist and Engineer* [The Report of the Committee for the Study of the Visual Arts of the Massachusetts Institute of Technology, 1952–1954] (Cambridge, MA: MIT, 1957), 16. See also the discussion of this report in Vallye, "Design and the Politics of Knowledge in America," 309.

121 See the transcript of "The Gropius Symposium" (held at the American Academy of Arts and Sciences, Boston, 31 January 1952), Walter Gropius Papers, Houghton Library, Harvard University, folder 142.

122 Ibid., 32.

123 For other readings of silence as a political and aesthetic strategy, see Caroline A. Jones, "Finishing School: John Cage and the Abstract Expressionist Ego," *Critical Inquiry* (Summer 1993): 643–647; and Jonathan D. Katz, "John Cage's Queer Silence; Or, How to Avoid Making Matters Worse," *GLQ* 5, no. 2 (1999): 231–252.

Chapter 3

1 Walter Benjamin, "On the Mimetic Faculty" (1933), in *Reflections: Essays, Aphorisms, Autobiographical Writings*, trans. Edmund Jephcott, ed. Peter Demetz (New York: Schocken, 1978), 336.

2 Jonathan Swift, *Gulliver's Travels* (London: George Bell and Sons, 1914), 192; originally published in 1726. Morrison also quotes this passage. The image used by Morrison appears in this edition of the book.

3 Philip Morrison, "The Modularity of Knowing," in Gyorgy Kepes, ed., *Module, Proportion, Symmetry, Rhythm* (New York: George Braziller, 1966), 1.

4 Ibid.

5 Swift, *Gulliver's Travels*, 190.

6 For a fascinating history of visual knowledge, see Susanna Berger, *The Art of Philosophy: Visual Thinking in Europe from the Late Renaissance to the Early Enlightenment* (Princeton: Princeton University Press, 2017).

7 *Arts of the Environments* is smaller than the other volumes. As Kepes explains, "due to the recession the publisher had to cut down both the size and number of pages of the book." Kepes to Annikki Luukela, 1 February 1972, Kepes papers, AAA, reel 5308, frame 0031.

8   Some of the planned essays were included in the *Arts of the Environment* instead, although there are separate materials for the unpublished book in Kepes's papers.

9   For an exemplary analysis of Moholy's books, see Pepper Stetler, *Stop Reading! Look! Modern Vision and the Weimar Photographic Book* (Ann Arbor: University of Michigan Press, 2015). For a reading of the *Vision + Value* series that emphasizes their role in perceptual training, see Bill Arning, "Gyorgy Kepes' *Vision + Value* 1965–1972: A Practical Training Program to Create a New Super-Perceiving Problem-Solving Artist" (PhD dissertation, Tufts University, 2008).

10   See a copy of the contract, dated 6 February 1961, in the George Braziller Papers, Princeton University, box 18, folder 8.

11   Gyorgy Kepes, introduction to Kepes, ed., *Structure in Art and in Science* (New York: George Braziller, 1965), iv.

12   Both were also key figures in the formation of systems discourse. Intriguingly, Kepes also mentions Ludwig Wittgenstein as an invited speaker, though I find no other evidence of Wittgenstein's visit to Kepes's courses. Kepes explains that the success of these seminars was not always obvious. "Sometimes it was an utter confusion. Sometimes it was bumbled." The courses nonetheless "stimulated" students. Oral history interview with Gyorgy Kepes, 7 March 1972–11 January 1973, Archives of American Art, Smithsonian Institution, page 29 of transcript.

13   "In this issue of *Daedalus* various disciplines are brought to bear on our visual culture," Kepes writes. See his introduction to the issue "The Visual Arts Today,'" *Daedalus* 89, no. 1 (Winter 1960): 3. This issue was later published in book form as Gyorgy Kepes, ed., *The Visual Arts Today* (Middletown, CT: Wesleyan University Press, 1960).

14   Quoted in Abraham H. Maslow, preface to Maslow, ed., *New Knowledge in Human Values* (New York: Harper and Row, 1959), viii.

15   Ibid., viii–ix.

16   Ibid., vii–viii.

17   Pitirim A. Sorokin, "Foreword from the Society," in Maslow, *New Knowledge in Human Values*, xii.

18   Ibid., xiii, xiv; emphasis in original.

19   Gyorgy Kepes, "Comments on Art," in Maslow, *New Knowledge in Human Values*.

20   Gyorgy Kepes, undated note titled "light introduction," Kepes papers, Stanford, box 82, folder 4.

21   Gyorgy Kepes, undated and untitled note, Kepes papers, Stanford, box 9, folder 6.

22   Gyorgy Kepes, undated and untitled note, Kepes papers, Stanford, box 82, folder 4. Emphasis in original.

23   Gyorgy Kepes, introduction to Kepes, ed., *The Nature and Art of Motion* (New York: George Braziller, 1965), v.

24 Gyorgy Kepes, undated and untitled note, Kepes papers, Stanford, box 82, folder 5.

25 Gyorgy Kepes, undated note titled "light," Kepes papers, Stanford, box 83, folder 1.

26 Gyorgy Kepes, undated and untitled note, Kepes papers, Stanford, box 82, folder 6.

27 Gyorgy Kepes, undated note titled "light scale XX century," Kepes papers, Stanford, box 83, folder 1.

28 Gyorgy Kepes, undated and untitled note, Kepes papers, Stanford, box 82, folder 6.

29 Kepes, introduction to *The Nature and Art of Motion*, v.

30 Ibid., vi.

31 Systems theory and cybernetics are distinct; the former derives from biology and is organismic, while the latter derives from engineering and is mechanical. Despite these demarcations, Kepes did not readily distinguish between the two discourses. Such distinctions may be more obvious now, in historical retrospect. George Braziller tended to publish works in systems theory while the Technology Press, now The MIT Press, tended to publish works in cybernetics; but Kepes had actually hoped to work with both presses. For another discussion of these two iterations of systems discourse in the arts, see Caroline A. Jones, "Hans Haacke 1967," in *Hans Haacke 1967* (Cambridge, MA: MIT List Visual Arts Center, 2011), 6–27.

32 See Rudolf Modley, "Graphic Symbols for World-Wide Communication," in Gyorgy Kepes, ed., *Sign, Image, Symbol* (New York: George Braziller, 1966), 108–125.

33 From the copy on the dustjackets for the first six *Vision + Value* volumes; the 1965 and 1966 texts are slightly different.

34 Ibid.

35 Gyorgy Kepes, undated and untitled note, Kepes papers, Stanford, box 83, folder 1.

36 Gyorgy Kepes, introduction to *Structure in Art and in Science*, i.

37 Gyorgy Kepes, undated and untitled note, Kepes papers, Stanford, box 81, folder 3.

38 Gyorgy Kepes, undated and untitled note, Kepes papers, Stanford, box 81, folder 3.

39 Gyorgy Kepes, undated and untitled note, Kepes papers, Stanford, box 82, folder 5.

40 Gyorgy Kepes, undated note titled "Braziller intro," Kepes papers, Stanford, box 82, folder 5.

41 Gyorgy Kepes, undated note titled "XX century," Kepes papers, Stanford, box 4, folder 2.

42 Kepes, introduction to *Structure in Art and in Science*, i.

43 See George Gaylord Simpson, *The Meaning of Evolution: A Study of the History of Life and Its Significance for Man* (New Haven: Yale University Press, 1949). See

also Anna Vallye's discussion of this concept in "Design and the Politics of Knowledge in America, 1937–1967: Walter Gropius, Gyorgy Kepes" (PhD dissertation, Columbia University, 2011), 301.

44   Kepes, introduction to *The Nature and Art of Motion*, xi.

45   Gyorgy Kepes, undated and untitled note, Kepes papers, Stanford, box 83, folder 4.

46   Rudolf Arnheim, "Visual Thinking," in Gyorgy Kepes, ed., *Education of Vision* (New York: George Braziller, 1965), 1.

47   Ibid.

48   Ibid., 2.

49   François Molnar, "The Unit and the Whole: Fundamental Problem of the Plastic Arts," in Kepes, *Module, Proportion, Symmetry, Rhythm*, 211.

50   Heinz von Foerster, "From Stimulus to Symbol: The Economy of Biological Computation," in Kepes, *Sign, Image, Symbol*, 54.

51   Ibid.

52   George Rickey, "The Morphology of Movement: A Study of Kinetic Art," in Kepes, *The Nature and Art of Motion*, 85.

53   Cyril Stanley Smith, "Structure Substructure Superstructure," in Kepes, *Structure in Art and in Science*, 31.

54   Ibid., 32.

55   Ibid.

56   Ibid., 40. See also Smith's volume *A Search for Structure: Selected Essays on Science, Art, and History* (Cambridge, MA: MIT Press, 1981) and his contribution to Judith Wechsler, ed., *On Aesthetics in Science* (Cambridge, MA: MIT Press, 1978). Wechsler's volume acknowledges Kepes's influence; it also emerged from her teaching at MIT in the early 1970s.

57   Richard P. Lohse, "Standard, Series, Module: New Problems and Tasks of Painting," in Kepes, *Module, Proportion, Symmetry, Rhythm*, 161.

58   As recounted by Cage in *A Year from Monday: New Lectures and Writings* (Middletown, CT: Wesleyan University Press, 1967), 120.

59   John Cage, "Rhythm Etc.," in Kepes, *Module, Proportion, Symmetry, Rhythm*, 195.

60   Max Bill, "Structure as Art? Art as Structure?," in Kepes, *Structure in Art and in Science*, 150; and Richard Lippold, "Illusion as Structure," in Kepes, *Structure in Art and in Science*, 153.

61   Ad Reinhardt, "Art in Art is Art as Art," in Kepes, *Sign, Image, Symbol*, 180.

62   Robert Osborn, "The Hangover," in Kepes, *Sign, Image, Symbol*, 184.

63   Lippold, "Illusion as Structure," 162–163.

64 Kepes, introduction to *Education of Vision*, v.

65 Kepes, introduction to *The Nature and Art of Motion*, xi.

66 See Geoffrey C. Bowker, "How to Be Universal: Some Cybernetic Strategies, 1943–1970," *Social Studies of Science* 23, no. 1 (February 1993): 116. Others have offered variations of this concept. Peter Galison defines the "trading zone" as a location where exchange occurs; Susan Leigh Star and James Griesemer define the "boundary object" as a discourse tangential to a field of study; and Fred Turner defines the "network forum" as physical sites that bridge the boundary object with the trading zone. See Peter Galison, "Trading Zone: Coordinating Action and Belief," in Mario Biagioli, ed., *The Science Studies Reader* (New York: Routledge, 1999), 137–160; Susan Leigh Star and James Griesemer, "Institutional Ecology, Translations, and Boundary Objects," *Social Studies of Science* 19, no. 3 (August 1989): 387–420; and Fred Turner, *From Counterculture to Cyberculture: Stewart Brand, the Whole Earth Network, and the Rise of Digital Utopianism* (Chicago: University of Chicago Press, 2006), 72.

67 As noted in von Foerster, "From Stimulus to Symbol," 61.

68 Stanislaw Ulam, "Patterns of Growth of Figures: Mathematical Aspects," in Kepes, *Module, Proportion, Symmetry, Rhythm*, 64.

69 As described in Ulam's biographical sketch in Kepes, *Module, Proportion, Symmetry, Rhythm*, 233. Ulam also invented the "Monte Carlo Method, a procedure for finding solutions to mathematical and physical problems by random sampling"— specifically problems related to weapons design, like calculating potential casualties.

70 See Abraham H. Maslow, "The Authoritarian Character Structure," *Journal of Social Psychology* 18, no. 2 (November 1943): 401–411.

71 Gyorgy Kepes, undated and untitled note, Kepes papers, Stanford, box 83, folder 4.

72 Gyorgy Kepes, undated note titled "art as model," Kepes papers, Stanford, box 9, folder 6.

73 George Gaylord Simpson to Gyorgy Kepes, 18 December 1961. George G. Simpson Papers, American Philosophical Society, no. 31, folder "Kepes, Gyorgy."

74 See her *Native Genius in American Architecture* (New York: Horizon Press, 1957).

75 Sibyl Moholy-Nagy, review of "The New Landscape in Art and Science," *College Art Journal* 18, no. 2 (Winter 1959): 193.

76 For example, she explains that Kepes's images were valuable only if one really believed that "an N-lexaltriaconstane crystal looks like a fine ceramic tile, or that micro abrasions conjure up a moon-lit lake." Moholy-Nagy's comments are ironic: she refers to the apparent irrelevance of the "N-lexaltriaconstane" crystal (actually an n-hexatriacontane crystal; both Kepes and Moholy-Nagy misspell it) to the artist, but the image would later inspire Robert Smithson. For Smithson's use of such images, see Jennifer L. Roberts, *Mirror-Travels: Robert Smithson and*

History (New Haven: Yale University Press, 2004). Smithson studied the same form of crystal growth, as captured in photographs produced by the same scientist, while developing the crystallographic concepts that led to *Spiral Jetty*. Smithson used Charles William Bunn's *Crystals: Their Role in Nature and Science* (1964) in his research; he tore a page from the book that included electron microscope photographs created by the crystallographer I. M. Dawson, who created the photograph in Kepes's book.

77   Kepes separated these letters from the papers that he donated to the Archives of American Art beginning in 1974, presumably because they embarrassed him. They are now located in the Kepes papers, Stanford, box 50, folder 11.

78   Sibyl Moholy-Nagy to Gyorgy Kepes, 15 March 1965, Kepes papers, Stanford, box 50, folder 11.

79   Ibid.

80   Kepes must have assumed that Moholy-Nagy would not want to participate, given her critical assessment of *The New Landscape in Art and Science*; he makes this point in his reply to her, explaining that "these books were formed from essays by people who share a conviction.... I know this is not your thought." Gyorgy Kepes to Sibyl Moholy-Nagy, 31 March 1965, Kepes papers, Stanford, box 50, folder 11.

81   In an earlier letter to Walter Gropius, Moholy-Nagy writes:

> Everyone has forgotten by now that Kepes left Moholy under extremely unpleasant circumstances, accusing him of the most absurd things and challenging his art convictions in every respect. This happened at a time (in 1943) when nothing would have been more vital than a front of unanimity; and when it happened it hurt Moholy deeper than anything that ever happened to him in his professional life.
>
> Then Kepes put out "Language of Vision" in which he carefully and small-mindedly refrained from any reference to his ten years of apprenticeship with Moholy, and which – in many ways – is a direct summary of the ideas he received in those ten years. Kepes himself is very much aware of his deep unfairness to Moholy which came out in a violent and almost tragic conversation we had a few months after Moholy's death. Kepes knows to the bottom of his heart that he did Moholy wrong.

Sibyl Moholy-Nagy to Walter Gropius, 21 October 1948, Walter Gropius Papers, Houghton Library, Harvard University, folder 1222 2/3. Moholy-Nagy wrote to Gropius to discourage the selection of Kepes as the exhibition designer for a 1950 show of Moholy's art at Harvard's Fogg Museum; despite Moholy-Nagy's protest, Kepes ultimately designed the exhibition.

82   Sibyl Moholy-Nagy to Gyorgy Kepes, 15 March 1965, Kepes papers, Stanford, box 50, folder 11.

83   In a 1947 letter to Robert Jay Wolff, she describes a lecture presented by Kepes at the Institute of Design in Chicago: there was "no new thought" in Kepes's talk, just a "regression into a very individualistic responsibility of 'the artist.'" Sibyl Moholy-Nagy to Robert J. Wolff, 12 November 1947, ID Collection, UIC, box 7,

folder 210. Kepes's lecture was later published in *Arts and Architecture* 65 (July/August 1948) as "Form and Motion."

84 Sibyl Moholy-Nagy to Gyorgy Kepes, 15 March 1965, Kepes papers, Stanford, box 50, folder 11.

85 Moholy-Nagy uses the phrase "book puzzle" in a letter commenting on an exhibition of Kepes's paintings. She claims, through a well-crafted backhanded compliment, to have enjoyed the paintings much more than *Vision + Value.* "You have in them [the paintings] a gift for original vision which one would hardly suspect from the book puzzles." Sibyl Moholy-Nagy to Gyorgy Kepes, 29 April 1966, Kepes papers, Stanford, box 50, folder 11.

86 Sibyl Moholy-Nagy to Gyorgy Kepes, 15 March 1965, Kepes papers, Stanford, box 50, folder 11.

87 Ibid.

88 Ibid.

89 Ibid. Moholy-Nagy is also responding to the work of MIT computer engineer Steven A. Coons, with whom Kepes did not directly collaborate.

90 Moholy-Nagy knew she had a reputation for hyperbole: "Any negative comment is brushed off with the remark: 'Ah, yes, Sibyl, she's against anything anyhow.'" Ibid. It is important to emphasize the gendered dimension of such a characterization; in fact, gender structured Moholy-Nagy's exchanges with all the male figures of the Bauhaus diaspora, all of whom dominated the field professionally, often to the exclusion of women connected to the Bauhaus diaspora.

91 That said, the letter is tinged with a dry sarcasm: "though it frequently bothers me, your extra-critical stand in life is a rewarding experience." I think Kepes is sincere despite his tone here. He respected Moholy-Nagy's deft argumentation, even if he disagreed with her arguments. Gyorgy Kepes to Sibyl Moholy-Nagy, 31 March 1965, Kepes papers, Stanford, box 50, folder 11. Moholy-Nagy considered Kepes to be typically reserved; she describes his "reticence" and "the perpetual solemnity of his mien." Sibyl Moholy-Nagy, *Moholy-Nagy: Experiment in Totality* (Cambridge, MA: MIT Press, 1969), 62; originally published in 1950.

92 Gyorgy Kepes to Sibyl Moholy-Nagy, 31 March 1965, Kepes papers, Stanford, box 50, folder 11.

93 Ibid.

94 Arthur M. Schlesinger, Jr., *The Vital Center: The Politics of Freedom* (Boston: Houghton Mifflin, 1949). The book advocated moderation as a means of attaining social stability, which is not unlike Kepes's ambition.

95 Gyorgy Kepes to Sibyl Moholy-Nagy, 31 March 1965, Kepes papers, Stanford, box 50, folder 11.

96 Gyorgy Kepes, undated note titled "Education of vision," Kepes papers, Stanford, box 81, folder 2.

97   Gyorgy Kepes, undated note titled "Braziller intro," Kepes papers, Stanford, box 82, folder 5. I believe Kepes is completely genuine in his fear of disorder. For an inverse argument that contrasts Kepes's fear with an embrace of disorder's critical potential, see Felicity D. Scott's intriguing study *Disorientation: Bernard Rudofsky in the Empire of Signs* (Berlin: Sternberg Press, 2016).

98   Kepes's daughter describes her father's predicament with a simple turn of phrase: "He was caught." Juliet Kepes Stone, interview with the author, November 2017.

99   Sibyl Moholy-Nagy to Gyorgy Kepes, 25 February 1968, Kepes papers, Stanford, box 50, folder 11.

100   Gyorgy Kepes to Sibyl Moholy-Nagy, 1 March 1968, Kepes papers, Stanford, box 50, folder 11.

101   Kepes and Sibyl remained at odds, according to Moholy-Nagy's daughter Claudia Moholy-Nagy, who contacted the Kepeses more than a year later, in 1969: "I know you and Mother have had a few confrontations. In that regard I am totally apolitical and 'side' with no-one. From what she has said to me, I gather that you and Sibyl are on opposite sides of an artistic fence." Claudia Moholy-Nagy to Gyorgy and Juliet Kepes, 29 October 1969, Kepes papers, Stanford, box 50, folder 11.

102   Sibyl Moholy-Nagy, "Hitler's Revenge," *Art in America* (September–October 1968): 42–43. See also Despina Stratigakos's helpful introduction to the essay in "Hitler's Revenge," *Places Journal*, March 2015, https://doi.org/10.22269/150316.

103   There were actual connections between the Bauhaus and fascism, though those are not the focus of Moholy-Nagy's critique. There were also connections between the systems thinkers in Kepes's *Vision + Value* volumes—namely Ludwig von Bertalanffy—and fascism; Bertalanffy joined the Nazi Party to maintain a professorship in Vienna.

104   Moholy-Nagy, "Hitler's Revenge." Breuer and Gropius also participated in Kepes's visual design program; Breuer contributed to the *Vision + Value* series and Gropius contributed to *The New Landscape in Art and Science*.

105   Moholy-Nagy, "Hitler's Revenge."

106   Ibid.

107   Ibid. Breuer won the American Institute of Architects Gold Medal in 1968; Gropius won it in 1959 and Mies in 1960.

108   Walter Gropius to Robert Jay Wolff, 29 October 1968. UIC, box 7, folder 209.

109   Robert Jay Wolff to Walter Gropius, 16 December 1968. UIC, box 7, folder 209.

110   Ibid.

111   The religious references in Kepes's work are not theological so much as metaphysical. For a provocative study that asserts the significance of religiosity in contemporary art, see Thomas Crow, *No Idols: The Missing Theology of Art* (Sydney.

Power Publications, 2017). I also thank Martin Kemp for discussing the thread in Leonardo's print with me.

112 Gyorgy Kepes, *The New Landscape in Art and Science* (Chicago: Paul Theobald, 1956), 327.

Chapter 4

1 Theodor W. Adorno, *Minima Moralia: Reflections from Damaged Life* (London: Verso, 2005), 247; originally published in 1951.

2 Gyorgy Kepes, "Light and Color: Tools (2)," typescript dated 3 November 1964, Kepes papers, Stanford, box 4, folder 7. Kepes's text is heavily annotated, with one set of corrections in pencil and another in pen; my transcription strives for readability over adherence to any single version.

3 For the cultural resonance of the bomb, see Paul Boyer's classic text: *By the Bomb's Early Light: American Thought and Culture at the Dawn of the Atomic Age* (Chapel Hill: University of North Carolina Press, 1985). For a fascinating account of the nuclear imagination, see Spencer R. Weart, *Nuclear Fear: A History of Images* (Cambridge, MA: Harvard University Press, 1988).

4 For a study of "technics-out-of-control," see Langdon Winner's *Autonomous Technology: Technics-out-of-Control as a Theme in Political Thought* (Cambridge, MA: MIT Press, 1977).

5 Kepes, "Light and Color: Tools (2)," typescript dated 3 November 1964, Kepes papers, Stanford, box 4, folder 7.

6 Ibid.

7 Kepes anticipates the argument of Wolfgang Schivelbusch in his *Disenchanted Night: The Industrialization of Light*, trans. Angela Davies (Berkeley: University of California Press, 1983). Schivelbusch explains light as historically contingent—a cultural, social, and political phenomenon rather than a neutral force of nature. He explores the negative role light played in both surveillance and spectacle. For a historical account that instead embraces light's positive attributes, see Chris Otter, *The Victorian Eye: A Political History of Light and Vision in Britain, 1800–1910* (Chicago: University of Chicago Press, 2008). Kate Flint offers a history of photography through photography's light in her *Flash! Photography, Writing, and Surprising Illumination* (Oxford: Oxford University Press, 2017). For a study of darkness, rather than light, see Noam Elcott, *Artificial Darkness: An Obscure History of Modern Art and Media* (Chicago: University of Chicago Press, 2016).

8 Kepes attributes his description of light as the "begetter of vision" to Johann Wolfgang von Goethe; in the margin of his typescript, he roughly transcribes a quotation from Goethe's *Theory of Colours*: "The eye is created by light and for light. Were the eye not sunlike how could we behold light?" Goethe in turn quotes the ancient philosopher Plotinus.

The test Kepes cites actually involved a hydrogen bomb, not an atomic bomb. For the official history of these tests, see *The Effects of Nuclear Weapons*, rev. ed. (Washington, DC: United States Atomic Energy Commission, 1962), 572. See also the description of Operation Hardtack in "Operation Hardtack I 1958," technical report, available from the Defense Technical Information Center: http://www.dtic.mil/dtic/tr/fulltext/u2/a136819.pdf.

9   Kepes, "Light and Color: Tools (2)," typescript dated 3 November 1964, Kepes papers, Stanford, box 4, folder 7.

10   This experiment is described in "Operation Hardtack I 1958," technical report, available from the Defense Technical Information Center: http://www.dtic .mil/dtic/tr/fulltext/u2/a136819.pdf. See also David M. Blades and Joseph M. Siracusa, *A History of U.S. Nuclear Testing and Its Influence on Nuclear Thought, 1945–1963* (Lanham, MD: Rowman and Littlefield, 2014). There are also declassified, though sanitized, films in the public domain that show the caged rabbits during the test: https://www.youtube.com/watch?v=ygQgWKtNFhg.

11   See John Hersey, *Hiroshima* (New York: Alfred A. Knopf, 1946). See also Alexander Nemerov, *Wartime Kiss: Visions of the Moment in the 1940s* (Princeton: Princeton University Press, 2013), 11.

12   There is an undated text labeled "Fiat Lux" with Kepes's Light Book materials in the Kepes papers, Stanford. Kepes elaborates on this reference in "Modulation of Light," in the exhibition catalog *Light as a Creative Medium* (Cambridge, MA: Harvard University, 1965), 12. A footnote in the catalog explains that the text is a "chapter from the forthcoming book 'Light in Art and Architecture,' by Gyorgy Kepes"—in other words, the Light Book.

13   Kepes, "Light and Color: Tools (2)," typescript dated 3 November 1964, Kepes papers, Stanford, box 4, folder 7.

14   In reconstructing Kepes's Light Book, I draw from Susan Buck-Morss's account of Benjamin's Arcades Project in her masterful study *The Dialectics of Seeing: Walter Benjamin and the Arcades Project* (Cambridge, MA: MIT Press, 1989).

15   Gyorgy Kepes, undated and untitled note, Kepes papers, Stanford, box 16, folder 1. Emphasis in original.

16   Gyorgy Kepes, undated and untitled note, Kepes papers, Stanford, box 82, folder 6.

17   Gyorgy Kepes, undated and untitled note, Kepes papers, Stanford, box 82, folder 6.

18   Gyorgy Kepes, untitled [Introduction second version] and undated typescript, 4, Kepes papers, Stanford, box 83, folder 4.

19   Ibid.

20   Gaston Bachelard, *The Flame of a Candle*, trans. Joni Caldwell (Dallas: Dallas Institute Publications, 1988), 64; originally published in 1961. Bachelard's volume is roughly concurrent with Kepes's Light Book.

21 Gyorgy Kepes, "Light-Art on a New Scale," *The Structurist* 13 / 14 (1973 / 1974): 76. The text is a revised version of the text I quote above, published about a decade later.

22 For discussions of Kepes's photography and light commissions, see Elizabeth Finch, "Languages of Vision: Gyorgy Kepes and the 'New Landscape' of Art and Science" (PhD dissertation, City University of New York, 2005), and especially Márton Orosz, "Light as a Creative Medium in the Art of György Kepes," in *A Fényjátékosok: Kepes György és Frank J. Malina a tudomány és a művészet metszéspontján* [*The Pleasure of Light: Gyorgy Kepes and Frank J. Malina at the Intersection of Science and Art*], exh. cat. (Budapest: Ludwig Museum Budapest, 2010), 32–89.

23 Gyorgy Kepes, undated note titled "Introduction Light-Mythology," Kepes papers, Stanford, box 4, folder 7.

24 John F. Kennedy, Radio and Television Address to the American People on the Nuclear Test Ban Treaty, 26 July 1963, transcript available at http://www. jfklibrary.org/Research/Research-Aids/JFK-Speeches/Nuclear-Test-Ban-Treaty _19630726.aspx.

25 The ad is available at https://www.youtube.com/watch?v=dDTBnsqxZ3k.

26 Gyorgy Kepes, undated note titled "Visual Fundamentals," Kepes papers, Stanford, box 83, folder 1. See also Henri Focillon, *The Life of Forms in Art*, trans. Charles Beecher Hogan and George Kubler (New York: Wittenborn, Schultz, 1948).

27 Typed quotation from the Kepes papers, Stanford, box 25, folder 8.

28 Typed quotation from the Kepes papers, Stanford, box 25, folder 8.

29 Gyorgy Kepes, undated note titled "Light Scale XX Century Background," Kepes papers, Stanford, box 83, folder 1. I have added minimal punctuation for readability.

30 See Norbert Wiener, *Cybernetics: Or Control and Communication in the Animal and the Machine* (Cambridge, MA: Technology Press, 1948); and John von Neumann and Oskar Morgenstern, *The Theory of Games and Economic Behavior* (Princeton: Princeton University Press, 1944).

31 See Herman Kahn's *On Thermonuclear War* (Princeton: Princeton University Press, 1960); and his *Thinking about the Unthinkable* (New York: Horizon Press, 1962).

32 The phrase comes from Marcus G. Raskin, "The Megadeath Intellectuals," *New York Review of Books* 1, no. 6 (14 November 1963).

33 For the history of Killian's and Wiesner's government involvements, see Sarah Bridger, *Scientists at War: The Ethics of Cold War Weapons Research* (Cambridge, MA: Harvard University Press, 2015).

34 See the lists of Boston-area owners of Kepes's painting in the Kepes papers, Stanford, box 78, folders 2 and 3.

35 A.D., "Kepes Looks at Light; Sees Spots," *Interiors and Industrial Design* 110, no. 6 (January 1951): 78.

36   Bacon's *New Atlantis* quoted in ibid.

37   Milton C. Nahm, "The Functions of Art and Fine Art in Communication," *Journal of Aesthetics and Art Criticism* 5, no. 4 (June 1947): 273. This issue also contained the artist Thomas Wilfred's article "Light and the Artist." Kepes participated on a panel with Nahm that year at the annual conference of the College Art Association.

38   The folder is in the Kepes papers, Stanford, box 50.

39   Oral history interview with Gyorgy Kepes, 7 March and 30 August 1972, 11 January 1973, Archives of American Art, Smithsonian Institution, page 15 of typescript.

40   See the bibliographies compiled in the Kepes papers, Stanford, box 39, folder 2. One was prepared by a reference librarian at the Library of Congress at Kepes's request in 1945, and two others were prepared by reference librarians at MIT in 1960 and 1965. Of the works I discuss below, William Bragg's *The Universe of Light*, Charles Wheatstone's *Scientific Papers*, John Tyndall's *Six Lectures on Light*, and works by Hamilton Hartridge and Matthew Luckiesh appear in these bibliographies.

41   The diagrams I discuss appear in multiple books by Luckiesh, making it impossible to know with certainty from which source Kepes copied them.

42   The term photogenics is explained in Anne Hoy, "Chronology," in *Gyorgy Kepes: Lightgraphics* (New York: International Center of Photography, 1984), unpaginated, note 7. Kepes apparently asked Hoy to describe his photograms this way.

43   See John Tyndall, *Six Lectures on Light: Delivered in America in 1872–1873* (London: Longmans, Green, 1875).

44   See Hamilton Hartridge, *Colours and How We See Them* (London: Bell, 1949).

45   For more on the technique, see Elizabeth Glassman and Marilyn F. Symmes, *Cliché-verre: Hand-drawn, Light-painted: A Survey of the Medium from 1839 to the Present* (Detroit: Detroit Institute of Arts, 1980).

46   See Charles Wheatstone, *The Scientific Papers of Sir Charles Wheatstone* (London: Taylor and Francis, 1879).

47   The light box is understood as Lerner's invention. When Lerner asked Kepes to author a catalog essay for a 1976 retrospective exhibition of Lerner's photography, Kepes's first draft of the essay explained that it was "mainly through Lerner's ingenuity" that the light box came to be. See the draft in the Nathan Lerner Papers, Chicago History Museum, box 12, folder 6. Lerner was unhappy with the word "mainly," and he asked Kepes to revise the passage. Kepes responded: "If I made some chronological misjudgment I am sorry about it, but as I am sure you know every idea has some parents and sometimes it is difficult to trace the birth certificate with all the legitimate and correct data." Gyorgy Kepes to Nathan Lerner, 9 April 1976, Nathan Lerner Papers, Chicago History Museum, box 12, folder 6. For the exhibition catalog, see *Nathan Lerner: Photographer*

(Chicago: Alan Frumkin Gallery, 1976). See also Lloyd Engelbrecht, "Educating the Eye: Photography and the Founding Generation at the Institute of Design, 1937–46," in David Travis and Elizabeth Siegel, eds., *Taken by Design: Photographs from the Institute of Design, 1937–1971* (Chicago: Art Institute of Chicago, 2002).

48 Gyorgy Kepes to Marc Freidus, 18 July 1978, Kepes papers, Stanford, box 25, folder 9. Kepes also designed an exhibition about Tintoretto at the Art Institute of Chicago in collaboration with Katherine Kuh.

49 Bragg, *The Universe of Light* (London: G. Bell and Sons, 1933), 1.

50 Ibid., 2. Kepes quotes this passage in the transcript of a dictated text made in preparation of his Light Book. Typescript titled "Disc A—Side 1," 1, Kepes papers, Stanford, box 81, folder 2.

51 Gyorgy Kepes, undated note titled "Light," Kepes papers, Stanford, box 16, folder 1; and Gyorgy Kepes, undated note titled "Light," Kepes papers, Stanford, box 83, folder 3.

52 Gyorgy Kepes, undated note titled "Light," Kepes papers, Stanford, box 83, folder 3.

53 Gyorgy Kepes, undated note titled "Light," Kepes papers, Stanford, box 83, folder 3.

54 See the categories listed on Gyorgy Kepes, undated note titled "Light," Kepes papers, Stanford, box 83, folder 3.

55 Gyorgy Kepes, undated note titled "Light," Kepes papers, Stanford, box 77, folder 4.

56 This comment appears throughout Kepes's Light Book manuscripts. It is also published in Gyorgy Kepes, "Light and Design," *Design Quarterly* 68 (1967): 4.

57 Gyorgy Kepes, undated and untitled note, Kepes papers, Stanford, box 82, folder 5.

58 Gyorgy Kepes, undated and untitled note, Kepes papers, Stanford, box 6, folder 1. Kepes's conception of a total situation resonates with the immersive media spaces of Ray and Charles Eames, and with expanded cinema; see Beatriz Colomina, "Enclosed by Images: The Eameses' Multimedia Architecture," *Grey Room* 2 (Winter 2001): 6–29; and Andrew V. Uroskie, *Between the Black Box and the White Cube: Expanded Cinema and Postwar Art* (Chicago: University of Chicago Press, 2014). It also evokes what Fred Turner terms the "democratic surround" in *The Democratic Surround: Multimedia and American Liberalism from World War II to the Psychedelic Sixties* (Chicago: University of Chicago Press, 2013).

59 From a set of notes for a lecture likely delivered at his exhibition "Light as a Creative Medium" at the Carpenter Center for the Visual Arts at Harvard University in 1965. Kepes papers, AAA, reel 5312, frame 1215.

60 Gyorgy Kepes, Memorandum on Color Illumination, dated 20 August 1947, Kepes papers, Stanford, box 83, folder 5, page 1 of typescript. There are two drafts of the memo in this folder.

61 Gyorgy Kepes, "Kinetic Light as a Creative Medium," *Technology Review* 70, no. 2 (December 1967): 25. Part of the problem was that Kepes had no technical understanding of Sylvania's equipment. He explains: "They knew very well that they could not expect from me, a painter, expert comment on technical matters." Ibid.

62 Gyorgy Kepes, Memorandum on Color Illumination, dated 20 August 1947, Kepes papers, Stanford, box 83, folder 5.

63 Ibid. The first term is the original one Kepes uses in his Memorandum on Color Illumination, though he quotes his memo later and revises it with his preferred term, "projected kinetic light forms." See Kepes, "Kinetic Light as a Creative Medium," 25.

64 Tom Johnston of the advertising firm Cecil & Presbrey, as quoted in "Light Mobile—Partial Chronology," undated document, Kepes papers, Stanford, box 4, folder 10.

65 Kepes, "Kinetic Light as a Creative Medium," 26. The project also reflects the influence of Moholy's Light-Space Modulator on Kepes's thinking.

66 See "Light Mobile—Partial Chronology," undated document, Kepes papers, Stanford, box 4, folder 10.

67 Gyorgy Kepes, undated note titled "Light," Kepes papers, Stanford, box 77, folder 5.

68 Sylvania's annual reports reveal their increasing involvement in military projects through the 1950s.

69 Gyorgy Kepes, undated note titled "Sylvania Light Mobile," Kepes papers, Stanford, box 83, folder 3.

70 Gyorgy Kepes, undated note titled "Randomness," Kepes papers, Stanford, box 4, folder 2.

71 Ralph K. Potter, "New Scientific Tools for the Arts," *Journal of Aesthetics and Art Criticism* 10, no. 2 (December 1951): 127. Kepes cites Potter on an undated note titled "Light, Mobile Light, Color Music," Kepes papers, Stanford, box 82, folder 5.

72 Bainbridge Bishop, *A Souvenir of the Color Organ, with Some Suggestions in Regard to the Soul of the Rainbow and the Harmony of Light* (New Russia, NY: De Vinne Press, 1893). Kepes also lists this text in his bibliographies on light.

73 Gyorgy Kepes, undated note titled "Palette of Light Technology," Kepes papers, Stanford, box 4, folder 2.

74 Gyorgy Kepes, undated note titled "Modulators," Kepes papers, Stanford, box 83, folder 1.

75 Gyorgy Kepes, undated and untitled note, Kepes papers, Stanford, box 4, folder 2.

76 Kepes, "Kinetic Light as a Creative Medium," 26.

77   For an exception, see Márton Orosz, "A látás revíziója. Művészet mint humaniesta tudomány Kepes György korai életművében" (PhD dissertation, Eötvös Loránd University, 2013).

78   Otto Piene, untitled statement in *Gyorgy Kepes: Works in Review* (Boston: Museum of Science, 1973), 23.

79   See the announcements of Kepes's retirement in the *The Tech* (10 July 1974) and the *Christian Science Monitor* (28 June 1974).

80   The existence of two archives of Kepes's papers, one in the Archives of American Art [AAA] at the Smithsonian Institution and a second in the Department of Special Collections at Stanford University, is a consequence of Kepes's Light Book. Kepes donated the material to the AAA in his lifetime, beginning in 1974, but withheld material related to the Light Book with the hope of completing the project. The Gyorgy Kepes papers at Stanford comprise material organized by Kepes's estate following the artist's death in 2001; the bulk of the archive contains this material.

81   See his proposal to the Guggenheim Foundation in the Kepes papers, Stanford, box 83, folder 5.

82   The Light Book material supported Kepes's exhibitions, lectures, essays, reports on the CAVS, and his *Vision + Value* volumes. Kepes constantly revised past material for future projects. For example, he often scrawled "light" on top of notes originally labeled "Visual Fundamentals," "Education of Vision," and "New Landscape," subsuming them into the project. In an attempt to differentiate old purposes from new, he created additional labels in bright marker in the 1970s: (T) for technology, (P) for pollution or participation, and (H) for homeostasis, often written over other labels.

83   Kepes's idea for a study of light may have emerged from collaborations with László Moholy-Nagy. In a letter to Kepes from the 1980s, Hattula Moholy-Nagy, Moholy's daughter, asks about a similar project: "I came across the mention of a 'Lichtbuch' [Light Book] that you and my father had begun to put together in Berlin." Hattula Moholy-Nagy to Gyorgy Kepes, 28 November 1988, Kepes papers, Stanford, box 36, folder 6. Mohly-Nagy was inquiring about an "incomplete 'contact book' of photographs" by her father that she suggested might be the actual "Lichtbuch." Given the nature of Kepes's archive, this contact book does not appear to be related to Kepes's project. In response to Hattula Moholy-Nagy's inquiry, Kepes neither denied nor affirmed the existence of a Lichtbuch (though he did invoke the metaphor of light): "I wish I could bring some light to this misty past. I have a vague recollection that Moholy-Nagy and I planned a book on light, but the book never happened." Gyorgy Kepes to Hattula Moholy-Nagy, 12 January 1989, Kepes papers, Stanford, box 36, folder 6. Kepes may have purposely disavowed Moholy's influence.

84   Gyorgy Kepes, untitled [Introduction third version] and undated typescript, page 1, Kepes papers, Stanford, box 16, folder 2.

85   Gyorgy Kepes, undated and untitled note, Kepes papers, Stanford, box 4, folder 6.

86   Kepes was not alone in attempting to chronicle light art. See Nan R. Piene, "Light Art," *Art in America* (May–June 1967). Kepes apparently assisted Piene as she researched her article; he was recruiting her then-husband, Otto Piene, to be an inaugural fellow at CAVS. See also Willoughby Sharp, "Luminism and Kineticism," in Gregory Battcock, ed., *Minimal Art: A Critical Anthology* (New York: E. P. Dutton, 1968), 317–358. The portion of Sharp's essay on light first appeared in the catalog *Light, Motion, Space* (Minneapolis: Walker Art Center, 1967).

      For a recent attempt to historicize these trends, see Andrea von Hülsen-Esch and Dirk Pörschmann, eds., *The Medium of Light in the Context of the Neo-Avant-Garde of the 1950s and 1960s* (Düsseldorf: Düsseldorf University Press, 2013); and especially the research of Tina Rivers Ryan.

87   These are the names Kepes listed on folders.

88   Gyorgy Kepes, typed application, page 4, Kepes papers, Stanford, box 83, folder 5.

89   Gyorgy Kepes, untitled [Introduction second version] and undated typescript, page 5, Kepes papers, Stanford, box 83, folder 4.

90   For a meditation on the strangeness of Edgerton's images, see James Elkins, "The Rapatronic Camera," in *What Photography Is* (London: Routledge, 2011), 160–176.

91   Gyorgy Kepes to Harold Edgerton, Harold Edgerton Papers, MC 25, MIT, box 24, folder 5. These images later appeared in Gyorgy Kepes, ed., *Arts of the Environment* (New York: George Braziller, 1972).

92   Gyorgy Kepes, "The Artist's Response to the Scientific World," in *Technology and Culture in Perspective: An Occasional Paper Published by the Church Society for College Work* (Cambridge, MA: Church Society for College Work, 1966), 36.

93   See Sam Hunter, "The Romantic Outlook," *Art in America* 48, no. 1 (1960): 20–21.

94   Kepes learned of the window tax from J. L. Hammond and Barbara Hammond, *The Bleak Age* (West Drayton, England: Penguin Books, 1947); originally published in 1934. See the Kepes papers, Stanford, box 25, folder 8.

95   Gyorgy Kepes, "The Nineteenth Century," undated typescript, Kepes papers, Stanford, box 82, folder 1.

96   Ibid.

97   Ibid.

98   Gyorgy Kepes, undated note titled "Light 19 C.," Kepes papers, Stanford, box 83, folder 5.

99   Gyorgy Kepes, undated note titled "Light 19 C.," Kepes papers, Stanford, box 83, folder 5.

100   Gyorgy Kepes, undated note titled "Light 19 C.," Kepes papers, Stanford, box 83, folder 5.

101 Gyorgy Kepes, undated note titled "Light 19 C.," Kepes papers, Stanford, box 83, folder 5.

102 Gyorgy Kepes, undated and untitled note, Kepes papers, Stanford, box 77, folder 5.

103 Kepes may have pulled the pages from Klingender's volume in order to hide the book; it would have been an incriminating title in the 1950s, when Kepes was worried about FBI investigation. The very same speech also inspired the name of the Cambridge-based underground newspaper the *Old Mole*; see chapter 5.

104 Some appear in Kepes, *Arts of the Environment*.

105 Gyorgy Kepes, undated note titled "Introduction," Kepes papers, Stanford, box 77, folder 5.

106 Gyorgy Kepes, "The Twentieth Century," undated typescript, Kepes papers, Stanford, box 83, folder 1.

107 Gyorgy Kepes, undated note titled "Introduction," Kepes papers, Stanford, box 77, folder 5.

108 Gyorgy Kepes, undated note titled "Introduction," Kepes papers, Stanford, box 77, folder 4.

109 Gyorgy Kepes, untitled [Introduction first version] and undated typescript, page 6, Kepes papers, Stanford, box 83, folder 1.

110 Max Horkheimer and Theodor W. Adorno, *Dialectic of Enlightenment*, ed. Gunzelin Schmid Noerr, trans. Edmund Jephcott (Stanford: Stanford University Press, 2002), 187; originally published in 1947.

111 Gyorgy Kepes to Seymour Lawrence, 20 November 1962, Kepes papers, Stanford, box 5, folder 7.

112 Gyorgy Kepes to Paul Elek, 15 October 1972, Kepes papers, Stanford, box 37, folder 3.

113 Gyorgy Kepes, transcript of lecture, Kepes papers, AAA, reel 5312, frame 1215.

114 Kepes repeats the aphorism often; it also appears in a typed list of quotations from Eckhart in the Kepes papers, Stanford, box 26, folder 3.

Chapter 5

1 Herbert Marcuse, *One-Dimensional Man: Studies in the Ideology of Advanced Industrial Society* (Boston: Beacon Press, 1964), 65.

2 It was held at MIT's Hayden Gallery from 28 February to 29 March 1970. It was held at the National Collection of Fine Arts from 4 April to 10 May 1970.

3 Other participants included Stephan Antonakos, Billy Apple, Charles Frazier, and John Goodyear. At the National Collection of Fine Arts, they also included the artists Lila Katzen, Preston McClanahan, Gary Thomas Rieveschl, Charles Ross, Friedrich St. Florian, James Seawright, Vera Simons, and William H. Wainwright,

and the engineers Richard L. Venezky and Jerry A. Erdman. See the exhibition catalog for *Explorations*. A poster for the exhibition at MIT's Hayden Gallery also lists affiliated artists Frank Carlton and David Morris, and technical collaborators Albert S. Makas, William M. Murray, Giorgio Trapani, Paul Earls, Paul Hughes, Jeffrey Millman, Phil Smith, William L. Verplank, Clifford A. Henricksen, and Henry H. Kolm. See the poster for *Exploration* in the Center for Advanced Visual Studies (CAVS) Special Collection, Massachusetts Institute of Technology [hereafter CAVS Special Collection, MIT]. I viewed this archive before all materials were assigned box or folder numbers.

4   Kepes quoted in "São Paulo, No! Smithsonian, Yes! Conversations with Gyorgy Kepes," *Arts Magazine* 44, no. 7 (May 1970): 16, 18.

5   Ibid., 16.

6   As per a visitor quoted in Grace Glueck, "'Explorations' Spotlights Use of Technology in Art," *New York Times* (6 April 1970): 48.

7   Ibid.

8   Kepes quoted in Meryle Secrest, "'Explorations': Art Weds Technology," *Washington Post* (4 April 1970): C1.

9   Gyorgy Kepes, "Toward Civic Art," *Explorations*, exh. Cat. (Washington, DC: National Collection of Fine Arts, 1970), unpaginated. Kepes later published this essay, in slightly revised form, in *Leonardo* 4, no. 1 (Winter 1971): 69–73.

10   Ibid.

11   As claimed by Gyorgy Kepes to Lois Bingham, 18 March 1970, CAVS Special Collection, MIT.

12   Gyorgy Kepes to John E. Burchard, 17 April 1970, CAVS Special Collection, MIT.

13   Gyorgy Kepes, "Center for Advanced Visual Studies," report in *Massachusetts Institute of Technology Bulletin* 106, no. 2 (September 1971): 35.

14   Lawrence Alloway, "Art," *The Nation* (20 April 1970): 477; emphasis in original.

15   Paul Richard, "Making a Work of Art at the Flip of a Poker Chip," *Washington Post* (10 May 1970): F8; and Glueck, "'Explorations' Spotlights Use of Technology in Art," 48.

16   Douglas Davis, "Improbable Marriage," *Newsweek* 75, no. 16 (20 April 1970): 101.

17   Emily Wasserman, "Washington," *Artforum* 8, no. 10 (June 1970): 89.

18   Ibid., 87.

19   For a comparison of Kepes's Center to these and other examples, see Anne Collins Goodyear, "The Relationship of Art to Science and Technology in the United States, 1957–1971: Five Case Studies" (PhD dissertation, University of Texas at Austin, 2002). For more on the A&T program at LACMA, see Pamela M. Lee,

"Eros and Technics and Civilization," in *Chronophobia: On Time in the Art of the 1960s* (Cambridge, MA: MIT Press, 2004).

20 Anne Collins Goodyear, "From Technophilia to Technophobia: The Impact of the Vietnam War on the Reception of 'Art and Technology,'" *Leonardo* 41, no. 1 (2008): 169–173. See also her essay "Gyorgy Kepes, Billy Klüver, and American Art of the 1960s: Defining Attitudes toward Science and Technology," *Science in Context* 17, no. 4 (2204): 611–635.

21 Kepes, "Toward Civic Art," no pagination. Kepes's concept anticipates Nicolas Bourriaud's notion of relational aesthetics. See Bourriaud's *Relational Aesthetics*, trans. Simon Pleasance and Fronza Woods (Dijon: Les Presses du Réel, 1998). See also an important critique of Bourriaud: Claire Bishop, "Antagonism and Relational Aesthetics," *October* 110 (Fall 2004): 51–79. Bishop argues instead for "relational antagonism."

22 Chryssa instead represented the United States, for which she was widely condemned. For an account of the tenth São Paulo Biennial from the perspective of Brazilian artists, see Claudia Calirman, *Brazilian Art under Dictatorship: Antonio Manuel, Art Barrio, and Cildo Meireles* (Durham: Duke University Press, 2012), 10–36. Isobel Whitelegg discusses the longer history of the Biennial under the reign of Brazil's military regime in "The Bienal de São Paulo: Unseen/Undone (1969–1981)," *Afterall* 22 (Autumn/Winter 2009), available at http://www.afterall.org/journal/issue.22/the.bienal.de.so.paulo.unseenundone.19691981. See also Gloria Sutton, *The Experience Machine: Stan VanDerBeek's Movie-Drome and Expanded Cinema* (Cambridge, MA: MIT Press, 2015), 58–61.

23 On the *coup d'état*, see Phyllis R. Parker, *Brazil and the Quiet Intervention, 1964* (Austin: University of Texas Press, 1979). Parker's account was the first to substantiate suspicions about "Operation Brother Sam" with archival evidence obtained from the files of the Johnson administration. For a study of how Americans viewed the events, see James N. Green, *We Cannot Remain Silent: Opposition to the Brazilian Military Dictatorship in the United States* (Durham: Duke University Press, 2010).

24 Kepes, "Toward Civic Art," no pagination.

25 Again, I borrow the phrase from Rebecca Lowen; see her *Creating the Cold War University: The Transformation of Stanford* (Berkeley: University of California Press, 1997). Her description of Stanford as the archetypal "Cold War University" is equally appropriate for MIT.

26 Janet Brown et al., "War Incorporated: The Complete Picture of the Congressional-Military-Industrial-University Complex," booklet published by the Student Research Facility, Berkeley, c. 1970.

27 Kepes is comparing students at MIT to those from Harvard who also took his courses, and had a broader "cultural orientation." Oral history interview with Gyorgy Kepes, 7 March 1972–11 January 1973, Archives of American Art, Smithsonian Institution, page 28 of transcript.

28 "M.I.T. and the WARFARE STATE," 19, photocopied pamphlet in the Pounds Panel Papers, Massachusetts Institute of Technology, folder 21, box 1. See also Kelly Moore, *Disrupting Science: Social Movements, American Scientists, and the Politics of the Military, 1945–1975* (Princeton: Princeton University Press, 2008), 137; and Sarah Bridger, *Scientists at War: The Ethics of Cold War Weapons Research* (Cambridge, MA: Harvard University Press, 2015), 162–163.

29 "M.I.T. and the WARFARE STATE," 19.

30 Ibid., 22.

31 Reinhold Martin, "Organicism's Other," *Grey Room* 4 (Summer 2001): 46. Martin expands this reading of Kepes in *The Organizational Complex: Architecture, Media, and Corporate Space* (Cambridge, MA: MIT Press, 2003). See also Melissa Ragain, "From Organization to Network: MIT's Center for Advanced Visual Studies," *X-tra* 14, no. 3 (Spring 2012), available at http://x-traonline.org/article/from -organization-to-network-mits-center-for-advanced-visual-studies/. Anne Collins Goodyear and Elizabeth Finch offer more general histories of the Center; see Goodyear, "The Relationship of Art to Science and Technology in the United States," and Finch, "Languages of Vision: Gyorgy Kepes and the 'New Landscape' of Art and Science" (PhD dissertation, City University of New York, 2005). Matthew Wisnioski considers what the Center offered scientists and engineers, particularly after Kepes's tenure as Director. See his "Centerbeam: Art of the Environment," in Arindam Dutta, ed., *A Second Modernism: MIT, Architecture, and the "Techno-Social" Moment* (Cambridge, MA: MIT Press, 2013), 188–225. See also Matthew Wisnioski, "Why MIT Institutionalized the Avant-Garde: Negotiating Aesthetic Virtue in the Postwar Defense Institute," *Configurations* 21, no. 1 (Winter 2013): 85–116.

32 Kepes, "The Visual Arts and the Sciences: A Proposal for Collaboration," *Daedalus* 94, no. 1 (Winter 1965): 145. Published in book form as Gerlad Holton, ed., *Science and Culture: A Study of Cohesive and Disjunctive Forces* (Boston: Beacon Press, 1965).

33 Stefan P. Munsing, Director, Art in Embassies Program, US Department of State, to Gyorgy Kepes, 24 April 1969, Kepes papers, Stanford, box 26, folder 16. Kepes's paintings were also "on duty" at MIT; the Institute's James R. Killian, Jr., recalls that he found "inspiration and delight" from the art by Kepes that hung on his office wall while Killian went about his "day-to-day duties." James R. Killian, Jr., *The Education of a College President: A Memoir* (Cambridge, MA: MIT Press, 1985), 236.

34 The cartoon is reprinted with a brief commentary titled "To See Ourselves as Others See Us," *Technology Review* (May 1958): 296; originally published in the 18 January 1958 issue of the *New Yorker*.

35 Jonathan Benthall, "Kepes's Center at M.I.T.," *Art International* 19, no. 1 (January 1974): 28.

36 Robert Reinhold, "M.I.T. Center Seeks to Wed Esthetics and Technology," *New York Times* (26 December 1969): 31.

37   A statement signed by Kepes on this letterhead is included in the Artists and Writers for Hughes Records, Houghton Library, Harvard University, box 1, folder 28.

38   "March on Washington," advertisement, *New York Times* (23 November 1965): 28.

39   Todd Gitlin explains how SANE's march also effectively coopted the radical elements of the antiwar movement, like SDS. See Gitlin's *The Whole World Is Watching: Mass Media and the Making and Unmaking of the New Left* (Berkeley: University of California Press, 1980).

40   "A Reply to Secretary Rusk on Vietnam," advertisement, *New York Times* (9 May 1965): E6.

41   "Scientists and Engineers for McCarthy," advertisement, *Boston Globe* (15 December 1967): 35.

42   "The War in Indochina must be stopped!," advertisement, *The Tech* (23 April 1971): 5.

43   A copy of the letter is in the Kepes papers, Stanford, box 32, folder 5.

44   For more on this portfolio, see Makeda Best, "Twenty-Five Prints for Artists against Racism and the War, 1968," *Art in Print* 3, no. 6 (2014): 15–19.

45   Gyorgy Kepes to Joan Seeman Robinson, 13 January 1988, Kepes papers, Stanford, box 36, folder 5. Kepes was replying to a query from Robinson about art and the Vietnam War, a topic Robinson was then researching.

46   Gyorgy Kepes, "The Visual Arts and Sciences: A Proposal for Collaboration," *Daedalus* 94, no. 1 (Winter 1965): 122. It is worth noting that the journal frequently considered these topics; see also the later "Art and Science" issue: *Daedalus* 115, no. 3 (Summer 1985).

47   László Moholy-Nagy, *Vision in Motion* (Chicago: Paul Theobald, 1947), 284; 359.

48   The diagram first appeared in a 1958 report but was also reprinted in *Technology Review* (December 1960).

49   Kepes, "The Visual Arts and Sciences: A Proposal for Collaboration," 123.

50   "A Center for Visual Studies," Introduction, 1, MIT, AC 134, box 29, folder 17. This introductory text is uncredited but likely authored by Lawrence B. Anderson, who probably draws from Kepes's own published proposal.

51   Lawrence B. Anderson, "The Visual Arts in the M.I.T. Community," dated 8 March 1963, MIT, AC 134, box 29, folder 17.

52   Gyorgy Kepes, interviewed by Douglas M. Davis, "Art & Technology—Conversations," *Art in America* 56, no. 1 (January–February 1968): 40. Daniel Bell elaborates on "intellectual technology" in multiple writings from the 1960s.

53   Kepes also imagined a symposium that would put these scientists in dialogue with figures from the arts, including art historians like Herbert Read, Kenneth

Clark, Rudolf Wittkower, Erwin Panofsky, and James Ackerman. Gyorgy Kepes to Julius A. Stratton, 17 June 1965, AC 31, MIT, box 1, folder "Center for Advanced Visual Studies."

54   Ibid.

55   The Andrew W. Mellon Foundation was formed in 1969 through the consolidation of two philanthropic organizations established by the children of Andrew W. Mellon: The Old Dominion Foundation and the Avalon Foundation.

56   Gyorgy Kepes, "A Collaborative Approach at the Center for Advanced Visual Studies: Prepared for the Old Dominion Foundation," typescript dated April 1966, AC 8, box 193, folder "Center for Advanced Visual Studies 3/3."

57   Gyorgy Kepes, "The Center for Advanced Visual Studies," brochure, no pagination, emphasis in original.

58   Grace Marmor Spruch, "A Report on a Symposium on Art and Science," *Artforum* 7, no.1 (January 1969), 32.

59   Ibid., 32.

60   Kepes, "The Visual Arts and the Sciences: A Proposal for Collaboration," 124.

61   Gyorgy Kepes, memorandum, 5 May 1969; Gyorgy Kepes, memorandum, 27 May 1971; and Gyorgy Kepes, memorandum, 22 November 1968, CAVS Special Collection, MIT.

62   Gyorgy Kepes, memorandum, 21 January 1972, CAVS Special Collection, MIT.

63   Gyorgy Kepes, memorandum, 4 February 1969, CAVS Special Collection, MIT.

64   Gyorgy Kepes, memorandum, 7 July 1969, CAVS Special Collection, MIT.

65   Gyorgy Kepes to Kay Stratton, 24 April 1969, CAVS Special Collection, MIT.

66   Gyorgy Kepes to James R. Killian, Jr., 3 March 1969, CAVS Special Collection, MIT.

67   Gyorgy Kepes, memorandum, 13 May 1970, CAVS Special Collection, MIT.

68   Charles Frazier to Gyorgy Kepes, 19 December 1969, CAVS Special Collection, MIT.

69   Kepes makes this "gas station" allusion in a later memorandum to the fellows: "Artists should not come to MIT simply to use it as a filling station of technical knowledge or help." Gyorgy Kepes, memorandum, 20 July [1971], CAVS Special Collection, MIT. This document is a transcript from a dictated memo; a version was circulated to fellows on 10 September 1971.

70   Gyorgy Kepes to Charles Frazier, 23 December 1969, CAVS Special Collection, MIT.

71   Daniel Bell, *The Reforming of General Education: The Columbia Experience in Its National Setting* (New Brunswick, NJ: Transaction Publishers, 2011), 76; originally published in 1966.

72 See Peter Galison, "Ontology of the Enemy: Norbert Wiener and the Cybernetic Vision," *Critical Inquiry* 21, no. 1 (Autumn 1994): 260.

73 The line derives from Marx's "Speech on the Anniversary of the *People's Paper*," an appropriate source for a newspaper. See *Karl Marx: Selected Writings*, ed. David McLellan (Oxford: Oxford University Press, 2000), 369. Ironically, this speech also appears in the materials Kepes compiled for his Light Book.

74 Gyorgy Kepes to Julius A. Stratton, 17 June 1965, MIT, AC 31, box 1.

75 For primary documents relating to the March 4th movement, see Jonathan Allen's edited volume *March 4: Scientists, Students, and Society* (Cambridge, MA: MIT Press, 1970); see also Dorothy Nelkin, *The University and Military Research: Moral Politics at M.I.T.* (Ithaca: Cornell University Press, 1972). For a history of the movement, see Stuart Leslie, "'Time of Troubles' for the Special Laboratories," in David Kaiser, ed., *Becoming MIT: Moments of Decision* (Cambridge, MA: MIT Press, 2010), 123–143. See also the accounts in Stuart W. Leslie, *The Cold War and American Science: The Military-Industrial-Academic Complex at MIT and Stanford* (New York: Columbia University Press, 1993), 233–256; Kelly Moore, *Disrupting Science*, 137–146; and Bridger, *Scientists at War*, 155–196. For an important study of the impact of these events on MIT's Urban Systems Laboratory, rather than on the CAVS, see Felicity D. Scott, "DISCOURSE, SEEK, INTERACT," in her *Outlaw Territories: Environments of Insecurity/Architectures of Counterinsurgency* (New York: Zone Books, 2016).

76 Howard Zinn, "The Academic Community and Governmental Power," in Allen, *March 4*, 60. Allen's volume includes transcripts of all the 4 March talks.

77 Review Panel on Special Laboratories, "Final Report" [Pounds Panel Report] (October 1969), 6. A copy of this report is available in the Institute Archives and Special Collections at MIT.

78 Howard Wesley Johnson, *Holding the Center: Memories of a Life in Higher Education* (Cambridge, MA: MIT Press, 1999), 169.

79 See Kepes's handwritten note in a folder labeled "Sao Paulo Log." His secretary later transcribed the note into a typed logbook labeled "Sao Paulo Contacts Made." The entry is dated 31 March 1969, CAVS Special Collection, MIT.

80 Ibid. This entry is dated 16 April 1969. Although the logbook does not indicate Wen-Ying Tsai's attendance at the meeting, the "cybernetic form" under discussion may involve technology later used in Tsai's *Cybernetic Sculpture System*, which employed a "feedback controlled oscillator" to vary the rate of a strobe light in response to ambient noise, thus making the work's vibrating rods appear to undulate in response to viewers.

81 C. S. Draper, "Technology, Engineering, Science and Modern Education," *Leonardo* 2, no. 2 (April 1969): 147–153.

82 Ibid., 147.

83 Ibid., 152. Kepes would also write for *Leonardo* and was a member of the journal's editorial board. In a published letter to the journal, he praised it: "I received my copy of *Leonardo* and want to tell you how impressed I am by it. I know too well the bitter experience of bringing people together to perform; how many headaches and disappointments are inevitably involved. You must feel happy about and proud of the final result." Gyorgy Kepes, letter to the editor, *Leonardo* 1, no. 3 (July 1968): 343.

84 Jack Nolan to Gyorgy Kepes, 14 December 1967, AC 8, MIT, box 193, folder: "Center for Advanced Visual Studies 3/3 1966–1980."

85 Gyorgy Kepes to Jack Nolan, 20 December 1967, AC 8, MIT, box 193, folder: "Center for Advanced Visual Studies 3/3 1966–1980."

86 See Nolan's obituary: Bryan Marquard, "Jack Nolan, Unusual Blend of an Artist, Scientist; at 83," *Boston Globe* (4 December 2007): A15. See also the recollections written by his colleague at Lincoln Laboratory, Amedio Armenti, who remembers Nolan's disaffection with the Lab due to its military involvements: https://alarmenti.wordpress.com/non-fiction/part-iv/.

87 See Clement Greenberg, "Avant-Garde and Kitsch," *Partisan Review* 6, no. 5 (Fall 1939): 34–49.

88 Gyorgy Kepes, memorandum, 4 March 1969, CAVS Special Collection, MIT.

89 *Explorations* exhibition catalog, no pagination. The Smithsonian staff argued that sponsors would be "attracted to the project if they consider the enormous prestige which would accrue" from supporting MIT. David W. Scott to Gyorgy Kepes, 14 February 1969, AC 8, MIT, box 193, folder "Center for Advanced Visual Studies 3/3 1966–1980."

90 Review Panel on Special Laboratories, "Final Report."

91 From the industrial sponsors listed in the *Explorations* catalog.

92 Gyorgy Kepes to Kenneth Germeshausen, E. G. & G., 30 April 1969, CAVS Special Collection, MIT.

93 See Gyorgy Kepes to Emilio G. Collado, 11 July 1969; Gyorgy Kepes to Emilio G. Collado, 8 August 1969; and John C. Haas to Vince Fulmer, 11 June 1969, CAVS Special Collection, MIT.

94 Executives from these companies likewise attended the exhibition opening at the National Collection of Fine Arts.

95 Steve Kaiser, "Alumni Confront SACC; Johnson Cites 'the Other 99%'," *The Tech* (23 July 1969): 3.

96 "M.I.T. 1969 Alumni Day Homecoming Afternoon Program—'The Human Purpose.'" Kepes papers, AAA, reel 5306, frame 0484–0485.

97 Ibid.

98   Claude W. Brenner to Gyorgy Kepes, 20 January 1969; Claude W. Brenner to Gyorgy Kepes, 19 May 1969, Kepes papers, AAA. Brenner recounts his work on weapons testing in an interview with Forrest Larson, 1 May 2009, available at https://libraries.mit.edu/music-oral-history/files/2013/05/BRE20090501.pdf. Brenner was also a witness at Operation Hardtack, the nuclear test that fascinated Kepes for its production of light. See chapter 4.

99   Quoted in Kaiser, "Alumni Confront SACC; Johnson Cites 'the Other 99%'," 3.

100   Saverio G. Greco, "Another Hippie College," letter to the editor, *The Tech* 90, no. 30 (15 September 1970): 4.

101   Review Panel on Special Laboratories, "Final Report," 17.

102   Ibid., 85–86. Chomsky's position relative to MIT's military involvements obviously represents a stark contrast with Kepes's more muted response, but it must be emphasized that even figures like Chomsky were themselves entangled in the institution's defense agenda. See Chris Knight, *Decoding Chomsky: Science and Revolutionary Politics* (New Haven: Yale University Press, 2016).

103   Review Panel on Special Laboratories, "Final Report," 51.

104   Gyorgy Kepes, memorandum, 15 October 1969, CAVS Special Collection, MIT.

105   William Irwin Thompson, *At the Edge of History* (New York: Harper and Row, 1971), 62, 67. Kepes retained an underlined fragment from Thompson's book in his papers, suggesting he was aware of, and bothered by, this critique. See the Kepes papers, Stanford, box 8, folder 4.

106   Thompson, *At the Edge of History*, 60.

107   Ibid., 68. Thompson's citation of Cardinal Newman at Oxford is a direct rebuttal to Clark Kerr, who also cites Newman in *The Uses of the University* (Cambridge, MA: Harvard University Press, 1963).

108   Benthall, "Kepes's Center at M.I.T.," 49.

109   The Academy's founding charter even declared that the institution's mission was "to cultivate every art and science." The line is printed in the journal's front matter.

110   See Stephen R. Graubard, "Note on the Tenth Anniversary," *Daedalus* 97, Tenth Anniversary Index: 1958–1968 (1968): 2; and Stephen R. Graubard, "*Daedalus*: Forty Years On," *Daedalus* 128 (1999): 9.

111   Herbert Marcuse, "Remarks on a Redefinition of Culture," *Daedalus* 94, no. 1 (Winter 1965): 193. Paradoxically, Kepes actually saw himself aligned with Marcuse; in a notebook, he alludes to Marcuse's *One-Dimensional Man*: "my paintings are wish action against pragmatic technological onesidedness—against one dimension man." From one of Kepes's notebooks in the Kepes papers, Stanford, box 20.

112   Ibid., 195.

113   Ibid., 205.

114   Ibid., 203.

115   Ibid., 206.

116   Herbert Marcuse, *An Essay on Liberation* (Boston: Beacon Press, 1969), 12.

117   Otto Piene, Kepes's successor as director of the Center, also described the Center in terms that resonate with conversion. Piene states that the values motivating the Center were *"love* and *peace,* and *interaction* among well-meaning creative people.... And these peace-loving people were not only among the artists, they were also among the scientists.... Some of them had turned peace activists, and some of them had turned lovers of art. And had to develop a rather different profile from than [*sic*] they had been made to adhere to during the war when very very [*sic*] serious stuff was being developed at MIT." Otto Piene and Matthew Wisnioski, "Art / Science / Technology," in Dutta, *A Second Modernism,* 772–773. Emphasis in original. Wisnioski also asked Piene whether there was "talk among the fellows about CAVS as a 'Cold War' institution," a question that Piene rejects as "a pretty bold concept!"

118   For more about Universitas, see "Designing Environments" in Felicity D. Scott, *Architecture or Techno-utopia: Politics after Modernism* (Cambridge, MA: MIT Press, 2007).

119   Gyorgy Kepes, from the transcript of the Third Working Session, Sunday 9 January 1972, in Emilio Ambasz, ed., *The Universitas Project: Solutions for a Post-Technological Society* (New York: Museum of Modern Art, 2006), 403–404.

120   Hans Haacke to Gyorgy Kepes, 22 April 1969, AC 118, MIT, box 39. See also the original in the CAVS Special Collection, MIT. This letter was circulated by Haacke and Clay in French; a French copy is included among the materials Kepes received from Clay. There are also two versions in English in MIT's archives, suggesting that this copy was subsequently translated back into English.

121   Gyorgy Kepes to Hans Haacke, 28 April 1969, AC 118, MIT, box 39.

122   Dore Ashton to Gyorgy Kepes, 5 June 1969, CAVS Special Collection, MIT.

123   Jean Clay to Gyorgy Kepes, 12 June 1969, CAVS Special Collection, MIT.

124   The documents within included a letter from Hélio Oiticica that initiated protest against French representation: see Oiticica, statement titled "Addressed to the French representation comittee [*sic*] in S. Paulo's 10th Biennal," typescript dated 10 June 1969, CAVS Special Collection, MIT.

125   Kepes first circulated the statement to Howard W. Johnson, who rejected it as inappropriately political.

126   Gyorgy Kepes, "X Bienal de Sao Paulo," typescript dated 30 June 1969, AC 118, Box 39. There are multiple versions of this statement in the archives.

127   See Grace Glueck, "No Rush for Reservation," *New York Times* (6 July 1969): D21.

128 Memo from Howard Johnson to JBW [Jerome B. Wiesner], 8 July 1969.

129 Jean Clay to Gyorgy Kepes, 9 July 1969, CAVS Special Collection, MIT.

130 Months earlier, Takis posted a flier critical of museum displays at the Museum of Modern Art in New York. The flier evoked the rhetoric of Kepes's Center: "Let us unite, artists with scientists, students with workers, to change these anachronistic situations into information centres for all artistic activities." Quoted in Julia Bryan-Wilson, *Art Workers: Radical Practice in the Vietnam War Era* (Berkeley: University of California Press, 2009), 13. Bryan-Wilson's important study explores the wider context of protest in the arts during this period.

131 Gyorgy Kepes to Takis, 20 June 1969, CAVS Special Collection, MIT.

132 Gyorgy Kepes to Dore Ashton, 11 June 1969, CAVS Special Collection, MIT. Ashton would have been sympathetic to Kepes; she participated in his *Vision + Value* volume *The Man-Made Object* (New York: George Braziller, 1966).

133 Jerome Rubin, an attorney and scientist who would later work at MIT's Media Lab, wrote a letter directly to Kubitschek on Kepes's behalf. The letter emphasized Kepes's background: "Professor Kepes, who was born in Hungary and suffered at the hands of a fascist government because of his outspoken devotion to democracy, is eager to go ahead with the exhibition." Of course, Kepes had actually been devoted to communism. Jerome S. Rubin to President Juscelino Kubitschek de Oliveira, 24 June 1969, AC 118, MIT, box 39.

134 Juscelino Kubitschek to Jerome S. Rubin, 11 July 1969, CAVS Special Collection, MIT. See also Roberto Burle Marx to Jerome S. Rubin, 3 July 1969, CAVS Special Collection, MIT.

135 The text of the telegram was published in *Trans-action: Social Science and Modern Society* 6, no. 8 [incorrectly printed as no. 7] (June 1969): 7.

136 Robert Smithson to Gyorgy Kepes, 3 July 1969. Published as "Letter to Gyorgy Kepes (1969)" in *Robert Smithson: The Collected Writings*, ed. Jack Flam (Berkeley: University of California Press, 1996), 369. This important letter is first discussed in Caroline A. Jones, *Machine in the Studio* (Chicago: University of Chicago Press, 1996), 330.

137 John Goodyear to Gyorgy Kepes, 4 July 1969, CAVS Special Collection, MIT.

138 Newton Harrison to Gyorgy Kepes, 9 July 1969, CAVS Special Collection, MIT.

139 Charles Frazier to Gyorgy Kepes, 4 July 1969, CAVS Special Collection, MIT.

140 Les Levine to Gyorgy Kepes, 17 July 1969, CAVS Special Collection, MIT.

141 Vera Simons to Gyorgy Kepes, 6 July 1969, CAVS Special Collection, MIT.

142 Gyorgy Kepes to Rubin Gerscham [Rubens Gerchman], 16 July 1969, CAVS Special Collection, MIT.

143 Gyorgy Kepes, statement, 14 July 1969, CAVS Special Collection, MIT.

144 Gyorgy Kepes to Jerome Rubin, 7 August 1969, CAVS Special Collection, MIT. This was a concern, Kepes explains, that he shared with Howard Johnson.

145 Gyorgy Kepes to Howard W. Johnson, 8 October 1969, CAVS Special Collection, MIT. It is not clear whom Kepes had in mind.

146 Gyorgy Kepes to Ida Rubin, 7 August 1969, CAVS Special Collection, MIT.

147 Many artists shared this disappointment. John Goodyear states that the pressure to withdrawal "has literally devastated me and my family." John Goodyear to Virginia Gunter, 4 July 1969, CAVS Special Collection, MIT.

148 Gyorgy Kepes to Jerome Wiesner, 7 August 1969, CAVS Special Collection, MIT.

149 Gyorgy Kepes, memorandum, 15 October 1969, CAVS Special Collection, MIT.

150 Gyorgy Kepes, "Center for Advanced Visual Studies," *Massachusetts Institute of Technology Bulletin* 106, no. 2 (September 1971): 35.

151 Matthew Wisnioski first draws attention to this pamphlet in "*Centerbeam*: Art of the Environment," in Dutta, *A Second Modernism*, 188–225.

152 See Walter H. G. Lewin, "Three Decades with Otto Piene," in Susanne Rennert and Stephan von Wiese, eds., *Otto Piene: Sky Art 1968–1996* (Cologne: Wienand Verlag, 1999), 37–41.

153 Pamphlet produced by the Council for Conscious Existence, Kepes papers, AAA, reel 5306, frame 0801.

154 See Donald Nicholson-Smith's version of the same passage used above: Raoul Vaneigem, *The Revolution of Everyday Life*, trans. Donald Nicholson-Smith (Oakland: PM Press, 2012), 169; first English translation published in 1983.

155 I pull this information from recollections about the CCE by former members posted online at www.greenmac.com/CCE/RAC_01.html.

156 The *Harvard Crimson* reported these events. To come full circle: the professor was Alex Inkeles, who organized the telegram signed by Kepes and other members of the Center and sent to Costa e Silva of Brazil. See the letter from Alex Inkeles to signatories to the Brazil telegram, 16 June 1969, CAVS Special Collection, MIT.

157 King Collins, personal communication to the author, November 2015.

158 Wilma Lewis [Mya Shone] expressed to me her regret for the action, and her later respect for Kepes. Personal communication to the author, November 2015.

159 Pamphlet produced by the Council for Conscious Existence, Kepes papers, AAA, reel 5306, frame 0803.

160 Ibid., frame 0802.

161 Ibid.

162 Gyorgy Kepes to Sid Lewis, Council for Conscious Existence, 24 November 1969, Kepes papers, AAA, reel 5306, frame 0806.

163 Otto Piene to Sid Lewis, Council for Conscious Existence, 21 November 1969, Kepes papers, AAA, reel 5306, frame 0801.

164 Wilma Lewis, Council for Conscious Existence, to Gyorgy Kepes, 28 November 1969, Kepes papers, AAA, reel 5306, frame 0807.

Chapter 6

1 Georg Lukács, *The Theory of the Novel: A Historico-philosophical Essay on the Forms of Great Epic Literature*, trans. Anna Bostock (Cambridge, MA: MIT Press, 1971), 63–64; first published 1920.

2 Gyorgy Kepes, "The Visual Arts and Sciences: A Proposal for Collaboration," *Daedalus* 94, no. 1 (Winter 1965): 123. The other two areas were "the creative use of light" and "the role of visual signs in artistic communication."

3 For an excellent study relating land art to ecology, see James Nisbet, *Ecologies, Environments, and Energy Systems in Art of the 1960s and 1970s* (Cambridge, MA: MIT Press, 2014). Nisbet also makes a visual comparison that evokes Kepes's picture essay in *Arts of the Environment*; see Nisbet's "Environmental Abstraction and the Polluted Image," *American Art* 31, no. 1 (Spring 2017): 114–131. See also Felicity D. Scott, *Architecture or Techno-Utopia: Politics after Modernism* (Cambridge, MA: MIT Press, 2007).

4 Lowry Burgess interviewed by Joan Brigham, 27 and 28 February 2003, transcription of interview in the CAVS Special Collection, MIT.

5 See Gyorgy Kepes, *The New Landscape in Art and Science* (Chicago: Paul Theobald, 1956); and Gyorgy Kepes to Brian O'Doherty, 6 September 1972, CAVS Special Collection, MIT, folder Charles River Project, 1971–1973. Kepes recycled portions of *The New Landscape* in essays like his "Art and Ecological Consciousness," in Gyorgy Kepes, ed., *Arts of the Environment* (New York: George Braziller, 1972).

6 For "tree of light," see George Kostritsky to Gyorgy Kepes, undated; for "garden of light," see the memorandum from George Kostritsky to Gyorgy Kepes, 11 September 1963; both are in the Kepes papers, Stanford, box 39, folder 1.

7 Kepes, "The Visual Arts and Sciences: A Proposal for Collaboration," 130.

8 Gyorgy Kepes, undated note titled "natural environment presents us with some luminous mobile kinetic spectacles," Kepes papers, Stanford, box 83, folder 4.

9 The Times Square Project was canceled by municipal agencies following the 1973 energy crisis, as the project's lighting elements would have been costly to operate. Such projects echo Kepes's involvement in urban camouflage as well as his KLM mural; he used language first written about his KLM mural in reports on the Times Square Project. See the undated report "Times Square Project," CAVS Special Collection, MIT, folder "Times Square project 1973–1974."

10 Otto Piene to Gyorgy Kepes, 28 January 1966, Kepes papers, Stanford, box 39, folder 6.

11   Otto Piene, "Attempt to list major proposal for common CENTER activities …," 19 November 1968, CAVS Special Collection, MIT, folder "CAVS Fellows proposals 1968–1972."

12   Gyorgy Kepes to Nishan Bichajian, 10 September 1968 (the letter was retyped in 1972), CAVS Special Collection, MIT, folder "Bichajian, Nishan."

13   As per the caption in Gyorgy Kepes, "The Artist's Role in Environmental Self-Regulation," in Kepes, *Arts of the Environment*, 180.

14   See the description in Gyorgy Kepes, "The Lost Pageantry of Nature," *Arts Canada* 25 (December 1968): 36.

15   Gyorgy Kepes to Brian O'Doherty, 6 September 1972, CAVS Special Collection, MIT, folder "Charles River Project, 1971–1973."

16   Gyorgy Kepes, memo to the fellows, 11 January 1972, CAVS Special Collection, MIT, folder "Charles River Project, 1971–1973."

17   Gyorgy Kepes to Brian O'Doherty, 6 September 1972, CAVS Special Collection, MIT, folder "Charles River Project, 1971–1973."

18   Ibid.

19   Gyorgy Kepes, undated and untitled text, CAVS Special Collection, MIT, folder "Charles River Project, 1971–1974," page 7 of typescript. Portions of this text were incorporated into Kepes's essay "The Artist's Role in Environmental Self-Regulation."

20   See Prince's proposal in the CAVS Special Collection, MIT, folder "Charles River Project, 1971–1973."

21   Keiko Prince, "Charles River Project," CAVS Special Collection, MIT, folder "Prince, Keiko."

22   Ted Kraynik, memorandum to Tom McNulty, CAVS Special Collection, MIT, folder "Charles River Project, 1971–1973."

23   Kepes, "The Artist's Role in Environmental Self-Regulation," 170.

24   Ibid, 178.

25   I thank Sabine Eckmann for discussing the connection to Speer with me.

26   Gyorgy Kepes to Philip Zeigler of the Boston Redevelopment Authority, 4 December 1968, Kepes papers, Stanford, box 27, folder 9.

27   Robert H. Lurcott of the Boston Redevelopment Authority to Gyorgy Kepes, 12 February 1969, Kepes papers, Stanford, box 27, folder 9. The Boston Redevelopment Authority financially supported the Center's later proposals for Boston's Long Wharf, which were more feasible.

28   Friedrich St. Florian to Bill N. Lacy, 10 October 1974, CAVS Special Collection, MIT, folder "Boston Harbor Islands Project."

29 Gyorgy Kepes to Brian O'Doherty, 6 September 1972, CAVS Special Collection, MIT, folder "Charles River Project, 1971–1973."

30 Gyorgy Kepes, undated note titled "desalination," Kepes papers, Stanford, box 82, folder 4.

31 Kepes, "The Artist's Role in Environmental Self-Regulation," 171.

32 Gyorgy Kepes, undated note titled "environmental-art," Kepes papers, Stanford, box 82, folder 4.

33 Gyorgy Kepes, undated note titled "1 technology," Kepes papers, Stanford, box 81, folder 1.

34 Gyorgy Kepes to Irwin Mill, 5 October 1970, CAVS Special Collection, MIT, folder "Kepes, Gyorgy."

35 Gyorgy Kepes to Eli Goldstone, 27 February 1968; and Gyorgy Kepes to Electro-Optical Systems, Inc., 29 May 1970. Both letters are in the CAVS Special Collection, MIT, folder "Kepes, Gyorgy."

36 Kepes had more luck with scientists at MIT, although these exchanges were not necessarily substantive. Kepes made contact, for example, with Frank Press, an MIT geophysicist; Thomas B. Sheridan, an MIT mechanical engineer; Waclaw Zalewski, an MIT structural engineer; and Louis D. Smullin, an MIT electrical engineer. Smullin previously ran the Radar and Weapons Division at Lincoln Laboratory, making him another controversial contact. See letters between Kepes and these scientists in the CAVS Special Collection, MIT, folders "Charles River Project, 1971–1973"; "Kepes, Gyorgy"; and "Kraynik, Ted."

37 Gyorgy Kepes to Brian O'Doherty, 6 September 1972, CAVS Special Collection, MIT, folder "Charles River Project, 1971–1973."

38 Gyorgy Kepes, undated note titled "Homeostasis← →pseudo-homeostasis," Kepes papers, Stanford, box 25, folder 6.

39 Gyorgy Kepes, undated note titled "culture shock," Kepes papers, Stanford, box 83, folder 4.

40 Gyorgy Kepes, undated and untitled note, Kepes papers, Stanford, box 83, folder 3.

41 Gyorgy Kepes, undated note titled "P" [pollution or participation], Kepes papers, Stanford, box 25, folder 1.

42 The text is infrequently cited in connection to land art or ecological practices of the 1970s; for an exception, see T. J. Demos, *Decolonizing Nature: Contemporary Art and the Politics of Ecology* (Berlin: Sternberg Press, 2016), 50. Demos cites Kepes's text as an example of "cultural homeostasis."

43 Quoted in Kepes, "Art and Ecological Consciousness," 1.

44 Quoted in ibid., 2.

45 Ibid., 12.

46   Ibid., 7.

47   See the article with Kepes's marginalia in the Kepes papers, Stanford, box 83, folder 5.

48   Judith Nottingham, "On Chemical and Biological Warfare," pamphlet published by the Women's International League for Peace and Freedom, c. 1968, Kepes papers, Stanford, box 83, folder 5.

49   "Chemical Biological Radiological WARFARE," pamphlet published by the Committee for World Development and World Disarmament, Kepes papers, Stanford, box 83, folder 5.

50   Gyorgy Kepes, undated note titled "elements + −," Kepes papers, Stanford, box 9, folder 6.

51   Gyorgy Kepes, undated note titled "elements," Kepes papers, Stanford, box 9, folder 6.

52   Kepes, "The Artist's Role in Environmental Self-Regulation," 184.

53   Ibid., 192. Baldeweg's proposal is consistent with the era's fascination for dome architecture. See especially Scott, *Architecture or Techno-Utopia*, 151–183.

54   Kepes, "The Artist's Role in Environmental Self-Regulation," 184.

55   Ibid., 186.

56   Ted Kraynik to Gyorgy Kepes, 7 October 1968, CAVS Special Collection, MIT, folder "Kraynik, Ted."

57   Ted Kraynik to Gyorgy Kepes, 16 September 1968, CAVS Special Collection, MIT, folder "CAVS Fellows Proposals 1968–1972."

58   A copy of the paper is included in the CAVS Special Collection, MIT, folder "Kraynik, Ted."

59   Kepes, "The Artist's Role in Environmental Self-Regulation," 184.

60   Ibid., 184–185.

61   Gyorgy Kepes, undated note titled "P" [pollution or participation], Kepes papers, Stanford, box 25, folder 1.

62   Gyorgy Kepes, undated and untitled note, Kepes papers, Stanford, box 4, folder 2.

63   See the drawings by Kepes in CAVS Special Collection, MIT, folder "Charles River Project, 1971–1973."

64   For the wider context of these architectural experiments, see especially Larry Busbea, *Topologies: The Urban Utopia in France, 1960–1970* (Cambridge, MA: MIT Press, 2007). Kepes studied Schöffer's designs; see the report about Schöffer's Cybernetic Light Tower annotated by Kepes in the Kepes papers, Stanford, box 4, folder 20.

65   Gyorgy Kepes, undated note titled "environmental art," Kepes papers, Stanford, box 9, folder 6.

66   Kepes, "The Artist's Role in Environmental Self-Regulation," 184.

67   Gyorgy Kepes, undated note titled "environmental art happening," Kepes papers, Stanford, box 9, folder 6.

68   Gyorgy Kepes, undated and untitled text, CAVS Special Collection, MIT, folder "Charles River Project, 1971–1974."

69   Friedrich St. Florian to Gyrogy Kepes, 19 July 1972, CAVS Special Collection, MIT, folder "Charles River Project, 1971–1973."

70   See Keiko Prince's proposal in the CAVS Special Collection, MIT, folder "Charles River Project, 1971–1973."

71   Ted Kraynik to Gyorgy Kepes, 16 September 1968, CAVS Special Collection, MIT, folder "CAVS Fellows Proposals, 1968–1972."

72   Kepes, "The Artist's Role in Environmental Self-Regulation," 184.

73   Gyorgy Kepes, undated and untitled note on light, Kepes papers, Stanford, box 82, folder 6.

74   Kepes, "The Lost Pageantry of Nature," 32.

75   Gyorgy Kepes, "The Nineteenth Century," undated typescript, Kepes papers, Stanford, box 82, folder 1. This concept is elaborated in chapter 4.

76   Gyorgy Kepes to Walter Rosenblith, 17 February 1972, AC 7, MIT, box 5, folder "Center for Advanced Visual Studies 1972–1974," pages 2 and 7 of typescript.

77   Gyorgy Kepes, undated and untitled text, CAVS Special Collection, MIT, folder "Charles River Project, 1971–1974," pages 2 and 7 of typescript.

78   Notes made by Walter Rosenblith in a memo dated 16 December 1973, AC 7, MIT, box 5, folder "Center for Advanced Visual Studies 1972–1974."

79   For more on Burnham, see the introduction by Melissa Ragain in Jack Burnham, *Dissolve into Comprehension*, ed. Melissa Ragain (Cambridge, MA: MIT Press, 2015); and Luke Skrebowski, "Jack Burnham Redux: The Obsolete in Reverse?," *Grey Room* 65 (Autumn 2016): 88–113. Burnham claims his disagreements with Kepes were a result of *Explorations* and the São Paulo Biennial.

80   Jack Burnham to Gyorgy Kepes, 9 June 1967, CAVS Special Collection, MIT.

81   Gyorgy Kepes to Robert H. Strotz, 10 March 1969, CAVS Special Collection, MIT. Kepes was endorsing Burnham for promotion at Northwestern University.

82   Jack Burnham, "Art and Technology: The Panacea That Failed," in Kathleen Woodward, ed., *The Myths of Information: Technology and Post-Industrial Culture* (Madison, WI: Coda Press, 1980), 208.

83   Gyorgy Kepes to John W. Sears, 8 December 1971, CAVS Special Collection, MIT.

84   Ted Kraynik, memorandum to Tom McNulty, 8 June 1971, CAVS Special Collection, MIT, folder "Charles River Project, 1971–1973." Emphasis in original.

85   Gyorgy Kepes to Ted Kraynik, 27 July 1972, CAVS Special Collection, MIT, folder "Kraynik, Ted."

86   Gyorgy Kepes to Ted Kraynik, 21 August 1972, CAVS Special Collection, MIT, folder "Kraynik, Ted."

87   Burnham, "Art and Technology," 208–209.

88   Friedrich St. Florian in a memorandum to Otto Piene and the Fellows, 3 September 1974, CAVS Special Collection, MIT, folder "Boston Harbor Islands Project, 1974–1975." Burnham was no longer affiliated with the Center, however.

89   "Plug-In" is described in Kepes, "The Lost Pageantry of Nature," 34.

90   Jack Burnham interviewed by Joan Brigham, 13 February 2004, transcription of interview from the CAVS Special Collection, MIT, reprinted in Burnham, *Dissolve into Comprehension.*

91   Harold Tovish interviewed by Joan Brigham, 18 September 2003, transcription of interview from the CAVS Special Collection, MIT.

92   Jack Burnham interviewed by Joan Brigham, 13 February 2004, transcription of interview from the CAVS Special Collection, MIT.

93   Roger Kizik to the Center for Advanced Visual Studies, undated [1976], CAVS Special Collection, MIT. These critiques were anticipated by the fellows; regarding the Charles River project, Paul Earls declares: "No individual nor collective lasting domination of the river." See the statement by Earls titled "Riverthoughts" [*sic*], 28 October 1971. Kepes even warns that the Center's projects might have "possible negative fallouts," including the creation of their own pollution, even "visual pollution." See his "First Draft Memo re: Charles River Project," 28 June 1971. Both documents are in the CAVS Special Collection, MIT, folder "Charles River Project, 1971–1973."

94   Quoted in Gyorgy Kepes, undated note titled "light," Kepes papers, Stanford, box 28, folder 14.

95   From one of Kepes's notebooks in the Kepes papers, Stanford, box 20. Kepes's notation reads: "after my radio talk I gave on color I received a letter dictated by a blind person please give us rainbow." The radio program may have been part of a series on WGBH Boston introduced by Lyman Bryson on various design topics. See https://www.youtube.com/watch?v=oBCYQNYrhh0.

96   From one of Kepes's notebooks in the Kepes papers, Stanford, box 20.

Epilogue

1   Sigfried Giedion, *Mechanization Takes Command: A Contribution to Anonymous History* (New York: W. W. Norton, 1948), 720.

2   "The Teaching of the Arts and Humanities at Harvard College: Mapping the Future" (Harvard University, 2013), 7. See the American Academy of Arts and Sciences "Humanities Indicators" project, available at http://www.humanities indicators.org/.

3   See the special issue of *Daedalus* 138, no. 1 (Winter 2009), devoted to "Reflecting on the Humanities"; the special issue of *Representations* 116, no. 1 (Fall 2011), on "The Humanities and the Crisis of the Public University"; and the special issue of *Critical Inquiry* 35, no. 4 (Summer 2009), on "The Fate of Disciplines."

4   Patricia Cohen, "A Rising Call to Promote STEM Education and Cut Liberal Arts Funding," *New York Times* (21 February 2016).

5   "Arts at MIT" (MIT, 2011), 10, available at http://orgchart.mit.edu/sites/default /files/reports/20110628_Provost_ArtsatMITFinal6-20-2011.pdf.

6   Hilarie M. Sheets, "At M.I.T., Science Embraces a New Chaos Theory: Art," *New York Times* (4 March 2016).

7   Though there are now critical evaluations of MIT's institutional history in the arts; see especially Arindam Dutta, ed., *A Second Modernism: MIT, Architecture, and the "Techno-Social" Moment* (Cambridge, MA: MIT Press, 2013).

8   Gyorgy Kepes, "Kinetic Light as a Creative Medium," *Technology Review* 70, no. 2 (December 1970): 31.

9   Kepes quoted in Rudolph de Harak, "Gyorgy Kepes Revisits the Visual Landscape," in Steven Heller and Marie Finamore, eds., *Design Culture: An Anthology of Writing from the AIGA Journal of Graphic Design* (New York: Allworth Press, 1997), 148; originally published in the *AIGA Journal of Graphic Design* 5, no. 2 (1987).

10   Gyorgy Kepes, undated note titled "visual fundamentals," Kepes papers, Stanford, box 83, folder 6.

Index